前言

✤ 為何撰寫本書

從事機器學習教育訓練已屆四年，其間也在『IT 邦幫忙』撰寫上百篇的文章 (https://ithelp.ithome.com.tw/users/20001976/articles)，從學員及讀者的回饋獲得許多寶貴意見，期望能將整個歷程集結成冊，同時，相關領域的進展也在飛速變化，過往的文章內容需要翻新，因此藉機再重整思緒，想一想如何能將演算法的原理解釋得更簡易清晰，協助讀者跨入 AI 的門檻，另外，也避免流於空談，盡量增加應用範例，希望能達到即學即用，不要有過多理論的探討。

AI 是一個將資料轉化為知識的過程，演算法就是過程中的生產設備，最後產出物是模型，再將模型植入各種硬體裝置，例如電腦、手機、智慧音箱、自駕車、醫療診斷儀器等，這些裝置就擁有特殊專長的智慧，再進一步整合各項技術就構建出智慧製造、智慧金融、智慧交通、智慧醫療、智慧城市、智慧家庭等應用系統。AI 的應用領域如此的廣闊，個人精力有限，當然不可能具備十八般武藝，樣樣精通，惟有從基礎紮根，再擴及有興趣的領域，因此，筆者撰寫這本書的初衷，非常單純，就是希望讀者在紮根的過程中，貢獻一點微薄的力量。

✤ 本書主要的特點

1. 由於筆者身為統計人，希望能『以統計 / 數學為出發點』，介紹深度學習必備的數理基礎，但又不希望內文有太多數學公式的推導，讓離開校園已久的在職者看到一堆數學符號就心生恐懼，因此，嘗試以『程式設計取代定理證明』，縮短學習歷程，增進學習樂趣。

2. TensorFlow 2.x 版有巨大的變動，預設模式改為 Eager Execution，並以 Keras 為主力，整合 TensorFlow 其他模組，形成完整的架構，本書期望對 TensorFlow 架構作完整性的介紹，並非只是介紹 Keras 而已。

3. 演算法介紹以理解為主，輔以大量圖表說明，摒棄長篇大論。

4. 完整的範例程式及各種演算法的延伸應用，以實用為要，希望能觸發讀者靈感，能在專案或產品內應用。

5. 介紹日益普及的演算法與相關套件的使用，例如 YOLO(物件偵測)、GAN(生成對抗網路)/DeepFake(深度偽造)、OCR(辨識圖像中的文字)、臉部辨識、BERT/Transformer、ChatBot、強化學習、自動語音辨識 (ASR) 等。

✣ 目標對象

1. 深度學習的入門者：必須熟悉 Python 程式語言及機器學習基本概念。

2. 資料工程師：以應用系統開發為職志，希望能應用各種演算法，進行實作。

3. 資訊工作者：希望能擴展深度學習知識領域。

4. 從事其他領域的工作，希望能一窺深度學習奧秘者。

✣ 閱讀重點

1. 第一章介紹 AI 的發展趨勢，鑑古知今，瞭解前兩波 AI 失敗的原因，比較第三波發展的差異性。

2. 第二章介紹深度學習必備的統計 / 數學基礎，不僅要理解相關知識，也力求能撰寫程式解題。

3. 第三章介紹 TensorFlow 基本功能，包括張量 (Tensor) 運算、自動微分及神經網路模型的組成，並說明梯度下降法求解的過程。

4. 第四章開始實作，依照機器學習 10 項流程，撰寫完整的範例，包括 Web、桌面程式。

5. 第五章介紹 TensorFlow 進階功能，包括各種工具，如 TensorBoard、TensorFlow Serving、Callbacks。

6. 第六～十章介紹圖像 / 視訊的演算法及各式應用。

7. 第十一～十四章介紹自然語言處理、語音及各式應用。

8. 第十五章介紹 AlphaGo 的基礎 --『強化學習』演算法。

本書範例程式碼全部收錄在 https://github.com/mc6666/DL_Book 。

✤ 致謝

原本只想整理過往文章集結成書，沒想到相關技術發展太快，幾乎全部重寫，結果比預計時程多花了三倍的時間才完成，但因個人能力有限，還是有許多議題成為遺珠之憾，仍待後續的努力，過程中要感謝冠瑀、琮文在編輯 / 校正 / 封面構想的盡心協助，也感謝深智出版社的大力支援，使本書得以順利出版，最後要藉此書，紀念一位摯愛的親人。

內容如有疏漏、謬誤或有其他建議，歡迎來信指教 (mkclearn@gmail.com)。

目錄

第三篇 進階的影像應用

12 自然語言處理的演算法

13 聊天機器人 (ChatBot)

14 語音辨識

第五篇 強化學習 (Reinforcement Learning)

15 強化學習

第一篇｜深度學習導論

在進入深度學習的殿堂前，學生常會想問以下的問題：

1. 人工智慧已歷經三波浪潮，這一波是否又將進入寒冬？
2. 人工智慧、資料科學、資料探勘、機器學習、深度學習到底有何關聯？
3. 機器學習開發流程與一般應用系統開發有何差異？
4. 深度學習的學習路徑為何？建議從哪裡開始？
5. 為什麼要先學習數學與統計，才能學好深度學習？
6. 先學哪一種深度學習框架比較好？TensorFlow 或 PyTorch？
7. 如何準備開發環境？

第一章就先針對以上的問題作一簡單的介紹，第二章則介紹深度學習必備的數學與統計基礎知識，有別於學校的教學，我們會偏重在程式的實作。

深度學習（Deep Learning）導論

1-1 人工智慧的三波浪潮

人工智慧(Artificial Intelligence，AI) 並不是最近幾年才興起的，其實它已經是第三波熱潮了，前兩波都經歷 10 餘年，就邁入寒冬，這一波熱潮至今也將近十年(2010~)，會不會又將邁入寒冬呢？我們就先來重點回顧一下過往的歷史吧。

人工智慧的三波浪潮

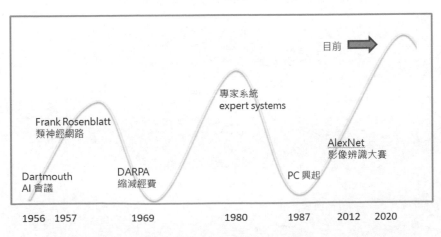

圖 1.1 人工智慧的三波浪潮

- 1956 年在達特茅斯(Dartmouth) 學院舉辦 AI 會議,確立第一波浪潮的開始。

- 1957 年 Frank Rosenblatt 創建感知器(Perceptron),即簡易的神經網路,然而當時並無法解決複雜多層的神經網路,直至 1980 年代才被想出解決辦法。

- 1969 年美國國防部(DARPA)基於投資報酬率過低,決定縮減 AI 研究經費,AI 邁入第一波寒冬。

- 1980 年專家系統(Expert System)興起,企圖將各行各業專家的內隱知識,外顯為一條條的規則,從而建立起專家系統,不過因陳義過高,且需要使用大型且昂貴的電腦設備,才能夠建構相關系統,又不巧適逢個人電腦(PC)的流行,相較之下,鋒頭就被掩蓋下去了,至此,AI 邁入第二波寒冬。

- 2012 年多倫多大學 Geoffrey Hinton 研發團隊利用分散式計算環境及大量影像資料,結合過往的神經網路知識,開發了 AlexNet 神經網路,參加 ImageNet 影像辨識大賽,大放異彩,一下子把錯誤率降低了十幾個百分比,就此興起 AI 第三波浪潮,至今仍方興未艾。

第三波熱潮至今也將近十年(2010~2021),會不會又將邁入寒冬呢?觀察這一波熱潮,相較於過去前兩波,具備了以下幾項的優勢:

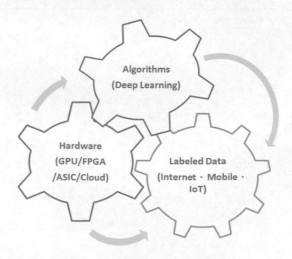

圖 1.2 第三波 AI 浪潮的觸媒

1. 先架構基礎功能，從影像、語音、文字辨識開始，再逐步往上構建各式的應用，例如自駕車(Self Driving)、對話機器人(ChatBot)、機器人(Robot)、智慧醫療、智慧城市等，這種由下往上的發展方式比較扎實。

2. 硬體的快速發展：

- 摩爾定律的發展速度：IC 上可容納的電晶體數目，約每隔 18 個月至兩年便會增加一倍，簡單來說，就是 CPU 每隔兩年便會增快一倍，過去 50 年均循此軌跡發展，此定律在未來十年也應該會繼續適用，之後也許量子電腦(Quantum Computing)等新科技會繼續接棒，它號稱目前電腦要計算幾百年的工作，量子電腦只要 30 分鐘就搞定了，如果成真，那時可能又是另一番光景了。

- 雲端資料中心的建立：大型 IT 公司在世界各地興建大型的資料中心，採取「用多少付多少」(Pay as you go)。由於模型訓練時，通常需要大量運算，採用雲端方案，一般企業就不須在前期購買昂貴設備，僅需支付必要運算的費用，也不需冗長的採購流程，只要幾分鐘就能開通(Provisioning)所需設備，省錢省時。

- GPU/ NPU 的開發：深度學習主要是以矩陣運算，GPU 這方面比 CPU 快上很多倍，專門生產 GPU 的美商輝達(NVidia)公司因而大放異彩，市值超越 Intel [1]。當然其他硬體及系統廠商不會錯失如此龐大的商機，因此各式的 NPU (Neural-network Processing Unit)或 xPU 紛紛出籠，積極搶食這塊大餅，各項產品使得運算越來越快，模型訓練的時間大幅縮短，通常模型調校需要反覆訓練，如果能在短時間得到答案，對資料科學家而言，是一大福音。另外，連接現場裝置的電腦(Raspberry pi、Jetson Nano、Auduino)，體積越來越小，運算能力越來越強，對於「邊緣運算」也有莫大的助益，例如路口監視器、無人機等。

3. 演算法推陳出新：過去由於計算能力受限，許多無法在短時間訓練完成的演算法一一解封，尤其是神經網路，現在已經可以建構上百層的模型，運算的參數也可以高達上兆個，都能在短時間內調校出最佳模型，因此，模型設計就可以更複雜，演算法邏輯也能夠更完備。

4. 大量資料的蒐集及標註(Labeling)：人工智慧必須仰賴大量資料，讓電腦學習，從中挖掘知識，近年來網際網路(Internet)及手機(Mobile)盛行，企業可以透過社群媒體蒐集到大量資料，再加上物聯網(IoT)也可藉由感測器產生源源不斷的資料，作為深度學習的養份(訓練資料)，而這些大型的網路公司銀彈充足，可以雇用大量的人力，進行資料標註，確保資料的品質，使得訓練出來的模型越趨精準。

因此，根據以上的趨勢發展，筆者猜測第三波的熱潮在短期內應該不會邁入寒冬吧。

1-2　AI 的學習地圖

AI 劃分三個階段，每一階段的重點為「人工智慧」(Artificial Intelligence)、「機器學習」(Machine Learning)、「深度學習」(Deep Learning)，每一階段都在縮小範圍，聚焦在特定的演算法，機器學習是人工智慧的部份領域，而深度學習又屬於機器學習的部份演算法。

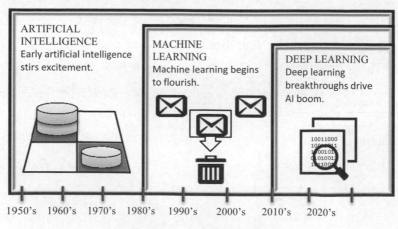

圖 1.3　AI 三波熱潮的重點

而一般教育機構規劃 AI 的學習地圖即依照這個軌跡，逐步深入各項技術，通常分為四個階段：

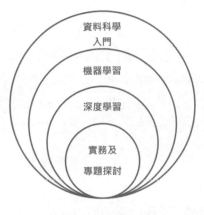

圖 1.4 AI 學習地圖

1. 資料科學(Data Science)入門：內容包括 Python/R 程式語言、數據分析(Data Analysis)、大數據平台(Hadoop、Spark) 等。
2. 機器學習(Machine Learning)：包含一些典型的演算法，如迴歸、羅吉斯迴歸、支援向量機(SVM)、K-means 集群等，這些演算法雖然簡單，但卻非常實用，較容易在一般企業內普遍性的導入。通常機器學習的大致分類如下圖：

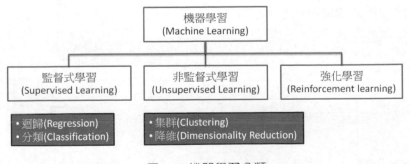

圖 1.5 機器學習分類

最新的發展還有半監督學習(Semi-supervised Learning)、自我學習(Self Learning)、聯合學習(Federated Learning) 等，不一而足，千萬不要被分類限制了你的想像。

另外，資料探勘(Data Mining)與機器學習的演算法大量重疊，其間的差異，有一說是資料探勘是著重挖掘資料的隱藏樣態(Pattern)，而機器學習則著重於預測。

3. 深度學習(Deep Learning)：深度學習屬於機器學習中的一環，所謂深度(Deep)是指多層式架構的模型，例如各種神經網路(Neural Network)模型、強化學習(Reinforcement learning, RL)演算法等，以多層的神經層或 try-and-error 的方式，以優化(Optimization)或反覆的方式求解。

4. 實務及專題探討(Capstone Project)：將各種演算法應用於各類領域/行業，強調專題探討及產業應用實作。

1-3 機器學習應用領域

機器學習的應用其實早已不知不覺間融入到我們的生活中了，例如：

1. 社群軟體大量運用 AI，預測使用者的行為模式、過濾垃圾信件。

2. 電商運用 AI，依據每位消費者的喜好推薦合適的商品。

3. 還有各式的 3C 產品，包括手機、智慧音箱，以人臉辨識取代登入，語音辨識代替鍵盤輸入，結合智慧家庭/保全。

4. 聊天機器人(ChatBot)取代客服人員，提供行銷服務。

5. 製造機器人(Robot) 提供智慧製造、老人照護、兒童陪伴，凡此種種，不及備載。

AI 相關的應用領域如下圖：

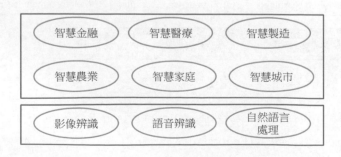

圖 1.6 AI 應用領域

目前相對熱門的研發領域包括：

1. 各種疾病的診斷 (Medical Diagnostics)及新藥的開發。
2. 聊天機器人(ChatBot)：包括行銷、銷售及售後服務的支援。
3. 物件辨識(Object Detection) / 人臉辨識 (Facial Recognition)。
4. 自駕車(Self Driving)。
5. 製造機器人(Robot)。

1-4 機器學習開發流程

一般來說，機器學習開發流程(Machine learning workflow)，有許多種建議的模型，例如資料探勘(Data Mining)流程，包括 CRISP-DM (cross-industry standard process for data mining,)、Google Cloud 建議的流程[2]等，個人偏好的流程如下：

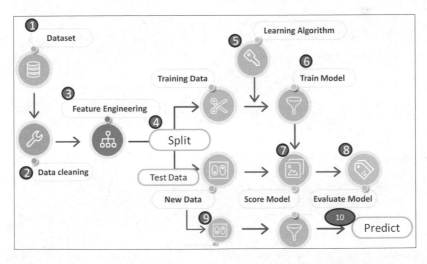

圖 1.7 機器學習開發流程(Machine Learning Workflow)

概分為 10 個步驟，不含較高層次的企業需求瞭解(Business Understanding)，只包括實際開發的步驟：

1. 蒐集資料，彙整為資料集(Dataset)。

2. 資料清理(Data Cleaning)、資料探索與分析(Exploratory Data Analysis, EDA)：EDA 通常是以描述統計量及統計圖觀察資料的分佈，瞭解資料的特性、極端值(Outlier)、變數之間的關聯性。

3. 特徵工程(Feature Engineering)：原始蒐集的資料未必是影響預測目標的關鍵因素，有時候需要進行資料轉換，以找到關鍵的影響變數。

4. 資料切割(Data Split)：切割為訓練資料(Training Data)及測試資料(Test Data)，一份資料提供模型訓練之用，另一份資料則用在衡量模型效能，例如準確度，切割的主要原因是確保測試資料不會參與訓練，以維持其公正性，即 Out-of-Sample Test。

5. 選擇演算法(Learning Algorithms)：依據問題的類型選擇適合的演算法。

6. 模型訓練(Model Training)：以演算法及訓練資料，進行訓練產出模型。

7. 模型計分(Score Model)：計算準確度等效能指標，評估模型的準確性。

8. 模型評估(Evaluate Model)：比較多個參數組合、多個演算法的準確度，找到最佳參數與演算法。

9. 佈署(Deploy)：複製最佳模型至正式環境(Production Environment)，製作使用介面或提供 API，通常以網頁服務(Web Services) 作為預測的 API。

10. 預測(Predict)：用戶端傳入新資料或檔案，系統以模型進行預測，傳回預測結果。

機器學習開發流程與一般應用系統開發有何差異？最大的差別如下圖：

1. 一般應用系統則利用輸入資料與轉換邏輯產生輸出，例如撰寫報表，根據轉換規則將輸入欄位轉換為輸出欄位，但機器學習會先產生模型，再根據模型進行預測，重用性(Reuse)高。

2. 機器學習除了輸入資料外，還會蒐集大量歷史資料或在網際網路中爬一堆資料，作為塑模的飼料。

3. 新產生的資料可回饋入模型，重新訓練，自我學習，使模型更聰明。

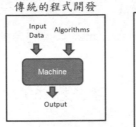 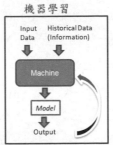

圖 1.8 機器學習與一般應用系統開發流程的差異

▌ 1-5 開發環境安裝

Python 是目前機器學習主流的程式語言，可以直接在本機安裝開發環境，亦能使用雲端環境，首先介紹本機安裝的程序，建議依照以下順序安裝：

1. 安裝 Anaconda：建議安裝此軟體，它內含 Python 及上百個常用套件。先至 https://www.anaconda.com/products/individual 下載安裝檔，在 Windows 作業系統安裝時，建議執行到下列畫面時，兩者都勾選，就可將安裝路徑加入至環境變數 Path 內，這樣就能在任何目錄下均可執行 python 程式。Mac/Linux 則須自行修改登入檔(profile)，增加 Anaconda 安裝路徑。

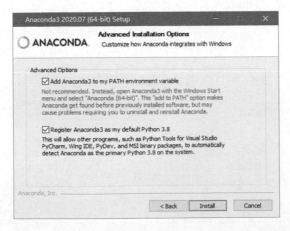

圖 1.9 Anaconda 安裝注意事項，將安裝路徑加入至環境變數 Path 內

2. 安裝 TensorFlow 最新版：Windows 作業系統安裝時，在檔案總管的路徑輸入 cmd，開啟 DOS 視窗，Mac/Linux 則須開啟終端機，輸入：

```
pip install TensorFlow
```

** 此指令同時支援 CPU 及 GPU，但要支援 **GPU 須另外安裝驅動程式及 SDK**。

3. 測試：安裝完，在 DOS 視窗或終端機內，輸入 python，進入互動式環境，再輸入以下指令測試：

```
> import tensorFlow
> exit()
```

▪ 若出現下列錯誤：

```
from tensorflow.python._pywrap_tensorflow_internal import *
ImportError: DLL load failed: The specified module could not be
found.
```

可能是 Windows 作業系統較舊，須安裝 MSVC 2019 runtime (https://aka.ms/vs/16/release/vc_redist.x64.exe)。

▪ 如果還是出現 Dll cannot be loaded. 也可使用 Anaconda 提供的指令來安裝。

```
conda install tensorFlow
```

▪ 注意，若安裝 Python V3.7 或更舊版，只能安裝 TensorFlow 2.0.0 版或以下版本。

4. 若要支援 GPU，還需安裝 CUDA Toolkit、cuDNN SDK，只能安裝 TensorFlow 支援的版本，請參考 TensorFlow 官網說明[3]，例如下表：

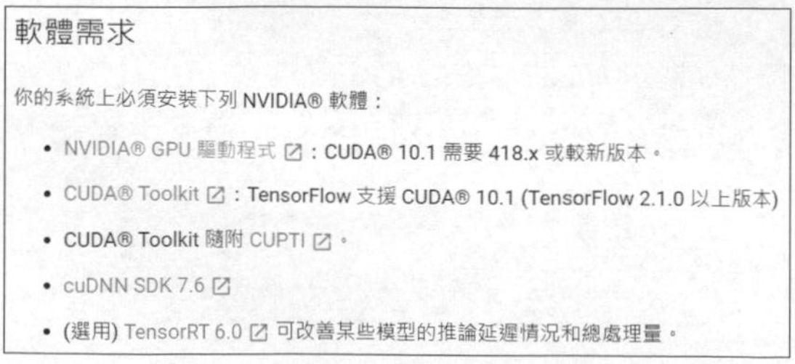

軟體需求

你的系統上必須安裝下列 NVIDIA® 軟體：

- NVIDIA® GPU 驅動程式 ☑ ：CUDA® 10.1 需要 418.x 或較新版本。
- CUDA® Toolkit ☑ ：TensorFlow 支援 CUDA® 10.1 (TensorFlow 2.1.0 以上版本)
- CUDA® Toolkit 隨附 CUPTI ☑ 。
- cuDNN SDK 7.6 ☑
- (選用) TensorRT 6.0 ☑ 可改善某些模型的推論延遲情況和總處理量。

圖 1.10 TensorFlow 支援的 CUDA Toolkit 版本

注意，目前只支援 **NVidia** 獨立顯卡，若是較舊型的顯卡必須查閱驅動程式搭配的版本資訊，請參考 NVIDIA 官網說明[4]，如下表：

Table 2. CUDA Toolkit and Compatible Driver Versions

CUDA Toolkit	Linux x86_64 Driver Version	Windows x86_64 Driver Version
CUDA 11.1.1 Update 1	>=455.32	>=456.81
CUDA 11.1 GA	>=455.23	>=456.38
CUDA 11.0.3 Update 1	>= 450.51.06	>= 451.82
CUDA 11.0.2 GA	>= 450.51.05	>= 451.48
CUDA 11.0.1 RC	>= 450.36.06	>= 451.22
CUDA 10.2.89	>= 440.33	>= 441.22
CUDA 10.1 (10.1.105 general release, and updates)	>= 418.39	>= 418.96
CUDA 10.0.130	>= 410.48	>= 411.31
CUDA 9.2 (9.2.148 Update 1)	>= 396.37	>= 398.26
CUDA 9.2 (9.2.88)	>= 396.26	>= 397.44
CUDA 9.1 (9.1.85)	>= 390.46	>= 391.29
CUDA 9.0 (9.0.76)	>= 384.81	>= 385.54
CUDA 8.0 (8.0.61 GA2)	>= 375.26	>= 376.51
CUDA 8.0 (8.0.44)	>= 367.48	>= 369.30
CUDA 7.5 (7.5.16)	>= 352.31	>= 353.66
CUDA 7.0 (7.0.28)	>= 346.46	>= 347.62

圖 1.11 CUDA Toolkit 版本與驅動程式的搭配

5. 奉勸各位讀者，太舊型的顯卡，若不能安裝 TensorFlow 支援的版本，就不用安裝 CUDA Toolkit、cuDNN SDK 了，因為顯卡記憶體過小，執行 TensorFlow 常常會發生記憶體不足(OOM)的錯誤，徒增困擾。

6. CUDA Toolkit、cuDNN SDK 安裝成功後，將安裝路徑下的 bin、libnvvp 加入至環境變數 Path。

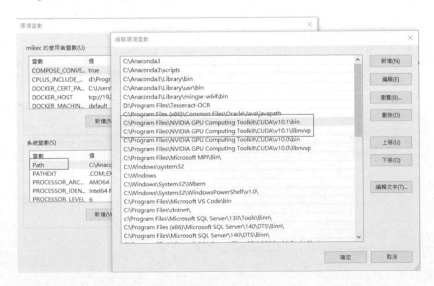

7. 安裝若有其他問題,可參考筆者的部落文「Day 01:輕鬆掌握 Keras」
   [5]。

再來我們談談雲端環境的開通,幾乎 Google GCP、AWS、Azure 都有提供
機器學習的開發環境,這裡介紹免費的 Google 雲端環境 – Colaboratory,
須具備 Gmail 帳號才能使用,具備以下特點:

1. 常用的套件均已預先安裝,包括 TensorFlow。
2. 免費的 GPU:為 NVIDIA Tesla K80 GPU 顯卡,含 12GB 記憶體,真
   是佛心啊。
3. 它在使用時即時開通 Docker container,限連續使用 12 小時,逾時的話
   虛擬環境會被回收,所有程式、資料一律會被刪除。

開通程序:

1. 使用 Google Chrome 瀏覽器,進入雲端硬碟(Google drive)介面。
2. 建立一個目錄,例如「0」,並切換至該目錄。

3. 在螢幕中間按滑鼠右鍵，點選「更多」>「連結更多應用程式」。

4. 在搜尋欄輸入「Colaboratory」，找到後點擊該 App，按「Connect」按鈕即可開通。

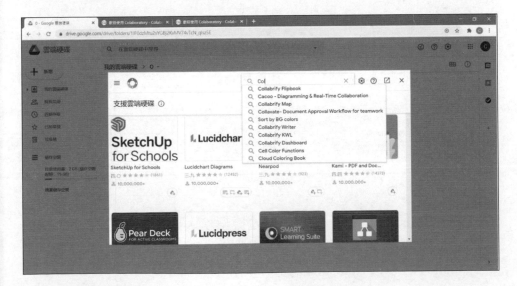

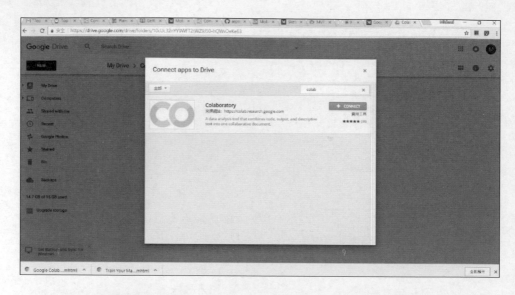

5. 開通後，即可開始使用。可新增一個「Colaboratory」的檔案。

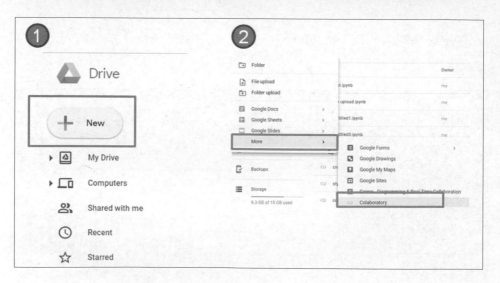

6. Google Colaboratory 會自動開啟虛擬環境，建立一個空白的 Jupyter
   Notebook 檔案，附檔名為 ipynb，幾乎所有的雲端環境及大數據平台
   Databricks，都以 Notebook 為主要使用介面。

7. 或者直接以滑鼠雙擊(Double click) Notebook 檔案，也可自動開啟虛擬環境，進行編輯與執行。本機的 Notebook 檔案也可上傳至雲端硬碟，點擊使用，完全不用轉換，非常方便。

8. 若要支援 GPU 可設定運行環境使用 GPU 或 TPU，TPU 為 Google 發明的 NPU。

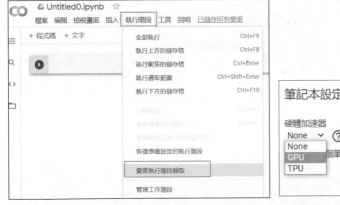

9. 「Colaboratory」相關操作，可參考官網說明[6]。

本書所附的範例程式，一律為 Notebook 檔案，因為 Notebook 可以使用 Markdown 語法撰寫美觀的說明，包括數學式，另外，程式也可以分格，單獨執行，便於講解，相關的用法可以參考「Jupyter Notebook: An Introduction」[7]。

參考資料 (References)

[1] Dylan Yeh、陳建鈞,《市值首度超越 Intel！NVIDIA 贏在哪裡？》, 2020
(https://www.bnext.com.tw/article/58410/nvidia-valuation-soars-past-intel-on-graphics-chip-boom)

[2] Google Cloud 官網指南
(https://cloud.google.com/ai-platform/docs/ml-solutions-overview)

[3] TensorFlow 官網說明
(https://www.tensorflow.org/install/gpu？hl=zh-tw)

[4] NVIDIA 官網說明
(https://developer.nvidia.com/cuda-toolkit-archive)

[5] 陳昭明,《Day 01：輕鬆掌握 Keras》, 2020
(https://ithelp.ithome.com.tw/articles/10233272)

[6] Colaboratory 官網說明
(https://colab.research.google.com/notebooks/intro.ipynb)

[7] Mike Driscoll, Jupyter Notebook: An Introduction
(https://realpython.com/jupyter-notebook-introduction/)

神經網路（Neural Network）原理

▎2-1 必備的數學與統計知識

現在每天幾乎都會看到幾則有關人工智慧(AI)的新聞，介紹 AI 的各式研發成果，一般人也許會基於好奇想一窺究竟，了解背後運用的技術與原理，就會發現一堆數學符號及統計公式，可能就會產生疑問：要從事 AI 系統開發，非要搞定數學、統計不可嗎？答案是肯定的，我們都知道機器學習是從資料中學習到知識(Knowledge Discovery from Data, KDD)，而演算法就是從資料中萃取出知識的果汁機，它必須以數學及統計為理論基礎，才能證明其解法具有公信力與精準度，然而數學/統計理論都有侷限，只有在假設成立的情況下，演算法才是有效的，因此，如果不瞭解演算法各個假設，隨意套用公式，就好像無視交通規則，在馬路上任意飆車一樣的危險。

圖片來源：Decision makers need more math[1]

因此，以深度學習而言，我們至少需要熟悉以下學科：

1. 線性代數(Linear Algebra)
2. 微積分(Calculus)
3. 統計與機率(Statistics and Probability)
4. 線性規劃(linear programming)

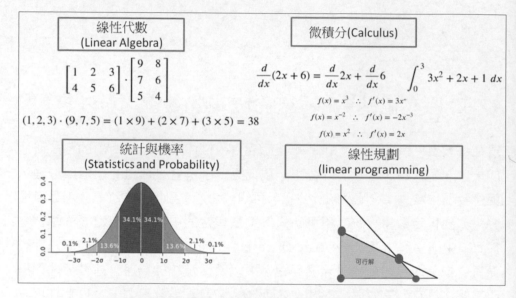

圖 2.1 必備的數學與統計知識

以神經網路優化求解的過程為例，4 門學科就全部用上了，如圖 2.2：

1. 正向傳導：藉由「線性代數」計算誤差及損失函數。
2. 反向傳導：透過「偏微分」計算梯度，同時，利用「線性規劃」優化技巧尋找最佳解。
3. 「統計」則串聯整個環節，例如資料的探索與分析、損失函數定義、效能衡量指標，通通都基於統計的理論架構而成的。
4. 深度學習的推論以「機率」為基礎，預測目標值。

四項學科相互為用，貫穿整個求解過程，因此，要通曉深度學習的運作原理，並且正確選用各種演算法，甚至進而能夠修改或創新演算法，都必須對其背後的數學和統計，有一定基礎的認識，以免誤用/濫用。

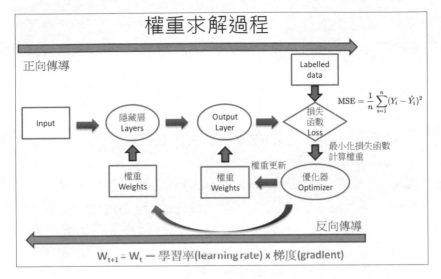

圖 2.2 神經網路權重求解過程

四門學科在大學至少都是兩學期的課程，加起來要總共要 24 個學分，對已經離開大學殿堂很久的工程師而言，在上班之餘，還要重修上述課程，相信大部份的人都會萌生退意了吧！想想看我們是不是有速成的捷徑呢？

筆者在這裡借用一個概念 Statistical Programming， 原意是「以程式解決統訊問題」，換個角度想，我們是不是也能**「以程式設計的方式學統計」**，以縮短學習歷程。

通常要按部就班學習數學及統計，都是從瞭解「假設」➔「定義/定理」➔「證明」➔「應用」，一步一步學習：

- 「假設」是「定義/定理」成立的前提。
- 「證明」是「定理」的驗證。
- 「應用」是「定義/定理」的實踐。

由於「證明」都會有一堆的數學符號及公式推導，常會讓人頭暈腦脹，降低學習的動力，因此，筆者大膽建議，工程師將心力著重在假設、定義、定理的理解與應用，並利用程式進行大量個案的驗證，對已進入職場的工程師會是一條較可行的捷徑，雖然忽略證明的作法，會讓學習無法徹底的融會貫通。

接下來我們就以上述的作法，對四項學科作一重點式的介紹，除了說明深度學習需要理解的知識外，更強調如何「**以撰寫程式實現相關理論及解題方法**」。

2-2 線性代數(Linear Algebra)

張量(Tensor)是描述向量空間(vector space)中物體的特徵，包括零維的純量、一維的向量(Vector)、二維的矩陣(Matrix)或更多維度的張量，線性代數則是說明張量如何進行各種運算，它被廣泛應用於各種數值分析的領域。以下就以實例說明張量的概念與運算。

2-2-1 向量(Vector)

向量(Vector)是一維的張量，它與線段的差別是除了長度(Magnitude)以外，還有方向(direction)，數學表示法為：

$$\vec{v} = \begin{bmatrix} 2 \\ 1 \end{bmatrix}$$

以圖形表示如下：

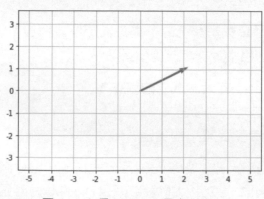

圖 2.3 向量(Vector) 長度與方向

以下使用程式計算向量的長度與方向，請參閱 02_2_1_線性代數_向量.ipynb。

1. 長度(magnitude)：計算公式為歐幾里德距離(Euclidean distance)。

$$\| \vec{v} \| = \sqrt{v_1^2 + v_2^2} = \sqrt{5}$$

程式撰寫如下：

```
1  # 向量(Vector)
2  v = np.array([2,1])
3
4  # 向量長度(magnitude)計算
5  (v[0]**2 + v[1]**2) ** (1/2)
```

也可以使用 np.linalg.norm() 計算向量長度：

```
1  # 使用 np.linalg.norm() 計算向量長度(magnitude)
2  import numpy as np
3
4  magnitude = np.linalg.norm(v)
5  print(magnitude)
```

2. 方向(direction)：使用 $\tan^{-1}()$ 函數計算。

$$\tan(\theta) = \frac{1}{2}$$

移項如下：

$$\theta = \tan^{-1}(\frac{1}{2}) \approx 26.57°$$

```
1   import math
2   import numpy as np
3
4   # 向量(Vector)
5   v = np.array([2,1])
6
7   vTan = v[1] / v[0]
8   print ('tan(θ) = 1/2')
9
10  theta = math.atan(vTan)
11  print('弧度(radian) =', round(theta,4))
12  print('角度(degree) =', round(theta*180/math.pi, 2))
13
14  # 也可以使用 math.degrees() 轉換角度
15  print('角度(degree) =', round(math.degrees(theta), 2))
```

3. 向量四則運算規則
   - 加減乘除一個常數：常數直接對每個元素作加減乘除。
   - 加減乘除另一個向量：兩個向量的相同位置的元素作加減乘除，所以兩個向量的元素個數須相等。

4. 向量加減法：加減一個常數，長度、方向均改變。

程式撰寫如下：

```
1   # 載入套件
2   import numpy as np
3   import matplotlib.pyplot as plt
4
5   # 向量(Vector) + 2
6   v = np.array([2,1])
7   v1 = np.array([2,1]) + 2
8   v2 = np.array([2,1]) - 2
9
10  # 原點
11  origin = [0], [0]
12
13  # 畫有箭頭的線
14  plt.quiver(*origin, *v1, scale=10, color='r')
15  plt.quiver(*origin, *v, scale=10, color='b')
16  plt.quiver(*origin, *v2, scale=10, color='g')
17
18  plt.annotate('orginal vector',(0.025, 0.01), xycoords='data'
19              , fontsize=16)
20
21  # 作圖
22  plt.axis('equal')
23  plt.grid()
24
25  plt.xticks(np.arange(-0.05, 0.06, 0.01), labels=np.arange(-5, 6, 1))
26  plt.yticks(np.arange(-3, 5, 1) / 100, labels=np.arange(-3, 5, 1))
27  plt.show()
```

執行結果：

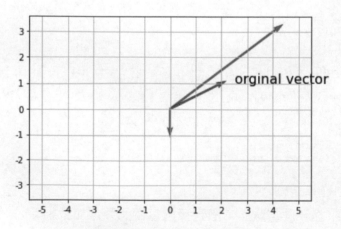

圖 2.4 向量加減一個常數，長度、方向均改變

5. 向量乘除法：乘除一個常數，長度改變、方向不改變。

```
1  # 載入套件
2  import numpy as np
3  import matplotlib.pyplot as plt
4
5  # 向量(Vector) * 2
6  v = np.array([2,1])
7  v1 = np.array([2,1]) * 2
8  v2 = np.array([2,1]) / 2
9
10 # 原點
11 origin = [0], [0]
12
13 # 畫有箭頭的線
14 plt.quiver(*origin, *v1, scale=10, color='r')
15 plt.quiver(*origin, *v, scale=10, color='b')
16 plt.quiver(*origin, *v2, scale=10, color='g')
17
18 plt.annotate('orginal vector',(0.025, 0.008), xycoords='data'
19             , color='b', fontsize=16)
20
21 # 作圖
22 plt.axis('equal')
23 plt.grid()
24
25 plt.xticks(np.arange(-0.05, 0.06, 0.01), labels=np.arange(-5, 6, 1))
26 plt.yticks(np.arange(-3, 5, 1) / 100, labels=np.arange(-3, 5, 1))
27 plt.show()
```

執行結果：

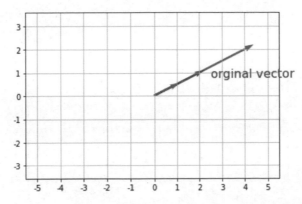

圖 2.5 向量乘除一個常數，長度改變、方向不改變

6. 向量加減乘除另一個向量：兩個向量的相同位置的元素作加減乘除。

```
1  # 載入套件
2  import numpy as np
3  import matplotlib.pyplot as plt
4
5  # 向量(Vector) * 2
6  v = np.array([2,1])
7  s = np.array([-3,2])
8  v2 = v+s
9
10 # 原點
11 origin = [0], [0]
12
13 # 畫有箭頭的線
14 plt.quiver(*origin, *v, scale=10, color='b')
15 plt.quiver(*origin, *s, scale=10, color='b')
16 plt.quiver(*origin, *v2, scale=10, color='g')
17
18 plt.annotate('orginal vector',(0.025, 0.008), xycoords='data'
19            , color='b', fontsize=16)
20
21 # 作圖
22 plt.axis('equal')
23 plt.grid()
24
25 plt.xticks(np.arange(-0.05, 0.06, 0.01), labels=np.arange(-5, 6, 1))
26 plt.yticks(np.arange(-3, 5, 1) / 100, labels=np.arange(-3, 5, 1))
27 plt.show()
```

執行結果：

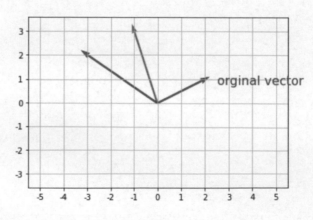

圖 2.6 向量加另一個向量

7. 「內積」或稱「點積」(Dot Product)。

$$\vec{v} \cdot \vec{s} = (v_1 \cdot s_1) + (v_2 \cdot s_2) \ldots + (v_n \cdot s_n)$$

NumPy 是以 **@** 作為內積的運算符號，而非 *。

```
1   # 載入套件
2   import numpy as np
3
4   # 向量(Vector)
5   v = np.array([2,1])
6   s = np.array([-3,2])
7
8   # 『內積』或稱『點積乘法』(Dot Product)
9   d = v @ s
10
11  print (d)
```

8. 計算兩個向量的夾角，公式如下：

$$\vec{v} \cdot \vec{s} = \|\vec{v}\| \|\vec{s}\| \cos(\theta)$$

移項：

$$\cos(\theta) = \frac{\vec{v} \cdot \vec{s}}{\|\vec{v}\| \|\vec{s}\|}$$

再利用 $\cos^{-1}()$ 計算夾角 θ。

```
1   # 載入套件
2   import math
3   import numpy as np
4
5   # 向量(Vector)
6   v = np.array([2,1])
7   s = np.array([-3,2])
8
9   # 計算長度(magnitudes)
10  vMag = np.linalg.norm(v)
11  sMag = np.linalg.norm(s)
12
13  # 計算 cosine(θ)
14  cos = (v @ s) / (vMag * sMag)
15
16  # 計算 θ
17  theta = math.degrees(math.acos(cos))
18
19  print(theta)
```

2-2-2 矩陣(Matrix)

矩陣是二維的張量，擁有列(Row)與行(Column)，可用以表達一個平面 N 個點(Nx2)、或一個 3D 空間 N 個點(Nx3)，例如：

$$A = \begin{bmatrix} 1 & 2 & 3 \\ 4 & 5 & 6 \end{bmatrix}$$

矩陣加法/減法與向量相似，相同位置的元素作運算即可，但乘法運算通常是指內積(dot product)，使用@。

➢ **以下程式請參考 02_2_2_線性代數_矩陣.ipynb。**

1. 試對兩個矩陣相加：

$$\begin{bmatrix} 1 & 2 & 3 \\ 4 & 5 & 6 \end{bmatrix} + \begin{bmatrix} 6 & 5 & 4 \\ 3 & 2 & 1 \end{bmatrix} = \begin{bmatrix} 7 & 7 & 7 \\ 7 & 7 & 7 \end{bmatrix}$$

程式撰寫如下：

```
1   # 載入套件
2   import numpy as np
3
4   # 矩陣
5   A = np.array([[1,2,3],
6                 [4,5,6]])
7   B = np.array([[6,5,4],
8                 [3,2,1]])
9
10  # 加法
11  print(A + B)
```

2. 試對兩個矩陣相乘

$$\begin{bmatrix} 1 & 2 & 3 \\ 4 & 5 & 6 \end{bmatrix} \cdot \begin{bmatrix} 9 & 8 \\ 7 & 6 \\ 5 & 4 \end{bmatrix} = \quad ?$$

解題：左邊矩陣的第 2 維須等於右邊矩陣的第 1 維，即(m, k) x (k, n) = (m, n)

$$\begin{bmatrix} 1 & 2 & 3 \\ 4 & 5 & 6 \end{bmatrix} \cdot \begin{bmatrix} 9 & 8 \\ 7 & 6 \\ 5 & 4 \end{bmatrix} = \begin{bmatrix} 38 & 32 \\ 101 & 86 \end{bmatrix}$$

其中左上角的計算過程為 $(1,2,3) \cdot (9,7,5)=(1\times9)+(2\times7)+(3\times5)=38$，
右上角的計算過程為 $(1,2,3) \cdot (8,6,4)=(1\times8)+(2\times6)+(3\times4)=32$，
以此類推，如下圖。

$$\left[\begin{array}{ccc} 1 & 2 & 3 \\ 4 & 5 & 6 \end{array}\right] \cdot \left[\begin{array}{cc} 9 & 8 \\ 7 & 6 \\ 5 & 4 \end{array}\right] = \left[\begin{array}{cc} \textcircled{38} & 32 \\ 101 & 86 \end{array}\right]$$

$$\left[\begin{array}{ccc} 1 & 2 & 3 \\ 4 & 5 & 6 \end{array}\right] \cdot \left[\begin{array}{cc} 9 & 8 \\ 7 & 6 \\ 5 & 4 \end{array}\right] = \left[\begin{array}{cc} 38 & \textcircled{32} \\ 101 & 86 \end{array}\right]$$

$$\left[\begin{array}{ccc} 1 & 2 & 3 \\ 4 & 5 & 6 \end{array}\right] \cdot \left[\begin{array}{cc} 9 & 8 \\ 7 & 6 \\ 5 & 4 \end{array}\right] = \left[\begin{array}{cc} 38 & 32 \\ \textcircled{101} & 86 \end{array}\right]$$

$$\left[\begin{array}{ccc} 1 & 2 & 3 \\ 4 & 5 & 6 \end{array}\right] \cdot \left[\begin{array}{cc} 9 & 8 \\ 7 & 6 \\ 5 & 4 \end{array}\right] = \left[\begin{array}{cc} 38 & 32 \\ 101 & \textcircled{86} \end{array}\right]$$

圖 2.7 矩陣相乘

```
1   # 載入套件
2   import numpy as np
3
4   # 矩陣
5   A = np.array([[1,2,3],
6                 [4,5,6]])
7   B = np.array([[9,8],
8                 [7,6],
9                 [5,4],
10                ])
11
12  # 乘法
13  print(A @ B)
```

3. 矩陣(A、B)相乘，A x B 是否等於 B x A ？

```
1   # 乘法：A x B != B x A
2
3   A = np.array([[1,2],
4                 [4,5]])
5   B = np.array([[9,8],
6                 [7,6],
7                 ])
8
9   print(A @ B)
10  print()
11  print(B @ A)
12  print()
13  print('A x B != B x A')
```

執行結果：A x B 不等於 B x A。

```
[[23 20]
 [71 62]]

[[41 58]
 [31 44]]

A x B != B x A
```

4. 矩陣在運算時，除了一般的加減乘除外，還有一些特殊的矩陣，包括轉置矩陣(Transpose)、反矩陣(Inverse)、對角矩陣(Diagonal matrix)、單位矩陣(Identity matrix) 等。

   • 轉置矩陣：列與行互換。

$$\begin{bmatrix} 1 & 2 & 3 \\ 4 & 5 & 6 \end{bmatrix}^T = \begin{bmatrix} 1 & 4 \\ 2 & 5 \\ 3 & 6 \end{bmatrix}$$

$(A^T)^T = A$：進行兩次轉置，會回復成原來的矩陣。

5. 對上述矩陣作轉置。

```
1  import numpy as np
2
3  A = np.array([[1,2,3],
4                [4,5,6]])
5
6  # 轉置矩陣
7  print(A.T)
```

也可以使用 np.transpose(A)。

6. 反矩陣(A^{-1})：A 必須為方陣，即列數與行數須相等，且必須是非奇異方陣(non-singular)，即每一列或行之間不可以相依於其他列或行。

```
1  import numpy as np
2
3  A = np.array([[1,2,3],
4                [4,5,6],
5                [7,8,9],
6                ])
7  print(np.linalg.inv(A))
```

執行結果：

```
[[ 3.15251974e+15 -6.30503948e+15  3.15251974e+15]
 [-6.30503948e+15  1.26100790e+16 -6.30503948e+15]
 [ 3.15251974e+15 -6.30503948e+15  3.15251974e+15]]
```

7. 若 A 為非奇異(Non-singular)矩陣，則 A @ A^{-1}=單位矩陣(I)。所謂的非奇異矩陣是任一列不能為其他列的倍數或多列的組合，包括各種四則運算。矩陣的行也須符合相同的規則。

8. 試對下列矩陣驗算 A @ A^{-1} 是否等於單位矩陣(I)。

$$A = \begin{bmatrix} 9 & 8 \\ 7 & 6 \end{bmatrix}$$

```
1  # A @ A反矩陣 = 單位矩陣(I)
2  A = np.array([[9,8],
3                [7,6],
4                ])
5
6  print(np.around(A @ np.linalg.inv(A)))
```

執行結果：

```
[[1. 0.]
 [0. 1.]]
```

結果為單位矩陣，表示 A 為非奇異(Non-singular)矩陣。

9. 試對下列矩陣驗算 A @ A^{-1} 是否等於單位矩陣(I)。

$$A = \begin{bmatrix} 1 & 2 & 3 \\ 4 & 5 & 6 \\ 7 & 8 & 9 \end{bmatrix}$$

```
1   # A @ A反矩陣 != 單位矩陣(I)
2   # A 為 singular 矩陣
3   # 第二行 = 第一行 + 1
4   # 第三行 = 第一行 + 2
5   A = np.array([[1,2,3],
6                 [4,5,6],
7                 [7,8,9],
8                 ])
9
10  print(np.around(A @ np.linalg.inv(A)))
```

執行結果：

```
[[ 0.  1. -0.]
 [ 0.  2. -1.]
 [ 0.  3.  2.]]
```

A 為奇異(Singular)矩陣,因為

第二行 = 第一行 + 1

第三行 = 第一行 + 2

故 A @ A^{-1} 不等於單位矩陣。

2-2-3 聯立方程式求解

在國中階段,我們通常會以高斯消去法(Gaussian Elimination)解聯立方程式。以下列方程式為例:

$$x + y = 16$$
$$10x + 25y = 250$$

第一個方程式乘以-10,加上第二個方程式,即可消去 x,變成:

$$-10(x + y) = -10(16)$$
$$10x + 25y = 250$$

簡化為:

$$15y = 90$$

得到 y=6,再代入任一方程式,得到 x=10。

以上過程,如果以線性代數求解就簡單多了:

- 以矩陣表示如下,A:方程式中未知數(x, y)的係數,B 為等號右邊的常數:

  AX = B
- 其中 A = $\begin{bmatrix} 1 & 1 \\ 10 & 25 \end{bmatrix}$ X = $\begin{bmatrix} x \\ y \end{bmatrix}$ B = $\begin{bmatrix} 16 \\ 250 \end{bmatrix}$

  則 X = A^{-1} B

證明如下:

- 兩邊各乘 A^{-1}:

  - A^{-1}AX = A^{-1} B
- A^{-1}A 等於單位矩陣,且任一矩陣乘以單位矩陣,還是等於原矩陣,故

  - X = A^{-1} B
- 以上式求得(x, y),注意,前提是 A 須為非奇異(Non-singular)矩陣。

> ➤ **以下程式均收錄在 02_2_3_聯立方程式求解.ipynb。**

1. 以 NumPy 套件求解上述聯立方程式。

```python
 1  # x+y=16
 2  # 10x+25y=250
 3
 4  # 載入套件
 5  import numpy as np
 6
 7  # 定義方程式的 A、B
 8  A = np.array([[1 , 1], [10, 25]])
 9  B = np.array([16, 250])
10  print('A=')
11  print(A)
12  print('')
13  print('B=')
14  print(B.reshape(2, 1))
15
16  # np.linalg.solve : 線性代數求解
17  print('\n線性代數求解：')
18  print(np.linalg.inv(A) @ B)
```

- inv(A)：A 的反矩陣。
- 執行結果：x=10，y=6。
- 也可以直接呼叫 np.linalg.solve() 函數求解。

```python
 1  print(np.linalg.solve(A, B))
```

2. 畫圖，交叉點即聯立方程式的解。

```python
 1  # x+y=16
 2  # 10x+25y=250
 3
 4  # 載入套件
 5  import numpy as np
 6  from matplotlib import pyplot as plt
 7
 8  # 取第一個方程式的兩端點
 9  A1 = [16, 0]
10  A2 = [0, 16]
11
12  # 取第二個方程式的兩端點
13  B1 = [25,0]
14  B2 = [0,10]
15
16  # 畫線
17  plt.plot(A1,A2, color='blue')
18  plt.plot(B1, B2, color="orange")
19  plt.xlabel('x')
20  plt.ylabel('y')
21  plt.grid()
```

```
22
23  # 交叉點 (10, 6)
24  plt.scatter([10], [6], color="red", s=100)
25  plt.show()
```

執行結果:

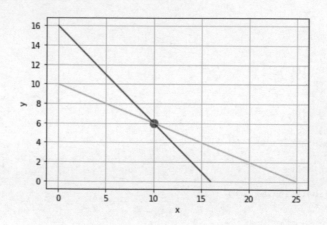

圖 2.8 交叉點為聯立方程式的解

3. 以 NumPy 套件求解下列聯立方程式。

$$-1x + 3y = -72$$

$$3x + 4y - 4z = -4$$

$$-20x - 12y + 5z = -50$$

```
1   # 載入套件
2   import numpy as np
3
4   # 定義方程式的 A、B
5   A = np.array([[-1, 3, 0], [3, 4, -4], [-20, -12, 5]])
6   B = np.array([-72, -4, -50])
7   print('A=')
8   print(A)
9   print('')
10  print('B=')
11  print(B.reshape(3, 1))
12
13  # np.linalg.solve : 線性代數求解
14  print('\n線性代數求解:')
15  print(np.linalg.solve(A, B))
```

- 執行結果:(x, y, z) = (12, -20, -10)。

- 也可以使用 SymPy 套件求解,直接將聯立方程式整理在等號左方,呼叫 solve()函數,參數內的多項式均假設等號(=)右方為 0。

```
1  from sympy.solvers import solve
2  from sympy import symbols
3
4  # 設定變數
5  x, y, z = symbols('x y z')
6
7  # 解題：設定聯立方程式，等號右邊預設為 0
8  solve([-1*x + 3*y + 72, 3*x + 4*y - 4*z + 4,-20*x + -12*y + 5*z + 50])
```

2-3 微積分(Calculus)

微積分包括微分(Differentiation)與積分(Integration)，微分是描述函數某一點的變化率，藉由微分可以得到特定點的斜率(Slope)或梯度(Gradient)，即變化的速度，二次微分可以得到加速度。積分則是微分的相反，互為逆運算，積分可以計算長度、面積、體積，也可以用來計算累積的機率密度函數(Probability density function)。

在機器學習中，微分在求解的過程中佔有舉足輕重的地位，因此 TensorFlow 套件就提供自動微分(Automatic Differentiation)的功能，用以計算各特徵變數的梯度(gradient)，進而求得模型的權重(Weight)參數。

2-3-1 微分(Differentiation)

微分是描述函數的變化率(Rate of Change)，例如 y=2x+5，表示 x 每增加一單位，y 會增加 2，因此，變化率就等於 2，也稱為「斜率」(Slope)，5 為「截距」(Intercept)，或稱「偏差」(Bias)。

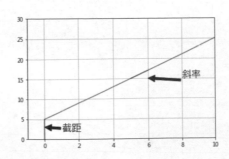

圖 2.9 斜率(Slope)與截距(Intercept)

我們先不管截距,只看斜率,算法如下,取非常相近的兩個點(距離 h 趨近於 0),y 座標值之差(Δy)除以 x 座標值之差(Δx):

$$f'(x) = \frac{\Delta y}{\Delta x} = \lim_{h \to 0} \left(\frac{f(x+h)-f(x)}{h} \right)$$

這就是微分的定義,但上述極限值(lim)不一定存在,存在的要素如下:

1. h 為正值時的極限值 = h 為負值時的極限值,亦即函數在該點時是連續 (Continuity)的。

2. 上述極限值不等於無窮大(∞)或負無窮大(-∞)。

以下函數在 x=5 的地方是連續的,由上方(5.25)逼近,或由下方(4.75)逼近是相等的,相關彩色圖形可參考 02_3_1_微分.ipynb。

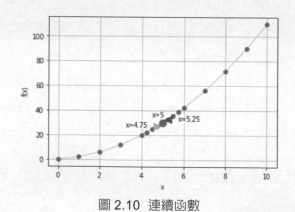

圖 2.10 連續函數

相反的,下圖在 x=0 時,是不連續的,逼近 x=0 時有兩個解。

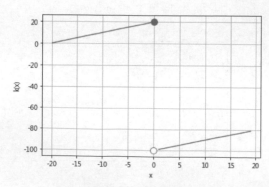

圖 2.11 不連續函數

接著來看看幾個應用實例。

➤ **以下程式請參考 02_3_1_微分**.ipynb。

1. 試繪製一次方函數 $f(x)=2x+5$。

```python
1  # 載入套件
2  import numpy as np
3  import matplotlib.pyplot as plt
4
5  # 樣本點
6  x = np.array(range(0, 11))
7  y = 2 * x + 5
8
9  # 作圖
10 plt.grid()
11 plt.plot(x, y, color='g')
12 pll.xlim(-1, 10)
13 plt.ylim(0, 30)
14
15 # 截距 (Intercept)
16 x = [0, 0]
17 y = [0, 5]
18 plt.plot(x, y, color="r")
19 plt.annotate("截距", xy=(0,3), xytext=(1, 2), xycoords='data',
20               arrowprops=dict(facecolor='black'), fontsize=18)
21
22 # 斜率(Slope)
23 x = np.array([5, 6])
24 y = 2 * x + 5
25 plt.plot(x, y, color="r")
26 plt.annotate("斜率", xy=(6,15), xytext=(8, 15), xycoords='data',
27               arrowprops=dict(facecolor='black'), fontsize=18)
28
29 plt.show()
```

執行結果：

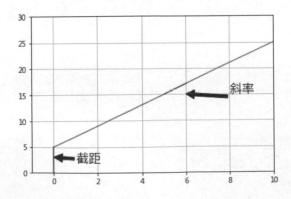

一次方函數每一點的斜率均相同。

2. 試繪製二次方曲線 $f(x)=-10x^2+100x+5$，求最大值。

```python
1  from matplotlib import pyplot as plt
2
3  # 二次曲線
4  def f(x):
5      return -10*(x**2) + (100*x)  + 5
6
7  # 一階導數
8  def fd(x):
9      return -20*x + 100
10
11 # 設定樣本點
12 x = list(range(0, 11))
13 y = [f(i) for i in x]
14
15 # 一階導數的樣本點
16 yd = [fd(i) for i in x]
17
18 # 畫二次曲線
19 plt.plot(x,y, color='green', linewidth=1)
20
21 # 畫一階導數
22 plt.plot(x,yd, color='purple', linewidth=3)
23
24 # 畫三個點的斜率 x = (2, 5, 8)
25 x1 = 2
26 x2 = 5
27 x3 = 8
28 plt.plot([x1-1,x1+1],[k(x1)-(fd(x1)),f(x1)+(fd(x1))], color='red', linewidth=3)
29 plt.plot([x2-1,x2+1],[k(x2)-(fd(x2)),f(x2)+(fd(x2))], color='red', linewidth=3)
30 plt.plot([x3-1,x3+1],[k(x3)-(fd(x3)),f(x3)+(fd(x3))], color='red', linewidth=3)
31
32 # 最大值
33 plt.axvline(5)
34
35 plt.grid()
36 plt.show()
```

執行結果：

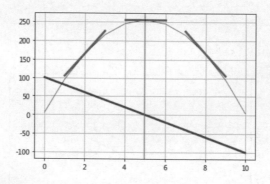

■ 一次方函數整條線上的每一個點的斜率都相同，但是二次方曲線上的每一個點的斜率就都不一樣了，如上圖，相關彩色圖形可參考 02_3_1_微分.ipynb：

• 綠線(細拋物線)：二次曲線，是一條對稱的拋物線。

- 紫線(斜線)：拋物線的一階導數。
- 紅線(拋物線的切線)：三個點(2, 5, 8)的斜率。
- 每一個點的斜率即該點與二次曲線的切線(紅線)，均不相同，斜率值可透過微分求得一階導數(圖中的斜線)，隨著 x 變大，斜率越來越小，二次曲線的最大值就發生在斜率=0 的地方，當 x=5 時，f(x)=255。

3. 試繪製二次方曲線 $f(x)=x^2+2x+7$，求最小值。

```
 1  from matplotlib import pyplot as plt
 2
 3  # 二次曲線
 4  def f(x):
 5      return (x**2) + (2*x) + 7
 6
 7  # 一階導數
 8  def fd(x):
 9      return 2*x + 2
10
11  # 設定樣本點
12  x = list(range(-10, 11))
13  y = [f(i) for i in x]
14
15  # 一階導數的樣本點
16  yd = [fd(i) for i in x]
17
18  # 畫二次曲線
19  plt.plot(x,y, color='green')
20
21  # 畫一階導數
22  plt.plot(x,yd, color='purple', linewidth=3)
23
24  # 畫三個點的斜率 x = (-7, -1, 5)
25  x1 = 5
26  x2 = -1
27  x3 = -7
28  plt.plot([x1-1,x1+1],[f(x1)-(fd(x1)),f(x1)+(fd(x1))], color='red', linewidth=3)
29  plt.plot([x2-1,x2+1],[f(x2)-(fd(x2)),f(x2)+(fd(x2))], color='red', linewidth=3)
30  plt.plot([x3-1,x3+1],[f(x3)-(fd(x3)),f(x3)+(fd(x3))], color='red', linewidth=3)
31
32  # 最大值
33  plt.axvline(-1)
34
35  plt.grid()
36  plt.show()
```

執行結果：

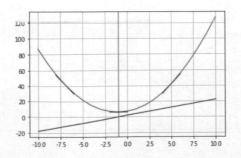

■ 斜率值可透過微分求得一階導數(圖中的斜線)，隨著 x 變大，斜率越來越大，二次曲線的最小值就發生在斜率=0 的地方，當 x=-1 時，f(x)=6。

綜合範例 2、3，可以得知微分兩次的**二階導數(f''(x))為常數**，且為正值時，函數有最小值，反之，為負值時，函數有最大值。但若是 **f(x) 為三次方(以上)的函數**，一階導數=0 的點，可能只是區域的最佳解(Local Minimum/Maximum)，而不是全局最佳解(Global Minimum/Maximum)。

4. 試繪製三次方曲線 $f(x) = x^3 - 2x + 100$，求最小值。

```
1   from matplotlib import pyplot as plt
2   import numpy as np
3
4   # f(x)= x^3-2x+100
5   def f(x):
6       return (x**3) - (2*x) + 100
7
8   # 一階導數
9   def fd(x):
10      return 3*(x**2) - 2
11
12  # 設定樣本點
13  x = list(range(-10, 11))
14  y = [f(i) for i in x]
15
16  # 一階導數的樣本點
17  yd = [fd(i) for i in x]
18
19  # 畫二次曲線
20  plt.plot(x,y, color='green')
21
22  # 畫一階導數
23  plt.plot(x,yd, color='purple', linewidth=3)
24
25  # 最小值
26  x1=np.array([sqrt(6)/3, -sqrt(6)/3])
27  plt.scatter([x1], [f(x1)])
28  plt.scatter([-sqrt(6)/3], [f(-1)])
29
30  plt.grid()
31  plt.show()
```

執行結果：

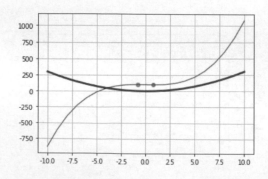

- 三次方曲線 $f(x)= x^3−2x+100$ 在斜率=0 的點只是區域的最佳解。
- 一般為凸函數才有全局最佳解。

2-3-2 微分定理

上述程式中的 kd() 函數為一階導數，它們是如何求得？只要運用以下微分的定理，就可以輕易解出上述範例的一階導數，相關定理整理如下：

1. $f(x)$ 一階導數的表示法為 $f'(x)$ 或 $\frac{dy}{dx}$。

2. $f(x)$ 為常數(C) ➜ $f'(x) = 0$

3. $f(x) = Cg(x)$ ➜ $f'(x) − Cg'(x)$

4. $(x)=g(x)+h(x)$ ➜ $f'(x)=g'(x)+h'(x)$

5. 次方的規則：$(x)=xn$ ➜ $f'(x)=nx^{n-1}$

6. 乘積的規則：$\frac{d[f(x)g(x)]}{dx} = f'(x)(x)+f(x)g'(x)$

7. 商的規則：若 $r(x)=s(x) / t(x)$

$$r'(x) = \frac{s'(x)t(x) - s(x)t'(x)}{[t(x)]^2}$$

8. 連鎖律(Chain Rule)

$$\frac{d}{dx}[o(i(x))] = o'(i(x)) \cdot i'(x)$$

以範例 2 為例 $f(x)=−10x^2 +100x+5$ ，針對多項式的每一項個別微分再相加，就得到 $f(x)$ 的一階導數：

$$f'(x) = -20x + 100$$

SymPy 套件直接支援微積分函數的計算，可以驗證定理，接下來我們就寫一些程式來練習一下。

➢ **以下程式請參考 02_3_1_微分.ipynb。**

1. $f(x)$ 為常數(C) ➜ $f'(x) = 0$

```
1  # 常數微分 f(x) = C ==> f'(x) = 0
2  from sympy import *
3
4  x = Symbol('x')
5  # f(x) 為常數
6  y = 0 * x + 5
7  yprime = y.diff(x)
8  yprime
```

執行結果：0。

2. f(x) = Cg(x) ➜ f'(x) = Cg'(x)

```
1  # f(x) = Cg(x) ==> f'(x) = Cg'(x)
2
3  from sympy import *
4
5  x = Symbol('x')
6
7  # Cg(x)
8  y1 = 5 * x ** 2
9  yprime1 = y1.diff(x)
10 print(yprime1)
11
12 # g(x)
13 y2 = x ** 2
14 # Cg'(x)
15 yprime2 = 5 * y2.diff(x)
16 print(yprime2)
17
18 # 比較
19 yprime1 == yprime1
```

執行結果均為 10x。

乘積的規則：$\dfrac{d[f(x)g(x)]}{dx} = f'(x)(x)+f(x)g'(x)$

```
1  # (d[f(x)g(x)])/dx = f'(x)g(x)+f(x)g'(x)
2
3  from sympy import *
4
5  x = Symbol('x')
6
7  # d[f(x)g(x)])/dx
8  f = x ** 2
9  g = x ** 3
10 y1 = f*g
11 yprime1 = y1.diff(x)
12 print(yprime1)
13
14 # f'(x)g(x)+f(x)g'(x)
15 yprime2 = f.diff(x) * g + f * g.diff(x)
16 print(yprime2)
17
18 # 比較
19 yprime1 == yprime1
```

執行結果：$5x^4$。

連鎖律(Chain Rule)：

$$\frac{d}{dx}[o(i(x))] = o'(i(x)) \cdot i'(x)$$

```
1   from sympy import *
2
3   x = Symbol('x')
4
5   # d[f(g(x))])/dx
6   g = x ** 3
7   f = g ** 2
8   yprime1 = f.diff(x)
9   print(yprime1)
10
11  # f'(g(x))g'(x)
12  g = Symbol('g')
13  f = g ** 2
14  g1 = x ** 3
15  yprime2 = f.diff(g) * g1.diff(x)
16  # 將 f'(g(x)) 的 g 以 x ** 3 取代
17  print(yprime2.subs({g:x ** 3}))
18
19  # 比較
20  yprime1 == yprime1
```

執行結果：$6x^5$。

驗證 $f(x)=-10x^2+100x+5$

```
1   from sympy import *
2
3   x = Symbol('x')
4   # f(x)=-10x2 +100x+5
5   y = -10 * x**2 + 100 * x + 5
6   yprime = y.diff(x)
7   yprime
```

執行結果：$100 - 20x$。

接著，利用一階導數=0，求最大值。

```
1   from sympy.solvers import solve
2
3   # 一階導數=0
4   dict1 = solve([yprime])
5   print(dict1)
6   x1 = dict1[x]
7   print(f'x={x1}, 最大值={-10 * x1**2 + 100 * x1 + 5}')
```

執行結果：x=5, 最大值=255。

f(x)=$x2+2x+7$

```
1  from sympy import *
2
3  x = Symbol('x')
4  # f(x)=x2+2x+7
5  y = (x**2) + (2*x) + 7
6  yprime = y.diff(x)
7  yprime
```

執行結果：2x + 2。

接著，利用一階導數=0，求最小值。

```
1  from sympy.solvers import solve
2
3  # 一階導數=0
4  dict1 = solve([yprime])
5  print(dict1)
6  x1 = dict1[x]
7  print(f'x={x1}, 最小值={(x1**2) + (2*x1) + 7}')
```

執行結果：x=-1, 最小值=6。

2-3-2 偏微分(Partial Differentiation)

在機器學習中，偏微分(Partial Differentiation) 在求解的過程中佔有舉足輕重的地位，常用來計算各特徵變數的梯度(gradient)，進而求得最佳權重(Weight)。梯度與斜率相似，單一變數的變化率，稱為斜率，多變數的斜率我們稱之為梯度。

在上一節中，f(x)只有單變數，如果 x 是多個變數 x_1、x_2、x_3……，要如何找到最小值或最大值呢？這時，我們就可以使用偏微分求取每個變數的梯度，讓函數沿著特定方向找最佳解，如下圖，即沿著等高線(Contour)，逐步向圓心逼近，這就是所謂的「梯度下降法」(Gradient Descent)。

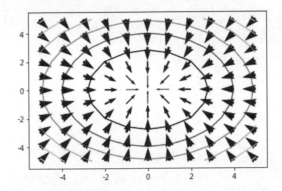

圖 2.12 梯度下降法(Gradient Descent) 圖解

> **以下程式請參考 02_3_2_偏微分**.ipynb。

1. 假設 $f(x,y)=x^2+y^2$，請分別對 x、y 作偏微分。

解題：先針對 x 作偏微分：

$$\frac{\partial f(x, y)}{\partial x} = \frac{\partial (x^2 + y^2)}{\partial x}$$

→ 將其他變數視為常數，對每一項個別微分：

$$\frac{\partial x^2}{\partial x} = 2x$$

$$\frac{\partial y^2}{\partial x} = 0$$

→ 加總：

$$\frac{\partial f(x, y)}{\partial x} = 2x + 0 = 2x$$

→再針對 y 作偏微分，同樣將其他變數視為常數

$$\frac{\partial f(x, y)}{\partial y} = 0 + 2y = 2y$$

→ 總結：

$$\frac{\partial f(x, y)}{\partial x} = 2x$$

$$\frac{\partial f(x, y)}{\partial y} = 2y$$

```
1  # f(x, y) = x^2 + y^2
2
3  from sympy import *
4  x, y = symbols('x y')
5
6  # 對 x 偏微分
7  y1 = x**2 + y**2
8  yprime1 = y1.diff(x)
9  print(yprime1)
10
11  # 對 y 偏微分
12  yprime2 = y1.diff(y)
13  print(yprime2)
14
```

第 8、11 行：y1.diff(x)、y1.diff(y) 分別對 x、y 作偏微分。

2. 假設 $f(x)=x^2$，請使用「梯度下降法」(Gradient Descent) 找最小值。

解題：程式邏輯如下。

- 任意設定一起始點(x_start)。
- 計算該點的梯度 fd(x)。
- 沿著梯度更新 x，逐步逼近最佳解，幅度大小以學習率控制。新的 x = x ─ 學習率(learning rate) * 梯度(gradient)
- 重複步驟 2、3，判斷梯度是否接近於 0，若已很逼近，即找到最佳解。

```
1  import numpy as np
2  import matplotlib.pyplot as plt
3
4  # 函數 f(x)=x^2
5  def f(x): return x ** 2
6
7  # 一階導數:dy/dx=2*x
8  def fd(x): return 2 * x
9
10  def GD(x_start, df, epochs, lr):
11      xs = np.zeros(epochs+1)
12      w = x_start
13      xs[0] = w
14      for i in range(epochs):
15          dx = df(w)
16          # 權重的更新
17          # W_NEW = W ─ 學習率(learning rate) x 梯度(gradient)
18          w += - lr * dx
19          xs[i+1] = w
20      return xs
21
22
```

```
23  # 超參數(Hyperparameters)
24  x_start = 5      # 起始權重
25  epochs = 25      # 執行週期數
26  lr = 0.1         # 學習率
```

```
27
28  # 梯度下降法, 函數 fd 直接當參數傳遞
29  w = GD(x_start, fd, epochs, lr=lr)
30  # 顯示每一執行週期得到的權重
31  print (np.around(w, 2))
32
33  # 畫圖
34  color = 'r'
35  from numpy import arange
36  t = arange(-6.0, 6.0, 0.01)
37  plt.plot(t, f(t), c='b')
38  plt.plot(w, f(w), c=color, label='lr={}'.format(lr))
39  plt.scatter(w, f(w), c=color, )
40
41  # 設定中文字型
42  from matplotlib.font_manager import FontProperties
43  font = FontProperties(fname=r"c:\windows\fonts\msjhbd.ttc", size=20)
44  plt.title('梯度下降法', fontproperties=font)
45  plt.xlabel('w', fontsize=20)
46  plt.ylabel('Loss', fontsize=20)
47
48  # 矯正負號
49  plt.rcParams['axes.unicode_minus'] = False
50
51  plt.show()
```

執行結果：

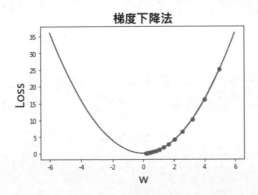

圖 2.13　梯度下降法(Gradient Descent) 實作

得到 x 每一點的座標如下：

```
[5.    4.    3.2   2.56 2.05 1.64 1.31 1.05 0.84 0.67 0.54 0.43 0.34 0.27
 0.22 0.18 0.14 0.11 0.09 0.07 0.06 0.05 0.04 0.03 0.02 0.02]
```

我們可以改變第 24~26 行的參數，觀察執行結果。

1. 改變起始點 x_start = -5，依然可以找到最小值。

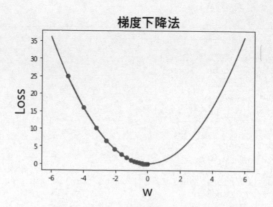

2. 設定學習率 lr = 0.9：如果函數較複雜，可能會跳過最小值。

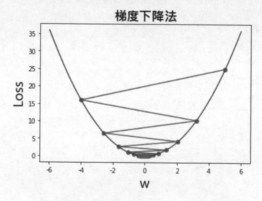

3. 設定學習率 lr = 0.01：還未逼近最小值，就提早停止了，可以增加「執行週期數」，解決問題。

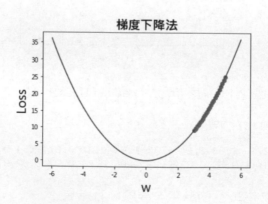

上述程式是神經網路優化器求解的簡化版，在後續的章節會詳細的探討，目前只聚焦在說明偏微分在深度學習的應用。

2-3-3 簡單線性迴歸求解

在優化求解中，也常利用一階導數等於 0 的特性，求取最佳解，例如，以「最小平方法」(Ordinary Least Square, OLS)對簡單線性迴歸 $y = wx + b$ 求解。首先定義「目標函數」 (Object Function) 或稱「損失函數」 (Loss Function) 為「均方誤差」(MSE)，即預測值與實際值差距的平方和，MSE 當然愈小愈好，所以它是一個最小化的問題，我們可以利用偏微分推導出公式，過程如下：

1. $MSE = \sum \varepsilon^2 / n = \sum (y - \hat{y})^2 / n$
   其中 ε：誤差，即實際值(y)與預測值($\hat{y}$)之差
   　　　n：為樣本個數

2. $MSE = SSE / n$，我們忽略常數 n，可以只考慮 SSE：
   $SSE = \sum \varepsilon^2 = \sum (y - \hat{y})^2 = \sum (y - wx - b)^2$

3. 分別為 w 及 b 作偏微分，並且令一階導數=0，可以得到兩個聯立方程式，進而求得 w 及 b 的解。

4. 對 b 偏微分，又因
   $f'(x) = g(x)g(x) = g'(x)g(x) + g(x)g'(x) = 2\,g(x)g'(x)$：

   $$\frac{dSSE}{db} = -2 \sum_{i=1}^{n} (y - wx - b) = 0$$

   → 兩邊同除 -2

   $$\sum_{i=1}^{n} (y - wx - b) = 0$$

   → 分解

   $$\sum_{i=1}^{n} y - \sum_{i=1}^{n} wx - \sum_{i=1}^{n} b = 0$$

→ 除以 n，$\bar{x}$、$\bar{y}$ 為 x、y 的平均數

$$\bar{y} - w\bar{x} - b = 0$$

→ 移項

$$b = \bar{y} - w\bar{x}$$

5. 對 w 偏微分：

$$\frac{dSSE}{dw} = -2\sum_{i=1}^{n}(y - wx - b)x = 0$$

→ 兩邊同除 -2

$$\sum_{i=1}^{n}(y - wx - b)x = 0$$

→ 分解

$$\sum_{i=1}^{n}yx - \sum_{i=1}^{n}wx - \sum_{i=1}^{n}bx = 0$$

→ 代入 4 的計算結果 $b = \bar{y} - w\bar{x}$

$$\sum_{i=1}^{n}yx - \sum_{i=1}^{n}wx - \sum_{i=1}^{n}(\bar{y} - w\bar{x})x = 0$$

→ 化簡

$$\sum_{i=1}^{n}(y - \bar{y})x - w\sum_{i=1}^{n}(x^2 - \bar{x}x) = 0$$

$$w = \sum_{i=1}^{n}(y - \bar{y})x \,/\, \sum_{i=1}^{n}(x^2 - \bar{x}x)$$

$$w = \sum_{i=1}^{n}(y - \bar{y})x \,/\, \sum_{i=1}^{n}(x - \bar{x})^2$$

結論：

$$w = \sum_{i=1}^{n}(y - \bar{y})x \,/\, \sum_{i=1}^{n}(x - \bar{x})^2$$
$$b = \bar{y} - w\bar{x}$$

> **以下程式請參考 02_3_3_線性迴歸.ipynb。**

現有一個世界人口統計資料集，以年度(year)為 x，人口數為 y，依上述公式計算迴歸係數 w、b，撰寫程式如下：

```python
1  # 使用 OLS 公式計算 w、b
2  # 載入套件
3  import matplotlib.pyplot as plt
4  import numpy as np
5  import math
6  import pandas as pd
7
8  # 載入資料集
9  df = pd.read_csv('./data/population.csv')
10
11 w = ((df['pop'] - df['pop'].mean()) * df['year']).sum() \
12     / ((df['year'] - df['year'].mean())**2).sum()
13 b = df['pop'].mean() - w * df['year'].mean()
14
15 print(f'w={w}, b={b}')
```

執行結果：

```
w=0.061159358661557375,  b=-116.35631056117687
```

使用 NumPy 的現成函數 polyfit() 驗算：

```python
1  # 使用 NumPy 的現成函數 polyfit()
2  coef = np.polyfit(df['year'], df['pop'], deg=1)
3  print(f'w={coef[0]}, b={coef[1]}')
```

執行結果：

```
w=0.061159358661554586,  b=-116.35631056117121
```

答案相去不遠，Yeah!

2-3-4 積分(Integration)

積分則是微分的相反，互為逆運算，微分用於斜率或梯度的計算，積分則可以計算長度、面積、體積，也可以用來計算累積的機率密度函數 (Probability density function)。

1. 積分一般數學表示式如下：

$$\int_0^n f(x)dx$$

2. 若 f(x) = x，積分結果為 $\frac{1}{2}x^2$。

3. 限定範圍的積分：先求積分，再將上、下限代入多項式，相減可得結果。

$$\int_0^3 x\,dx = \frac{1}{2}x^2\Big|_0^3$$

➔ $((1/2) * 3^2) - ((1/2) * 0^2) = 4.5$

接下來撰寫程式，看看積分如何計算。

➢ **以下程式請參考 02_3_4_積分.ipynb。**

【範例 1】積分計算並作圖。

$$\int_0^3 x\,dx$$

1. 積分運算：SciPy 套件支援積分運算

```python
1  # 載入套件
2  import numpy as np
3  import scipy.integrate as integrate
4  import numpy as np
5  import math
6
7
8  f = lambda x: x
9  i, e = integrate.quad(f, 0, 3)
10
11 print('積分值: ' + str(i))
12 print('誤差: ' + str(e))
```

執行結果如下：

```
積分值: 4.5
誤差: 4.9960036108132044e-14
```

程式說明：

■ 第 3 行載入 SciPy 套件。

■ 第 9 行呼叫 integrate.quad()，作限定範圍的積分，參數如下：

　● 函數 f(x)

　● 範圍下限：設為負無窮大(-∞)

　● 範圍上限：設為正無窮大(∞)

　● 輸出：含積分結果及誤差值

2. 作圖

```
1   import matplotlib.pyplot as plt
2
3   # 樣本點
4   x = range(0, 11)
5
6   # 作圖
7   plt.plot(x, f(x), color='purple')
8
9   # 積分面積
10  area = np.arange(0, 3, 1/20)
11  plt.fill_between(area, f(area), color='green')
12
13  # 設定圖形屬性
14  plt.xlabel('x')
15  plt.ylabel('f(x)')
16  plt.grid()
17
18  plt.show()
```

執行結果：

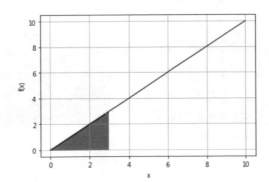

程式説明：

- 第 11 行 fill_between()會填滿整個積分區域。

【範例 2】 計算常態分配(Normal Distribution)的機率並作圖。

1. 常態分配的機率密度函數(Probability Density Function, 簡稱 pdf) 如下：

$$f(x; \mu, \sigma) = \frac{1}{\sqrt{2\pi * \sigma^2}} * e^{-\frac{1}{2} * (\frac{x-\mu}{\sigma})^2}$$

計算常態分配$(-\infty, \infty)$ 的機率。

```
1   # 載入套件
2   import scipy.integrate as integrate
3   import numpy as np
4   import math
5
6
7   # 平均數(mean)、標準差(std)
8   mean = 0
9   std = 1
10  # 常態分配的機率密度函數(Probability Density Function, pdf)
11  f = lambda x: (1/((2*np.pi*std**2) ** .5)) * np.exp(-0.5*((x-mean)/std)**2)
12
13  # 積分，從負無窮大至無窮大
14  i, e = integrate.quad(f, -np.inf, np.inf)
15
16  print('累積機率:', round(i, 2))
17  print('誤差:', str(e))
```

執行結果：

```
累積機率: 1.0
誤差: 1.0178191437091558e-08
```

程式說明：

- 第 7、8 行定義平均數(mean)為 0、標準差(std)為 1，也就是「標準」常態分配，又稱 Z 分配。

- 第 11 行定義常態分配的機率密度函數(pdf)。

- 注意，任何機率分配的總和必然為 1，即所有事件發生的機率總和必然為 100%。

2. 常態分配常見的信賴區間包括正負 1、2、3 倍標準差，我們可以計算其機率。有關信賴區間的定義會在機率與統計的章節說明。

- ±1 倍標準差的信賴區間機率= 68.3%。

```
1   # 1倍標準差區間之機率
2   i, e = integrate.quad(f, -1, 1)
3
4   print('累積機率:', round(i, 3))
5   print('誤差:', str(e))
```

執行結果如下：

```
累積機率: 0.683
誤差: 7.579375928402476e-15
```

integrate.quad() 參數設為 -1 及 1。

■　±2 倍標準差的信賴區間機率= 95.4%。

```
1  # 2倍標準差區間之機率
2  i, e = integrate.quad(f, -2, 2)
3
4  print('累積機率:', round(i, 3))
5  print('誤差:', str(e))
```

執行結果：

```
累積機率: 0.954
誤差: 1.8403560456416134e-11
```

integrate.quad() 參數設為 -2 及 2。

■　±3 倍標準差的信賴區間機率= 99.7%。

```
1  # 3倍標準差區間之機率
2  i, e = integrate.quad(f, -3, 3)
3
4  print('累積機率:', round(i, 3))
5  print('誤差:', str(e))
```

執行結果：

```
累積機率: 0.997
誤差: 1.1072256503105314e-14 程式說明:
```

integrate.quad() 參數設為 -3 及 3。

■　我們常用 ±1.96 倍標準差的信賴區間，機率=95%，剛好是個整數。

```
1  # 95%信賴區間之機率
2  i, e = integrate.quad(f, -1.96, 1.96)
3
4  print('累積機率:', round(i, 3))
5  print('誤差:', str(e))
```

執行結果：

```
累積機率: 0.95
誤差: 1.0474096492701325e-11
```

integrate.quad() 參數設為 -1.96 及 1.96。

■　另外，可以使用隨機亂數及直方圖，繪製標準常態分配圖。

```
1  import matplotlib.pyplot as plt
2  import numpy as np
3  import seaborn as sns
4
5  # 使用 randn 產生標準常態分配的亂數 10000 個樣本點
6  x = np.random.randn(10000)
7
8  # 直方圖，參數 hist=False：不畫階梯直方圖，只畫平滑曲線
9  sns.distplot(x, hist=False)
10
11 # 設定圖形屬性
12 plt.xlabel('x')
13 plt.ylabel('f(x)')
14 plt.grid()
15
16 # 對正負1、2、3倍標準差畫虛線
17 plt.axvline(-3, c='r', linestyle=':')
18 plt.axvline(3, c='r', linestyle=':')
19 plt.axvline(-2, c='g', linestyle=':')
20 plt.axvline(2, c='g', linestyle=':')
21 plt.axvline(-1, c='b', linestyle=':')
22 plt.axvline(1, c='b', linestyle=':')
23
24 # 1 倍標準差機率
25 plt.annotate(text='', xy=(-1,0.25), xytext=(1,0.25), arrowprops=dict(arrowstyle='<->'))
26 plt.annotate(text='68.3%', xy=(0,0.26), xytext=(0,0.26))
27 # 2 倍標準差機率
28 plt.annotate(text='', xy=(-2,0.05), xytext=(2,0.05), arrowprops=dict(arrowstyle='<->'))
29 plt.annotate(text='95.4%', xy=(0,0.06), xytext=(0,0.06))
30 # 3 倍標準差機率
31 plt.annotate(text='', xy=(-3,0.01), xytext=(3,0.01), arrowprops=dict(arrowstyle='<->'))
32 plt.annotate(text='99.7%', xy=(0,0.02), xytext=(0,0.02))
33
34 plt.show()
```

執行結果如下：

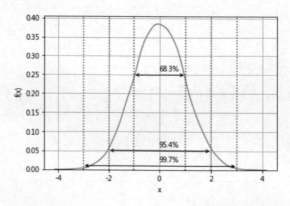

圖 2.14　常態分配 1、2、3 倍標準差的信賴區間

程式說明：

■ 第 9 行 sns.distplot(x, hist=False)：使用 Seaborn 畫直方圖，參數 hist=False 表示不畫階梯直方圖，只畫平滑曲線。

■ 第 17 行 axvline()：畫垂直線，標示正負 1、2、3 倍標準差。

2-4 機率(Probability)與統計(Statistics)

統計是從資料所推衍出來的資訊，包括一些描述統計量、機率分佈等，例如平均數、標準差。而我們蒐集到的資料統稱為「資料集」(dataset)，它包括一些觀察值(Observations)或案例(Cases)，即資料集的列，以及一些特徵(features)，或稱屬性(attributes)，即資料集的欄位。譬如，要預測下一任台北市長的選舉，我們做一次問卷調查，相關定義如下：

1. 資料集：全部問卷調查資料。
2. 觀察值：每張問卷。
3. 特徵或屬性：每一個問題，通常以 X 表示。
4. 屬性值或特徵值：每一個問題的回答。
5. 目標(Target)欄位：市長候選人，通常以 y 表示。

以下我們會依序介紹以下內容：

1. 抽樣
2. 描述統計量
3. 機率
4. 機率分配函數
5. 假設檢定

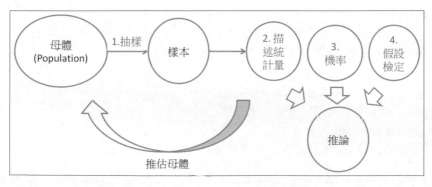

圖 2.15 基礎統計介紹的範圍及其關聯

2-4-1　資料型態

依照特徵的資料型態，分為定性(qualitative)及定量(quantitative)的欄位。定性(qualitative)欄位為非數值型的欄位，通常包含有限類別。又可以分為名目資料(Nominal Data)及有序資料(Ordinal Data)。定量(quantitative)欄位為數值型欄位，又可以分為離散型資料(Discrete Data)及連續型資料(Continuous Data)。預測的目標(Target)欄位如為離散型變數，適用分類(Classification)的演算法，反之，目標欄位為連續型變數，適用迴歸(Regression) 的演算法。

1. 名目資料(Nominal Data)：欄位值並沒有順序或大小的隱含意義，例如顏色，紅、藍、綠並沒有誰大誰小的隱含意義。轉換為代碼時，應以 One-hot encoding 處理，將每一類別轉為個別的啞變數(Dummy variable)，每個啞變數只有 True/False 或 1/0 兩種值。例如下圖：

	color	is_blue	is_green	is_red
0	green	0	1	0
1	red	0	0	1
2	green	0	1	0
3	blue	1	0	0

Color 特徵有三種類別，經過 One-hot encoding 處理，會轉換為三個啞變數。

➤ **以下程式請參考 02_4_1_特徵轉換.ipynb。**

```
1  df2 = pd.get_dummies(df["color"], columns=["color"],
2                       prefix='is', prefix_sep='_')
3
4  # 連結轉換前後的欄位，相互比較
5  pd.concat((df["color"], df2), axis=1)
```

資料框(Data Frame)、要轉換的欄位、前置符號(prefix)、分隔符號(prefix_sep)。

2. 有序資料(Ordinal Data)：欄位值有順序或大小的隱含意義，例如衣服尺寸，XL > L > M > S。

	color	size	price	classlabel
0	green	XL	10.1	class1
1	red	M	10.1	class1
2	green	L	13.5	class2
3	blue	S	15.3	class1

size 欄位依尺寸大小，分別編碼為 XL(4)、L(3)、M(2)、S(1)，轉換後如下。

	color	size	price	classlabel
0	green	4	10.1	class1
1	red	2	10.1	class1
2	green	3	13.5	class2
3	blue	1	15.3	class1

```
1  # 以字典定義轉換規則
2  size_mapping = {'XL': 4,
3                  'L': 3,
4                  'M': 2,
5                  'S': 1}
6
7  # 使用 map() 轉換
8  df['size'] = df['size'].map(size_mapping)
9  df
```

程式說明：

- 第 2 行：以字典定義轉換規則，key 為原值，value 為轉換後的值。
- 第 4 行 map()：以字典轉換 size 欄位的每一個值。

2-4-2 抽樣(Sampling)

再以預測下一任台北市長的選舉為例，囿於人力、時間及經費的限制，不太可能調查所有市民的投票傾向，通常我們只會隨機抽樣 1000 份或更多一點的樣本進行調查，這種方式稱之為抽樣(Sampling)，相關名詞定義如下：

1. 母體(Population)：全體有投票權的市民。
2. 樣本(Sample)：被抽中調查的市民。
3. 分層抽樣(Stratified Sampling)：依母體某些屬性的比例，進行相同比例的抽樣，希望能充分代表母體的特性，例如政黨、年齡、性別比例。

照例我們看看程式如何撰寫，請參閱 02_4_2_抽樣.ipynb。

簡單抽樣，從一個集合中隨機抽出 n 個。

```
1  import random
2  import numpy as np
3
4  # 1~10 的集合
5  list1 = list(np.arange(1, 10 + 1))
6
7  # 隨機抽出 5 個
8  print(random.sample(list1, 5))
```

執行結果：[1, 6, 2, 8, 5]

程式說明：

- 第 8 行 random.sample()：自集合中隨機抽樣。
- 每次執行結果均不相同，且每個項目不重複，此種抽樣方法稱為「不放回式抽樣」(Sampling Without Replacement)。

放回式抽樣(Sampling With Replacement)。

```
1  import random
2  import numpy as np
3
4  # 1~10 的集合
5  list1 = list(np.arange(1, 10 + 1))
6
7  # 隨機抽出 5 個
8  print(random.choices(list1, k=5))
```

執行結果：[8, 4, 7, 4, 6]

程式說明：

- 第 8 行 random.choices()：自集合中隨機抽樣。
- 每次執行結果均不相同，但項目會重複，如上，4 被重複抽出兩次，此種抽樣方法稱為「放回式抽樣」。

以 Pandas 套件進行抽樣。

```
1  from sklearn import datasets
2  import pandas as pd
3
4  # 載入鳶尾花(iris)資料集
5  ds = datasets.load_iris()
6
7  # x, y 合成一個資料集
8  df = pd.DataFrame(data=ds.data, columns=ds.feature_names)
9  df['y'] = ds.target
10
11 # 隨機抽出 5 個
12 print(df.sample(5))
```

執行結果如下：

	sepal length (cm)	sepal width (cm)	petal length (cm)	petal width (cm)	y
102	7.1	3.0	5.9	2.1	2
40	5.0	3.5	1.3	0.3	0
32	5.2	4.1	1.5	0.1	0
48	5.3	3.7	1.5	0.2	0
77	6.7	3.0	5.0	1.7	1

程式說明：

- 第 12 行 df.sample(5)：自集合中隨機抽樣 5 筆。
- 為「不放回式抽樣」(Sampling Without Replacement)。

以 Pandas 套件進行分層抽樣(Stratified Sampling)。

```
1  from sklearn import datasets
2  from sklearn.model_selection import StratifiedShuffleSplit
3  import pandas as pd
4
5  # 載入鳶尾花(iris)資料集
6  ds = datasets.load_iris()
7
8  # x, y 合成一個資料集
9  df = pd.DataFrame(data=ds.data, columns=ds.feature_names)
10 df['y'] = ds.target
11
12 # 隨機抽出 6 個
13 stratified = StratifiedShuffleSplit(n_splits=1, test_size=6)
14 x = list(stratified.split(df, df['y']))
15
16 print('重新洗牌的全部資料:')
17 print(x[0][0])
18
19 print('\n抽出的索引值:')
20 print(x[0][1])
```

執行結果：

```
重新洗牌的全部資料：
[ 38 115 136  80 111  94   1   0  48 100 108 104   8  51 131  78   9 142
 112  11 126  79  95   2  46 128 125  65  55  10  72 145 130  56 138  96
  88  19   7  43   4  82  32  91 127  87 133  73  85  62 129  42  57  84
  40 105  49  75 113 147  99  27   6 135  58  35  26 124  92  70  69 139
  66 101  74  60 110  15  39  59   3  53  89 107  61 143 118  86  71  98
  50  41  34  12 149  77  23  21 117 121  97  54 119  64 120  45  81 141
 122 114  20 144 134 132  17  24  13  22  44 123  31 116  76  18  47 137
  63  83  29  25  36 102  28  37  33  93 148  14 146  16 103  68  90 140]

抽出的索引值：
[ 52   5  30 109 106  67]
```

程式說明：

- 第 13 行 StratifiedShuffleSplit(test_size=6)：將集合重新洗牌，並從中隨機抽樣 6 筆。
- 第 14 行 stratified.split(df, df['y'])：以 y 欄位為 key，分層抽樣，得到的結果是 generator 資料型態，須轉為 list，才能一次取出。
- 第 1 個輸出是重新洗牌的全部資料，第 2 個是抽出的索引值。
- 利用 df.iloc[x[0][1]] 指令，取得資料如下，觀察 y 欄，每個類別各有兩筆與母體比例相同。

	sepal length (cm)	sepal width (cm)	petal length (cm)	petal width (cm)	y
52	6.9	3.1	4.9	1.5	1
5	5.4	3.9	1.7	0.4	0
30	4.8	3.1	1.6	0.2	0
109	7.2	3.6	6.1	2.5	2
106	4.9	2.5	4.5	1.7	2
67	5.8	2.7	4.1	1.0	1

母體比例可以 df['y'].value_counts() 指令取得資料如下，0/1/2 類別各 50 筆。

```
2    50
1    50
0    50
```

以 Pandas 套件進行不分層抽樣。

```
1  from sklearn import datasets
2  from sklearn.model_selection import train_test_split
3  import pandas as pd
4
5  # 載入鳶尾花(iris)資料集
6  ds = datasets.load_iris()
7
8  # x, y 合成一個資料集
9  df = pd.DataFrame(data=ds.data, columns=ds.feature_names)
10 df['y'] = ds.target
11
12 # 隨機抽出 6 個
13 train, test = train_test_split(df, test_size=6)
14 x = list(stratified.split(df, df['y']))
15
16 print('\n抽出的資料:')
17 print(test)
```

執行結果：

	sepal length (cm)	sepal width (cm)	petal length (cm)	petal width (cm)	y
141	6.9	3.1	5.1	2.3	2
147	6.5	3.0	5.2	2.0	2
97	6.2	2.9	4.3	1.3	1
80	5.5	2.4	3.8	1.1	1
22	4.6	3.6	1.0	0.2	0
61	5.9	3.0	4.2	1.5	1

程式說明：

- 第 13 行 train_test_split(test_size=6)：將資料集合重新洗牌，並從中隨機抽樣 6 筆為測試資料，其餘為訓練資料。
- 取得資料如上，觀察 y 欄，類別 0 有 1 筆，類別 1 有 3 筆，類別 2 有 2 筆，與母體比例不相同。

2-4-3 基礎統計(Statistics Fundamentals)

經過各種方式蒐集到一組樣本後，我們會先對樣本進行探索，通常會先衡量其集中趨勢(Central Tendency)及資料離散的程度(Measures of Variance)，這些指標稱之為描述統計量(Descriptive Statistics)，較常見的描述統計量如下：

1. 集中趨勢(Central Tendency)

   平均數 (Mean)：母體以 μ 表示，樣本以 $\bar{x}$ 表示。

   $$\mu = \frac{\sum_{i=1}^{n} x_i}{n}$$

   中位數 (Median)：將所有樣本由小排到大，以中間的樣本值為準。若樣本數為偶數，則取中間樣本值的平均數。中位數可以避免離群值(Outlier)的影響，例如，統計平均年收入，如果加入一位超級富豪，則平均數將大幅提高，但中位數毫不受影響。假設有一組樣本：

   [100, 200, 300, 400, 500]
   平均數 = 中位數 = 300

   現在把 500 改為 50000：

   [100, 200, 300, 400, 50000]

   平均數 = 100200，被 50000 影響，大幅提升，無法反應大多數資料的的樣態。

   中位數 = 300，仍然不變。

   眾數 (Mode)：頻率發生最高的數值，以大多數的資料為主。

2. 資料離散的程度(Measures of Variance)

   級距(Range)：最大值 - 最小值。

   百分位數(Percentiles)與四分位數(Quartiles)：例如 100 為 75 百分位數，表示有 75%的樣本小於 100。

   變異數(Variance)：母體以 δ 表示，樣本以 s 表示。

   $$\delta = \sqrt{\frac{\sum (x - \mu)^2}{n}}$$

   $$s = \sqrt{\frac{\sum (x-\mu)^2}{n-1}}$$

3. 可以使用箱形圖(box plot) 或稱盒鬚圖，直接顯示上述相關的統計量。

➤ **以下進行實作，請參閱 02_4_3_基礎統計**.ipynb。

1. 以美國歷屆總統的身高資料，計算各式描述統計量。

```
1  # 集中趨勢(Central Tendency)
2  print(f"平均數={df['height'].mean()}")
3  print(f"中位數={df['height'].median()}")
4  print(f"眾　數={df['height'].mode()[0]}")
5  print()
6
7  # 資料離散的程度(Measures of Variance)
8  from scipy import stats
9
10 print(f"級距(Range)={df['height'].max() - df['height'].min()}")
11 print(f"182cm 百分位數={stats.percentileofscore(df['height'], 182, 'strict')}")
12 print(f"變異數={df['height'].std():.2f}")
```

執行結果：

- 平均數=179.73809523809524
- 中位數=182.0
- 眾　數=183
- 級距(Range)=30
- 182cm 的百分位數=47.61904761904761
- 變異數=7.02

2. 以美國歷屆總統的身高資料，計算四分位數(Quartiles)。

```
1  print(f"四分位數\n{df['height'].quantile([0.25,0.5,0.75,1])}")
```

執行結果：

- 四分位數
- 0.25　174.25
- 0.50　182.00
- 0.75　183.00
- 1.00　193.00

```
1  print(f"{df['height'].describe()}")
```

執行結果：

- count　42.000000
- mean　179.738095
- std　　7.015869

- min　　163.000000
- 25%　　174.250000
- 50%　　182.000000
- 75%　　183.000000
- max　　193.000000

3. 繪製箱形圖(box plot)或稱盒鬚圖,定義如下。

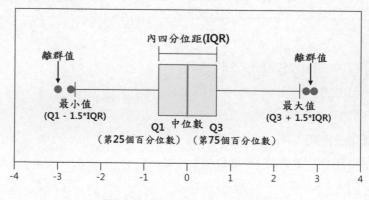

圖 2.16　箱形圖(box plot)定義

可觀察下列特性:

1. 中間的線:中位數。
2. 中間的箱子:Q1~Q3 共 50%的資料集中在箱子內。
3. 是否存在離群值(Outlier),注意,離群值發生的可能原因包括記錄或登打錯誤、感測器失常造成量測錯誤或是重要的警訊(Influential Sample),應仔細探究為何發生,不要直接剔除掉。一般而言,會更進一步收集更多樣本,來觀察離群值是否再次出現,另一方面,更多的樣本也可以稀釋離群值的影響力。

程式碼如下:

```
1  import matplotlib.pyplot as plt
2
3  df['height'].plot(kind='box', title='美國歷屆總統的身高資料', figsize=(8,6))
4  plt.show()
```

執行結果：

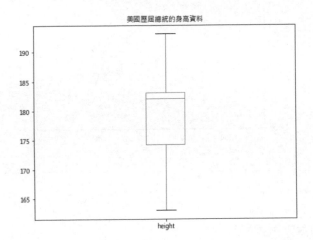

除了單變數的統計量，也會做多變數的分析，了解變數之間的關聯度 (Correlation)，在資料探索與分析(Exploratory Data Analysis, 簡稱 EDA)的 階段，我們還會依據資料欄位的屬性進行不同面向的觀察，如下圖：

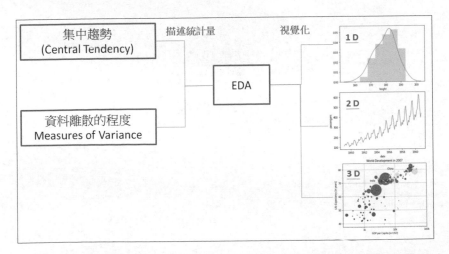

圖 2.17 資料探索與分析(Exploratory Data Analysis)不同面向的觀察

常用的統計圖功能簡要説明如下，同時也使用程式實作相關圖表：

1. 直方圖(Histogram)：觀察資料集中趨勢、離散的程度、機率分配及偏 態(Skewness)/峰態(Kurtosis)。

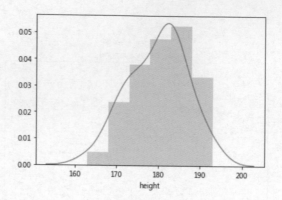

```
1  # 直方圖
2  import seaborn as sns
3
4  # 繪圖
5  sns.distplot(df['height'])
6  plt.show()
```

2. 圓餅圖(Pie Chart)：顯示單一變數的各類別所佔比例。

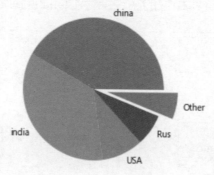

```
1   # 餅圖(Pie Chart)
2   import matplotlib.pyplot as plt
3   import pandas as pd
4   import numpy as np
5
6   # 讀取資料檔
7   df = pd.read_csv('./data/gdp.csv')
8
9   # 轉為整數欄位
10  df.pop = df['pop'].astype(int)
11
12  # 取最大 5 筆
13  df2 = df.nlargest(5, 'pop')
14
15  # 散佈圖(Scatter Chart)
16  plt.pie(df2['pop'], explode=[0, 0,0,0,0.2], labels=['china', 'india', 'USA', 'Rus', 'Other'])
17
18  plt.show()
```

3. 折線圖(Line Chart)：觀察趨勢，尤其是時間的趨勢，譬如股價、氣象溫度、營收、運量分析等。

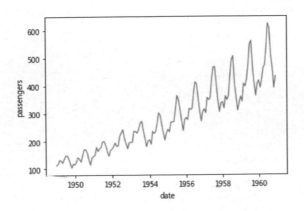

```
1   # 線圖(Line Chart)
2   import matplotlib.pyplot as plt
3   import pandas as pd
4   import seaborn as sns
5
6   # 讀取資料檔
7   df = pd.read_csv('./data/airline.csv')
8
9   # 轉為日期欄位
10  df['date'] = pd.to_datetime(df['date'])
11
12  # 繪圖
13  sns.lineplot(df['date'], df['passengers'])
14
15  plt.show()
```

4. 散佈圖(Scatter Chart)：顯示兩個變數的關聯，亦用於觀察是否有離群值。

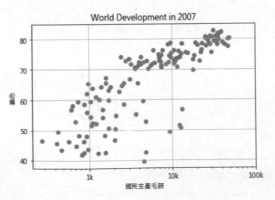

```
1   # 散佈圖(Scatter Chart)
2   import matplotlib.pyplot as plt
3   import pandas as pd
4   import numpy as np
5
6   # 讀取資料檔
7   df = pd.read_csv('./data/gdp.csv')
8
9   # 轉為整數欄位
10  df.pop = df['pop'].astype(int)
11
12  # 散佈圖(Scatter Chart)
13  plt.scatter(x = df.gdp, y = df.life_exp)
14
15  # 設定繪圖屬性
16  plt.xscale('log')
17  plt.xlabel('國民生產毛額')
18  plt.ylabel('壽命')
19  plt.title('World Development in 2007')
20  plt.xticks([1000,10000,100000], ['1k','10k','100k'])
21  plt.grid()
22
23  plt.show()
```

5. 氣泡圖(Bubble Chart)：散佈圖再加一個變數，以該變數值作為點的大小，並且做 3 個維度的分析。

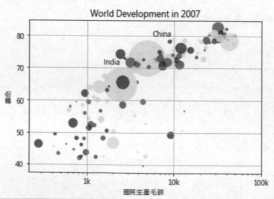

```
1   # 氣泡圖(Bubble Chart)
2   import matplotlib.pyplot as plt
3   import pandas as pd
4   import numpy as np
5
6   # 讀取資料檔
7   df = pd.read_csv('./data/gdp.csv')
8
9   # 轉為整數欄位
10  df.pop = df['pop'].astype(int)
11
```

```
12  # 散佈圖(Scatter Chart) + 加一個變數 pop，作為點的大小
13  col=np.resize(['red', 'green', 'blue', 'yellow', 'lightblue'], df.shape[0])
14  plt.scatter(x = df.gdp, y = df.life_exp, s = df.pop * 2, c = col, alpha = 0.8)
15
16  # 設定繪圖屬性
17  plt.xscale('log')
18  plt.xlabel('國民生產毛額')
19  plt.ylabel('壽命')
20  plt.title('World Development in 2007')
21  plt.xticks([1000,10000,100000], ['1k','10k','100k'])
22  plt.grid()
23
24  # 加註
25  plt.text(1550, 71, 'India')
26  plt.text(5700, 80, 'China')
27
28  plt.show()
```

6. 熱力圖(Heatmap)：顯示各變數之間的關聯度(Correlation)，得以輕易辨識出較高的關聯度。輸入資料須為各變數之間的關聯係數(Correlation coefficient)，公式如下，可使用 Pandas 的 corr() 函數計算。

$$r_{x,y} = \frac{\sum_{i=1}^{n}(x_i - \bar{x})(y_i - \bar{y})}{\sqrt{\sum_{i=1}^{n}(x_i - \bar{x})^2(y_i - \bar{y})^2}}$$

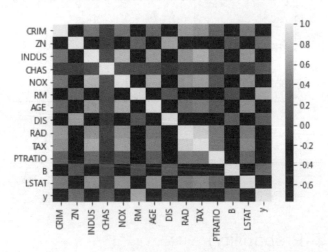

```
 1  # 熱力圖(Heatmap)
 2  import matplotlib.pyplot as plt
 3  import pandas as pd
 4  import seaborn as sns
 5  from sklearn import datasets
 6
 7  # 讀取 sklearn 內建資料檔
 8  ds = datasets.load_boston()
 9
10  df = pd.DataFrame(ds.data, columns=ds.feature_names)
11  df['y'] = ds.target
12
13
14  # 繪製熱力圖(Heatmap)
15  # df.corr()：關聯係數(Correlation coefficient)
16  sns.heatmap(df.corr())
17
18  plt.show()
```

7. 另外，Seaborn 套件還提供許多指令，使用一個指令就可以產生多張的
   圖表，例如 pairplot(下圖)、facegrid……，可參閱 Seaborn 官網[2]。

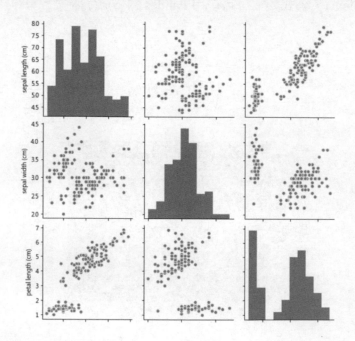

2-4-4　機率(Probability)

以台北市長選舉為例，當我們收集到問卷調查後，接著就可以建立模型來
預測誰會當選，預測結果通常是以機率表示各候選人當選的可能性。

2-4-4-1 機率的定義與定理

接下來我們先瞭解機率的相關術語(Terminology)及定義。

1. 實驗(Experiment)或稱試驗(Trial)：針對計畫目標進行一連串的行動，例如擲硬幣、擲骰子、抽撲克牌、買樂透等。
2. 樣本空間(Sample Space)：一個實驗會出現的所有可能結果，例如擲硬幣一次的樣本空間為{正面、反面}，擲硬幣兩次的樣本空間為{正正、正反、反正、反反}。
3. 樣本點(Sample Point)：樣本空間內任一可能的結果，例如擲硬幣兩次會有四種樣本點，抽一張撲克牌，有 52 種樣本點。
4. 事件(Event)：某次實驗發生的結果，一個事件可能含多個樣本點，例如，抽一張撲克牌，出現紅色的事件，該事件就包含 26 種樣本點。
5. 機率：某次事件發生的可能性，因此

$$機率 = \frac{發生此事件的樣本點}{樣本空間的所有樣本點}$$

    例如，擲硬幣兩次出現兩個正面為{正正} 1 個樣本點，樣本空間所有樣本點為{正正、正反、反正、反反} 4 個樣本點，故機率等於 1/4。同樣的，擲硬幣兩次出現一正一反的機率 = {正反、反正} / {正正、正反、反正、反反} = 1/2。

6. 「條件機率」(Conditional Probability)與相依性(Dependence)

    事件「獨立」(Independent)：A 事件發生，不會影響 B 事件出現的機率，則稱 A、B 兩事件獨立。例如，擲硬幣兩次，第一次出現正面或反面，都不會影響第二次的擲硬幣。

    事件「相依」(Dependent)：抽一張撲克牌，不放回，再抽第二張，機率會受到第一張抽出結果所影響。例如：

    *   第一張抽到紅色的牌，第二張再抽到紅色的機率
        第一張抽到紅色 ➔ 紅色牌剩 25 張，黑色牌剩 26 張。
        第二張抽到紅色機率 = **25** / (25 + 26)。
    *   第一張抽到紅色的牌，第二張再抽到黑色的機率
        第二張抽到紅色機率 = **26** / (25 + 26)。

兩個機率是不相等的,故兩次抽牌的事件是「相依」的。

事件「互斥」(Mutually Exclusive):A 事件發生,就不會出現 B 事件,兩事件不可能同時發生。次日天氣出現晴天的機率為 1/2,陰天的機率為 1/4,雨天的機率為 1/4,次日出現晴天,就不可能出現陰天,故次日「不是雨天」的機率為 1/2 + 1/4 = 3/4,互斥事件的機率等於個別事件的機率直接相加。

依照上述定義,衍生的相關定理如下:

1. A、B 事件獨立,則同時發生 A 及 B 事件的機率 $P(A \cap B) = P(A)$ x $P(B)$。

2. A、B 事件互斥,則發生 A 或 B 事件的機率 $P(A \cup B) = P(A) + P(B)$。

3. A、B 相依事件,則發生 A 或 B 事件的機率 $P(A \cup B) = P(A) + P(B) - P(A \cap B)$。

4. 樣本空間內所有互斥事件的機率總和 = 1。

以上條件機率可以使用集合論(Set Theory)表示。

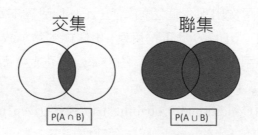

圖 2.18 交集(Intersection)及聯集(Union)

2-4-4-2 排列與組合

有些事件我們會關心發生的順序,相反的,有些時候則不關心事件發生的順序,計算的公式會有所不同,結果也不一樣,關心事件發生的順序稱為「排列」(Permutation),反之,不關心事件發生順序稱為「組合」(Combination)。例如,擲硬幣兩次的「排列」樣本空間為{正正、正反、反正、反反},但「組合」樣本空間為{正正、一正一反、反反},因為「正反」、「反正」都是「一正一反」。

以下列舉各種案例說明排列與組合的相關計算公式及程式撰寫，**請參閱02_4_4_機率.ipynb**。

【範例 1】有三顆球，標號各為 1、2、3，它們排列的事件共有幾種？

程式碼：

```
1   from itertools import permutations
2
3   # 測試資料
4   list1 = [1, 2, 3]
5
6   # 宣告排列的類別
7   perm = permutations(list1)
8
9   # 印出所有事件
10  print('所有事件:')
11  list_output = list(perm)
12  for i in list_output:
13      print(i)
14
15  print()
16  print(f'排列數={len(list_output)}')
```

執行結果：所有事件如下。

```
(1, 2, 3)
(1, 3, 2)
(2, 1, 3)
(2, 3, 1)
(3, 1, 2)
(3, 2, 1)
排列數=6
```

【範例 2】有三顆球，抽出兩顆球排列的事件共有幾種？

程式碼：

```
1   from itertools import permutations
2
3   # 測試資料
4   list1 = [1, 2, 3]
5
6   # 宣告排列的類別，三個球抽出兩個球
7   perm = permutations(list1, 2)
8
9   # 印出所有事件
10  print('所有事件:')
11  list_output = list(perm)
12  for i in list_output:
13      print(i)
```

```
14
15  print()
16  print(f'排列數={len(list_output)}')
```

執行結果：所有事件如下。

```
(1, 2)
(1, 3)
(2, 1)
(2, 3)
(3, 1)
(3, 2)
排列數=6
```

出現(1, 2)的機率=[(1, 2), (2, 1)] / 6 = 2/6 = 1/3

【範例 3】有三顆球，抽出兩顆球組合的事件共有幾種？

程式碼：

```
1   from itertools import combinations
2
3   # 測試資料
4   list1 = [1, 2, 3]
5
6   # 宣告排列的類別，三個球抽出兩個球
7   comb = combinations(list1, 2)
8
9   # 印出所有事件
10  print('所有事件:')
11  list_output = list(comb)
12  for i in list_output:
13      print(i)
14
15  print()
16  print(f'組合數={len(list_output)}')
```

執行結果：所有事件如下。

```
(1, 2)
(1, 3)
(2, 3)
組合數=3
```

出現(1, 2)的機率=[(1, 2)] / 3 = 1/3

【範例 4】再進一步衍生，袋子中有 10 顆不同顏色的球，抽出 3 顆共有幾種排列方式？

解題：排列計算公式如下：

$$排列 = \frac{n!}{(n-k)!}$$

n: 樣本點數 10

k: 3

```
1  import math
2
3  n=10
4  k=3
5
6  # math.factorial: 階乘
7  perm = math.factorial(n) / math.factorial(n-k)
8
9  print(f'排列方式-{perm:.0f}')
```

執行結果：排列方式共 720 種。

公式由來如下:

抽第一次有 10 種選擇，抽第二次有 9 種選擇，抽第三次有 8 種選擇，故

$$10 \times 9 \times 8 = \frac{10!}{(10-3)!}$$

【範例 5】擲硬幣 3 次，出現 0、1、2、3 次正面的組合共有幾種？

解題：因為只考慮出現正面的次數，不管出現的順序，故屬於組合的問題，公式如下，因為出現的順序不同，仍視為同一事件，故要多除於 k!：

$$組合 = \frac{n!}{k!\,(n-k)!}$$

n: 實驗次數，4 次

k: 出現正面次數，0、1、2、3 次

```
1  import math
2
3  n=3
4
5  # math.factorial: 階乘
6  for k in range(4):
7      # 組合公式
8      comb = math.factorial(n) / (math.factorial(k) * math.factorial(n-k))
9      print(f'出現 {k} 次正面的組合共 {comb:.0f} 種')
```

執行結果：

出現 0 次正面的組合共 1 種

出現 1 次正面的組合共 3 種
出現 2 次正面的組合共 3 種
出現 3 次正面的組合共 1 種

2-4-4-3 二項分配(Binomial Distribution)

擲硬幣、擲骰子、抽撲克牌一次,均假設每個單一樣本點發生的機率都均
等,如果不是均等,上述的排列/組合的公式需要再考慮機率 p,例如擲硬
幣出現正面的機率=0.4,p^k 代表出現 k 次正面的機率, $(1-p)^{(n-k)}$ 代表出現
n-k 次反面的機率。

$$P(x = k) = \frac{n\,!}{k!\,(n-k)!}\,p^k\,(1-p)^{(n-k)}$$

【範例 6】擲硬幣出現正面的機率=0.4,擲硬幣 3 次,出現 0、1、2、3 次
正面的機率為何?

```
1  import math
2
3  n=3
4  p=0.4
5
6  # math.factorial: 階乘
7  for k in range(4):
8      # 組合公式
9      comb = math.factorial(n) / (math.factorial(k) * math.factorial(n-k))
10     prob = comb * (p**k) * ((1-p)**(n-k))
11     print(f'出現 {k} 次正面的機率 = {prob:.4f}')
```

執行結果:

出現 0 次正面的機率 = 0.2160
出現 1 次正面的機率 = 0.4320
出現 2 次正面的機率 = 0.2880
出現 3 次正面的機率 = 0.0640

使用 SciPy comb 組合函數,執行結果同上。

```
1  # 使用 scipy comb 組合函數
2  from scipy import special as sps
3
4  n=3
5  p=0.4
6
7  for k in range(4):
8      prob = sps.comb(n, k) * (p**k) * ((1-p)**(n-k))
9      print(f'出現 {k} 次正面的機率 = {prob:.4f}')
10
```

直接使用 SciPy 的統計模組 binom.pmf(k,n,p)，執行結果同上。

```python
1  from scipy.stats import binom
2
3  n=3
4  p=0.4
5
6  # binom
7  for k in range(4):
8      print(f'出現 {k} 次正面的機率 = {binom.pmf(k,n,p):.4f}')
```

【範例 7】試算今彩 539 平均報酬率[3]。

基本玩法：從 39 個號碼中選擇 5 個號碼。

各獎項的中獎方式如下表：

獎項	中獎方式	中獎方式圖示
頭獎	與當期五個中獎號碼完全相同者	●●●●●
貳獎	對中當期獎號之其中任四碼	●●●●
參獎	對中當期獎號之其中任三碼	●●●
肆獎	對中當期獎號之其中任二碼	●●

各獎項金額：

獎項	頭獎	貳獎	參獎	肆獎
單注獎金	$8,000,000	$20,000	$300	$50

解題：

1. 假定每個號碼出現的機率是相等的(1/39)，固定選 5 個號碼，故計算機率時不需考慮 p。

2. 計算從 39 個號碼選 5 個號碼，總共有幾種組合。

```python
1  # 從39個號碼選5個號碼，總共有幾種
2  from scipy import special as sps
3
4  print(f'從39個號碼選5個號碼，總共有{int(sps.comb(39,5))}種')
5
6  # 頭獎號碼的個數，5個中獎號碼全中，不含非中獎號碼
7  first_price_count = int(sps.comb(5,5) * sps.comb(34,0))
8  print(f'頭獎號碼的個數: {first_price_count}')
9
10  # 頭獎號碼的個數，5個中獎號碼中4個，有1個非中獎號碼
11  second_price_count = int(sps.comb(5,4) * sps.comb(34,1))
```

```
12  print(f'二獎號碼的個數: {second_price_count}')
13
14  # 頭獎號碼的個數，5個中獎號碼中3個，有3個非中獎號碼
15  third_price_count = int(sps.comb(5,3) * sps.comb(34,2))
16  print(f'三獎號碼的個數: {third_price_count}')
17
18  # 頭獎號碼的個數，5個中獎號碼中2個，有3個非中獎號碼
19  fourth_price_count = int(sps.comb(5,2) * sps.comb(34,3))
20  print(f'四獎號碼的個數: {fourth_price_count}')
21
```

執行結果如下：

```
從 39 個號碼選 5 個號碼，總共有 575757 種
頭獎號碼的個數: 1
二獎號碼的個數: 170
三獎號碼的個數: 5610
四獎號碼的個數: 59840
```

計算平均中獎金額

```
1  # 平均中獎金額=(頭獎金額 * 頭獎組數  + 二獎金額 * 二獎組數  +
2  #               三獎金額 * 三獎組數  + 四獎金額 * 四獎組數) / 全部組合數
3  average_return = (8000000 * 1 + 20000 * 170 + 300 * 5610 + 50 * 59840) / 575757
4  print(f'平均中獎金額: {average_return:.2f}')
5
6  print(f'今彩539平均報酬率={((average_return / 50) - 1) * 100:.2f}%')
```

執行結果如下：

```
平均中獎金額: 27.92
今彩 539 平均報酬率=-44.16%
```

排列與組合的理論看似簡單，然而實務上的應用千變萬化，建議讀者多找
一些案例實作，才能運用自如。

2-4-5 機率分配(Distribution)

機率分配(Distribution)結合了描述統計量及機率的概念，對資料作進一
步的分析，希望推測母體是呈現何種形狀的分佈，例如常態分配、均勻分
配、卜瓦松(Poisson)分配或二項分配等，並且依據樣本推估機率分配相關
的母數，譬如平均數、變異數等。有了完整機率分配資訊後，就可以進行
預測、區間估計、假設檢定等。

機率分配的種類非常多，如下圖，這裡我們僅介紹幾個常見的機率分配。

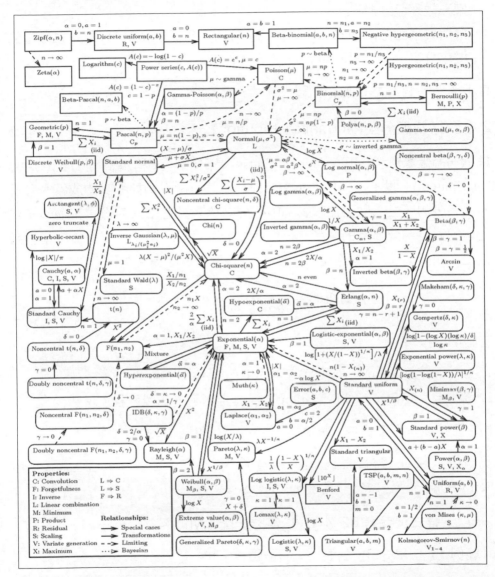

圖 2.19 各種機率分配及其關聯，圖片來源：Univariate Distribution Relationships [4]

首先介紹幾個專有名詞：

1. 機率密度函數(Probability Density Function, 簡稱 pdf)：發生各種事件的機率。

2. 累積分配函數(Cumulative Distribution Function, 簡稱 cdf)：等於或低於某一觀察值的機率。

3. 機率質量函數(Probability Mass Function, 簡稱 pmf)：如果是離散型的機率分配，pdf 改稱為 pmf。

接下來看幾個常見的機率分配。

2-4-5-1 常態分配(Normal Distribution)

常態分配因為是 Carl Friedrich Gauss 提出的，又稱為高斯分配(Gauss Distribution)，世上大部份的事件都屬於常態分配，例如考試的成績，考低分及高分的學生人數會比較少，中等分數的人數會占大部份。其他例子如全年的溫度、業務員的業績、一堆紅豆的重量等。常態分配的機率密度函數定義如下：

$$f(x; \mu, \sigma) = \frac{1}{\sqrt{2\pi * \sigma^2}} * e^{-\frac{1}{2} * (\frac{x - \mu}{\sigma})^2}$$

有兩個參數：

1. μ：平均數(Mean)，是全體樣本的平均數。
2. δ：標準差(Standard Deviation)，描述全體樣本的離散程度。

機率密度函數可簡寫成 $N(\mu, \delta)$。

我們在「2-3-4 積分」節已經介紹過了，並撰寫一些程式，這裡直接使用 SciPy 的統計模組做一些實驗。

➤ **以下程式請參考 02_4_5_1_常態分配**.ipynb。

【範例 1】使用 SciPy 繪製機率密度函數(pdf)。

```
1  import matplotlib.pyplot as plt
2  import numpy as np
3  from scipy.stats import norm
4
5  # 觀察值範圍
6  z1, z2 = -4, 4
7
8  # 樣本點
9  x = np.arange(z1, z2, 0.001)
```

```
10  y = norm.pdf(x,0,1)
11
12  # 繪圖
13  plt.plot(x,y,'black')
14
15  # 填色
16  plt.fill_between(x,y,0, alpha=0.3, color='b')
```

執行結果：

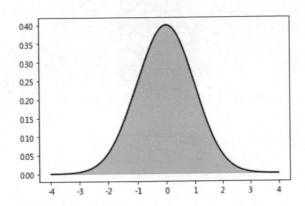

【**範例 2**】繪製常態分配的正負 1、2、3 倍標準差區間及其機率。

```
1   import matplotlib.pyplot as plt
2   import numpy as np
3   from scipy.stats import norm
4
5   z1, z2 = -1, 1
6   x1 = np.arange(z1, z2, 0.001)
7   y1 = norm.pdf(x1,0,1)
8
9   # 1倍標準差區域
10  plt.fill_between(x1,y1,0, alpha=0.3, color='b')
11  plt.ylim(0,0.5)
12
13  # 觀察值範圍
14  z1, z2 = -4, 4
15
16  # 樣本點
17  x2 = np.arange(z1, z2, 0.001)
18  y2 = norm.pdf(x2,0,1)
19
20  # 繪圖
21  plt.plot(x2,y2,'black')
22
23  # 1倍標準差機率
24  plt.annotate(text='', xy=(-1,0.25), xytext=(1,0.25), arrowprops=dict(arrowstyle='<->'))
25  plt.annotate(text='68.3%', xy=(0,0.26), xytext=(0,0.26))
26
27  # 填色
28  plt.fill_between(x2,y2,0, alpha=0.1, color='b')
29  plt.show()
```

執行結果：

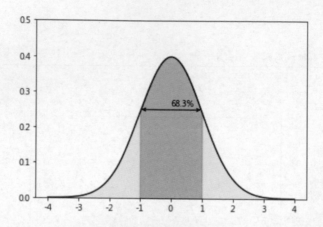

第 5 行改為(-2, 2)、(-3, 3)，結果如下：

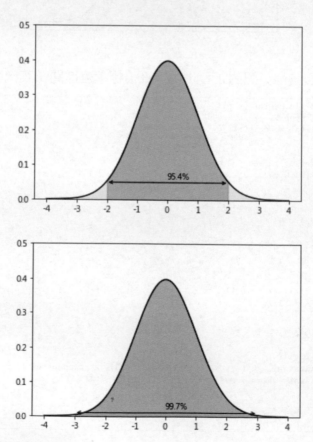

【範例 3】使用 SciPy 繪製累積分配函數(cdf)。

```python
1   import matplotlib.pyplot as plt
2   import numpy as np
3   from scipy.stats import norm
4
5   # 觀察值範圍
6   z1, z2 = -4, 4
7
8   # 樣本點
9   x = np.arange(z1, z2, 0.001)
10
11  # norm.cdf：累積機率密度函數
12  y = norm.cdf(x,0,1)
13
14  # 繪圖
15  plt.plot(x,y,'black')
16
17  plt.show()
```

第 11 行從 pdf 改為 cdf。

執行結果：

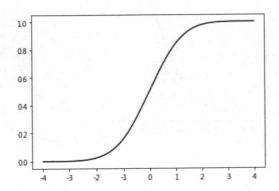

有一個 Student's t 分配，很像常態分配，常被用來作假設檢定(Hypothesis Test)，SciPy 也有支援。

【範例 4】使用 SciPy 繪製 Student's t 分配的機率密度函數(pdf)，並與常態分配比較。

```python
1   import matplotlib.pyplot as plt
2   import numpy as np
3   from scipy.stats import t, norm
4
5   # 觀察值範圍
6   z1, z2 = -4, 4
7
```

```
 8  # 樣本點
 9  x = np.arange(z1, z2, 0.1)
10  dof = 10 #len(x) - 1
11  y = t.pdf(x, dof)
12
13  # 繪圖
14  plt.plot(x,y,'red', label='t分配', linewidth=5)
15
16  # 繪製常態分配
17  y2 = norm.pdf(x)
18  plt.plot(x,y2, 'green', label='常態分配')
19
20  plt.legend()
21  plt.show()
```

第 11 行為 t 為 Student's t 分配。

執行結果：

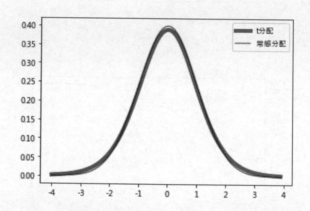

較粗的線(紅色)：t 為 Student's t 分配。

2-4-5-2 均勻分配(Uniform Distribution)

這也是一種常見的分配，通常是屬於離散型的資料，所有樣本點發生的機率都相同，例如擲硬幣、擲骰子，出現每一面的機會都一樣。均勻分配的機率密度函數如下：

$$f(x) = \frac{1}{(b-a)} \quad , a \le x \le b$$

累積分配函數如下：

【範例 1】繪製擲骰子的機率密度函數。

```
 1  import numpy as np
 2  import matplotlib.pyplot as plt
 3
 4  # 點數與機率
 5  face = [1,2,3,4,5,6]
 6  probs = np.full((6), 1/6)
 7
 8  # 繪製長條圖
 9  plt.bar(face, probs)
10  plt.ylabel('機率', fontsize=12)
11  plt.xlabel('點數', fontsize=12)
12  plt.title('擲骰子', fontsize=12)
13  plt.ylim([0,1])
14  plt.show()
```

執行結果：

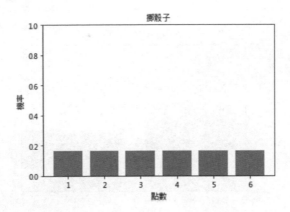

可以使用隨機亂數模擬，程式碼為 np.random.randint(1, 6)。

2-4-5-3 二項分配(Binomial Distribution)

顧名思義二項分配就是只有兩種觀察值，成功/失敗、正面/反面、有/沒有等，類似的分配有三種：

1. 伯努利(Bernoulli)分配：做一次二分類的實驗，機率密度函數如下：
   $f(x) = p^x(1-p)^{1-x}$, x=0 或 1, p: 成功的機率

2. 二項(Binomial)分配：做多次二分類的實驗。
   $f(k, n, p) = \binom{n}{k}p^k(1-p)^{n-k}$, n：試驗次數, k：成功次數, p:成功的機率

3. 多項(Multinomial)分配：做多次多分類的實驗。
   $F(x_1, x_2, ... x_k, n, p_1, p_2 ..., p_k) = \frac{n!}{x_1! x_2! ... x_k!} p_1^{x_1} p_2^{x_2} ... p_k^{x_k}$, $\sum x_i = n$

【範例 2】繪製擲硬幣的機率密度函數。

```
1   import matplotlib.pyplot as plt
2   import numpy as np
3
4   # 正面/反面機率
5   probs = np.array([0.6, 0.4])
6   face = [0, 1]
7   plt.bar(face, probs)
8   plt.title('伯努利(Bernoulli)分配', fontsize=16)
9   plt.ylabel('機率', fontsize=14)
10  plt.xlabel('正面/反面', fontsize=14)
11  plt.ylim([0,1])
12
13  plt.show()
```

執行結果:

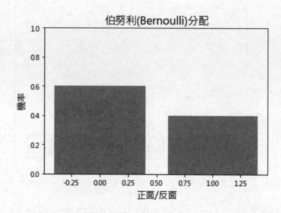

【範例 3】繪製擲硬幣 10 次的機率密度函數。

```
1   import matplotlib.pyplot as plt
2   import numpy as np
3   from scipy.stats import binom
4
5   # 正面/反面機率
6   n = 10
7   p = 0.6
8
9   probs=[]
10  for k in range(11):
11      p_binom = binom.pmf(k,n,p)
12      probs.append(p_binom)
13
14  face=np.arange(11)
15  plt.bar(face, probs)
16  plt.title('二項(Binomial)分配', fontsize=16)
17  plt.ylabel('機率', fontsize=14)
18  plt.xlabel('正面次數', fontsize=14)
19
20  plt.show()
```

執行結果：

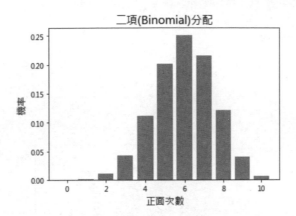

【範例 4】 繪製擲骰子 10 次的機率。

```
1   import matplotlib.pyplot as plt
2   import numpy as np
3   from scipy.stats import multinomial
4
5   # 擲骰子次數
6   n = 6
7
8   # 假設擲出各種點數的機率
9   p = [0.1, 0.2, 0.2, 0.2, 0.2, 0.1]
10
11  probs=[]
12
13  # 擲骰子1~6點各出現1次的機率
14  p_multinom = multinomial(n,p)
15  p_multinom.pmf([1,1,1,1,1,1])
```

執行結果：0.011520000000000013

若擲骰子 60 次，1~6 點各出現 10 次的機率，執行結果：4.006789054168284e-06，機率小很多，由於，擲骰子 60 次出現的樣本空間有更多的組合。

```
1   # 擲骰子60次
2   n = 60
3
4   # 擲骰子1~6點各出現10次的機率
5   p_multinom = multinomial(n,p)
6   p_multinom.pmf([10,10,10,10,10,10])
```

擲骰子 1000 次，1~6 點各出現的次數。

```
1  # 隨機抽樣1000次
2  y = np.random.choice(np.arange(1,7), size=1000, p=p)
3
4  # np.bincount：0~6點各出現的次數，故要扣掉第一個元素
5  times = np.bincount(y)[1:]
6
7  # 繪圖
8  plt.bar(np.arange(1,7), times)
9  plt.show()
```

執行結果：

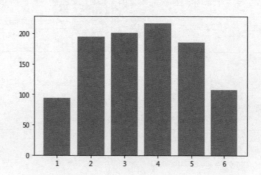

2-4-5-4　卜瓦松分配(Poisson Distribution)

卜瓦松分配經常運用在「給定的時間內發生 k 次事件」的機率分配函數，我們得以建立一套顧客服務的模型，用來估計一個服務櫃台的等候人數，進而計算出需要安排幾個服務櫃台，來達成特定的服務水準(Service Level Agreement, SLA)，應用層面非常廣，例如售票櫃台、便利商店結帳櫃台、客服中心的電話線數、依個人發生車禍的次數決定車險定價等。

卜瓦松分配的機率密度函數如下：

$$f(k, \lambda) = \frac{\lambda^k e^{-\lambda}}{k!} \quad , \lambda: 平均發生次數, e: 自然對數, k: 發生次數$$

【範例 5】繪製各種 λ 值的卜瓦松分配機率密度函數。

```
1  import matplotlib.pyplot as plt
2  import numpy as np
3  from scipy.stats import poisson
4
5  # λ = 2、4、6、8
6  for lambd in range(2, 10, 2):
7      k = np.arange(0, 10)
8      poisson1 = poisson.pmf(k, lambd)
9
```

```
10      # 繪圖
11      plt.plot(n, poisson1, '-o', label=f"λ = {lambd}")
12
13  # 設定圖形屬性
14  plt.xlabel('發生次數', fontsize=12)
15  plt.ylabel('機率', fontsize=12)
16  plt.title("Poisson Distribution (各種λ值)", fontsize=16)
17  plt.legend()
```

執行結果：

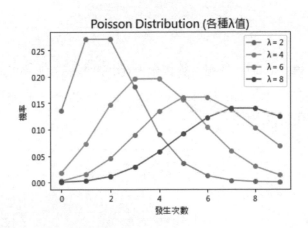

2-4-6　假設檢定(Hypothesis Testing)

根據維基百科的定義[5]，假設檢定（Hypothesis Testing）是推論統計中用於檢定統計假設的一種方法。而「統計假設」可以通過觀察一組隨機變數的模型進行檢定的科學假說。通常判定一種新藥是否有效，並不希望因為隨機抽樣的誤差，而造成判定的錯誤，所以，治癒人數的比例必須超過某一顯著水準，才能夠認定為具有醫療效果。在談到假設檢定之前，我們先來了解什麼是「信賴區間」(Confidence Interval)。

2-4-6-1　「信賴區間」(Confidence Interval)

之前我們談到平均數、中位數都是以單一值表達樣本的集中趨勢，這稱為「點估計」，不過，這種表達方式並不精確，以常態分配而言，如下圖機率密度函數，樣本剛好等於平均數的機率也只有 0.4，因此，我們改以區間估計會是一個比較穩健的做法，例如，估計 95%的樣本會落在特定區間內，這個區間我們稱為「信賴區間」(Confidence Interval)。

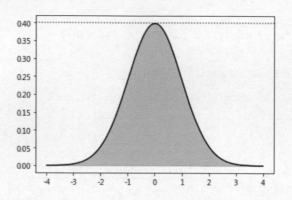

以常態分配而言，1 倍的標準差信賴水準(Confidence Level)約為 68.3%，2 倍的標準差信賴水準(Confidence Level)約為 95.4%，3 倍的標準差信賴水準(Confidence Level)約為 99.7%。

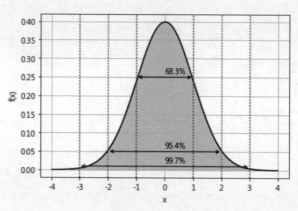

以業界常用的 95%信賴水準為例，就是「我們確信 95%的樣本會落在 $(-1.96\delta, 1.96\delta)$ 區間內」，一般醫藥有效性的檢定也是以此概念表達，例如疫苗。

➤ **以下程式請參考 02_4_6_1_信賴區間.ipynb。**

【範例 1】以美國歷屆總統的身高資料，計算各式描述統計量及 95%的信賴區間。

解題：

```
1  import random
2  import pandas as pd
3  import numpy as np
4
5  # 讀取檔案
6  df = pd.read_csv('./data/president_heights.csv')
7  df.rename(columns={df.columns[-1]:'height'}, inplace=True)
8
9  # 計算信賴區間
10 m = df['height'].mean()
11 sd = df['height'].std()
12 print(f'平均數={m}, 標準差={sd}, 信賴區間=({m-2*sd}, {m+2*sd})')
```

執行結果：平均數=179.74, 標準差=7.02, 信賴區間=(165.71, 193.77)。

如上所述，如果單純以平均數 179.74 說明美國總統的身高並不完整，因為其中有多位總統的身高低於 170，若再加上「95%美國總統的身高介於 (165.71, 193.77)」，會讓資訊更加完整。

【範例 2】利用隨機亂數產生常態分配的樣本，再使用 SciPy 的統計模組 (stats)計算信賴區間。

解題：

1. 利用隨機亂數產生 10000 筆樣本，樣本來自常態分配 N(5, 2)，即平均數為 5，標準差=2，使用 norm.interval()計算信賴區間，參數分別為信賴水準、平均數、標準差。

```
1  import matplotlib.pyplot as plt
2  import numpy as np
3  from scipy.stats import norm
4
5  # 產生隨機亂數的樣本
6  mu = 5       # 平均數
7  sigma = 2    # 標準差
8  n = 10000    # 樣本數
9  data = np.random.normal(mu, sigma, n)
10
11 cl = .95     #信賴水準(Confidence Level)
12
13 # 計算信賴區間
14 m = data.mean()
15 sd = data.std()
16 y1 = norm.interval(cl, m, sd)
17 print(f'平均數={m}, 標準差={sd}, 信賴區間={y1}')
```

執行結果：

信賴區間 = (1.0487, 8.9326)，約為 $(\mu-1.96\delta , \mu+1.96\delta)$。

2. 繪圖。

```
1   import seaborn as sns
2
3   # 直方圖，參數 hist=False：不畫階梯直方圖，只畫平滑曲線
4   sns.distplot(data, hist=False)
5
6   # 畫信賴區間
7   plt.axvline(y1[0], c='r', linestyle=':')
8   plt.axvline(y1[1], c='r', linestyle=':')
9
10  # 標示 95%
11  plt.annotate(text='', xy=(y1[0],0.025), xytext=(y1[1],0.025),
12               arrowprops=dict(arrowstyle='<->'))
13  plt.annotate(text='95%', xy=(mu,0.03), xytext=(mu,0.03))
14
15  plt.show()
```

執行結果如下：

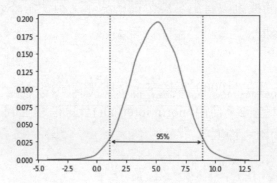

【範例 3】利用隨機亂數產生二項分配的樣本，再使用 SciPy 的統計模組 (stats)計算信賴區間。

解題：

```
1   import matplotlib.pyplot as plt
2   import numpy as np
3   from scipy.stats import binom
4
5   # 產生隨機亂數的樣本
6   trials = 1    # 實驗次數
7   p = 0.5       # 出現正面的機率
8   n = 10000     # 樣本數
9   data = np.random.binomial(trials, p, n)
```

```
10
11  cl = .95     #信賴水準(Confidence Level)
12
13  # 計算信賴區間
14  m = data.mean()
15  y1 = binom.interval(cl, n, m)
16  print(f'平均數={m}, 信賴區間={y1}')
```

執行結果：

平均數=0.5039, 標準差=0.4999847897686489, 信賴區間=(4941.0, 5137.0)，約為 $(\mu-1.96\,\delta\,,\,\mu+1.96\delta)$。

二項分配標準差公式 $=\sqrt{p(1-p)}$，可以驗算 $(p*(1-p))**.5 = 0.5$，隨機亂數的樣本與理論值相去不遠$(0.49998 \cong 0.5)$。

2-4-6-2 中央極限定理(Central Limit Theorem)

「中央極限定理」(Central Limit Theorem, CLT) 是指當樣本數愈大時，不管任何機率分配，每批樣本的平均數之機率分配會近似於常態分配。因此，我們就可以使用常態分配，估計任何樣本平均數之機率分配的信賴區間或做假設檢定。

然而，因為每一個樣本是一批的資料，假設個數為 n，則每批樣本的平均數所形成的標準差不是原樣本標準差，而是如下：

$$\delta_{\bar{x}} = \frac{\delta}{\sqrt{n}}$$

例如，我們有 1000 批的樣本，每批的樣本有 10 筆資料，則每批樣本的平均數所形成的標準差：

$$\delta_{\bar{x}} = \frac{\delta}{\sqrt{10}}$$

又例如，樣本來自二項分配，標準差 $=\sqrt{p(1-p)}$，則每批樣本的平均數所形成的標準差：

$$\delta_{\bar{x}} = \frac{\sqrt{p(1-p)}}{\sqrt{n}} = \sqrt{\frac{p(1-p)}{n}}$$

平均數的標準差稱為標準誤差(standard error)，有別於原樣本的標準差。

➢ **以下程式請參考 02_4_6_2_中央極限定理.ipynb。**

【**範例 1**】利用隨機亂數產生二項分配的 10000 批樣本，每批含 100 筆資料，請計算平均數、標準誤差，並繪圖驗證是否近似常態分配。

解題：

```python
import matplotlib.pyplot as plt
import numpy as np
from scipy.stats import binom
import seaborn as sns

# 產生隨機亂數的樣本
trials = 100 # 實驗次數
p = 0.5      # 出現正面的機率
n = 10000   # 樣本數
data = np.random.binomial(trials, p, n)

cl = .95    #信賴水準(Confidence Level)

# 計算信賴區間
data = data / trials
m = data.mean()
sd = data.std()
y1 = binom.interval(cl, n, m)
print(f'平均數={m}, 標準差={sd}, 信賴區間={y1}')

# 直方圖
sns.distplot(data, bins=20)
plt.show()
```

執行結果：平均數=0.5003, 標準差=0.04982, 信賴區間=(4906.0, 5102.0)。

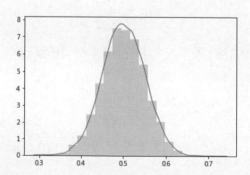

- 平均數與原樣本的平均數幾乎相同(0.5003 ≅ 0.5)。
- 標準差與理論值也很接近(0.04982 ≅ 0.05)，理論值 = (p*(1-p) / trials) **.5=0.04982。
- 直方圖近似常態分配。

2-4-6-3 假設檢定(Hypothesis Testing)

在我們設定特定的顯著水準(Significance Level)，以 α 表示，例如 5%，或者相對的信賴水準 95%，如果樣本的平均數落在信賴區間之外，我們就說新藥是顯著有效，這樣的判定只有 5% 機率是錯誤的，這種檢定方法就稱為「假設檢定」 (Hypothesis Testing)。當然，也可以用較嚴謹的顯著水準(0.3%)，即三倍標準差，檢定新藥是否顯著有效。

依檢定範圍會分為單邊(One-side)、雙邊(Two-side)，又依樣本的設計會分為單樣本(Single-sample)、雙樣本(Two-sample)，又針對檢定的統計量，可能是平均數檢定或標準差的檢定。

單邊(One-side)檢定是只關心機率分配的一邊，例如，A 班的成績是否優於 B 班($\mu_A > \mu_B$)，為右尾檢定(Right-tail Test)，病毒核酸檢測(PCR)，CT 值 30 以下即視為確診($\mu < 30$) ，為左尾檢定(Left-tail Test)，雙邊(Two-side)檢定則關心兩邊，例如，A 班的成績是否與 B 班有顯著差異($\mu_A \neq \mu_B$)。

單樣本(Single-sample)檢定，只有一份樣本，例如，抽樣一群顧客，調查是否喜歡公司特定的產品，雙樣本(Two-sample)則有兩份樣本互相作比較，例如，A 班的成績是否優於 B 班，或者，新藥有效性測試，通常會將實驗對象分為兩組，一組為實驗組(Treatment Group)，讓他們服用新藥，另一組為對照組，又稱控制組(Control Group)，讓他們服用安慰劑，這種設計又稱為 A/B Test。

檢定時會有兩個假設，虛無假設(Null Hypothesis)及對立假設(Alternative Hypothesis)：

虛無假設 H_0：$\mu = 0$
對立假設 H_1：$\mu \neq 0$
H_1 也可以寫成 H_A。

對立假設是虛無假設反面的條件，通常對立假設是我們希望的結果， 因此，檢定後，如果虛無假設為真時，我們會使用「不能拒絕」(fail to reject) 虛無假設，而不會說虛無假設成立。

以下就以實例説明各式的檢定。

> **以下程式請參考 02_4_6_3_假設檢定.ipynb。**

【**範例 1**】川普的身高 190 公分,是否比歷屆的美國總統平均身高有顯著的不同?假設歷屆的美國總統身高為常態分配,以顯著水準 5% 檢定。

解題:顯著水準 5% 時,信賴區間為$(\mu - 1.96\delta, \mu + 1.96\delta)$,計算如下:

```python
1   import random
2   import pandas as pd
3   import numpy as np
4   import seaborn as sns
5
6   # 讀取檔案
7   df = pd.read_csv('./data/president_heights.csv')
8   df.rename(columns={df.columns[-1]:'height'}, inplace=True)
9
10  # 計算信賴區間
11  m = df['height'].mean()
12  sd = df['height'].std()
13  print(f'平均數={m:.2f}, 標準差={sd:.2f},
14      信賴區間=({m-1.96*sd:.2f}, {m+1.96*sd:.2f})')
15
16  sns.distplot(df['height'])
17  plt.axvline(m, color='yellow', linestyle='dashed', linewidth=2)
18  plt.axvline(m-1.96*sd, color='magenta', linestyle='dashed', linewidth=2)
19  plt.axvline(m+1.96*sd, color='magenta', linestyle='dashed', linewidth=2)
20
21  # 川普的身高190公分
22  plt.axvline(190, color='red', linewidth=2)
23
24  plt.show()
```

執行結果:平均數=179.74, 標準差=7.02, 信賴區間=(165.99, 193.49),川普的身高 190 公分在信賴區間內,表示川普身高並不顯著比歷屆的美國總統高。

以圖形看,川普的身高 190 公分在左右兩條虛線(信賴區間)之間。

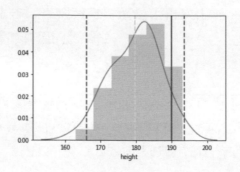

【**範例 2**】要調查顧客是否喜歡公司新上市的食品，公司進行問卷調查，取得客戶的評價，範圍介於 0~10 分，已知母體平均分數為 5 分，標準差 2 分，請使用假設檢定確認顧客是否喜歡公司新上市的食品。

解題：

虛無假設 $H_0：\mu \leq 5$

對立假設 $H_1：\mu > 5$

檢驗平均數，我們會使用 t 檢定，當樣本大於 30 時，會近似於常態分配。t 統計量如下：

$$t = \frac{\bar{x} - \mu}{s / \sqrt{n}}$$

1. 使用隨機亂數模擬問卷調查結果如下：

```
1  import numpy as np
2  import matplotlib.pyplot as plt
3
4  # 問卷調查，得到客戶的評價
5  np.random.seed(123)
6  lo = np.random.randint(0, 5, 6)      # 0~4分  6筆
7  mid = np.random.randint(5, 7, 38)   # 5~6分 38筆
8  hi = np.random.randint(7, 11, 6)    # 7~10分 6筆
9  sample = np.append(lo,np.append(mid, hi))
10
11 print(f"最小值:{sample.min()}, 最大值:{sample.max()}, 平均數:{sample.mean()}")
12
13 plt.hist(sample)   # 畫直方圖
14 plt.show()
```

執行結果：最小值:1, 最大值:9, 平均數:5.46。

2. 使用 stats.ttest_1samp()作假設檢定，參數 alternative='greater'為右尾檢定，SciPy 需 v1.6.0 以上才支援 alternative 參數。

```
1  from scipy import stats
2  import seaborn as sns
3
4  # t檢定，scipy 需 v1.6.0 以上才支援 alternative
5  # https://docs.scipy.org/doc/scipy/reference/generated/scipy.stats.ttest_1samp.html
6  t,p = stats.ttest_1samp(sample, 5, alternative='greater')
7  print(f"t統計量:{t:.4f}, p值:{p}")
8
9  # 單尾檢定，右尾顯著水準 5%，故取信賴區間 90%，兩邊各 5%
10 ci = stats.norm.interval(0.90, pop_mean, pop_std)
11 sns.distplot(pop, bins=100)
12
```

```
13  # 畫平均數
14  plt.axvline(pop.mean(), color='yellow', linestyle='dashed', linewidth=2)
15
16  # 畫右尾顯著水準
17  plt.axvline(ci[1], color='magenta', linestyle='dashed', linewidth=2)
18
19  # 畫t統計量
20  plt.axvline(pop.mean() + t*pop.std(), color='red', linewidth=2)
21
22  plt.show()
```

執行結果：t 統計量 2.0250, p 值 0.024。90%的信賴區間約在$(\mu-1.645\delta, \mu+1.645\delta)$之間，因 t 統計量(2.0250) >1.645，故對立假設為真，確認顧客喜歡公司新上市的食品。使用 t 統計量比較，需考慮單/雙尾檢定，須比較不同的值，有點麻煩，專家會改用 p 值與顯著水準比較，若小於顯著水準，則對立假設為真，反之，不能拒絕虛無假設，因為 p 值的計算公式=1-cdf(累積分配函數)。

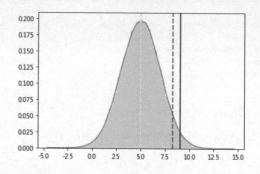

上圖實線為「母體平均數+ t 統計量 x 母體標準差」，旁邊虛線為信賴區間的右界，前者大於後者，表示效果顯著，有明顯差異。

若使用左尾檢定，也是呼叫 stats.ttest_1samp()，參數 alternative='less'。

若使用雙尾檢定，也是呼叫 stats.ttest_1samp()，刪除參數 alternative 即可。

【範例 3】要檢定新藥有效性，將實驗對象分為兩組，一組為實驗組 (Treatment Group)，讓他們服用新藥，另一組為對照組，又稱控制組 (Control Group)，讓他們服用安慰劑，檢驗兩組疾病復原狀況是否有明顯差異。

解題：

虛無假設 $H_0：\mu_1=\mu_2$

對立假設 $H_1：\mu_1\neq\mu_2$

1. 使用隨機亂數模擬復原狀況如下：

```
1  import numpy as np
2  import matplotlib.pyplot as plt
3  from scipy import stats
4
5  np.random.seed(123)
6  Control_Group = np.random.normal(66.0, 1.5, 100)
7  Treatment_Group = np.random.normal(66.55, 1.5, 200)
8  print(f"控制組平均數:{Control_Group.mean():.2f}")
9  print(f"實驗組平均數:{Treatment_Group.mean():.2f}")
```

執行結果：控制組平均數:66.04, 實驗組平均數:66.46。

2. 使用 stats.ttest_ind() 作假設檢定，參數放入兩組數據及 alternative= 'greater' 表示右尾檢定。

```
1  # t檢定，scipy 帶 v1.6.0 以上才支援 alternative
2  # https://docs.scipy.org/doc/scipy/reference/generated/scipy.stats.ttest_1samp.html
3  t,p = stats.ttest_ind(Treatment_Group, Control_Group, alternative='greater')
4  print(f"t統計量:{t:.4f}, p值:{p:.4f}")
5
6  # 單尾檢定，右尾顯著水準 5%，故取信賴區間 90%，兩邊各 5%
7  pop = np.random.normal(Control_Group.mean(), Control_Group.std(), 100000)
8  ci = stats.norm.interval(0.90, Control_Group.mean(), Control_Group.std())
9  sns.distplot(pop, bins=100)
10
11 # 畫平均數
12 plt.axvline(pop.mean(), color='yellow', linestyle='dashed', linewidth=2)
13
14 # 畫右尾顯著水準
15 plt.axvline(ci[1], color='magenta', linestyle='dashed', linewidth=2)
16
17 # 畫t統計量
18 plt.axvline(pop.mean() + t*pop.std(), color='red', linewidth=2)
19
20 plt.show()
```

執行結果：t 統計量 2.2390, p 值 0.0129，p 值若小於顯著水準(0.05)，對立假設為真，表示新藥有顯著療效。

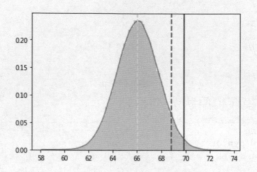

上圖實線為「母體平均數+ t 統計量 x 母體標準差」，旁邊虛線為信賴區間的右界，前者大於後者，表示效果顯著，有明顯差異。

【**範例 4**】另外有一種配對檢定(**Paired Tests**)，例如學生同時參加期中及期末考試，我們希望檢驗期末時學生是否有顯著進步，因兩次考試都是以同一組學生作實驗，故稱為配對檢定。

解題：

虛無假設 H_0：$\mu_1 = \mu_2$

對立假設 H_1：$\mu_1 \neq \mu_2$

1. 使用隨機亂數模擬學生兩次考試成績如下：

```
1  import numpy as np
2  import matplotlib.pyplot as plt
3  from scipy import stats
4
5  np.random.seed(123)
6  midTerm = np.random.normal(60, 5, 100)
7  endTerm = np.random.normal(61, 5, 100)
8  print(f"期中考:{midTerm.mean():.2f}")
9  print(f"期末考:{endTerm.mean():.2f}")
```

執行結果：期中考:60.14, 期末考:60.90。

2. 使用 stats.ttest_rel()作假設檢定，參數放入兩組數據及 alternative= 'greater' 表右尾檢定。

```
1  # t檢定，scipy 需 v1.6.0 以上才支援 alternative
2  # https://docs.scipy.org/doc/scipy/reference/generated/scipy.stats.ttest_1samp.html
3  t,p = stats.ttest_rel(endTerm, midTerm, alternative='greater')
4  print(f"t統計量:{t:.4f}, p值:{p:.4f}")
5
6  # 單尾檢定，右尾顯著水準 5%，故取信賴區間 90%，兩邊各 5%
```

```
 7  pop = np.random.normal(Control_Group.mean(), Control_Group.std(), 100000)
 8  ci = stats.norm.interval(0.90, Control_Group.mean(), Control_Group.std())
 9  sns.distplot(pop, bins=100)
10
11  # 畫平均數
12  plt.axvline(pop.mean(), color='yellow', linestyle='dashed', linewidth=2)
13
14  # 畫右尾顯著水準
15  plt.axvline(ci[1], color='magenta', linestyle='dashed', linewidth=2)
16
17  # 畫t統計量
18  plt.axvline(pop.mean() + t*pop.std(), color='red', linewidth=2)
19
20  plt.show()
```

執行結果：t 統計量 1.0159, p 值 0.1561，p 值大於顯著水準(0.05)，對立假設為假，表示學生成績沒有顯著進步。

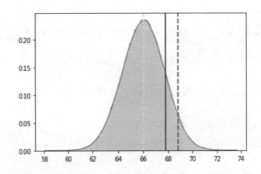

上圖實線為「母體平均數+ t 統計量 x 母體標準差」，右側虛線為信賴區間的右界，前者小於後者，表示效果不顯著。

以上基礎統計的介紹，環環相扣，以台北市長選舉為例：

1. 抽樣：使用抽樣方法，從有投票權的市民中隨機抽出一批市民意見作為樣本。
2. 計算描述統計量：推估母體的平均數、標準差、變異數等描述統計量，描繪出大部分市民的想法與差異。
3. 機率：推測候選人當選的機率，再進而推估機率分配函數。
4. 假設檢定：依據機率分配函數，進行假設檢定，推論候選人當選的可信度或確定性，這就是一般古典統計的基礎流程，與機器學習以資料為主的預測相輔相成，可彼此驗證對方。

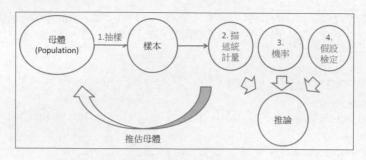

圖 2.20 上述介紹的基礎統計範圍及其關聯

2-5 線性規劃(Linear Programming)

線性規劃是「作業研究」(Operations Research, OR)中非常重要的領域,它是給定一些線性的限制條件,求取目標函數的最大值或最小值。

➤ **以下程式請參考 02_5_線性規劃.ipynb。**

【**範例 1**】最大化目標函數 $z = 3x+2y$

限制條件: $2x+y \leq 100$

$x+y \leq 80$

$x \leq 40$

$x \geq 0, y \geq 0$

解題:

1. 先畫個圖,塗色區域(黃色)為可行解(Feasible Solutions),即符合限制條件的區域:

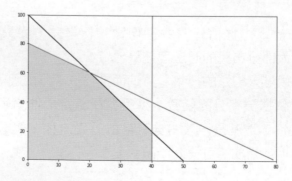

程式碼如下，程式太長分兩段截圖：

```python
1  import numpy as np
2  import matplotlib.pyplot as plt
3
4  plt.figure(figsize=(10,6))
5
6  # 限制式 2x+y = 100
7  x = np.arange(0,80)
8  y = 100 - 2 * x
9
10 # 限制式 x+y = 80
11 y_2 = 80 - x
12 plt.plot(x, y, 'black', x, y_2, 'g')
13
14 # 限制式 x = 40
15 plt.axvline(40)
16
17 # 座標軸範圍
18 plt.xlim(0,80)
19 plt.ylim(0,100)
```

```python
20
21 # 限制式 x+y = 80 取邊界線樣本點
22 x1 = np.arange(0,21)
23 y1 = 80 - x1
24
25 # 限制式 2x+y = 100 取邊界線樣本點
26 x2 = np.arange(20,41)
27 y2 = 100 - 2 * x2
28
29 # 限制式 x = 40 取邊界線樣本點
30 x3 = np.array([40]*20)
31 y3 = np.arange(0,20)
32
33 # 整合邊界線樣本點
34 x1 = np.concatenate((x1, x2, x3))
35 y1 = np.concatenate((y1, y2, y3))
36
37 # 可行解(Feasible Solutions)
38 plt.fill_between(x1, y1, color='yellow')
39
40 plt.show()
```

2. 以上的線性規劃求解可使用「單形法」(Simplex Method)，上述問題比較簡單，凸集合的最佳解發生在可行解的頂點(Vertex)，所以依上圖，我們只要求每一個頂點對應的目標函數值，比較並找到最大的數值即可。不過，深度學習的變數動輒幾百個，且神經層及神經元又很多，無法使用單形法求解，而是採取優化的方式，逐漸逼近找到近似解。因此，我們只運用程式來解題，不介紹單形法的原理。首先安裝套件 pulp：

pip install pulp

3. 以套件 pulp 求解，定義目標函數及限制條件。

```
1  from pulp import LpMaximize, LpProblem, LpStatus, lpSum, LpVariable
2
3  # 設定題目名稱及最大化(LpMaximize)或最小化(LpMinimize)
4  model  = LpProblem("範例1. 最大化目標函數", LpMaximize)
5
6  # 變數初始化，x >= 0, y >= 0
7  x = LpVariable(name="x", lowBound=0)
8  y = LpVariable(name="y", lowBound=0)
9
10 # 目標函數
11 objective_function = 3 * x + 2 * y
12
13 # 限制條件
14 constraint = 2 * x + 4 * y >= 8
15 model += (2 * x + y <= 100, "限制式1")
16 model += (x + y <= 80, "限制式2")
17 model += (x <= 40, "限制式3")
18
19 model += objective_function
20 model
```

執行結果如下，顯示定義內容：

```
範例1._最大化目標函數:
MAXIMIZE
3*x + 2*y + 0
SUBJECT TO
限制式1: 2 x + y <= 100

限制式2: x + y <= 80

限制式3: x <= 40

VARIABLES
x Continuous
y Continuous
```

4. 呼叫 model.solve()求解。

```
1  status = model.solve()
2  status = 'yes' if status == 1 else 'no'
3  print(f'有解嗎? {status}')
4
5  print(f"目標函數: {model.objective.value()}")
6  for var in model.variables():
7      print(f"{var.name}: {var.value()}")
8
9  print(f'\n限制式的值(不太重要)')
10 for name, constraint in model.constraints.items():
11     print(f"{name}: {constraint.value()}")
```

執行結果如下，當 x=20、y=60 時，目標函數最大值=180。

```
有解嗎? yes
目標函數: 180.0
x: 20.0
y: 60.0

限制式的值(不太重要)
限制式1: 0.0
限制式2: 0.0
限制式3: -20.0
```

【**範例 2**】來個實例應用，運用線性規劃來安排客服中心各時段的人力配置，請參考拙著的部落文[6]。

以上是簡單的線性規劃，還有其他的模型，例如整數規劃(Integer Programming)、二次規劃(Quadratic Programming)、非線性規劃(Non-linear Programming)，除了應用在機器學習求解外，也廣泛被應用於財務工程、工業工程等領域。

2-6 最小平方法(OLS) vs. 最大概似法(MLE)

最小平方法(Ordinary Least Squares, OLS)、最大概似法(Maximum Likelihood Estimation MLE)是常見的優化和估算參數值的方法，例如迴歸係數、常態分配的平均數/變異數。筆者戲稱兩者是優化器求解的倚天劍與屠龍刀，許多演算法憑藉這兩把寶刀迎刃而解。

2-6-1 最小平方法(OLS)

在「2-3-3 簡單線性迴歸求解」章節，利用一階導數等於 0 的特性，對簡單線性迴歸(y = wx + b)求取最佳解，使用的就是「最小平方法」(Ordinary Least Square, OLS)，這裡再強化一下，使用矩陣運算，讓解法可應用到多元迴歸、深度學習。

$$y = w_1x_1 + w_2x_2 + w_3x_3 + ... + w_nx_n + b$$

首先定義線性迴歸的「目標函數」(Objective Function) 或稱「損失函數」

(Loss Function) 為「均方誤差」(MSE)，公式為誤差平方和(SSE)，除以樣本數(n)，其中

$$誤差(\varepsilon) = 實際值(y) - 預測值(\hat{y})$$

MSE 當然愈小愈好，所以它是一個最小化的問題：

1. 目標函數 MSE $= \sum \varepsilon^2 / n = \sum(y - \hat{y})^2 / n$

   - MSE = SSE / n，我們忽略常數 n，可以只考慮 SSE：
     $$SSE = \sum \varepsilon^2 = \sum(y - \hat{y})^2 = \sum(y - wx - b)^2$$
   - 以矩陣表示線性迴歸 y = wx。
     其中 $b = b * x^0$，故將 b 也可視為 w 的一個項目。
   - 以矩陣表示 SSE：
     - ➔ $(y - wx)^T * (y - wx)$
     - ➔ $y^T y - 2w^T x^T y + w^T x^T wx$

2. 對 w 偏微分
   $$\frac{dSSE}{dw} = -2x^T y + 2x^T xw = 0$$

   $$w = (x^T x)^{-1} x^T y$$

結合矩陣、微分，我們就能夠輕鬆求出線性迴歸的係數，此原理是神經網路優化求解的基石。

➢ **以下程式請參考 02_4_8_1_最小平方法.ipynb。**

【範例 1】以最小平方法(OLS)建立線性迴歸模型，預測波士頓(Boston)房價。

解題：

1. 依上述公式計算迴歸係數。

```
1  # 載入套件
2  import numpy as np
3  import matplotlib.pyplot as plt
4  import pandas as pd
5  import seaborn as sns
6  from sklearn import datasets
7
```

```
 8  # 載入 sklearn 內建資料集
 9  ds = datasets.load_boston()
10
11  # 特徵變數
12  X=ds.data
13
14  # b = b * x^0
15  b=np.ones((X.shape[0], 1))
16  |
17  # 將 b 併入 w
18  X=np.hstack((X, b))
19
20  # 目標變數
21  y = ds.target
22
23  # 以公式求解
24  W = np.linalg.inv(X.T @ X) @ X.T @ y
25  print(f'W={W}')
```

執行結果：

```
W=[-1.08011358e-01  4.64204584e-02  2.05586264e-02  2.68673382e+00
   -1.77666112e+01  3.80986521e+00  6.92224640e-04 -1.47556685e+00
    3.06049479e-01 -1.23345939e-02 -9.52747232e-01  9.31168327e-03
   -5.24758378e-01  3.64594884e+01]
```

2. 計算相關效能衡量指標。

```
 1  # 計算效能衡量指標
 2  SSE = ((X @ W - y ) ** 2).sum()
 3  MSE = SSE / y.shape[0]
 4  RMSE = MSE ** (1/2)
 5  print(f'MSE={MSE}')
 6  print(f'RMSE={RMSE}')
 7
 8  # 計算判別係數(R^2)
 9  y_mean = y.ravel().mean()
10  SST = ((y - y_mean) ** 2).sum()
11  R2 = 1 - (SSE / SST)
12  print(f'R2={R2}')
```

3. 以 Scikit-learn 套件驗證。

```
 1  from sklearn.linear_model import LinearRegression
 2  from sklearn.metrics import r2_score, mean_squared_error
 3
 4  # 模型訓練
 5  lr = LinearRegression()
 6  lr.fit(X, y)
 7
 8  # 預測
 9  y_pred = lr.predict(X)
10
```

```
11  # 迴歸係數
12  print(f'W={lr.coef_},{lr.intercept_}\n')
13
14  # 計算效能衡量指標
15  print(f'MSE={mean_squared_error(y, y_pred)}')
16  print(f'RMSE={mean_squared_error(y, y_pred) ** .5}')
17  print(f'R2={r2_score(y, y_pred)}')
```

執行結果與公式計算一致，驗證無誤。

【範例 2】使用 SciPy 以最小平方法(OLS) 計算函數 x^2+5 的最小值。

1. 先將函數繪圖

```
1   # 函數繪圖
2   import numpy as np
3   import matplotlib.pyplot as plt
4   from scipy.optimize import leastsq
5
6   x=np.linspace(-5, 5, 11)
7   # x^2+5
8   def f(x):
9       return x**2+5
10
11  # 繪座標軸
12  plt.axhline()
13  plt.axvline()
14  # 繪圖
15  plt.plot(x, f(x), 'g')
16  plt.scatter([0],[5], color='r')
```

執行結果：

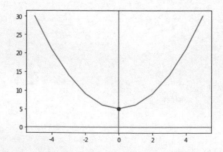

呼叫 scipy.optimizea 模組的 leastsq 函數進行優化求解。

```
1   import numpy as np
2   from scipy.optimize import leastsq
3
4   # x^2+5
5   def f(x):
6       return x**2+5
7
8   # 最小平方法
9   leastsq(f, 5, full_output=1) # full_output=1 ==> 顯示詳盡的結果
```

第一個參數是求解的函數。

- 第二個參數數是起始點。

- leastsq 是採逼近法，而非純數學公式求解，nfev 顯示它經過 22 次執行週期，才找到最小值 5 (fvec)，當時 x=1.72892379e-05 $\cong$ 0。

執行結果：

```
(array([1.72892379e-05]),
 None,
 {'fvec': array([5.]),
  'nfev': 22,
  'fjac': array([[-0.]]),
  'ipvt': array([1], dtype=int32),
  'qtf': array([5.])},
```

起始點可設為任意值，通常採隨機亂數或是直接給 0。指定值設定不佳的話，仍然可以找到最佳解，不過，需要較多次的執行週期，也就是所謂的較慢收斂(Convergence)。

當面對較複雜的函數或較多的變數，我們很難單純運用數學去求解，因此，逼近法會是一個比較務實的方法，深度學習的梯度下降法就是一個典型的例子，後面章節我們將使用 TensorFlow 程式來說明。

2-6-2 最大概似法(MLE)

最大概似法(Maximum Likelihood Estimation, MLE) 也是估算參數值的方法，目標是找到一個參數值，使出現目前事件的機率最大。如下圖，圓點是樣本點，曲線是四個可能的常態機率分配(平均數/變異數不同)，我們希望利用最大概似法找到最適配(fit)的一個常態機率分配。

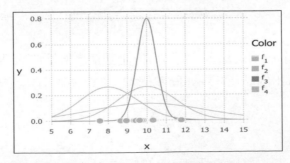

圖 2.21 最大概似法(Maximum Likelihood Estimation, MLE) 示意圖

以下使用最大概似法求算常態分配的參數 μ 及 δ。

常態分配

$$P(x; \mu, \sigma) = \frac{1}{\sigma\sqrt{2\pi}} \exp\left(-\frac{(x-\mu)^2}{2\sigma^2}\right)$$

假設來自常態分配的多個樣本互相獨立,聯合機率就等於個別的機率相乘。

$$\mathcal{L}(\mathcal{D}|\mu, \sigma^2) = \prod_{i=1}^{N} \frac{1}{\sqrt{2\pi\sigma^2}} e^{-\frac{(x_i-\mu)^2}{2\sigma^2}}$$

目標就是求取參數值 μ 及 δ,最大化機率值,即最大化目標函數。

$$\underset{\mu, \sigma^2}{\arg\max} \, \mathcal{L}(\mathcal{D}|\mu, \sigma^2)$$

N 個樣本機率全部相乘,不易計算,通常我們會取 log,變成一次方,所有的機率值取 log 後,大小順序並不會改變。

$$\underset{\mu, \sigma^2}{\arg\max} \, \log \mathcal{L}(\mathcal{D}|\mu, \sigma^2)$$

取 Log 後,化簡如下:

$$\log \mathcal{L}(\mathcal{D}|\mu, \sigma^2) = -\frac{N}{2}\log(2\pi\sigma^2) - \frac{1}{2\sigma^2}\sum_{i=1}^{N}(x_i-\mu)^2$$

對 μ 及 δ 分別偏微分。

$$\frac{\partial \log \mathcal{L}}{\partial \mu} = -\frac{1}{\sigma^2}\sum_{i=1}^{N}(x_i-\mu)$$

$$\frac{\partial \log \mathcal{L}}{\partial \sigma} = -\frac{N}{\sigma} + \frac{1}{\sigma^3}\sum_{i=1}^{N}(x_i-\mu)^2$$

一階導數為 0 時,有最大值,得到目標函數最大值下的 μ 及 δ。所以,使用最大概似法算出的 μ 及 δ 就如同我們常見的公式。

$$\mu = \frac{1}{N}\sum_{i=1}^{N} x_i$$

$$\sigma^2 = \frac{1}{N}\sum(x_i-\mu)^2$$

> **以下程式請參考 02_6_2_最大概似法.ipynb。**

【範例 1】如果樣本點 x=1, 計算來自常態分配 N(0,1)的機率。

1. 使用 NumPy 計算機率密度函數(pdf)。

```python
# 載入套件
import numpy as np
import math

# 常態分配的機率密度函數(Probability Density Function, pdf)
def f(x, mean, std):
    return (1/((2*np.pi*std**2) ** .5)) * np.exp(-0.5*((x-mean)/std)**2)

f(1, 0, 1)
```

執行結果：0.24。

2. 也可以使用 scipy.stats 模組計算。

```python
from scipy.stats import norm

# 平均數(mean)、標準差(std)
mean = 0
std = 1

# 計算來自常態分配N(0,1)的機率
norm.pdf(1, mean, std)
```

執行結果：0.24。

3. 繪製機率密度函數(pdf)。

```python
import matplotlib.pyplot as plt
import numpy as np
from scipy.stats import norm

# 觀察值範圍
z1, z2 = -4, 4

# 平均數(mean)、標準差(std)
mean = 0
std = 1

# 樣本點
x = np.arange(z1, z2, 0.001)
y = norm.pdf(x, mean, std)

# 繪圖
plt.plot(x,y,'black')

# 填色
```

```
20  plt.fill_between(x,y,0, alpha=0.3, color='b')
21
22  plt.plot([1, 1], [0, norm.pdf(1, mean, std)], color='r')
23  plt.show()
```

執行結果：

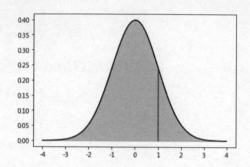

【**範例 2**】如果有兩個樣本點 x=1、3, 來自常態分配 N(1,1)及 N(2,3)的可能性，哪一個比較大？

1. 假設兩個樣本是獨立的，故聯合機率為兩個樣本機率相乘，使用 scipy.stats 模組計算機率。

```
1   # 載入套件
2   import numpy as np
3   import math
4   from scipy.stats import norm
5
6   # 計算來自常態分配 N(1,1)的機率
7   mean = 1    # 平均數(mean)
8   std = 1     # 標準差(std)
9   print(f'來自 N(1,1)的機率：{norm.pdf(1, mean, std) * norm.pdf(3, mean, std)}')
10
11
12  # 計算來自常態分配 N(2,3)的機率
13  mean = 2    # 平均數(mean)
14  std = 3     # 標準差(std)
15  print(f'來自 N(2,3)的機率：{norm.pdf(1, mean, std) * norm.pdf(3, mean, std)}')
```

執行結果如下，來自 N(1,1) 可能性比較大。

```
來自 N(1,1)的機率：0.021539279301848634
來自 N(2,3)的機率：0.01582423339377573
```

【**範例 3**】如果有一組樣本，計算來自哪一個常態分配 $N(\mu, \delta)$機率最大，請依上面證明計算 μ、δ？

解題：

1. 對常態分配的 pdf 取對數(log)：

```
1  # 載入套件
2  from scipy.stats import norm
3  from sympy import symbols, pi, exp, log
4  from sympy.stats import Probability, Normal
5
6  # 樣本
7  data = [1,3,5,3,4,2,5,6]
8
9  # x變數、平均數(m)、標準差(s)
10 x, m, s = symbols('x m s')
11
12 # 常態分配的機率密度函數(Probability Density Function, pdf)
13 pdf = (1/((2*pi*s**2) ** .5)) * exp(-0.5*((x-m)/s)**2)
14 # 顯示 log(pdf) 函數
15 log_p = log(pdf)
16 log_p
```

執行結果如下，Jupyter Notebook 可顯示 Latex 數學式。

$$\log\left(\frac{0.707106781186547 e^{-\frac{0.5(-m+x)^2}{s^2}}}{\pi^{0.5}\left(s^2\right)^{0.5}} \right)$$

2. 帶入樣本資料：

```
1  # 帶入資料
2  logP = 0
3  for xi in data:
4      logP += log_p.subs({x: xi})
5
6  logP
```

3. 上述函數使用 diff()各對平均數、變異數偏微分

```
1  from sympy import diff
2
3  logp_diff_m = diff(logP, m) # 對平均數(m)偏微分
4
5  logp_diff_s = diff(logP, s) # 對變異數(s)偏微分
6
7  print('m 偏導數:', logp_diff_m)
8  print('s 偏導數:', logp_diff_s)
```

4. 使用 simplify() 簡化偏導數

```
1  from sympy import simplify
2
3  # 簡化 m 偏導數
4  logp_diff_m = simplify(logp_diff_m)
5  print(logp_diff_m)
6  print()
7
8  # 簡化 s 偏導數
9  logp_diff_s = simplify(logp_diff_s)
10 logp_diff_s
```

5. 令一階導數=0，有最大值，可得到聯立方程式

```
1  from sympy import solve
2
3  funcs = [logp_diff_s, logp_diff_m]
4  solve(funcs, [m, s])
```

執行結果：[(3.62, -1.57), (3.62, 1.57)]。

6. 使用 NumPy 驗證。

```
1  np.mean(data), np.std(data)
```

執行結果：(3.62, 1.57)，與 MLE 求解的結果相符。

2-7 神經網路(Neural Network)求解

有了以上的基礎後，我們將綜合微積分、矩陣、機率等數學知識，對神經網路求解，這是進入深度學習領域非常重要的概念。

2-7-1 神經網路(Neural Network)

神經網路是深度學習最重要的演算法，它主要是模擬生物神經網路的傳導系統，希望透過層層解析，歸納出預測的結果。

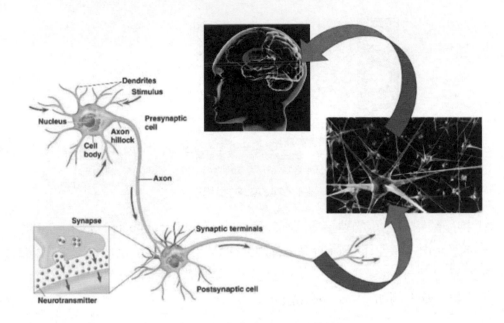

圖 2.22　生物神經網路的傳導系統

生物神經網路中表層的神經元接收到外界訊號，歸納分析後，再透過神經末梢，將分析結果傳給下一層的每個神經元，下一層神經元進行相同的動作，再往後傳導，最後傳至大腦，大腦作出最後的判斷與反應。

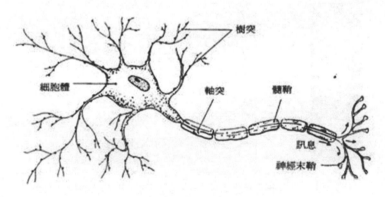

圖 2.23　單一神經元結構

於是，AI 科學家將上述生物神經網路簡化成下列的網路結構：

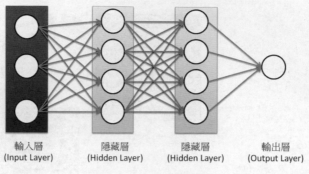

圖 2.24 AI 神經網路

AI 神經網路最簡單的連接方式稱為完全連接(Full connected, FC)，亦即每一神經元均連接至下一層的每個神經元，因此，我們可以把第二層以後的神經元均視為一條迴歸線的 y，它的特徵變數(x)就是前一層的每一個神經元，如下圖，例如 y_1、z_1 兩條迴歸線。

$$y_1 = w_1x_1 + w_2x_2 + w_3x_3 + b$$

$$z_1 = w_1y_1 + w_2y_2 + w_3y_3 + b$$

所以，簡單的講，一個神經網路可視為多條迴歸線組合而成的模型。

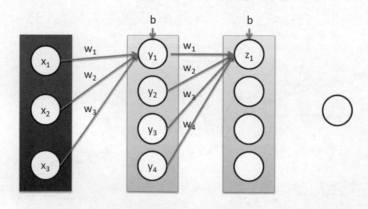

圖 2.25 一個神經網路可視為多條迴歸線組合而成的模型

以上的迴歸線是線性的，為了支援更通用性的解決方案(Generic Solution)，模型還會乘上一個非線性的函數，稱為「激勵函數」(Activation Function)，期望也能解決非線性的問題，如下圖。因激勵函數並不能明確表達原意，以下直接以 Activation Function 表示。

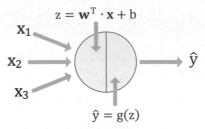

$$z = \mathbf{w}^{\mathrm{T}} \cdot \mathbf{x} + b$$

$$\hat{y} = g(z)$$

g：Activation Function

圖 2.26 激勵函數(Activation Function)

如果不考慮激勵函數，每一條線性迴歸線的權重(Weight)及偏差(Bias)可以透過最小平方法(OLS)求解，請參閱 2-4-8-1 說明，但如果乘上非線性的 Activation Function，就比較難用單純的數學求解了，因此，學者就利用優化(Optimization)理論，針對權重、偏差各參數分別偏微分，沿著切線(即梯度)逐步逼近，找到最佳解，這種演算法就稱為「梯度下降法」(Gradient Descent)。

有人用了一個很好的比喻，「當我們在山頂時，不知道下山的路，於是，就沿路往下走，遇到叉路時，就選擇坡度最大的叉路走，直到抵達平地為止。」，所以梯度下降法利用「偏微分」(Partial Differential)，求算斜率，依斜率的方向，一步步的往下走，逼近最佳解，直到損失函數沒有顯著改善為止，這時我們就認為已經找到最佳解了。

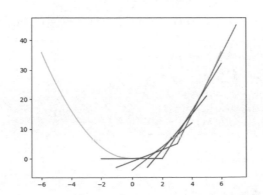

圖 2.26 梯度下降法(Gradient Descent)示意圖

➢ **上圖及以下程式請參考 02_7_梯度下降法**.ipynb。

2-7-2 梯度下降法(Gradient Descent)

梯度其實就是斜率，單變數迴歸線的權重稱為斜率，多變數迴歸線時，須個別作偏微分求取權重值，就稱為梯度。以下，先針對單變數求解，示範如何使用梯度下降法(Gradient Descent)求取最小值。

【範例 1】假定損失函數 $f(x) = x^2$，而非 MSE，請使用梯度下降法求取最小值。

注意，損失函數又稱為目標函數或成本函數，在神經網路相關文獻中大多稱為損失函數，本書從善如流，以下將統一以「損失函數」取代「目標函數」。

1. 定義函數(func)及其導數(dfunc)：

```
1  # 載入套件
2  import numpy as np
3  import matplotlib.pyplot as plt
4
5  # 目標函數(損失函數):y=x^2
6  def func(x): return x ** 2 #np.square(x)
7
8  # 目標函數的一階導數:dy/dx=2*x
9  def dfunc(x): return 2 * x
```

2. 定義梯度下降法函數，反覆更新 x，更新的公式如下，後面章節我們會推算公式的由來。

   新的 x = 目前的 x - 學習率(learning_rate) * 梯度(gradient)

```
1  # 梯度下降
2  # x_start: x的起始點
3  # df: 目標函數的一階導數
4  # epochs: 執行週期
5  # lr: 學習率
6  def GD(x_start, df, epochs, lr):
7      xs = np.zeros(epochs+1)
8      x = x_start
9      xs[0] = x
10     for i in range(epochs):
11         dx = df(x)
12         # x更新 x_new = x – learning_rate * gradient
13         x += - dx * lr
14         xs[i+1] = x
15     return xs
```

3. 設定起始點、學習率(lr)、執行週期數(epochs)等參數後，呼叫梯度下降法求解。

```
1  # 超參數(Hyperparameters)
2  x_start = 5       # 起始權重
3  epochs = 15       # 執行週期數
4  lr = 0.3          # 學習率
5
6  # 梯度下降法
7  # *** Function 可以直接當參數傳遞 ***
8  w = GD(x_start, dfunc, epochs, lr=lr)
9  print (np.around(w, 2))
10
11 color = 'r'
12 from numpy import arange
13 t = arange(-6.0, 6.0, 0.01)
14 plt.plot(t, func(t), c='b')
15 plt.plot(w, func(w), c=color, label='lr={}'.format(lr))
16 plt.scatter(w, func(w), c=color, )
17
18 # 設定中文字型
19 from matplotlib.font_manager import FontProperties
20 font = FontProperties(fname=r"c:\windows\fonts\msjhbd.ttc", size=20)
21 # 矯正負號
22 plt.rcParams['axes.unicode_minus'] = False
23
24 plt.title('梯度下降法', fontproperties=font)
25 plt.xlabel('X', fontsize=20)
26 plt.ylabel('損失函數', fontsize=20)
27 plt.show()
```

執行結果：

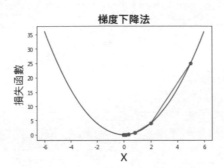

每一執行週期的損失函數如下，隨著 x 變化，損失函數逐漸收斂，即前後週期的損失函數差異逐漸縮小，最後當 x=0 時，損失函數 f(x)等於 0，為函數的最小值，與最小平方法(OLS)的計算結果相同。

[5.2, 0.8, 0.32, 0.13, 0.05, 0.02, 0.01, 0, 0, 0, 0, 0, 0, 0, 0]

如果改變起始點(x_start)為其他值，例如-5，依然可以找到相同的最小值。

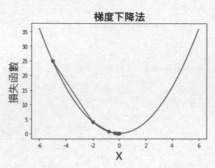

【範例 2】假定損失函數 $f(x) = 2x^4 - 3x - 20$，請使用梯度下降法求取最小值。

1. 定義函數及其微分：

```
1  # 損失函數
2  def func(x): return 2*x**4-3*x+2*x-20
3
4  # 損失函數一階導數
5  def dfunc(x): return 8*x**3-6*x+2
```

2. 繪製損失函數。

```
1   from numpy import arange
2   t = arange(-6.0, 6.0, 0.01)
3   plt.plot(t, func(t), c='b')
4
5   # 設定中文字型
6   from matplotlib.font_manager import FontProperties
7   font = FontProperties(fname=r"c:\windows\fonts\msjhbd.ttc", size=20)
8   # 矯正負號
9   plt.rcParams['axes.unicode_minus'] = False
10
11  plt.title('梯度下降法', fontproperties=font)
12  plt.xlabel('X', fontsize=20)
13  plt.ylabel('損失函數', fontsize=20)
14  plt.show()
```

執行結果：

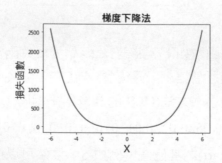

3. 梯度下降法函數(GD)不變，執行程式，如果學習率不變(lr = 0.3)，會出現錯誤訊息：「Result too large」，原因是學習率過大，梯度下降過程錯過最小值，往函數左方逼近，造成損失函數值越來越大，最後導致溢位。

```
OverflowError                          Traceback (most recent call last)
<ipython-input-20-e4de47c74444> in <module>
      6 # 梯度下降法
      7 # *** Function 可以直接當參數傳遞 ***
----> 8 w = GD(x_start, dfunc, epochs, lr=lr)
      9 print (np.around(w, 2))
     10

<ipython-input-3-32de2a0b3ea7> in GD(x_start, df, epochs, lr)
      9     xs[0] = w
     10     for i in range(epochs):
---> 11         dx = df(w)
     12         # 權重的更新 W_new = W - learning_rate * gradient
     13         w += - dx * lr

<ipython-input-17-2c5b4a8cd0cb> in dfunc(x)
      3
      4 # 損失函數一階導數
----> 5 def dfunc(x): return 8*x**3-6*x+2
      6

OverflowError: (34, 'Result too large')
```

4. 修改學習率(lr = 0.001)，同時增加執行週期數(epochs = 15000)，避免還未逼近到最小值，就先提早結束。

```
# 超參數(Hyperparameters)
x_start = 5      # 起始權重
epochs = 15000  # 執行週期數
lr = 0.001       # 學習率
```

執行結果：當 x=0.51 時，函數有最小值。

[5.   4.03 3.53 ... 0.51 0.51 0.51]

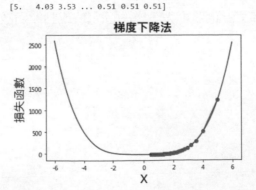

觀察上述範例，不管函數為何，我們以相同的梯度下降法(GD 函數)都能夠找到函數最小值，最重要的關鍵是「x 的更新公式」：

新的 x = 目前的 x - 學習率(learning_rate) * 梯度(gradient)

接著我們會說明此公式的由來，也就是神經網路求解的精華所在。

2-7-3 神經網路求解

神經網路求解是一個正向傳導與反向傳導反覆執行的過程，如下圖所示。

1. 由於神經網路是多條迴歸線的組合，建立模型的主要任務就是計算出每條迴歸線的權重(w)與偏差(b)。

2. 依上述範例的邏輯，一開始我們指定 w、b 為任意值，建立迴歸方程式 y=wx+b，將特徵值(x)帶入方程式，可以求得預測值(ŷ)，進而計算出損失函數，例如 MSE，這個過程稱為「正向傳導」(Forward Propagation)。

3. 透過最小化 MSE 的目標和偏微分，可以找到更好的 w、b，並依學習率來更新每一層神經網路的 w、b，此過程稱之為「反向傳導」(Backpropagation)。這部份可以藉由微分的連鎖率(Chain Rule)，一次逆算出每一層神經元對應的 w、b，公式為：

$$W_{t+1} = W_t - 學習率(learning\ rate) \times 梯度(gradient)$$

其中：

梯度 = - 2 * x * (y - ŷ) [後續證明]

學習率：優化器事先設定的固定值或動能函數。

4. 重複 2、3 步驟，一直到損失函數不再有明顯改善為止。

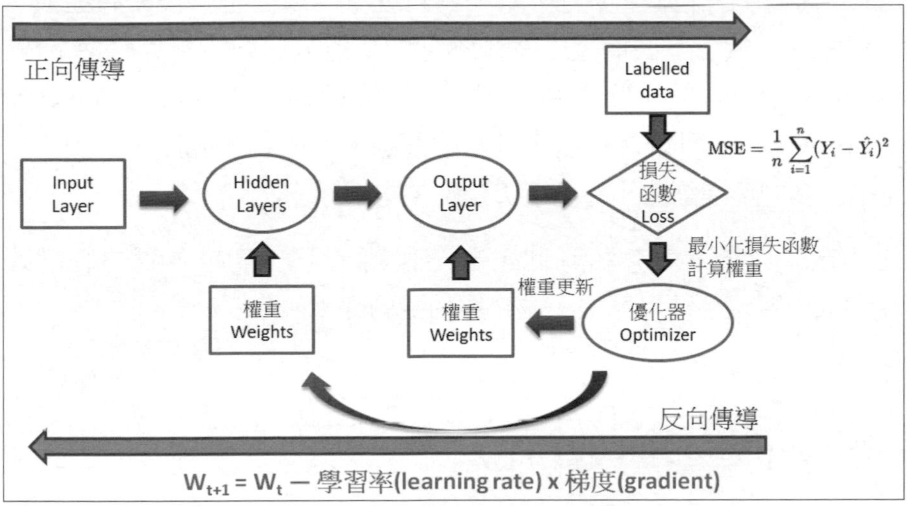

圖 2.27 神經網路權重求解過程

梯度(gradient)公式證明如下：

1. 損失函數 $MSE = \frac{\sum(y-\hat{y})^2}{n}$ ，因 n 為常數，故僅考慮分子，即 SSE

2. $SSE = \sum(y-\hat{y})^2 = \sum(y-wx)^2 = \sum(y^2 - 2ywx + w^2x^2)$

3. 以矩陣表示，$SSE = y^2 - 2ywx + w^2x^2$

4. $\frac{\partial SSE}{\partial w} = -2yx + 2wx^2 = -2\ x\ (y-wx) = -2\ x(y-\hat{y})$

5. 同理，$\frac{\partial SSE}{\partial b} = -2\ x^0(y-\hat{y}) = -2\ (y-\hat{y})$

6. 為了簡化公式，常把係數 2 拿掉

7. 最後公式為：

$$調整後權重 = 原權重 + (學習率 * 梯度)$$

8. 有些文章將梯度負號拿掉，公式就修正為：

$$調整後權重 = 原權重 - (學習率 * 梯度)$$

以上是以 MSE 為損失函數時的梯度計算公式，若是其他損失函數，梯度計算結果也會有所不同，如果再加上 Activation Function，梯度公式計算就更加複雜了，還好，深度學習框架均提供自動微分(Automatic Differentiation)、計算梯度的功能，我們就不用煩惱了。後續有些演算法會自訂損失函數，會因而產生意想不到的功能，例如「風格轉換」(Style Transfer)可以合成兩張圖像，生成對抗網路(Generative Adversarial Network, GAN)，可以產生幾可亂真的圖像。也因為如此關鍵，我們才花費了這麼多的篇幅鋪陳「梯度下降法」。

基礎的數學與統計就介紹到此，告一段落，下一章起，我們將開始以TensorFlow實現各種演算法，並介紹相關的應用。

參考資料 (References)

[1] Keith McNulty,《Decision makers need more math》, 2018
(https://towardsdatascience.com/decision-makers-need-more-math-ed4d4fe3dc09)

[2] Seaborn 官網(https://seaborn.pydata.org/examples/index.html)

[3] 台灣彩券官網(https://www.taiwanlottery.com.tw/DailyCash/index.asp)

[4] Univariate Distribution Relationships
(http://www.math.wm.edu/~leemis/chart/UDR/UDR.html)

[5] 維基百科中關於假設檢定的定義
(https://zh.wikipedia.org/wiki/%E5%81%87%E8%AA%AA%E6%AA%A2%E5%AE%9A)

[6] 陳昭明,《Day 14：客服人力規劃(Workforce Planning) -- 線性規劃求解》, 2019
(https://ithelp.ithome.com.tw/articles/10222877)

|第二篇| TensorFlow 基礎篇

TensorFlow 是 Google Brain 於 2015 年發布的深度學習框架，它是目前深度學習框架佔有率最高的套件，又於 1.4 版納入 Keras，兼顧簡易性與效能，是深度學習最佳的入門套件。

本篇將介紹 TensorFlow 的整體架構，包含下列內容：

1. 從張量(Tensor)運算、自動微分(Automatic Differentiation)、神經層，最後構建完整的神經網路。
2. 說明神經網路的各項函數，例如 Activation Function、損失函數(Loss Function)、優化器(Optimizer)、效能衡量指標(Metrics)，並介紹運用梯度下降法找到最佳解的原理與過程。
3. 示範 TensorFlow 各項工具的使用，包含回呼函數(Callback)、TensorBoard 視覺化工具、TensorFlow Dataset、TensorFlow Serving 佈署等。
4. 神經網路完整流程的實踐。
5. 卷積神經網路(Convolutional Neural Network, CNN)。
6. 預先訓練的模型(Pre-trained Model)。
7. 轉移學習(Transfer Learning)。

TensorFlow 架構與主要功能

▌ 3-1 常用的深度學習套件

維基百科[1]列舉近 20 種的深度學習套件，有許多已經逐漸被淘汰了，以
Python/C++語言的套件而言，目前比較流行的只剩以下這四種：

套件名稱	程式語言	優點	缺點
TensorFlow/Keras	Python, C++, js, Java, Go	1. Keras 簡單、易入門。支援多種程式語言。 2. 效能佳。	與 Python 未完全整合，例如模型結構不能直接使用 if。
PyTorch	Python, C++, Java	1. 與 Python 整合較好，有彈性。 2. 學界採用佔比有增加的趨勢。	執行速度較慢。
Caffe	C++	1. 執行速度非常快。 2. 專長於影像應用，NVidia 有一改良版。	1. 複雜。 2. 無 NLP 模組。
Apache MXNet	Python, Scala, and Julia	AWS/Azure 雲端支援。	複雜。

TensorFlow、PyTorch 分居佔有率的前兩名，其他套件均望塵莫及，如下
圖所示，因此推薦讀者若是使用 Python 語言，熟悉 TensorFlow、PyTorch
就綽綽有餘了，假如偏好 C/C++的話，則可以使用 Caffe 框架。

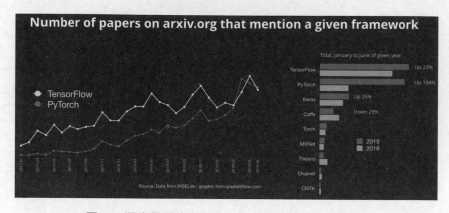

圖 3.1 深度學習套件在 Arxiv 網站的論文採用比率

圖片來源：Which deep learning framework is the best？[2]

本書的範例以 TensorFlow/Keras 為主，不會涉及其他深度學習套件，希望對 TensorFlow/Keras 作全面而深入的探討。

3-2 TensorFlow 架構

梯度下降法是神經網路主要求解的方法，計算過程需要使用大量的張量 (Tensor)運算，另外，在反向傳導的過程中，要進行偏微分，並須解決多層結構的神經網路，因此，大多數的深度學習套件至少都會具備下列功能：

1. 張量運算：包括各種向量、矩陣運算。
2. 自動微分(Auto Differentiation)
3. 神經網路(Neural Network)及各種神經層(Layers)

學習的路徑可以從簡單的張量運算開始，再逐漸熟悉高階的神經層函數，以奠定扎實的基礎。

接著再來瞭解 TensorFlow 執行環境(Runtime)，如下圖，支援 GPU、分散式運算、低階 C API 等，也支援多種網路協定。

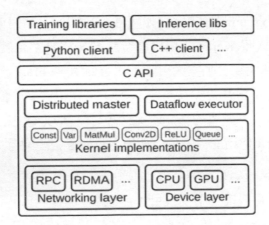

圖 3.2 TensorFlow 執行環境(Runtime)，圖片來源：TensorFlow GitHub[3]

TensorFlow 執行環境包括三種版本：

1. TensorFlow：一般電腦版本。
2. TensorFlowjs：網頁版本，適合邊緣運算的裝置及虛擬機裝置 (Docker/Kubernetes)。
3. TensorFlow Lite：輕量版，適合運算速度及記憶體有限的手機和物聯網 裝置。

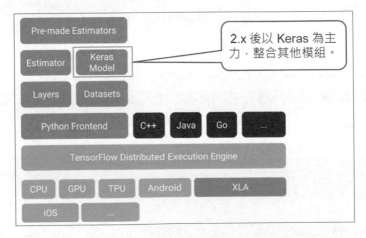

圖 3.3 TensorFlow 程式設計堆疊(Programming Stack)
圖片來源：Explained: Deep Learning in TensorFlow[4]

程式設計堆疊(Programming Stack)如圖 3.3，在 2.x 版後以 Keras 為主軸，TensorFlow 團隊依照 Keras 的規格重新開發，逐步整合其他模組，比獨立開發的 Keras 套件功能更加強大，因而 Keras 大神 François Chollet 也宣告 Keras 獨立套件也不再升級，甚至將 Keras 官網也改為介紹 TensorFlow 的 Keras 模組，而非自家開發的 Keras 獨立套件。所以，現在要撰寫 Keras 程式，應以 TensorFlow 為主，避免再使用 Keras 獨立套件了。

TensorFlow Keras 模組逐步整合其他模組及工具，如果，讀者之前是使用 Keras 獨立套件，可以補強下列項目：

1. 低階梯度下降的訓練(GradientTape)
2. 資料集載入(Dataset and Loader)
3. 回呼函數(Callback)
4. 估計器(Estimator)
5. 預訓模型(Keras Application)
6. TensorFlow Hub：進階的預訓模型
7. TensorBoard 視覺化工具
8. TensorFlow Serving 佈署工具

我們會在後面的章節陸續探討各項功能。

除了 Keras 的引進，特別要注意的是，**TensorFlow 2.x 版之後，預設模式改為 Eager Execution Mode**，捨棄 Sessionm 語法，若未調回原先的 Graph Execution Mode，1.x 版的程式執行時將會產生錯誤。由於目前網路上許多範例的程式新舊雜陳，還有很多 1.x 版的程式，讀者應特別注意。

▌3-3 張量(Tensor)運算

TensorFlow 顧名思義就是提供張量(Tensor)的定義、計算與變數值的傳遞，因此，它最底層的功能即支援各項張量的資料型態與其運算函數，基本上都遵循 NumPy 套件的設計理念與語法，包括傳播(Broadcasting)機制的支援。

以下我們直接以實作代替長篇大論，請參閱 **03_1_張量運算.ipynb**。

1. 顯示 TensorFlow 版本

```
1  # 載入套件
2  import tensorflow as tf
3
4  # 顯示版本
5  print(tf.__version__)
```

2. 檢查 GPU 是否存在

```
1  # check cuda available
2  tf.config.list_physical_devices('GPU')
```

執行結果如下，可顯示 GPU 簡略資訊：

```
[PhysicalDevice(name='/physical_device:GPU:0',device_type='GPU')]
```

3. 如果要知道 GPU 的詳細規格，可安裝 PyCuda 套件。請注意在 Windows 環境下，無法以 pip install pycuda 安裝，須至「Unofficial Windows Binaries for Python Extension Packages」(https://www.lfd.uci.edu/~gohlke/pythonlibs/?cm_mc_uid=08085305845514542921829&cm_mc_sid_50200000=1456395916#pycuda) 下載對應 Python、Cuda Toolkit 版本的二進位檔案。

   舉例來説，Python v3.8，且安裝 Cuda Toolkit v10.1 則須下載：pycuda-2020.1+cuda*101*-cp*38*-cp38-win_amd64.whl，並執行：

```
pip install pycuda-2020.1+cuda101-cp38-cp38-win_amd64.whl。
```

接著就可以執行本書所附的範例：
python GpuQuery.py。

執行結果如下，即可顯示 GPU 的詳細規格，重要資訊排列在前面：

```
偵測 1 個CUDA裝置

裝置 0: GeForce GTX 1050 Ti
         計算能力: 6.1
         GPU記憶體: 4096 MB
         6 個處理器, 各有 128 個CUDA核心數, 共 768 個CUDA核心數

         ASYNC_ENGINE_COUNT: 5
         CAN_MAP_HOST_MEMORY: 1
         CLOCK_RATE: 1392000
         COMPUTE_CAPABILITY_MAJOR: 6
         COMPUTE_CAPABILITY_MINOR: 1
         COMPUTE_MODE: DEFAULT
         CONCURRENT_KERNELS: 1
         ECC_ENABLED: 0
         GLOBAL_L1_CACHE_SUPPORTED: 1
         GLOBAL_MEMORY_BUS_WIDTH: 128
         GPU_OVERLAP: 1
         INTEGRATED: 0
         KERNEL_EXEC_TIMEOUT: 1
         L2_CACHE_SIZE: 1048576
         LOCAL_L1_CACHE_SUPPORTED: 1
         MANAGED_MEMORY: 1
         MAXIMUM_SURFACE1D_LAYERED_LAYERS: 2048
         MAXIMUM_SURFACE1D_LAYERED_WIDTH: 32768
```

4. 宣告常數(constant)，參數可以是常數、list、NumPy array。

```
1  # 宣告常數(constant)，參數可以是常數、list、numpy array
2  x = tf.constant([[1, 2]])
```

5. 支援四則運算。

```
1  print(x+10)
2  print(x-10)
3  print(x*2)
4  print(x/2)
```

執行結果如下：

```
tf.Tensor([[11 12]], shape=(1, 2), dtype=int32)
tf.Tensor([[-9 -8]], shape=(1, 2), dtype=int32)
tf.Tensor([[2 4]], shape=(1, 2), dtype=int32)
tf.Tensor([[0.5 1. ]], shape=(1, 2), dtype=float64)
```

如果要顯示數值，須轉為 NumPy array：

```
(x+10).numpy()
```

6. 四則運算也可以使用 TensorFlow 函數。

```
1  # 轉為負數
2  print(tf.negative(x))
3
4  # 常數、List、Numpy array 均可運算
5  print(tf.add(1, 2))
6  print(tf.add([1, 2], [3, 4]))
```

```
7
8   print(tf.square(5))
9   print(tf.reduce_sum([1, 2, 3]))
10
11  # 混用四則運算符號及TensorFlow 函數
12  print(tf.square(2) + tf.square(3))
```

reduce_sum 是沿著特定軸加總，輸出會少一維，故[1, 2, 3]套用函數運算後等於 6。

7. TensorFlow 會自動決定在 CPU 或 GPU 運算，可由下列指令偵測。一般而言，常數(tf.constant)放在 CPU，其他變數則放在 GPU 上，兩者加總時，會將常數搬到 GPU 上再加總，過程不需人為操作。但請注意，**PyTorch 套件稍嫌麻煩，必須手動將變數搬移至 CPU 或 GPU 運算，不允許一個變數在 CPU，另一個變數在 GPU。**

```
1   x1 = tf.constant([[1, 2, 3]], dtype=float)
2   print("x1 是否在 GPU #0 上: ", x1.device.endswith('GPU:0'))
3
4   # 設定 x 為均勻分配亂數 3x3
5   x2 = tf.random.uniform([3, 3])
6   print("x2 是否在 GPU #0 上: ", x2.device.endswith('GPU:0'))
7
8   x3=x1+x2
9   print("x3 是否在 GPU #0 上: ", x3.device.endswith('GPU:0'))
```

執行結果:

```
x1 是否在 GPU #0 上:     False
x2 是否在 GPU #0 上:     True
x3 是否在 GPU #0 上:     True
```

8. 也能夠使用 with tf.device("CPU:0")或 with tf.device("GPU:0")，強制指定在 CPU 或 GPU 運算。

```
1   import time
2
3   # 計算 10 次的時間
4   def time_matmul(x):
5       start = time.time()
6       for loop in range(10):
7           tf.matmul(x, x)
8
9       result = time.time()-start
10      print("{:0.2f}ms".format(1000*result))
```

```
11
12  # 強制指定在cpu運算
13  print("On CPU:", end='')
14  with tf.device("CPU:0"):
15      x = tf.random.uniform([1000, 1000])
16      assert x.device.endswith("CPU:0")
17      time_matmul(x)
18
19  # 強制指定在gpu運算
20  if tf.config.list_physical_devices("GPU"):
21      print("On GPU:", end='')
22      with tf.device("GPU:0"):
23          x = tf.random.uniform([1000, 1000])
24          assert x.device.endswith("GPU:0")
25          time_matmul(x)
```

第一次執行結果如下，CPU 運算比 GPU 快速:

```
On CPU: 64.00ms
On GPU: 311.49ms
```

多次執行後，GPU 反而比 CPU 運算快很多:

```
On CPU:58.00ms
On GPU:1.00ms
```

9. 稀疏矩陣(Sparse Matrix)運算：稀疏矩陣是指矩陣內只有很少數的非零元素，如果依一般的矩陣儲存會非常浪費記憶體，運算也是如此，因為大部份項目為零，不需浪費時間計算，所以，科學家針對稀疏矩陣設計特殊的資料儲存結構及運算演算法，TensorFlow 也支援此類資料型態。

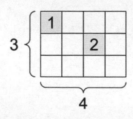

圖 3.4 稀疏矩陣(Sparse Matrix)

10. TensorFlow 稀疏矩陣只需設定有值的位置(indices)和數值(values)，並設定維度(dense_shape)如下：

```
1   # 稀疏矩陣只需設定有值的位置及數值
2   sparse_tensor = tf.sparse.SparseTensor(indices=[[0, 0], [1, 2]],
3                                          values=[1, 2],
4                                          dense_shape=[3, 4])
5   print(sparse_tensor)
```

執行結果：

```
SparseTensor(indices=tf.Tensor(
[[0 0]
 [1 2]], shape=(2, 2), dtype=int64), values=tf.Tensor([1 2], shape=(2,), dtype=int32), dense_shape=tf.Tensor([3 4], shape=(2,),
dtype=int64))
```

11. 轉為正常的矩陣格式。

```
1   # 轉為正常的矩陣格式
2   x = tf.sparse.to_dense(sparse_tensor)
3   print(type(x))
4
5   # 2.31 以前版本會出錯
6   x.numpy()
```

執行結果：

```
<class 'tensorflow.python.framework.ops.EagerTensor'>

array([[1, 0, 0, 0],
       [0, 0, 2, 0],
       [0, 0, 0, 0]])
```

12. 如果要執行 TensorFlow 1.x 版 Graph Execution Mode 的程式，須禁用
    (Disable)2.x 版的功能，並改變載入套件的命名空間(Namespace)。

```
1   if tf.__version__[0] != '1':         # 是否為 TensorFlow 1.x版
2       import tensorflow.compat.v1 as tf  # 改變載入套件的命名空間(Namespace)
3       tf.disable_v2_behavior()           # 使 2.x 版功能失效(Disable)
```

TensorFlow 大刀闊斧的改良反而造成與舊版相容性差的困擾，1.x 版的程
式在 2.x 版的預設模式下均無法執行，雖然可以把 2.x 版的預設模式切換
回 1.x 版的模式，但是要自行修改的地方還是蠻多的，而且也缺乏未來
性，因此，在這裡建議大家：

- Eager Execution Mode 已是 TensorFlow 的主流，不要再使用 1.x 版的
  Session 或 TFLearn 等舊的架構，套句電視劇對白，已經回不去了。
- 手上有許多 1.x 版的程式，如果很重要，非用不可，可利用官網移轉

(Migration)指南[5]進行修改，單一檔案比較可行，但若是複雜的套件，可能就要花上大半天的時間了。

- 官網也有提供指令，能夠一次升級整個目錄的所有程式，如下：
tf_upgrade_v2 --intree <1.x 版程式目錄> --outtree <輸出目錄>
詳細使用方法可參閱 TensorFlow 官網升級指南[6]。

13. 禁用 2.x 版的功能後，測試 1.x 版程式，Graph Execution Mode 程式須使用 tf.Session：

```
1  # 測試1.x版程式
2  x = tf.constant([[1, 2]])
3  neg_x = tf.negative(x)
4
5  with tf.Session() as sess:    # 使用 session
6      result = sess.run(neg_x)
7      print(result)
```

14. GPU 記憶體管理：由於 TensorFlow 對於 GPU 記憶體的垃圾回收 (Garbage Collection)機制並不完美，因此，常會出現下列 GEMM 錯誤：

```
InternalError:  Blas GEMM launch failed : a.shape=(32, 784), b.shape=(784, 256), m=32, n=256, k=784
        [[node sequential/dense/MatMul (defined at <ipython-input-3-9d42ad511782>:5) ]] [Op:__inference_train_function_581]
```

此訊息表示 GPU 記憶體不足，尤其是使用 Jupyter Notebook 時，因為 Jupyter Notebook 是一個網頁程式，關掉某一個 Notebook 檔案，網站仍然在執行中，所以不代表該檔案的資源會被回收，通常要選擇「**Kernel > Restart**」選單才會回收資源。另一個方法，就是限制 GPU 的使用配額，以下是 TensorFlow 2.x 版的方式，1.x 版並不適用：

```
1  # 限制 TensorFlow 只能使用 GPU 2GB 記憶體
2  gpus = tf.config.experimental.list_physical_devices('GPU')
3  if gpus:
4      try:
5          # 限制 第一顆 GPU 只能使用 2GB 記憶體
6          tf.config.experimental.set_virtual_device_configuration(gpus[0],
7              [tf.config.experimental.VirtualDeviceConfiguration(memory_limit=1024*2)])
8
9          # 顯示 GPU 個數
10         logical_gpus = tf.config.experimental.list_logical_devices('GPU')
11         print(len(gpus), "Physical GPUs,", len(logical_gpus), "Logical GPUs")
12     except RuntimeError as e:
13         # 顯示錯誤訊息
14         print(e)
```

15. 也可以不使用 GPU：

```
1  import os
2
3  os.environ["CUDA_VISIBLE_DEVICES"] = "-1"
```

3-4　自動微分(Automatic Differentiation)

反向傳導時，會更新每一層的權重，這時就輪到偏微分運算派上用場，所以，深度學習套件的第二項主要功能就是自動微分 (Automatic Differentiation)。

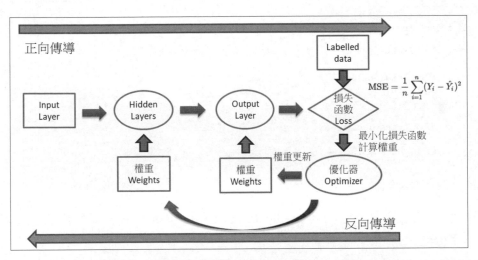

圖 3.5　神經網路權重求解過程

同樣地直接以實作代替長篇大論，請參閱 **03_1_張量運算.ipynb**。

1. 呼叫 tf.GradientTape()函數可自動微分，再呼叫 g.gradient(y, x)可取得 y 對 x 作偏微分的梯度。

```
1  import numpy as np
2  import tensorflow as tf
3
4  x = tf.Variable(3.0)            # 宣告TensorFlow變數(Variable)
5
```

```
6  with tf.GradientTape() as g: # 自動微分
7      y = x * x                # y = x^2
8
9  dy_dx = g.gradient(y, x)      # 取得梯度，f'(x) = 2x, x=3 ==> 6
10
11 print(dy_dx.numpy())          # 轉換為 NumPy array 格式
```

執行結果：

```
f(x) = x²
f'(x) = 2x
f'(3) = 2 * 3 = 6。
```

2. 宣告為變數(tf.Variable)時，該變數會自動參與自動微分，但宣告為常數(tf.constant)時，如欲參與自動微分，須額外設定 g.watch()。

```
1  import numpy as np
2  import tensorflow as tf
3
4  x = tf.constant(3.0)          # 宣告 TensorFlow 常數
5
6  with tf.GradientTape() as g: # 自動微分
7      g.watch(x)                # 設定常數參與自動微分
8      y = x * x                # y = x^2
9
10 dy_dx = g.gradient(y, x)      # 取得梯度，f'(x) = 2x, x=3 ==> 6
11
12 print(dy_dx.numpy())          # 轉換為 NumPy array 格式
```

執行結果：與上面相同。

3. 計算二階導數：呼叫 tf.GradientTape()、g.gradient(y, x)函數兩次，即能取得二階導數。

```
1  x = tf.constant(3.0)                  # 宣告 TensorFlow 常數
2  with tf.GradientTape() as g:
3      g.watch(x)
4      with tf.GradientTape() as gg: # 自動微分
5          gg.watch(x)                # 設定常數參與自動微分
6          y = x * x                # y = x^2
7
8      dy_dx = gg.gradient(y, x)      # 一階導數
9  d2y_dx2 = g.gradient(dy_dx, x)     # 二階導數
10
11 print(f'一階導數={dy_dx.numpy()}, 二階導數={d2y_dx2.numpy()}')
```

執行結果：一階導數=6.0, 二階導數=2.0

```
f(x) = x²
f'(x) = 2x
f''(x) = 2。
f''(3) = 2。
```

4. 多變數計算導數：各自呼叫 g.gradient(y, x)函數，可取得每一個變數的梯度。若呼叫 g.gradient()兩次或以上，tf.GradientTape()須加參數 persistent=True，使 tf.GradientTape()不會被自動回收，用完之後，可使用「del g」刪除 GradientTape 物件。

```
1  x = tf.Variable(3.0)              # 宣告 TensorFlow 常數
2  with tf.GradientTape(persistent=True) as g:  # 自動微分
3      y = x * x                     # y = x^2
4      z = y * y                     # z = y^2
5
6  dz_dx = g.gradient(z, x)      # 4*x^3
7  dy_dx = g.gradient(y, x)      # 2*x
8
9  del g                            # 不用時可刪除 GradientTape 物件
10
11 print(f'dy/dx={dy_dx.numpy()}, dz/dx={dz_dx.numpy()}')
```

執行結果：dy/dx=6, dz/dx=108。

```
z = f(x) = y² = x⁴
➔ f'(x) = 4x³
➔ f'(3) = 108。
```

5. 順便認識一下 PyTorch 自動微分的語法，與 TensorFlow 稍有差異。

```
1  import torch            # 載入套件
2
3  x = torch.tensor(3.0, requires_grad=True)   # 設定 x 參與自動微分
4  y=x*x                  # y = x^2
5
6  y.backward()           # 反向傳導
7
8  print(x.grad)          # 取得梯度
```

■ requires_grad=True 參數宣告 x 參與自動微分。

■ 呼叫 y.backward()，要求作反向傳導。

■ 呼叫 x.grad 取得梯度。

【範例】利用 TensorFlow 自動微分求解簡單線性迴歸的參數(w、b)。

程式:03_3_簡單線性迴歸.ipynb。

1. 載入套件。

```
1  # 載入套件
2  import numpy as np
3  import tensorflow as tf
```

2. 定義損失函數 $MSE = \dfrac{\sum(y - \hat{y})^2}{n}$。

```
1  # 定義損失函數
2  def loss(y, y_pred):
3      return tf.reduce_mean(tf.square(y - y_pred))
```

3. 定義預測值函數 $y = w\,x + b$。

```
1  # 定義預測值函數
2  def predict(X):
3      return w * X + b
```

4. 定義訓練函數:在自動微分中需重新計算損失函數值,assign_sub 函數相當於「-=」。

```
1   # 定義訓練函數
2   def train(X, y, epochs=40, lr=0.0001):
3       current_loss=0                           # 損失函數值
4       for epoch in range(epochs):              # 執行訓練週期
5           with tf.GradientTape() as t:         # 自動微分
6               t.watch(tf.constant(X))          # 宣告 TensorFlow 常數參與自動微分
7               current_loss = loss(y, predict(X))   # 計算損失函數值
8
9           dw, db = t.gradient(current_loss, [w, b]) # 取得 w, b 個別的梯度
10
11          # 更新權重:新權重 = 原權重 — 學習率(learning_rate) * 梯度(gradient)
12          w.assign_sub(lr * dw) # w -= lr * dw
13          b.assign_sub(lr * db) # b -= lr * db
14
15          # 顯示每一訓練週期的損失函數
16          print(f'Epoch {epoch}: Loss: {current_loss.numpy()}')
```

5. 產生隨機亂數作為資料集，進行測試。

```
1  # 產生線性隨機資料100筆，介於 0-50
2  n = 100
3  X = np.linspace(0, 50, n)
4  y = np.linspace(0, 50, n)
5
6  # 資料加一點雜訊(noise)
7  X += np.random.uniform(-10, 10, n)
8  y += np.random.uniform(-10, 10, n)
```

6. 執行訓練。

```
1  # w、b 初始值均設為 0
2  w = tf.Variable(0.0)
3  b = tf.Variable(0.0)
4
5  # 執行訓練
6  train(X, y)
7
8  # w、b 的最佳解
9  print(f'w={w.numpy()}, b={b.numpy()}')
```

執行結果：w=0.9464, b=0.0326。

損失函數值隨著訓練週期越來越小，如下圖。

```
Epoch 0: Loss: 890.1063232421875
Epoch 1: Loss: 607.071533203125
Epoch 2: Loss: 419.7763671875
Epoch 3: Loss: 295.8358459472656
Epoch 4: Loss: 213.81948852539062
Epoch 5: Loss: 159.5459747314453
Epoch 6: Loss: 123.63106536865234
Epoch 7: Loss: 99.86465454101562
Epoch 8: Loss: 84.1374282836914
Epoch 9: Loss: 73.7300033569336
Epoch 10: Loss: 66.84292602539062
Epoch 11: Loss: 62.28538513183594
Epoch 12: Loss: 59.269386291503906
Epoch 13: Loss: 57.27349090576172
Epoch 14: Loss: 55.95263671875
Epoch 15: Loss: 55.078487396240234
Epoch 16: Loss: 54.49993133544922
Epoch 17: Loss: 54.116981506347656
Epoch 18: Loss: 53.86347579956055
Epoch 19: Loss: 53.69562911987305
Epoch 20: Loss: 53.584468841552734
```

7. 顯示結果：迴歸線確實居於樣本點中線。

```
1  import matplotlib.pyplot as plt
2
3  plt.scatter(X, y, label='data')
4  plt.plot(X, predict(X), 'r-', label='predicted')
5  plt.legend()
```

執行結果：

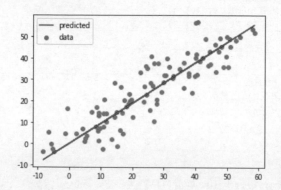

有了 TensorFlow 自動微分的功能，正向與反向傳導變的非常簡單，只要熟悉運作的架構，後續複雜的模型也可以運用自如。

3-5 神經層(Neural Network Layer)

上一節運用自動微分實現一條簡單線性迴歸線的求解，然而神經網路是多條迴歸線的組合，並且每一條迴歸線可能再乘上非線性的 Activation Function，假如使用自動微分函數逐一定義每條公式，層層串連，程式可能要很多個迴圈才能完成。所以為了簡化程式開發的複雜度，TensorFlow/Keras 直接建構各式各樣的神經層(Layer)函數，可以使用神經層組合神經網路的結構，我們只需要專注在演算法的設計即可，輕鬆不少。

神經網路是多個神經層組合而成的，如下圖，包括輸入層(Input Layer)、隱藏層(Hidden Layer)及輸出層(Output Layer)，其中隱藏層可以有任意多層，一般而言，隱藏層大於或等於兩層，即稱為「深度」(Deep)學習。

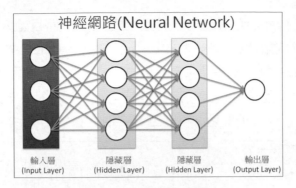

圖 3.6 神經網路示意圖

TensorFlow/Keras 提供數十種神經層，分成下列類別，可參閱 Keras 官網說明(https://keras.io/api/layers/)：

1. 核心類別(Core Layer)：包括完全連接層(Full Connected Layer)、Activation layer、嵌入層(Embedding layer) 等。

2. 捲積層(Convolutional Layer)。

3. 池化層(Pooling Layer)。

4. 循環層(Recurrent Layer)。

5. 前置處理層(Preprocessing layer)：提供 one-hot encoding、影像前置處理、資料增補(Data augmentation) 等。

我們先來看看兩個最簡單的「完全連接層」範例。

【範例 1】使用完全連接層估算簡單線性迴歸的參數(w、b)。

程式：03_4_簡單的完全連階層.ipynb。

1. 產生隨機資料，與上一節範例相同。

```
1  # 載入套件
2  import numpy as np
3  import tensorflow as tf
4
5  # 產生線性隨機資料100筆，介於 0-50
6  n = 100
7  X = np.linspace(0, 50, n)
8  y = np.linspace(0, 50, n)
9
10 # 資料加一點雜訊(noise)
11 X += np.random.uniform(-10, 10, n)
12 y += np.random.uniform(-10, 10, n)
```

2. 建立模型：神經網路僅使用一層完全連接層，而且輸入只有一個神經元，即 X，輸出也只有一個神經元，即 y。Dense 本身有一個參數 use_bias，即是否有偏差項，預設值為 True，除了一個神經元輸出外，還會有一個偏差項。以上設定其實就等於 y=wx+b。為聚焦概念的說明，暫時不解釋其他參數，後面章節會有詳盡說明。

```
1  # 定義完全連接層(Dense)
2  # units：輸出神經元個數，input_shape：輸入神經元個數
3  layer1 = tf.keras.layers.Dense(units=1, input_shape=[1])
4
5  # 神經網路包含一層完全連接層
6  model = tf.keras.Sequential([layer1])
```

3. 定義模型的損失函數(loss)及優化器(optimizer)。

```
1  # 定義模型的損失函數(loss)為 MSE，優化器(optimizer)為 adam
2  model.compile(loss='mean_squared_error',
3                optimizer=tf.keras.optimizers.Adam())
```

4. 模型訓練：只需一個指令 model.fit(X, y) 即可，訓練過程的損失函數變化都會存在 history 變數中。

```
1  history = model.fit(X, y, epochs=500, verbose=False)
```

5. 訓練過程繪圖。

```
1  import matplotlib.pyplot as plt
2
3  plt.xlabel('訓練週期', fontsize=20)
4  plt.ylabel("損失函數(loss)", fontsize=20)
5  plt.plot(history.history['loss'])
```

執行結果：損失函數值隨著訓練週期越來越小。

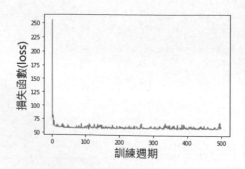

6. 取得模型參數 w=第一層的第一個參數，b 為輸出層的第一個參數。

```
1  w = layer1.get_weights()[0][0][0]
2  b = layer1.get_weights()[1][0]
3
4  print(f"w : {w:.4f} , b : {b:.4f}")
```

執行結果：w：0.8798，b：3.5052，因輸入資料為隨機亂數。

7. 繪圖顯示迴歸線。

```
1  import matplotlib.pyplot as plt
2
3  plt.scatter(X, y, label='data')
4  plt.plot(X, X * w + b, 'r-', label='predicted')
5  plt.legend()
```

執行結果：

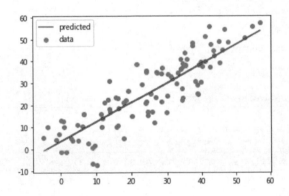

與自動微分比較，程式更簡單，只要設定模型結構、損失函數、優化器後，呼叫訓練(fit)函數即可。

再看一個有趣的例子，利用神經網路自動求出華氏與攝氏溫度的換算公式。

【範例 2】使用完全連接層推算華氏與攝氏溫度的換算公式。

$$華氏(F) = 攝氏(C)*(9/5) + 32$$

1. 利用換算公式，隨機產生 151 筆資料。

```python
1  # 載入套件
2  import numpy as np
3  import tensorflow as tf
4
5  # 隨機產生151筆資料
6  n = 151
7  C = np.linspace(-50, 100, n)
8  F = C * (9/5) + 32
9
10 for i, x in enumerate(C):
11     print(f"華氏(F)：{F[i]:.2f}，攝氏(C)：{x:.0f}")
```

2. 建立模型：神經網路只有一層完全連接層，而且輸入只有一個神經元，即攝氏溫度，輸出只有一個神經元，即華氏溫度。

```python
1  # 定義完全連接層(Dense)
2  # units：輸出神經元個數，input_shape：輸入神經元個數
3  layer1 = tf.keras.layers.Dense(units=1, input_shape=[1])
4
5  # 神經網路包含一層完全連接層
6  model = tf.keras.Sequential([layer1])
7
8  # 定義模型的損失函數(loss)為 MSE，優化器(optimizer)為 adam
9  model.compile(loss='mean_squared_error',
10               optimizer=tf.keras.optimizers.Adam(0.1))
```

3. 模型訓練。

```python
1  history = model.fit(X, y, epochs=500, verbose=False)
```

4. 訓練過程繪圖。

```python
1  import matplotlib.pyplot as plt
2
3  plt.xlabel('訓練週期', fontsize=20)
4  plt.ylabel("損失函數(loss)", fontsize=20)
5  plt.plot(history.history['loss'])
```

執行結果如下圖：損失函數值隨著訓練週期越來越小。

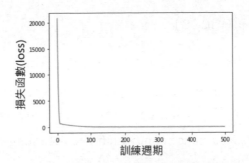

5. 測試：輸入攝氏 100 度及 0 度預測華氏溫度，答案完全正確。

```
1  y_pred = model.predict([100.0])[0][0]
2  print(f"華氏(F)：{y_pred:.2f} , 攝氏(C)：100")
3
4  y_pred = model.predict([0.0])[0][0]
5  print(f"華氏(F)：{y_pred:.2f} , 攝氏(C)：0")
```

執行結果：

```
華氏(F)：212.00 , 攝氏(C)：100
華氏(F)：32.00 , 攝氏(C)：0
```

6. 取得模型參數 w、b。

```
1  w = layer1.get_weights()[0][0][0]
2  b = layer1.get_weights()[1][0]
3
4  print(f"w：{w:.4f} , b：{b:.4f}")
```

執行結果：w：1.8000 , b：31.9999，近似下列華氏與攝氏溫度的換算公式，實在太神奇了。

$$華氏(F) = 攝氏(C)*(9/5) + 32$$

其實換算公式也是一條迴歸線，想通了，對於結果就不足為奇了。

讀者讀到這裡，應該會好奇想知道如何使用更多的神經元和神經層，甚至更複雜的神經網路結構，下一章我們將正式邁入深度學習的殿堂，學習如何用 TensorFlow 解決各種實際的案例，並且詳細剖析各個函數的用法及參數說明。

參考資料 (References)

[1] 維基百科中針對常見的深度學習套件之比較圖表
(https://en.wikipedia.org/wiki/Comparison_of_deep-learning_software)

[2] Amol Mavuduru ,《Which deep learning framework is the best?》, 2020
(https://towardsdatascience.com/which-deep-learning-framework-is-the-best-eb51431c39a)

[3] TensorFlow 官方 GitHub
(https://github.com/tensorflow/docs/blob/master/site/en/r1/guide/extend/architecture.md)

[4] Sonu Sharma ,《Explained: Deep Learning in TensorFlow》, 2019
(https://towardsdatascience.com/explained-deep-learning-in-tensorflow-chapter-1-9ab389fe90a1)

[5] TensorFlow 官網移轉指南
(https://www.tensorflow.org/guide/migrate)

[6] TensorFlow 官網升級指南
(https://www.tensorflow.org/guide/upgrade)

神經網路實作

接下來,將開始以神經網路實作各種應用,可以暫時跟數學/統計說再見,會著重於概念的澄清與程式的撰寫。筆者會盡可能的運用大量圖解,幫助讀者迅速掌握各種演算法的原理,避免流於長篇大論。

同時也會藉由「手寫阿拉伯數字辨識」的案例,實作機器學習流程的 10 大步驟,並詳細解說構建神經網路的函數用法及各項參數代表的意義,到最後我們會撰寫一個完整的視窗介面程式及網頁程式,讓終端使用者(End User)親身體驗 AI 應用程式,期望激發使用者對「企業導入 AI」有更多的發想。

4-1 撰寫第一支神經網路程式

手寫阿拉伯數字辨識,問題定義如下:

1. 讀取手寫阿拉伯數字的影像,將影像中的每個像素當成一個特徵。資料來源為 MNIST 機構所收集的 60000 筆訓練資料,另含 10000 筆測試資料,每筆資料是一個阿拉伯數字,寬高為(28, 28)的點陣圖形。
2. 建立神經網路模型,利用梯度下降法,求解模型的參數值,一般稱為權重(Weight)。
3. 依照模型推算每一個影像是 0~9 的機率,再以最大機率者為預測結果。

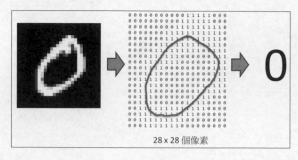

28 x 28 個像素

圖 4.1　手寫阿拉伯數字辨識，左圖為輸入的圖形，中間為圖形的像素，右圖為預測結果

4-1-1　最簡短的程式

TensorFlow 1.x 版使用會話(Session)及運算圖(Computational Graph)的概念
來編寫，光是要將兩個張量相加就要寫上一大段程式，被對手 Facebook
PyTorch 批評得體無完膚，於是 TensorFlow 2.x 為了回擊對手並一雪前
恥，官網直接出大招，在文件首頁展示一支超短程式，示範如何撰寫手寫
阿拉伯數字的辨識，要證明改版後的 TensorFlow 確實超好用，現在我們就
來看看這支程式。

【範例 1】TensorFlow 官網的手寫阿拉伯數字辨識。
程式：04_01_手寫阿拉伯數字辨識.ipynb：

```
1   import tensorflow as tf
2   mnist = tf.keras.datasets.mnist
3
4   # 匯入 MNIST 手寫阿拉伯數字 訓練資料
5   (x_train, y_train),(x_test, y_test) = mnist.load_data()
6
7   # 特徵縮放至 (0, 1) 之間
8   x_train, x_test = x_train / 255.0, x_test / 255.0
9
10  # 建立模型
11  model = tf.keras.models.Sequential([
12    tf.keras.layers.Flatten(input_shape=(28, 28)),
13    tf.keras.layers.Dense(128, activation='relu'),
14    tf.keras.layers.Dropout(0.2),
15    tf.keras.layers.Dense(10, activation='softmax')
16  ])
17
18  # 設定優化器(optimizer)、損失函數(loss)、效能衡量指標(metrics)
19  model.compile(optimizer='adam',
20                loss='sparse_categorical_crossentropy',
21                metrics=['accuracy'])
```

```
22
23  # 模型訓練，epochs：執行週期，validation_split：驗證資料佔 20%
24  model.fit(x_train, y_train, epochs=5, validation_split=0.2)
25
26  # 模型評估
27  model.evaluate(x_test, y_test)
```

執行結果如下：

```
Epoch 1/5
1500/1500 [==============================] - 5s 3ms/step - loss: 0.5336 - accuracy: 0.8432 - val_loss: 0.1558 - val_accuracy:
0.9555
Epoch 2/5
1500/1500 [==============================] - 5s 3ms/step - loss: 0.1676 - accuracy: 0.9505 - val_loss: 0.1134 - val_accuracy:
0.9657
Epoch 3/5
1500/1500 [==============================] - 5s 3ms/step - loss: 0.1203 - accuracy: 0.9646 - val_loss: 0.0975 - val_accuracy:
0.9704
Epoch 4/5
1500/1500 [==============================] - 5s 3ms/step - loss: 0.0981 - accuracy: 0.9703 - val_loss: 0.0968 - val_accuracy:
0.9717
Epoch 5/5
1500/1500 [==============================] - 5s 3ms/step - loss: 0.0786 - accuracy: 0.9758 - val_loss: 0.0958 - val_accuracy:
0.9713
313/313 [==============================] - 1s 3ms/step - loss: 0.0807 - accuracy: 0.9752

[0.08072374016046524, 0.9751999974250793]
```

上述的程式扣除註解，僅 10 多行，辨識的準確率高達 97~98%，
TensorFlow 成功扳回一城，真是厲害！

4-1-2 程式強化

上一節的範例「手寫阿拉伯數字辨識」是官網為了炫技刻意縮短了程式，
本節將會按照機器學習流程的 10 大步驟，撰寫完整程式，針對每個步驟
仔細解析，務必理解每一行程式背後代表的意涵。

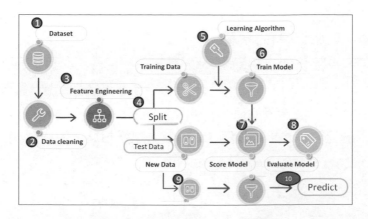

圖 4.2　機器學習流程 10 大步驟

【範例 2】依據上圖 10 大步驟撰寫手寫阿拉伯數字辨識。

程式：04_02_手寫阿拉伯數字辨識_完整版.ipynb。

1. 步驟 1：載入 MNIST 手寫阿拉伯數字資料集。

```
1  import tensorflow as tf
2  mnist = tf.keras.datasets.mnist
3
4  # 載入 MNIST 手寫阿拉伯數字資料
5  (x_train, y_train),(x_test, y_test) = mnist.load_data()
6
7  # 訓練/測試資料的 X/y 維度
8  print(x_train.shape, y_train.shape,x_test.shape, y_test.shape)
```

執行結果：取得 60000 筆訓練資料，10000 筆測試資料，每筆資料是一個阿拉伯數字，寬高各為(28, 28)的點陣圖形，要注意資料的維度及其大小，必須與模型的輸入規格契合。

```
(60000, 28, 28) (60000,) (10000, 28, 28) (10000,)
```

2. 步驟 2：EDA，對資料集進行探索與分析，首先觀察訓練資料的目標值 (y)，即影像的真實結果。

```
1  # 訓練資料前10筆圖片的數字
2  y_train[:10]
```

執行結果如下，每筆資料是一個阿拉伯數字。

```
array([5, 0, 4, 1, 9, 2, 1, 3, 1, 4], dtype=uint8)
```

3. 列印第一筆訓練資料的像素。

```
1  # 顯示第1張圖片內含值
2  x_train[0]
```

執行結果如下，每筆像素的值介於(0, 255)之間，為灰階影像，0 為白色，255 為最深的黑色，注意，這與 **RGB** 色碼剛好相反，**RGB** 黑色為 **0**，白色為 **255**。

```
[   0,    0,    0,    0,    0,    0,    0,    0,    0,    0,    0,    0,    0,
    0,    0,    0,    0,    0,    0,    0,    0,    0,    0,    0,    0,    0,
    0,    0],
[   0,    0,    0,    0,    0,    0,    0,    0,    0,    0,    0,    0,    3,
   18,   18,   18,  126,  136,  175,   26,  166,  255,  247,  127,    0,    0,
    0,    0],
[   0,    0,    0,    0,    0,    0,    0,    0,   30,   36,   94,  154,  170,
  253,  253,  253,  253,  253,  225,  172,  253,  242,  195,   64,    0,    0,
    0,    0],
[   0,    0,    0,    0,    0,    0,    0,   49,  238,  253,  253,  253,  253,
  253,  253,  253,  253,  251,   93,   82,   82,   56,   39,    0,    0,    0,
    0,    0],
[   0,    0,    0,    0,    0,    0,    0,    0,   18,  219,  253,  253,  253,
  253,  198,  182,  247,  241,    0,    0,    0,    0,    0,    0,    0,    0,
    0,    0],
[   0,    0,    0,    0,    0,    0,    0,    0,   80,  156,  107,  253,  253,
  205,   11,    0,   43,  154,    0,    0,    0,    0,    0,    0,    0,    0,
    0,    0],
```

4. 為了看清楚圖片的手寫的數字，將非 0 的數值轉為 1，變為黑白兩色的圖片。

```
1  # 將非0的數字轉為1，顯示第1張圖片
2  data = x_train[0].copy()
3  data[data>0]=1
4
5  # 將轉換後二維內容顯示出來，隱約可以看出數字為 5
6  text_image=[]
7  for i in range(data.shape[0]):
8      text_image.append(''.join(str(data[i])))
9  text_image
```

執行結果如下，筆者以筆描繪 1 的範圍，隱約可以看出是 5。

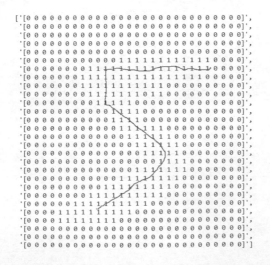

5. 顯示第一筆訓練資料圖像，確認是 5。

```
1  # 顯示第1張圖片圖像
2  import matplotlib.pyplot as plt
3
4  # 第一筆資料
5  X2 = x_train[0,:,:]
6
7  # 繪製點陣圖，cmap='gray':灰階
8  plt.imshow(X2.reshape(28,28), cmap='gray')
9
10 # 隱藏刻度
11 plt.axis('off')
12
13 # 顯示圖形
14 plt.show()
```

執行結果：

6. 步驟 3：進行特徵工程，將特徵縮放成(0, 1)之間，特徵縮放可提高模型準確度，並且可以加快收斂速度。特徵縮放採常態化(Normalization)公式：

   (x - 樣本最小值) / (樣本最大值 - 樣本最小值)。

```
1  # 特徵縮放，使用常態化(Normalization)，公式 = (x - min) / (max - min)
2  # 顏色範圍：0~255，所以，公式簡化為 x / 255
3  # 注意，顏色0為白色，與RGB顏色不同，(0,0,0) 為黑色。
4  x_train_norm, x_test_norm = x_train / 255.0, x_test / 255.0
5  x_train_norm[0]
```

執行結果：

```
[0.         , 0.         , 0.         , 0.         , 0.         ,
 0.         , 0.         , 0.         , 0.         , 0.         ,
 0.         , 0.         , 0.00392157, 0.00392157, 0.00392157,
 0.00392157, 0.00392157, 0.00392157, 0.00392157, 0.00392157,
 0.00392157, 0.00392157, 0.00392157, 0.00392157, 0.         ,
 0.         , 0.         , 0.         ],
[0.         , 0.         , 0.         , 0.         , 0.         ,
 0.         , 0.         , 0.         , 0.00392157, 0.00392157,
 0.00392157, 0.00392157, 0.00392157, 0.00392157, 0.00392157,
 0.00392157, 0.00392157, 0.00392157, 0.00392157, 0.00392157,
 0.00392157, 0.00392157, 0.00392157, 0.00392157, 0.         ,
 0.         , 0.         , 0.         ],
```

7. 步驟 4：資料分割為訓練及測試資料，此步驟無需進行，因為載入 MNIST 資料時，已經切割好了。

8. 步驟 5：建立模型結構如下：

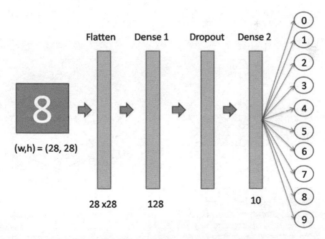

圖 4.3 手寫阿拉伯數字辨識的模型結構

Keras 提供兩類模型，包括順序型模型(Sequential Model)及 Functional API 模型，順序型模型函數為 tf.keras.models.Sequential，適用於簡單的結構，神經層一層接一層的順序執行，使用 Functional API 可以設計成較複雜的模型結構，包括多個輸入層或多個輸出層，也允許分叉，後續用到再詳細說明。這裡使用簡單的順序型模型，內含各種神經層如下：

```
1  # 建立模型
2  model = tf.keras.models.Sequential([
3    tf.keras.layers.Flatten(input_shape=(28, 28)),
4    tf.keras.layers.Dense(128, activation='relu'),
5    tf.keras.layers.Dropout(0.2),
6    tf.keras.layers.Dense(10, activation='softmax')
7  ])
```

- 扁平層(Flatten Layer)：將寬高各 28 個像素的圖壓扁成一維陣列 (28 x 28 = 784 個特徵)。

- 完全連接層(Dense Layer)：輸入為上一層的輸出，輸出為 128 個神經元，即構成 128 條迴歸線，每一條迴歸線有 784 個特徵。輸出通常訂為 4 的倍數，並無建議值，可經由實驗調校取得較佳的參數值。

- Dropout Layer：類似正則化(Regularization)，希望避免過度擬合，在訓練週期隨機丟棄一定比例(0.2)的神經元，一方面可以估計較少的參數，另一方面能夠取得多個模型的均值，避免受極端值影響，藉以矯正過度擬合的現象。通常會在每一層 Dense 後面加一個 Dropout，比例也無建議值。

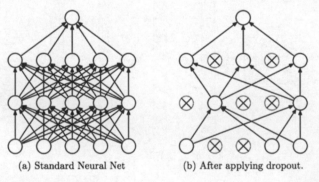

(a) Standard Neural Net   (b) After applying dropout.

圖 4.2　左邊為標準的神經網路，右邊是 Dropout Layer 形成的的神經網路。

- 第二個完全連接層(Dense)：為輸出層，因為要辨識 0~9 十個數字，故輸出要設成 10，透過 Softmax Activation Function，可以將輸出轉為機率形式，即預測 0~9 的個別機率，再從中選擇最大機率者為預測值。

9. 編譯指令(model.compile)需設定參數，優化器(optimizer)為 adam，損失函數(loss)為 sparse_categorical_crossentropy(交叉熵)，而非 MSE，相關參數後面章節會詳細説明。

```
1  # 設定優化器(optimizer)、損失函數(loss)、效能衡量指標(metrics)的類別
2  model.compile(optimizer='adam',
3                loss='sparse_categorical_crossentropy',
4                metrics=['accuracy'])
```

10. 步驟 6：結合訓練資料及模型結構，進行模型訓練。

```
1  # 模型訓練
2  history = model.fit(x_train_norm, y_train, epochs=5, validation_split=0.2)
```

- validation_split：將訓練資料切割一部份為驗證資料，目前設 0.2，即驗證資料佔 20%，在訓練過程中，會用驗證資料計算準確度及損失函數值，確認訓練過程有無異常。

- epochs：設定訓練要執行的週期數，所有訓練資料經過一次正向和反向傳導，稱為一個執行週期。

執行結果如下所示，每一個執行週期(epoch)，都包含訓練的損失(loss)、準確率(accuracy)及驗證資料的損失(val_loss)、準確率(val_accuracy)，這些資訊都會儲存在 history 變數內，為一字典(dict)資料型態。

```
Train on 48000 samples, validate on 12000 samples
Epoch 1/5
48000/48000 [==============================] - 4s 77us/sample - loss: 0.3264 - accuracy: 0.9055 - val_loss: 0.1576 - val_accura
cy: 0.9572
Epoch 2/5
48000/48000 [==============================] - 3s 71us/sample - loss: 0.1593 - accuracy: 0.9534 - val_loss: 0.1187 - val_accura
cy: 0.9654
Epoch 3/5
48000/48000 [==============================] - 3s 71us/sample - loss: 0.1188 - accuracy: 0.9649 - val_loss: 0.1039 - val_accura
cy: 0.9682
Epoch 4/5
48000/48000 [==============================] - 3s 72us/sample - loss: 0.0969 - accuracy: 0.9704 - val_loss: 0.1071 - val_accura
cy: 0.9668
Epoch 5/5
48000/48000 [==============================] - 3s 71us/sample - loss: 0.0829 - accuracy: 0.9740 - val_loss: 0.0876 - val_accura
cy: 0.9739
```

11. 對訓練過程的準確率繪圖：

```
1  # 對訓練過程的準確率繪圖
2  plt.figure(figsize=(8, 6))
3  plt.plot(history.history['accuracy'], 'r', label='訓練準確率')
4  plt.plot(history.history['val_accuracy'], 'g', label='驗證準確率')
5  plt.legend()
```

執行結果：隨著執行週期(epoch)次數的增加，準確率越來越高，且驗證資料與訓練資料的準確率應趨於一致，若不一致或準確率過低就要檢查每個環節是否出錯。

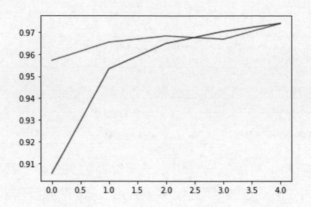

12. 對訓練過程的損失繪圖：

```
1   # 對訓練過程的損失繪圖
2   import matplotlib.pyplot as plt
3
4   plt.figure(figsize=(8, 6))
5   plt.plot(history.history['loss'], 'r', label='訓練損失')
6   plt.plot(history.history['val_loss'], 'g', label='驗證損失')
7   plt.legend()
```

執行結果：隨著執行週期(epoch)次數的增加，損失越來越低，驗證資料與
訓練資料的損失應趨於一致。

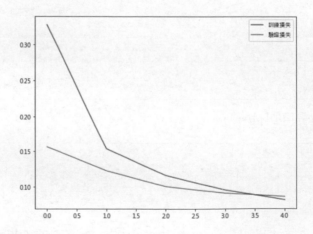

13. 步驟 7：評分(Score Model)，呼叫 evaluate()，輸入測試資料，會計算
    出損失及準確率。

```
1   # 評分(Score Model)
2   score=model.evaluate(x_test_norm, y_test, verbose=0)
3
4   for i, x in enumerate(score):
5       print(f'{model.metrics_names[i]}: {score[i]:.4f}')
```

執行結果：loss: 0.0833，accuracy: 0.9743。

14. 實際比對測試資料的前 20 筆，呼叫 predict_classes()，可以得到預測類別。

```
1   # 實際預測 20 筆資料
2   predictions = model.predict_classes(x_test_norm)
3
4   # 比對
5   print('actual    :', y_test[0:20])
6   print('prediction:', predictions[0:20])
```

執行結果如下，全部正確。

```
actual    : [7 2 1 0 4 1 4 9 5 9 0 6 9 0 1 5 9 7 3 4]
prediction: [7 2 1 0 4 1 4 9 5 9 0 6 9 0 1 5 9 7 3 4]
```

顯示第 9 筆的機率：呼叫 predict()，可以得到 0~9 預測機率各自的值。

```
1   # 顯示第 9 筆的機率
2   import numpy as np
3
4   predictions = model.predict(x_test_norm[8:9])
5   print(f'0~9預測機率: {np.around(predictions[0], 2)}')
```

執行結果：發現 5 及 6 的機率很相近，表示模型並不很肯定，所以，實務上，我們可以提高門檻，規定機率須超過規定的下限，例如 0.8，才算是辨識成功，以提高可信度，避免模擬兩可的預測。

```
0~9預測機率: [0.   0.   0.   0.   0.   0.59 0.41 0.   0.   0.  ]
```

第 9 筆圖像如下，像 5 又像 6。

16. 步驟 8：效能評估，暫不進行，之後可調校相關「超參數」
(Hyperparameter)及模型結構，尋找最佳模型和參數。超參數是指在模
型訓練前可以調整的參數，例如學習率、執行週期、權重初始值、訓
練批量等，但不含模型求算的參數如權重(Weight)或偏差項(Bias)。

17. 步驟 9：模型佈署，將最佳模型存檔，再開發使用者介面或提供 API，
連同模型檔一併佈署到上線環境(Production Environment)。

```
1  # 模型存檔
2  model.save('model.h5')
3
4  # 模型載入
5  model = tf.keras.models.load_model('model.h5')
```

18. 步驟 10：接收新資料預測，之前都是使用 MNIST 內建資料測試，嚴格
說並不可靠，因為這些都是出自同一機構所收集的資料，因此，建議
讀者自己利用繪圖軟體親自撰寫測試。我們準備一些圖檔，放在
myDigits 目錄內，讀者可自行修改，再利用下列程式碼測試，注意，
從圖檔讀入影像後要反轉顏色，顏色 0 為白色，與 RGB 色碼不同，它
的 0 為黑色。

```
1   # 使用小畫家，繪製 0~9，實際測試看看
2   from skimage import io
3   from skimage.transform import resize
4   import numpy as np
5
6   # 讀取影像並轉為單色
7   uploaded_file = './myDigits/8.png'
8   image1 = io.imread(uploaded_file, as_gray=True)
9
10  # 縮為 (28, 28) 大小的影像
11  image_resized = resize(image1, (28, 28), anti_aliasing=True)
12  X1 = image_resized.reshape(1,28, 28) #/ 255
13
14  # 反轉顏色，顏色0為白色，與 RGB 色碼不同，它的 0 為黑色
15  X1 = np.abs(1-X1)
16
17  # 預測
18  predictions = model.predict_classes(X1)
19  print(predictions)
```

19. 使用下列指令顯示模型彙總資訊(summary)。

```
1   # 顯示模型的彙總資訊
2   model.summary()
```

執行結果：包括每一神經層的名稱及輸出入參數的個數。

```
Model: "sequential_2"

Layer (type)                 Output Shape              Param #
=================================================================
flatten_2 (Flatten)          (None, 784)               0

dense_4 (Dense)              (None, 128)               100480

dropout_2 (Dropout)          (None, 128 )              0

dense_5 (Dense)              (None, 10 )               1290
=================================================================
Total params: 101,770
Trainable params: 101,770
Non-trainable params: 0
```

計算參數個數：舉例來説 dense_5，輸出參數為 1290，意思是共有 10 條迴歸線，每一條迴歸線都有 128 個特徵對應的權重(w)與一個偏差項(b)，所以總共有 10 x (128 +1) = 1290 個參數。

20. 繪製圖形，顯示模型結構：要繪製圖形顯示模型結構，需先完成以下步驟，才能順利繪製圖形。

安裝 graphviz 軟體，網址為 https://www.graphviz.org/download，再把安裝目錄下的 bin 路徑加到環境變數 Path 中。
安裝兩個套件：

pip install graphviz pydotplus

```
1   tf.keras.utils.plot_model(model, to_file='model.png')
```

執行 plot_model 指令，可以同時顯示圖形和存檔。

執行結果：

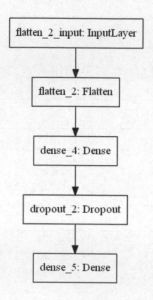

以上我們依機器學習流程的 10 大步驟撰寫了一支完整的程式，雖然篇幅很長，讀者應該還是有些疑問，許多針對細節的描述，將於下一節登場，我們會做些實驗來說明建構模型的考量，同時解答教學現場同學們常提出的問題。

4-1-3　實驗

前一節我們完成了第一支深度學習的程式，也見識到它的威力，扣除說明，短短 10 幾行的程式就能夠辨識手寫阿拉伯數字，且準確率達到97%。然而，仔細思考後會產生許多疑問：

1. 模型結構為什麼要設計成兩層 Dense？更多層準確率會提高嗎？
2. 第一層 Dense 輸出為什麼要設為 128？設為其他值會有何影響？
3. 目前第一層 Dense 的 Activation Function 設為 relu，代表什麼意義？設為其他值又會有何不同？
4. 優化器(optimizer)、損失函數(loss)、效能衡量指標(metrics)有哪些選擇？設為其他值會有何影響？

5. Dropout 比例為 0.2，設為其他值會更好嗎？

6. 影像為單色灰階，若是彩色可以辨識嗎？怎麼修改？

7. 執行週期(epoch)設為 5，設為其他值會更好嗎？

8. 準確率可以達到 100% 嗎？這樣企業才可以安心導入。

9. 如果要辨識其他物件，程式要修改那些地方？

10. 如果要辨識多個數字，例如輸入 4 位數，要如何辨識？

11. 希望了解更詳細的相關資訊，有哪些資源可以參閱？

以上問題是這幾年來授課時學員常提出的疑惑，我們就來逐一實驗，試著尋找答案。

【問題 1】模型結構為什麼要設計成兩層 Dense？更多層準確率會提高嗎？

解答：

1. 前面曾經説過，神經網路是多條迴歸線的組合，而且每一條迴歸線可能還會包在 Activation Function 內，變成非線性的函數，因此，要單純以數學求解幾乎不可能，只能以優化方法求得近似解，但是，只有凸集合的資料集，才保證有全局最佳解(Global Minimization)，以 MNIST 為例，總共有 784 個特徵，即 784 度空間，根本無法知道它是否為凸集合，因此嚴格來講，到目前為止，神經網路依然是一個黑箱(Black Box)科學，我們只知道它威力強大，但如何達到較佳的準確率，仍就需要經驗與實驗，因此，模型結構並沒有明確規定要設計成幾層，會隨著不同的問題及資料來進行測試，case by case 進行效能調校，找尋較佳的參數值。

2. 理論上，越多層架構，迴歸線就越多，預測應當越準確，像是 ResNet 模型就高達 150 層，但是，經過實驗證實，超過某一界限後，準確率可能會不升反降，這跟訓練資料量有關，如果只有少量的資料，要估算過多的參數(w、b)，自然準確率不高。

3. 我們就來小小實驗一下，多一層 Dense，準確率是否會提高？請參閱程式 **04_03_手寫阿拉伯數字辨識_實驗 1.ipynb**。

4. 修改模型結構如下，加一對 Dense/Dropout，其餘程式碼不變。

```
1  # 建立模型
2  model = tf.keras.models.Sequential([
3    tf.keras.layers.Flatten(input_shape=(28, 28)),
4    tf.keras.layers.Dense(128, activation='relu'),
5    tf.keras.layers.Dropout(0.2),
6    tf.keras.layers.Dense(64, activation='relu'),
7    tf.keras.layers.Dropout(0.2),
8    tf.keras.layers.Dense(10, activation='softmax')
9  ])
```

執行結果如下，準確率不見提升，反而微降。

```
loss: 0.0840
accuracy: 0.9733
```

【問題 2】第一層 Dense 輸出為什麼要設為 128？設為其他值會有何影響？

解答：

1. 輸出的神經元個數可以任意設定，一般來講，會使用 4 的倍數，以下我 們 修 改 為 256，參 閱 程 式 **04_03_手寫阿拉伯數字辨識_實驗 2.ipynb**。

```
2  model = tf.keras.models.Sequential([
3    tf.keras.layers.Flatten(input_shape=(28, 28)),
4    tf.keras.layers.Dense(256, activation='relu'),
5    tf.keras.layers.Dropout(0.2),
6    tf.keras.layers.Dense(10, activation='softmax')
7  ])
```

執行結果如下，準確率略為提高，但不明顯。

```
loss: 0.0775
accuracy: 0.9764
```

2. 同問題 1，照理來説，神經元個數越多，迴歸線就越多，特徵也越多，預測應該會越準確，但經過驗證，準確率並未顯著提高。依「Deep

Learning with TensorFlow 2.0 and Keras」一書測試如下圖，也是有一個極限，超過就會不升反降。

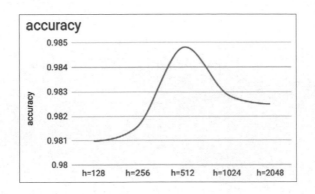

3. 神經元個數越多，訓練時間就越長，如下圖：

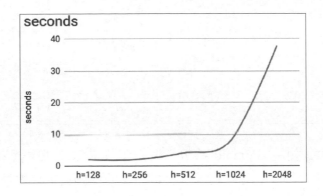

【問題 3】目前第一層 Dense 的 Activation Function 設為 relu，代表什麼意義？設為其他值會有何不同？

解答：

Activation Function 有很多種，後面會有詳盡介紹，可先參閱維基百科 [1]，部份表格擷取如下，包括函數的名稱、機率分配圖形、公式及一階導數：

Name ⬍	Plot	Function, $f(x)$ ⬍	Derivative of f, $f'(x)$
Identity		x	1
Binary step		$\begin{cases} 0 & \text{if } x < 0 \\ 1 & \text{if } x \geq 0 \end{cases}$	$\begin{cases} 0 & \text{if } x \neq 0 \\ \text{undefined} & \text{if } x = 0 \end{cases}$
Logistic, sigmoid, or soft step		$\sigma(x) = \dfrac{1}{1 + e^{-x}}$ [1]	$f(x)(1 - f(x))$
tanh		$\tanh(x) = \dfrac{e^x - e^{-x}}{e^x + e^{-x}}$	$1 - f(x)^2$
Rectified linear unit (ReLU)[11]		$\begin{cases} 0 & \text{if } x \leq 0 \\ x & \text{if } x > 0 \end{cases}$ $= \max\{0, x\} = x\mathbf{1}_{x>0}$	$\begin{cases} 0 & \text{if } x < 0 \\ 1 & \text{if } x > 0 \\ \text{undefined} & \text{if } x = 0 \end{cases}$
Gaussian error linear unit (GELU)[6]		$\dfrac{1}{2}x\left(1 + \text{erf}\left(\dfrac{x}{\sqrt{2}}\right)\right)$ $= x\Phi(x)$	$\Phi(x) + x\phi(x)$
Softplus[12]		$\ln(1 + e^x)$	$\dfrac{1}{1 + e^{-x}}$

早期隱藏層大都使用 Sigmoid 函數，近幾年發現 ReLU 準確率較高，我們先試著比較這兩種，請參閱程式 **04_03_手寫阿拉伯數字辨識_實驗 3.ipynb**。

將 relu 改為 sigmoid，如下所示：

```
1  # 建立模型
2  model = tf.keras.models.Sequential([
3    tf.keras.layers.Flatten(input_shape=(28, 28)),
4    tf.keras.layers.Dense(128, activation='relu'),
5    tf.keras.layers.Dropout(0.2),
6    tf.keras.layers.Dense(10, activation='sigmoid')
7  ])
```

執行結果如下，準確率確實略低於 ReLU。

```
loss: 0.0847
accuracy: 0.9762
```

【問題 4】優化器(Optimizer)、損失函數(Loss)、效能衡量指標(Metrics)有哪些選擇？設為其他值會有何影響？

解答：

1. 優化器有很多種，從最簡單的固定值學習率(SGD)，到很複雜的動態改變學習率，甚至是能夠自訂優化器。請參考[2]或[3]。優化器的選擇，主要會影響收斂的速度，大多數的狀況下，Adam 優化器都有不錯的表現。

2. 損失函數也種類繁多，包括常見的 MSE、Entropy，其他更多的請參考[4]或[5]。損失函數的選擇，主要也是影響著收斂的速度，另外，某些自訂損失函數有特殊功能，例如風格轉換(Style Transfer)，它能夠製作影像合成的效果，生成對抗網路(GAN)更是發揚光大，後面章節會有詳細的介紹。

3. 效能衡量指標(metrics)：除了準確率(Accuracy)，還可以計算精確率(Precision)、召回率(Recall)、F1，也可以同時設定多個效能衡量指標，請參考[6]，例如下面程式碼，完整程式可參閱程式 **04_03_手寫阿拉伯數字辨識_實驗 4.ipynb**。

```
10  model.compile(optimizer='adam',
11              loss='categorical_crossentropy',
12              metrics=[tf.keras.metrics.CategoricalAccuracy(),
13                      tf.keras.metrics.Precision(),
14                      tf.keras.metrics.Recall()])
```

注意，設定多個效能衡量指標時，準確率請不要使用 Accuracy，數值會非常低，需使用 CategoricalAccuracy，表示分類的準確率，而非迴歸的準確率。

執行結果如下：

```
loss: 0.0757
categorical_accuracy: 0.9781
precision_3: 0.9810
recall_3: 0.9751
```

【問題 5】Dropout 比例為 0.2,設為其他值會更好嗎?

解答:

Dropout 比例為 0.1,測試看看,參閱程式 **04_03_手寫阿拉伯數字辨識_實驗 5.ipynb**。

```
1   # 建立模型
2   model = tf.keras.models.Sequential([
3     tf.keras.layers.Flatten(input_shape=(28, 28)),
4     tf.keras.layers.Dense(128, activation='relu'),
5     tf.keras.layers.Dropout(0.1),
6     tf.keras.layers.Dense(10, activation='softmax')
7   ])
```

執行結果如下,準確率略為提高。

```
loss: 0.0816
accuracy: 0.9755
```

拋棄比例過高時,準確率會陡降。

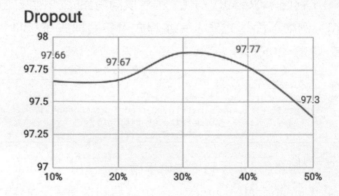

【問題 6】目前 MNIST 影像為單色灰階,若是彩色可以辨識嗎?怎麼修改?

解答:可以,若顏色有助於辨識,可以將 RGB 三通道分別輸入辨識,後面我們談到卷積神經網路(Convolutional Neural Networks,CNN)時會有範例説明。

【問題 7】執行週期(epoch)設為 5,設為其他值會更好嗎?

解答：執行週期(epoch)改為 10，參閱程式 **04_03_手寫阿拉伯數字辨識_實驗 6.ipynb**。

```
15 | history = model.fit(x_train_norm, y_train, epochs=10, validation_split=0.2)
```

執行結果如下，準確率略為提高。

```
loss: 0.0700
accuracy: 0.9785
```

理論上，訓練週期越多，準確率越高，但是，過多的訓練週期會造成過度擬合(Overfitting)，反而會使準確率降低。

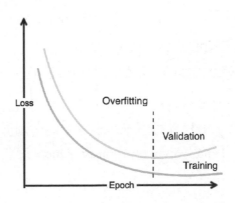

【問題 8】準確率可以達到 100% 嗎？

解答：很少模型準確率能夠達到 100%，除非是用數學證明，然而，神經網路是只是近似解而已，另一方面，神經網路是從訓練資料中學習到知識，但是，測試或預測資料並不參與訓練，若與訓練資料分佈有所差異，甚至來自不同的機率分配，很難確保準確率能達到 100%。

【問題 9】如果要辨識其他物件，程式要修改那些地方？

解答：我們只需修改很少的程式碼，就可以辨識其他物件，例如，MNSIT另一個資料集 FashionMnist，它包含女人身上的 10 種配件，請參閱 **04_03_FashionMnist_實驗.ipynb**，除了載入資料的指令不同之外，其他的程式碼幾乎不變。這也說明了一點，**神經網路並不是真的認識 0~9 或女人**

身上的 10 個配件，它只是從像素資料中推估出的模型，即所謂的 「從資料中學習到知識」(Knowledge Discovery from Data, KDD)，以 MNIST 而言，模型只是統計 0~9 個個數字，他們的像素大部份分布在那些位置而已。

【問題 10】如果要辨識多個數字，例如輸入 4 位數，要如何辨識？

解答：可以使用影像處理分割數字，再分別依序輸入模型預測即可，更簡單的方法，直接將視覺介面(UI)設計成 4 格，規定使用者只能在每格子內各輸入一個數字。

【問題 11】希望了解更詳細的相關資訊，有哪些資源可以參閱？

解答：可以參考 TensorFlow 官網[7]或 Keras 官網[8]，版本快速的更新已經使網路上的資訊新舊雜陳，官網才是最新資訊的正確來源。

以上的實驗大多只針對單一參數作比較，假如要同時比較多個變數，就必須跑遍所有參數組合，這樣程式豈不是很複雜嗎？別擔心，有一些套件可以幫忙，包括 Keras Tuner、hyperopt、Ray Tune、Ax…等，在後續「超參數調校」有較詳細的介紹。

由於這個模型的辨識率很高，要觀察超參數調整對模型的影響，並不容易，建議找一些辨識率較低的模型進行相關實驗，例如 FashionMnist[9]、CiFar 資料集，才能有比較顯著的效果，筆者針對 FashionMnist 作了另一次實驗 **04_03_FashionMnist_實驗.ipynb**。

4-2 Keras 模型種類

TensorFlow/Keras 提供兩類模型結構：

1. Sequential model：順序型的模型，按神經層的排列順序，由上往下執行，每一層的輸出都是下一層的輸入，所以，除了第一層要設定輸入的維度外，其他層都不需要設定。

2. Functional API：提供較有彈性的結構，允許非直線型的結構、共享的神經層及多輸入/輸出，亦即模型結構可以有分叉或合併(Split/Merge)。

4-2-1 Sequential model

程式：04_05_ Sequential_model.ipynb。

1. 模型內可包含各式的神經層，簡潔的寫法如下，以 List 包住神經層。

```
1  model = tf.keras.models.Sequential([
2    tf.keras.layers.Flatten(input_shape=(28, 28)),
3    tf.keras.layers.Dense(128, activation='relu'),
4    tf.keras.layers.Dropout(0.2),
5    tf.keras.layers.Dense(10, activation='softmax')
6  ])
```

除了第一層要設定輸入的維度(input_shape)外，其他層都不需要設定，只要在第一個參數指定輸出維度。

2. 可以變換另一種寫法，將 input_shape 拿掉，在 model 內設定輸入層及維度參數(shape)。

```
1   model = tf.keras.models.Sequential([
2     tf.keras.layers.Flatten(),
3     tf.keras.layers.Dense(128, activation='relu'),
4     tf.keras.layers.Dropout(0.2),
5     tf.keras.layers.Dense(10, activation='softmax')
6   ])
7
8   x = tf.keras.layers.Input(shape=(28, 28))
9   # 或 x = tf.Variable(tf.random.truncated_normal([28, 28]))
10  y = model(x)
```

3. 可以直接串連神經層：最後一行指令。

```
1  layer1 = tf.keras.layers.Dense(2, activation="relu", name="layer1")
2  layer2 = tf.keras.layers.Dense(3, activation="relu", name="layer2")
3  layer3 = tf.keras.layers.Dense(4, name="layer3")
4
5  # Call layers on a test input
6  x = tf.ones((3, 3))
7  y = layer3(layer2(layer1(x)))
```

4. 可以事後加減神經層，pop()會刪減最上層(Top)，注意，神經層是堆疊(Stack)，後進先出，即最後一層 Dense。

```
1  model = tf.keras.models.Sequential([
2    tf.keras.layers.Flatten(input_shape=(28, 28)),
3    tf.keras.layers.Dense(128, activation='relu'),
4    tf.keras.layers.Dropout(0.2),
5    tf.keras.layers.Dense(10, activation='softmax')
6  ])
7
8  # 刪減一層
9  model.pop()
10 print(f'神經層數: {len(model.layers)}')
11 model.layers
```

執行結果：

```
神經層數: 3

[<tensorflow.python.keras.layers.core.Flatten at 0x1826cfb8580>,
 <tensorflow.python.keras.layers.core.Dense at 0x1826cf94340>,
 <tensorflow.python.keras.layers.core.Dropout at 0x1826cf94b20>]
```

5. 增加一層神經層。

```
1  # 增加一層
2  model.add(tf.keras.layers.Dense(10))
3  print(f'神經層數: {len(model.layers)}')
4  model.layers
```

執行結果：

```
神經層數: 4

[<tensorflow.python.keras.layers.core.Flatten at 0x1826cfb8580>,
 <tensorflow.python.keras.layers.core.Dense at 0x1826cf94340>,
 <tensorflow.python.keras.layers.core.Dropout at 0x1826cf94b20>,
 <tensorflow.python.keras.layers.core.Dense at 0x18270b86040>]
```

6. 取得模型各神經層資訊。

```
1  # 建立 3 layers
2  layer1 = tf.keras.layers.Dense(2, activation="relu", name="layer1",
3                                 input_shape=(28, 28))
4  layer2 = tf.keras.layers.Dense(3, activation="relu", name="layer2")
5  layer3 = tf.keras.layers.Dense(4, name="layer3")
6
7  # 建立模型
8  model = tf.keras.models.Sequential([
9    layer1,
10   layer2,
11   layer3
12 ])
```

```
13
14   # 讀取模型權重
15   print(f'神經層參數類別總數: {len(model.weights)}')
16   model.weights
```

執行結果：有 3 層共 6 類資訊，包括 3 層權重(weight)及 3 層偏差(bias)，共 6 類。

```
神經層參數類別總數: 6

[<tf.Variable 'layer1/kernel:0' shape=(28, 2) dtype=float32, numpy=
 array([[-0.40281397,  0.2205016 ],
        [-0.03376389,  0.15561956],
        [ 0.4151038 ,  0.02849016],
        [-0.15993604,  0.03800485],
        [ 0.06474555,  0.39933097],
        [ 0.41804487, -0.35671836],
        [-0.09304568, -0.12987521],
        [-0.07738718,  0.4214089 ],
        [-0.03624141, -0.15986142],
        [ 0.41953433, -0.14494684],
        [-0.16713199, -0.36726573],
        [ 0.28147936, -0.07306954],
        [ 0.06702495,  0.29314017],
        [-0.4076838 , -0.22369348],
```

7   取得特定神經層資訊。

```
1   print(f'{layer2.name}: {layer2.weights}')
```

8. 取得模型彙總資訊。

```
1   model.summary()
```

9. 可以一邊加神經層，一邊顯示模型彙總資訊，這樣有利於除錯，查看中間處理結果。

```
1   from tensorflow.keras import layers
2
3   model = tf.keras.models.Sequential()
4   model.add(tf.keras.Input(shape=(250, 250, 3)))   # 250x250 RGB images
5   model.add(layers.Conv2D(32, 5, strides=2, activation="relu"))
6   model.add(layers.Conv2D(32, 3, activation="relu"))
7   model.add(layers.MaxPooling2D(3))
8
9   # 顯示目前模型彙總資訊
```

```
10  model.summary()
11
12  # The answer was: (40, 40, 32), so we can keep downsampling...
13
14  model.add(layers.Conv2D(32, 3, activation="relu"))
15  model.add(layers.Conv2D(32, 3, activation="relu"))
16  model.add(layers.MaxPooling2D(3))
17  model.add(layers.Conv2D(32, 3, activation="relu"))
18  model.add(layers.Conv2D(32, 3, activation="relu"))
19  model.add(layers.MaxPooling2D(2))
20
21  # 顯示目前模型彙總資訊
22  model.summary()
23
24  # Now that we have 4x4 feature maps, time to apply global max pooling.
25  model.add(layers.GlobalMaxPooling2D())
26
27  # Finally, we add a classification layer.
28  model.add(layers.Dense(10))
```

10. 取得每一層神經層的 output：可設定模型的 input/output。

```
1   # 設定模型
2   initial_model = tf.keras.Sequential(
3       [
4           tf.keras.Input(shape=(250, 250, 3)),
5           layers.Conv2D(32, 5, strides=2, activation="relu"),
6           layers.Conv2D(32, 3, activation="relu"),
7           layers.Conv2D(32, 3, activation="relu"),
8       ]
9   )
10
11  # 設定模型的input/output
12  feature_extractor = tf.keras.Model(
13      inputs=initial_model.inputs,
14      outputs=[layer.output for layer in initial_model.layers],
15  )
16
17  # 呼叫 feature_extractor 取得 output
18  x = tf.ones((1, 250, 250, 3))
19  features = feature_extractor(x)
20  features
```

11. 取得特定神經層的 output：設定模型的 output 為特定的神經層。

```
1   # 設定模型
2   initial_model = tf.keras.Sequential(
3       [
4           tf.keras.Input(shape=(250, 250, 3)),
5           layers.Conv2D(32, 5, strides=2, activation="relu"),
6           layers.Conv2D(32, 3, activation="relu", name="my_intermediate_layer"),
7           layers.Conv2D(32, 3, activation="relu"),
8       ]
9   )
```

```
10
11  # 設定模型的input/output
12  feature_extractor = tf.keras.Model(
13      inpuls=initial_model.inputs,
14      outputs=initial_model.get_layer(name="my_intermediate_layer").output,
15  )
16
17  # 呼叫 feature_extractor 取得 output
18  x = tf.ones((1, 250, 250, 3))
19  features = feature_extractor(x)
20  features
```

4-2-2 Functional API

Functional API 由於提供較有彈性的結構，適用於相對複雜的模型結構，允許非直線型的結構、共享的神經層及多輸入/輸出，結構可以分叉或合併 (Split/Merge)。

直接來看範例說明，請參閱程式 04_06_ Functional_API.ipynb。

【範例 1】先看一個簡單的程式語法，除了第一層之外，每一層均須設定前一層，同時，Model 函數必須指定輸入/輸出(input/output)是哪些神經層，都是 List 資料型態，允許多個輸入/輸出。

```
1   # Functional API
2
3   # 建立第一層 InputTensor
4   InputTensor = layers.Input(shape=(100,))
5
6   # H1 接在 InputTensor 後面
7   H1 = layers.Dense(10, activation='relu')(InputTensor)
8
9   # H2 接在 H1 後面
10  H2 = layers.Dense(20, activation='relu')(H1)
11
12  # Output 接在 H2 後面
13  Output = layers.Dense(1, activation='softmax')(H2)
14
15  # 建立模型，必須指定 inputs / outputs
16  model = tf.keras.Model(inputs=InputTensor, outputs=Output)
17
18  # 顯示模型彙總資訊
19  model.summary()
```

【範例 2】模型包括 3 個輸入、2 個輸出，先不管模型用途，只觀察程式語法，layers.concatenate()可合併神經層。

```
 6  # 建立第一層 InputTensor
 7  title_input = tf.keras.Input(shape=(None,), name="title")
 8  body_input = tf.keras.Input(shape=(None,), name="body")
 9  tags_input = tf.keras.Input(shape=(num_tags,), name="tags")
10
11  # 建立第二層
12  title_features = layers.Embedding(num_words, 64)(title_input)
13  body_features = layers.Embedding(num_words, 64)(body_input)
14
15  # 建立第三層
16  title_features = layers.LSTM(128)(title_features)
17  body_features = layers.LSTM(32)(body_features)
18
19  # 合併以上神經層
20  x = layers.concatenate([title_features, body_features, tags_input])
21
22  # 建立第四層，連接合併的 x
23  priority_pred = layers.Dense(1, name="priority")(x)
24  department_pred = layers.Dense(num_departments, name="department")(x)
25
26  # 建立模型，必須指定 inputs / outputs
27  model = tf.keras.Model(
28      inputs=[title_input, body_input, tags_input],
29      outputs=[priority_pred, department_pred],
30  )
31
32  # 繪製模型
33  # show_shapes=True：Layer 含 Input/Output 資訊
34  tf.keras.utils.plot_model(model, "multi_input_and_output model.png",
35                            show_shapes=True)
```

最後一行程式碼繪製的模型圖如下：

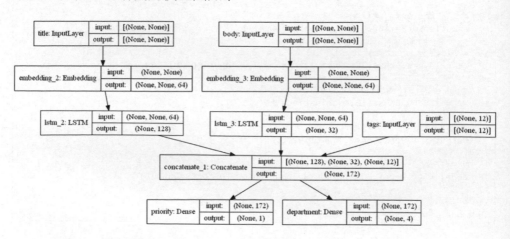

- concatenate()合併了 3 個 layers，它們的輸出維度大小分別為 128/32/12，
  故合併後，輸出維度大小=128+32+12=172。

■ 最後一行程式碼的參數 show_shapes=True，結構圖會額外添加含有 Input/Output 資訊。

4-3 神經層(Layer)

神經層是神經網路的主要成員，TensorFlow 有各式各樣的神經層，詳情可參閱 Keras 官網[10]，目前包括以下類別，隨著演算法的發明，會不斷的增加，就算 TensorFlow 的更新腳步跟不上你的需求，還可以自訂神經層 (Custom layer)。

■ 核心神經層(Core Layers)
■ 卷積神經層(Convolution layers)
■ 池化神經層(Pooling layers)
■ 循環神經層(Recurrent layers)
■ 前置神經層(Preprocessing layers)
■ 常態化神經層(Normalization layers)
■ 正則神經層(Regularization layers)
■ 注意力神經層(Attention layers)
■ 維度重置神經層(Reshaping layers)
■ 合併神經層(Merging layers)
■ 激勵神經層(Activation layers)

由於中文翻譯大部份都不能望文生義，後面的內容均使用英文術語。

現階段僅介紹之前用到的核心神經層，其他類型的神經層在後續演算法用到時再說明。

4-3-1 完全連接神經層(Dense Layer)

Dense 是最常見的神經層，每個輸入的神經元都會完全連接到輸出神經元。

直接來看範例說明，請參閱程式 **04_07_神經層.ipynb**。

【範例 1】計算 Dense 的參數個數。

1. 模型結構如下：

```python
import tensorflow as tf
from tensorflow.keras import layers

# 建立模型
model = tf.keras.models.Sequential([
  tf.keras.layers.Flatten(input_shape=(28, 28)),
  tf.keras.layers.Dense(128, activation='relu', name="layer1"),
  tf.keras.layers.Dropout(0.2),
  tf.keras.layers.Dense(10, activation='softmax', name="layer2")
])

# 設定優化器(optimizer)、損失函數(loss)、效能衡量指標(metrics)的類別
model.compile(optimizer='adam',
              loss='sparse_categorical_crossentropy',
              metrics=['accuracy'])

# 顯示模型彙總資訊
model.summary()
```

執行結果如下，顯示各層的 Output 及參數個數。

```
Model: "sequential_2"

Layer (type)              Output Shape          Param #
=================================================================
flatten_2 (Flatten)       (None, 784)           0

layer1 (Dense)            (None, 128)           100480

dropout_3 (Dropout)       (None, 128)           0

layer2 (Dense)            (None, 10)            1290
=================================================================
Total params: 101,770
Trainable params: 101,770
Non-trainable params: 0
```

2. 設定模型的 output 為第一層 Dense，顯示第一層 Dense output 個數 (128)。

```
1   # 設定模型的 input/output
2   feature_extractor = tf.keras.Model(
3       inputs=model.inputs,
4       outputs=model.get_layer(name="layer1").output,
5   )
6
7   # 呼叫 feature_extractor 取得 output
8   x = tf.ones((1, 28, 28))
9   features = feature_extractor(x)
10  features.shape
```

3. 第一層 Dense 的參數個數計算。

```
1   # 第一層 Dense 參數個數計算
2   parameter_count = (28 * 28) * features.shape[1] + features.shape[1]
3   print(f'參數(parameter)個數：{parameter_count}')
```

執行結果：參數個數共有 100,480 個與模型彙總資訊一致。

4. 第二層 Dense 的參數個數計算。

```
1   # 設定模型的 input/output
2   feature_extractor = tf.keras.Model(
3       inputs=model.inputs,
4       outputs=model.get_layer(name="layer2").output,
5   )
6
7   # 呼叫 feature_extractor 取得 output
8   x = tf.ones((1, 28, 28))
9   features = feature_extractor(x)
10
11  parameter_count = 128 * features.shape[1] + features.shape[1]
12  print(f'參數(parameter)個數：{parameter_count}')
```

執行結果：參數個數共有 1290 個與模型彙總資訊一致。

Dense 神經層的參數說明如下：

- units：輸出神經元個數。
- activation：指定要使用 activation function，也可以獨立使用 activation layer。
- use_bias：權重參數估計是否要含偏差項。
- bias_initializer：偏差初始值。
- kernel_initializer：權重初始值，預設值是 glorot_uniform，它是均勻分配的隨機亂數。

- kernel_regularizer：權重是否要使用防止過度擬合的正則函數 (regularizer)，預設值是無，也可設為 L1 或 L2。
- bias_regularizer：偏差是否要使用防止過度擬合的正則函數，預設值是無，也可設為 L1 或 L2。
- activity_regularize：activation function 是否要使用防止過度擬合的正則函數，預設值是無，也可設為 L1 或 L2。
- kernel_constraint：權重是否有限制範圍。
- bias_constraint：偏差是否有限制範圍。

4-3-2 Dropout Layer

Dropout layer 在每一 epoch/step 訓練時，會隨機丟棄設定比例的輸入神經元，避免過度擬合，只在訓練時運作，預測時會忽視 Dropout，不會有任何作用。參數說明如下：

- rate：丟棄的比例，介於(0, 1)之間。
- training：是否在訓練時運作。

根據大部份學者的經驗，在神經網路中使用 Dropout 會比 Regularizer 效果來的好。

4-4 激勵函數(Activation Function)

Activation Function 是將線性方程式轉為非線性，目的是希望能提供更通用的解決方案。

$$Output = activation\ function(x_1w_1 + x_2w_2 + \cdots + x_nw_n + bias)$$

Activation Function 有非常多種函數，可以參考維基百科表格[3]，如下圖：

表 4.1 Activation Function 列表，資料來源：維基百科

Name ⬍	Plot	Function, $f(x)$ ⬍	Derivative of f, $f'(x)$ ⬍	Range ⬍
Identity		x	1	$(-\infty, \infty)$
Binary step		$\begin{cases} 0 & \text{if } x < 0 \\ 1 & \text{if } x \geq 0 \end{cases}$	$\begin{cases} 0 & \text{if } x \neq 0 \\ \text{undefined} & \text{if } x = 0 \end{cases}$	$\{0, 1\}$
Logistic, sigmoid, or soft step		$\sigma(x) = \dfrac{1}{1 + e^{-x}}$[1]	$f(x)(1 - f(x))$	$(0, 1)$
tanh		$\tanh(x) = \dfrac{e^x - e^{-x}}{e^x + e^{-x}}$	$1 - f(x)^2$	$(-1, 1)$
Rectified linear unit (ReLU)[11]		$\begin{cases} 0 & \text{if } x \leq 0 \\ x & \text{if } x > 0 \end{cases}$ $= \max\{0, x\} = x\mathbf{1}_{x>0}$	$\begin{cases} 0 & \text{if } x < 0 \\ 1 & \text{if } x > 0 \\ \text{undefined} & \text{if } x = 0 \end{cases}$	$[0, \infty)$
Gaussian error linear unit (GELU)[6]		$\dfrac{1}{2}x\left(1 + \text{erf}\left(\dfrac{x}{\sqrt{2}}\right)\right)$ $= x\Phi(x)$	$\Phi(x) + x\phi(x)$	$(-0.17\ldots, \infty)$
Softplus[12]		$\ln(1 + e^x)$	$\dfrac{1}{1 + e^{-x}}$	$(0, \infty)$
Exponential linear unit (ELU)[13]		$\begin{cases} \alpha\left(e^x - 1\right) & \text{if } x \leq 0 \\ x & \text{if } x > 0 \end{cases}$ with parameter α	$\begin{cases} \alpha e^x & \text{if } x < 0 \\ 1 & \text{if } x > 0 \\ 1 & \text{if } x = 0 \text{ and } \alpha = 1 \end{cases}$	$(-\alpha, \infty)$
Scaled exponential linear unit (SELU)[14]		$\lambda\begin{cases} \alpha(e^x - 1) & \text{if } x < 0 \\ x & \text{if } x \geq 0 \end{cases}$ with parameters $\lambda = 1.0507$ and $\alpha = 1.67326$	$\lambda\begin{cases} \alpha e^x & \text{if } x < 0 \\ 1 & \text{if } x \geq 0 \end{cases}$	$(-\lambda\alpha, \infty)$
Leaky rectified linear unit (Leaky ReLU)[15]		$\begin{cases} 0.01x & \text{if } x < 0 \\ x & \text{if } x \geq 0 \end{cases}$	$\begin{cases} 0.01 & \text{if } x < 0 \\ 1 & \text{if } x \geq 0 \end{cases}$	$(-\infty, \infty)$
Parameteric rectified linear unit (PReLU)[16]		$\begin{cases} \alpha x & \text{if } x < 0 \\ x & \text{if } x \geq 0 \end{cases}$ with parameter α	$\begin{cases} \alpha & \text{if } x < 0 \\ 1 & \text{if } x \geq 0 \end{cases}$	$(-\infty, \infty)$[2]
ElliotSig,[17][18] softsign[19][20]		$\dfrac{x}{1 + |x|}$	$\dfrac{1}{(1 + |x|)^2}$	$(-1, 1)$
Square nonlinearity (SQNL)[21]		$\begin{cases} 1 & \text{if } x > 2.0 \\ x - \dfrac{x^2}{4} & \text{if } 0 \leq x \leq 2.0 \\ x + \dfrac{x^2}{4} & \text{if } -2.0 \leq x < 0 \\ -1 & \text{if } x < -2.0 \end{cases}$	$1 \mp \dfrac{x}{2}$	$(-1, 1)$

S-shaped rectified linear activation unit (SReLU)[22]		$\begin{cases} t_l + a_l(x - t_l) & \text{if } x \leq t_l \\ x & \text{if } t_l < x < t_r \\ t_r + a_r(x - t_r) & \text{if } x \geq t_r \end{cases}$ where t_l, a_l, t_r, a_r are parameters.	$\begin{cases} a_l & \text{if } x \leq t_l \\ 1 & \text{if } t_l < x < t_r \\ a_r & \text{if } x \geq t_r \end{cases}$	$(-\infty, \infty)$
Bent identity		$\dfrac{\sqrt{x^2 + 1} - 1}{2} + x$	$\dfrac{x}{2\sqrt{x^2 + 1}} + 1$	$(-\infty, \infty)$
Sigmoid linear unit (SiLU,[6] SiL,[23] or Swish-1[24])		$\dfrac{x}{1 + e^{-x}}$	$\dfrac{1 + e^{-x} + xe^{-x}}{(1 + e^{-x})^2}$	$[-0.278\ldots, \infty)$
Gaussian		e^{-x^2}	$-2xe^{-x^2}$	$(0, 1]$
SQ-RBF		$\begin{cases} 1 - \frac{x^2}{2} & \text{if } \lvert x \rvert \leq 1 \\ \frac{1}{2}(2 - \lvert x \rvert)^2 & \text{if } 1 < \lvert x \rvert < 2 \\ 0 & \text{if } \lvert x \rvert \geq 2 \end{cases}$	$\begin{cases} -x & \text{if } \lvert x \rvert \leq 1 \\ x - 2\,\text{sgn}(x) & \text{if } 1 < \lvert x \rvert < 2 \\ 0 & \text{if } \lvert x \rvert \geq 2 \end{cases}$	$[0, 1]$

TensorFlow 支援大部份的函數,如果找不到的話,也能夠自訂函數,它們可以直接設定在神經層的參數,也可以是獨立的函數。

【範例 1】實際測試常用的 Activation Function。

程式:04_08_Activation_Function.ipynb。

1. ReLU(Rectified linear unit):是目前隱藏層最常用的函數,公式請參考上表,函數名稱為 relu。

```
1  # 設定 x = -10, -9, ..., 10 測試
2  x= np.linspace(-10, 10, 21)
3  x_tf = tf.constant(x, dtype = tf.float32)
4
5  # ReLU
6  y = activations.relu(x_tf).numpy()
7
8  # 繪圖
9  plt.plot(x, y)
10 plt.show()
```

執行結果:會忽視過小的外部輸入,比如說,我們輕輕碰一下皮膚,大腦可能不會做出反應。

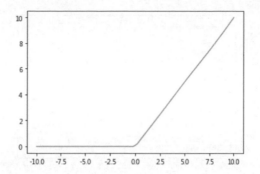

relu 函數有三個參數：

■ threshold：超過此門檻值，y 才會>0。例如 threshold=5，如下圖。

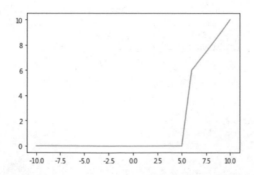

■ max_value：y 的上限。例如 max_value=5，如下圖。

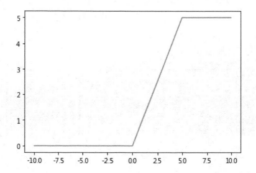

■ alpha：小於門檻值，y 會等於 x * alpha。例如 alpha=0.5，如下圖，又稱為 Parameteric rectified linear unit (PReLU)，若 alpha=0.01，則稱為 Leaky rectified linear unit (Leaky ReLU)。

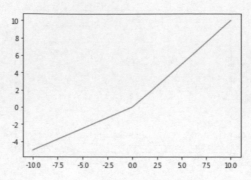

相關測試請參閱程式。

2. Sigmoid：即羅吉斯迴歸，因為函數為 S 型而得名，適用於二分類，可加在最後一層 Dense 內。

```
1  x= np.linspace(-10, 10, 21)
2  x_tf = tf.constant(x, dtype = tf.float32)
3
4  # sigmoid
5  y = activations.sigmoid(x_tf).numpy()
6
7  # 模糊地帶
8  plt.axvline(-4, color='r')
9  plt.axvline(4, color='r')
10
11 plt.plot(x, y)
12 plt.show()
```

執行結果：函數最小值為 0，最大值為 1，只有兩條直線中間是模糊地帶，但也是一個平滑改變的過程，而非階梯形的函數，可降低預測的變異性(Variance)。

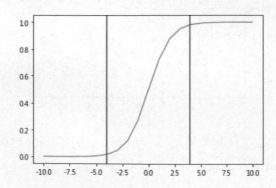

3. tanh：與 Sigmoid 類似，但最小值是-1。

```
1  x= np.linspace(-10, 10, 21)
2  x_tf = tf.constant(x, dtype = tf.float32)
3
4  # tanh
5  y = activations.tanh(x_tf).numpy()
6
7  # 模糊地帶
8  plt.axvline(-3, color='r')
9  plt.axvline(3, color='r')
10
11 plt.plot(x, y)
12 plt.show()
```

執行結果：函數最小值為-1，最大值為 1，只有兩條直線中間是模糊地帶，平滑改變的過程與 Sigmoid 相比較為陡峭。

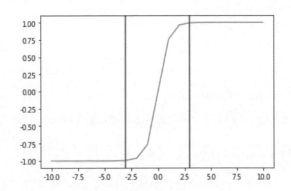

4. Softmax：這個函數會將輸入轉為機率，即所有值介於(0, 1)之間，總和為 1，適用於多分類，可加在最後一層 Dense 內。

```
1  # activations.softmax 輸入需為 2 維資料，設定 x 為均勻分配，轉換後每一列加總為 1
2  x = np.random.uniform(1, 10, 40).reshape(10, 4)
3  print('輸入：\n', x)
4  x_tf = tf.constant(x, dtype = tf.float32)
5
6  # Softmax
7  y = activations.softmax(x_tf).numpy()
8  print('加總：', np.round(np.sum(y, axis=1)))
```

執行結果：activations.softmax() 輸入必須是 2 維資料，設定 x 為均勻分配，轉換後，每一列總和為 1。

```
輸入：
[[1.87635979 5.65431617 5.0807942  7.84746665]
 [5.71430835 1.0332283  9.15220976 4.59581463]
 [2.91586061 1.01252361 9.21877442 5.51475778]
 [4.22146643 1.18823067 6.42331145 6.23998232]
 [2.61397256 7.75011603 2.62736731 9.18788281]
 [4.07362709 1.12741729 8.18113136 8.01828864]
 [9.20565064 7.19991452 1.14951624 6.12306693]
 [3.77131905 3.75599114 9.43088289 5.65098351]
 [7.56332186 3.4039989  2.22646493 8.14256405]
 [5.1090939  2.1567948  9.0736152  8.6324396 ]]
加總： [1. 1. 1. 1. 1. 1. 1. 1. 1. 1.]
```

使用 NumPy 計算 Softmax。

```
1  # 設定 x = 1, 2, ..., 10 測試
2  x= np.random.uniform(1, 10, 40)
3
4  # Softmax
5  y = np.e ** (x) / tf.reduce_sum(np.e ** (x))
6  print(sum(y))
```

5. 自訂函數，可以使用 TensorFlow 張量函數，只要傳回與輸入/輸出相符合的維度和資料型態即可，例如：

```
model.add(layers.Dense(64, activation=tf.nn.tanh))
```

6. 其他的函數請參見 Keras 官網，包括這二個網址：

- https://keras.io/api/layers/activations/ [11]。
- https://keras.io/api/layers/activation_layers/ [12]。

一般來説，Activation Function 會接在神經層後面，如下所示：

- x = layers.Dense(10)(x)
- x = layers.LeakyReLU()(x)

TensorFlow/Keras 為簡化語法，容許將 Activation Function 當作參數使用，直接包在神經層的定義中，如下：

- tf.keras.layers.Dense(128, activation='relu')
- tf.keras.layers.Dense(10, activation='softmax')

4-5 損失函數(Loss Functions)

損失函數(Loss Functions) 又稱為目標函數(Objective Function)、成本函數
(Cost Function)，模型以預測總誤差最小化為目標，因而，學者因應不同
的場域，以各種函數來定義「總誤差」。

TensorFlow 損失函數分成三類，請參閱官網[7]：

1. 機率相關的損失函數(Probabilistic Loss)：例如，二分類的交叉熵
   (BinaryCrossentropy)、多分類的交叉熵(CategoricalCrossentropy)。
2. 迴歸相關的損失函數(Regression Loss)：比如，均方誤差(MSE)。
3. 鉸鏈損失函數(Hinge Loss)：經常用在「最大間格分類」(Maximum-
   margin Classification)，適用於支援向量機(SVM)等演算法。

TensorFlow 損失函數一般在 model.compile()中設定，如下：

```
model.compile(loss='mean_squared_error', optimizer='sgd')
```

上面直接使用字串，如果擔心粗心大意拼錯字的話，也可以使用函數：

```
from keras import losses
model.compile(loss=losses.mean_squared_error, optimizer-'sgd')
```

【範例 1】實際測試幾個常用的損失函數。

程式：04_09_Loss_Function.ipynb。

- BinaryCrossentropy：二分類的交叉熵，熵(Entropy)是指不確定的程
  度，公式如下：
  $s = - \int p(x) \log p(x)\, dx$ ➔ 連續型分配
  $s = - \sum p(x) \log p(x)$ ➔ 離散型分配

- 若是二分類 y=0 或 1，則離散型分配的二分類交叉熵等於：
  $s = - y \log(p) - (1 - y)\log(1 - p)$ ➔ 公式 4.1
  當
  y=0 時 ➔ $s = -\log(1 - p)$

$$y=1 \text{ 時} \rightarrow s = -\log(p)$$

■ 使用 Sigmoid 計算 p，就可得到 BinaryCrossentropy。

1. 兩筆資料的實際值和預測值如下，計算 BinaryCrossentropy。

```
1  # 兩筆資料實際及預測值
2  y_true = [[0., 1.], [0., 0.]]     # 實際值
3  y_pred = [[0.6, 0.4], [0.4, 0.6]] # 預測值
4
5  # 二分類交叉熵(BinaryCrossentropy)
6  bce = tf.keras.losses.BinaryCrossentropy()
7  bce(y_true, y_pred).numpy()
```

執行結果：0.8149。

依照公式 4.1 驗算：

```
1  # 驗算
2  import math
3
4  ((0-math.log(1-0.6) - math.log(0.4)) + (0-math.log(1-0.6) - math.log(0.6)) )/4
```

執行結果：0.8149，與 BinaryCrossentropy()一致。

2. CategoricalCrossentropy：多分類的交叉熵。兩筆資料的實際值和預測值如下，計算 BinaryCrossentropy。

```
1  # 兩筆資料實際及預測值
2  y_true = [[0, 1, 0], [0, 0, 1]]     # 實際值
3  y_pred = [[0.05, 0.95, 0], [0.1, 0.8, 0.1]] # 預測值
4
5  # 多分類交叉熵(CategoricalCrossentropy)
6  cce = tf.keras.losses.CategoricalCrossentropy()
7  cce(y_true, y_pred).numpy()
```

執行結果：1.1769。

3. SparseCategoricalCrossentropy：稀疏矩陣的多分類交叉熵，預期的目標值是單一整數，而非 one-hot encoding 的資料型態，所以，使用此損失函數有個好處是，前置處理就可以省去 one-hot encoding 的轉換。兩筆資料的實際值和預測值如下，計算 SparseCategoricalCrossentropy。

```
1   # 兩筆資料實際及預測值
2   y_true = [1, 2]      # 實際值
3   y_pred = [[0.05, 0.95, 0], [0.1, 0.8, 0.1]] # 預測值
4
5   # 多分類交叉熵(CategoricalCrossentropy)
6   cce = tf.keras.losses.SparseCategoricalCrossentropy()
7   cce(y_true, y_pred).numpy()
```

執行結果：1.1769。

4. MeanSquaredError：計算實際值和預測值的均方誤差。

兩筆資料的實際值和預測值如下，計算 MeanSquaredError。

```
1   # 兩筆資料實際及預測值
2   y_true = [[0., 1.], [0., 0.]]      # 實際值
3   y_pred = [[1., 1.], [1., 0.]]      # 預測值
4
5   # 多分類父義熵(CategoricalCrossentropy)
6   mse = tf.keras.losses.MeanSquaredError()
7   mse(y_true, y_pred).numpy()
```

執行結果： $((1\text{-}1)^2+(0\text{-}1)^2) / 2 = 0.5$。

■ sample_weight：可加參數設定樣本類別的權重比例。

■ reduction=tf.keras.losses.Reduction.SUM：取總和，即 SSE，而非 MSE。

大部份損失函數也可以加這些參數。

5. 鉸鏈損失函數(Hinge Loss)：常用於支援向量機(SVM)，詳細可參閱 [13]：

total loss = $\sum$ maximum(1 - y_true * y_pred, 0)

不考慮負值的損失，所以也被稱作單邊損失函數，真實值(y_true)通常是-1 或 1，如果訓練資料的 y 是 0/1，Hinge Loss 會自動將其轉成-1/1，再計算損失。

兩筆資料的實際值和預測值如下，計算 Hinge Loss。

```
1   # 兩筆資料實際及預測值
2   y_true = [[0., 1.], [0., 0.]]      # 實際值
3   y_pred = [[0.6, 0.4], [0.4, 0.6]] # 預測值
4
5   # Hinge Loss
6   loss_function = tf.keras.losses.Hinge()
7   loss_function(y_true, y_pred).numpy()
```

執行結果：1.3。

驗算：

```
1  # 驗算
2  # loss = sum (maximum(1 - y_true * y_pred, 0))
3  (max(1 - (-1) * 0.6, 0) + max(1 - 1 * 0.4, 0) +
4      max(1 - (-1) * 0.4, 0) + max(1 - (-1) * 0.6, 0)) / 4
```

執行結果與 Hinge()相同。

6. 自訂損失函數(Custom Loss)：撰寫一個函數，輸入為 y 的實際值及預
   測值，輸出為常數即可。下列範例自訂損失為 MSE。

```
1  # 自訂損失函數(Custom Loss)
2  def my_loss_fn(y_true, y_pred):
3      # MSE
4      squared_difference = tf.square(y_true - y_pred)
5      return tf.reduce_mean(squared_difference, axis=-1)  # axis=-1 須設為 -1
6
7  model.compile(optimizer='adam', loss=my_loss_fn)
```

4-6 優化器(Optimizer)

優化器是神經網路中反向傳導的求解方法，著重在兩方面：

1. 設定學習率的變化，加速求解的收斂速度。
2. 避開馬鞍點(Saddle Point)等局部最小值，並且找到全局的最小值
   (Global Minimum)。

優化器的類別一樣在 model.compile()設定，TensorFlow 支援很多種不同的
優化器如下，可參閱 Keras 官網[5]。

- SGD
- RMSprop
- Adam
- Adadelta
- Adagrad

- Adamax
- Nadam
- Ftrl

【範例 1】實際操作幾個常用的優化器。

程式：04_10_Optimizer.ipynb。

1. 隨機梯度下降法(Stochastic Gradient Decent, SGD)
   依據權重更新的時機差別，梯度下降法分為下面三種:

- 批量梯度下降法 (Batch Gradient Descent, BGD)：以「全部」樣本計算梯度，更新權重。
- 隨機梯度下降法 (Stochastic Gradient Descent, SGD)：一次抽取「一個」樣本計算梯度，並立即更新權重。優點是更新速度快，但收斂會較曲折，因為訓練過程中，可能抽到好樣本，也可能抽到壞樣本。

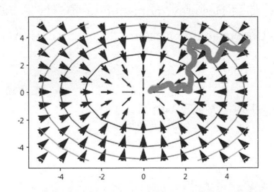

圖 4.2 隨機梯度下降法 (Stochastic Gradient Descent, SGD) 求解圖示

- 小批量梯度下降法 (Mini-batch Gradient Descent)：折衷前兩種做法，以「一批」樣本計算梯度，再更新權重。

小批量梯度下降法可涵蓋前兩種，批量等於全部樣本，即為 BGD；批量為 1，即為 SGD，故小批量梯度下降法通稱為隨機梯度下降法(SGD)。

SGD 語法:

```
1  # SGD
2  tf.keras.optimizers.SGD(
3      learning_rate=0.01, momentum=0.0, nesterov=False, name="SGD"
4  )
```

- 權重更新公式:w = w - learning_rate * g

  其中 w:權重,g:梯度,learning_rate:學習率

- 動能(momentum):公式為

  velocity = momentum * velocity - learning_rate * g

  w = w + velocity

  動能通常介於(0, 1),若等於 0,表示學習率為固定值。一般而言,剛開始訓練時,離最小值較遠的時候,學習率可以放膽邁大步,越接近最小值時,學習率變動幅度就要變小,以免錯過最小值,這種動態調整學習率的方式,能夠使求解收斂速度加快,又不會錯過最小值。

- nesterov:是否使用 Nesterov momentum,預設值是 False。要瞭解技術細節可參閱「Understanding Nesterov Momentum (NAG)」[14]。

2. 隨機梯度下降法的簡單測試。

```
1  # SGD
2  opt = tf.keras.optimizers.SGD(learning_rate=0.1)
3
4  # 任意變數
5  var = tf.Variable(1.0)
6
7  # 損失函數
8  loss = lambda: (var ** 2)/2.0
9
10 # step_count : 優化的步驟
11 for i in range(51):
12     step_count = opt.minimize(loss, [var]).numpy()
13     if i % 10 == 0 and i > 0:
14         print(f'優化的步驟:{step_count}, 變數:{var.numpy()}')
```

損失函數 $x^2/2$。

執行結果:每 10 步列印結果,越來越接近最小值 0。

```
優化的步驟:11, 變數:0.3138105869293213
優化的步驟:21, 變數:0.10941897332668304
優化的步驟:31, 變數:0.03815203905105591
優化的步驟:41, 變數:0.01330279465764761
優化的步驟:51, 變數:0.0046383971348404884
```

3. 優化三次測試隨機梯度下降法的動能。

```
1  opt = tf.keras.optimizers.SGD(learning_rate=0.1, momentum=0.9)
2  var = tf.Variable(1.0)
3
4  # 損失函數起始值
5  val0 = var.value()
6
7  # 損失函數
8  loss = lambda: (var ** 2)/2.0
9
10 # 優化第一次
11 step_count = opt.minimize(loss, [var]).numpy()
12 val1 = var.value()
13 print(f'優化的步驟:{step_count}, 變化值:{(val0 - val1).numpy()}')
14
15 # 優化第二次
16 step_count = opt.minimize(loss, [var]).numpy()
17 val2 = var.value()
18 print(f'優化的步驟:{step_count}, 變化值:{(val1 - val2).numpy()}')
19
20 # 優化第三次
21 step_count = opt.minimize(loss, [var]).numpy()
22 val3 = var.value()
23 print(f'優化的步驟:{step_count}, 變化值:{(val2 - val3).numpy()}')
```

執行結果：

```
val0:1.0
優化的步驟:1, val1:0.8999999761581421, 變化值:0.10000002384185791
優化的步驟:2, val2:0.7199999690055847, 變化值:0.18000000715255737
優化的步驟:3, val3:0.4860000014305115, 變化值:0.23399996757507324
```

4. Adam(Adaptive Moment Estimation)是常用的優化器，這裡就引用 Kingma 等學者於 2014 年發表的「Adam: A Method for Stochastic Optimization」[15]一文所作的評論「Adam 計算效率高、記憶體耗費少，適合大資料集及參數個數很多的模型」。

Adam 語法：

```
1  # Adam
2  tf.keras.optimizers.Adam(
3      learning_rate=0.001,
4      beta_1=0.9,
5      beta_2=0.999,
6      epsilon=1e-07,
7      amsgrad=False,
8      name="Adam",
9  )
```

- beta_1：一階動能衰減率(exponential decay rate for the 1st moment estimates)。
- beta_2：二階動能衰減率(exponential decay rate for the 2nd moment estimates)。
- epsilon：誤差值，小於這個值，優化即停止。
- amsgrad：是否使用 AMSGrad，預設值是 False。技術細節可參閱「一文告訴你 Adam、AdamW、Amsgrad 區別和聯繫」[16]。

5. Adam 簡單測試。

```
1  # Adam
2  opt = tf.keras.optimizers.Adam(learning_rate=0.1)
3
4  # 任意變數
5  var = tf.Variable(1.0)
6
7  # 損失函數
8  loss = lambda: (var ** 2)/2.0
9
10 # step_count：優化的步驟
11 for i in range(11):
12     step_count = opt.minimize(loss, [var]).numpy()
13     if i % 2 == 0 and i > 0:
14         print(f'優化的步驟:{step_count-1}, 變數:{var.numpy()}')
```

執行結果：SGD 執行 50 步，Adam 只需要執行 10 步，就已收斂。

```
優化的步驟:2, 變數:0.7015826106071472
優化的步驟:4, 變數:0.5079597234725952
優化的步驟:6, 變數:0.3234168291091919
優化的步驟:8, 變數:0.15358148515224457
優化的步驟:10, 變數:0.005128741264343262
```

3. 另外還有幾種常用的優化器：

- Adagrad(Adaptive Gradient-based optimization)：設定每個參數的學習率更新頻率不同，較常變動的特徵使用較小的學習率，較少調整，反之，使用較大的學習率，比較頻繁的調整，主要是針對稀疏的資料集。
- RMS-Prop：每次學習率更新是除以均方梯度(average of squared gradients)，以指數的速度衰減。
- ADAM：是 Adagrad 改良版，學習率更新會配合過去的平均梯度調整。

官網還有介紹其他的優化器，網路上也有許多優化器的比較和動畫，有興趣的讀者可自行搜尋。話説回來，雖然研發領域也是一個值得探究的小宇宙，但畢竟我們學習東西，能不能實際派上用場最重要(其實是筆者的腦容量有限)，所以本書的核心是以實務為主，論文研究就不是本書的重點。

4-7 效能衡量指標(Performance Metrics)

效能衡量指標是定義模型優劣的衡量標準，要了解各種效能衡量指標，先要理解混淆矩陣(Confusion Matrix)，以二分類而言，如下圖。

真實

		真(True)	假(False)
預測	陽性(Positive)	TP	FP
	陰性(Negative)	TN	FN

圖 4.3 混淆矩陣(Confusion Matrix)

橫軸為預測結果，分為陽性(Positive, 簡稱 P)、陰性(Negative, 簡稱 N)。縱軸為真實狀況，分為真(True, 簡稱 T)、假(False, 簡稱 F)。

依預測結果及真實狀況的組合，共分為四種狀況：

- TP(真陽性)：預測為陽性，且預測正確。
- TN(真陰性)：預測為陰性，且預測正確。
- FP(偽陽性)：預測為陽性，但預測錯誤，又稱型一誤差(Type I Error)，或 α 誤差。
- FN(偽陰性)：預測為陰性，但預測錯誤，又稱型二誤差(Type II Error)，或 β 誤差。

有了 TP/TN/FP/FN 之後，我們就可以定義各種效能衡量指標，常見的有四種：

- 準確率（Accuracy）= (TP+TN)/(TP+FP+FN+TN)，即
  「預測正確數 / 總數」。
- 精確率（Precision）= TP/(TP+FP)，即
  「正確預測陽性數 / 總陽性數」。
- 召回率（Recall）= TP/(TP+FN)，即
  「正確預測陽性數 / 實際為真的總數」。
- F1 = 精確率與召回率的調和平均數，即
  1 / ((1 / Precision) + (1 / Recall))。

FP(偽陽性)與 FN(偽陰性)是相衝突的，以 Covid-19 檢驗為例，如果降低陽性認定值，可以盡最大可能找到所有的確診者，減少偽陰性，避免傳染病擴散，但有些沒病的人會被誤判，偽陽性會因而相對增加，導致資源的浪費，更嚴重可能造成醫療體系崩潰，得不償失，所以，疾病管制署(CDC)會因應疫情的發展，隨時調整陽性認定值。

除了準確率之外，為什麼還需要參考其他指標？

- 以醫療檢驗設備來舉例，假設某疾病實際染病的比率為 1%，這時我們拿一個故障的檢驗設備，它不管有無染病，都判定為陰性，這時候計算設備準確率，結果竟然是 99%。會有這樣離譜的統計，是因為在此

案例中，驗了 100 個樣本，確實只錯一個。所以，碰到真假比例懸殊的不平衡(Imbalanced)樣本，必須使用其他指標來衡量效能。

- 精確率：再以醫療檢驗設備為例，我們只關心被驗出來的陽性病患，有多少比例是真的染病，而不去關心驗出為陰性者，因為驗出為陰性，通常不會再被複檢，或者不放心又跑到其他醫院複檢，醫院其實很難追蹤他們是否真的沒病。

- 召回率：比方 Covid-19，我們關心的是所有的染病者有多少比例被驗出陽性，因為一旦有漏網之魚(偽陰性)，可能就會造成重大的傷害，如社區傳染。

針對二分類，還有一種較客觀的指標稱為 ROC/AUC 曲線，它是在各種檢驗門檻值下，以假陽率為 X 軸，真陽率為 Y 軸，繪製出來的曲線，稱為 ROC。覆蓋的面積(AUC)越大，表示模型在各種門檻值下的平均效能越好，這個指標有別於一般預測固定以 0.5 當作判斷真假的基準。

有趣的是，損失函數也可視為效能衡量指標，因為當損失函數越小，就表示預測值與實際值越接近，所以，TensorFlow 也可以把損失函數當作效能衡量指標來使用。

TensorFlow 的效能衡量指標可參閱 Keras 官網[17]。

程式：04_11_Metrics.ipynb。

【範例 1】假設有 8 筆資料如下，請計算混淆矩陣(Confusion Matrix)。

實際值 = [0, 0, 0, 1, 1, 1, 1, 1]
預測值 = [0, 1, 0, 1, 0, 1, 0, 1]

1. 載入相關套件

```
1  import tensorflow as tf
2  from tensorflow.keras import metrics
3  import numpy as np
4  import matplotlib.pyplot as plt
5  from sklearn.metrics import accuracy_score, classification_report
6  from sklearn.metrics import precision_score, recall_score, confusion_matrix
```

2. Scikit-learn 提供混淆矩陣(Confusion Matrix)函數，程式碼如下。

```
1  from sklearn.metrics import confusion_matrix
2
3  y_true = [0, 0, 0, 1, 1, 1, 1, 1] # 實際值
4  y_pred = [0, 1, 0, 1, 0, 1, 0, 1] # 預測值
5
6  # 混淆矩陣(Confusion Matrix)
7  tn, fp, fn, tp  = confusion_matrix(y_true, y_pred).ravel()
8  print(f'TP={tp}, FP={fp}, TN={tn}, FN={fn}')
```

注意，Scikit-learn 提供的混淆矩陣，傳回值與圖 4.3 位置不同。

實際值與預測值上下比較，TP 為(1, 1)、FP 為(0, 1)、TN 為(0, 0)、FN 為(1, 0)。

執行結果：TP=3, FP=1, TN=2, FN=2。

3. 繪圖

```
1  # 顯示矩陣
2  fig, ax = plt.subplots(figsize=(2.5, 2.5))
3
4  # 1:藍色, 0:白色
5  ax.matshow([[1, 0], [0, 1]], cmap=plt.cm.Blues, alpha=0.3)
6
7  # 標示文字
8  ax.text(x=0, y=0, s=tp, va='center', ha='center')
9  ax.text(x=1, y=0, s=fp, va='center', ha='center')
10 ax.text(x=0, y=1, s=tn, va='center', ha='center')
11 ax.text(x=1, y=1, s=fn, va='center', ha='center')
12
13 plt.xlabel('實際', fontsize=20)
14 plt.ylabel('預測', fontsize=20)
15
16 # x/y 標籤
17 plt.xticks([0,1], ['T', 'F'])
18 plt.yticks([0,1], ['P', 'N'])
19 plt.show()
```

執行結果：

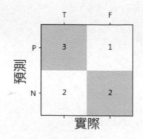

【範例 2】依上述資料計算效能衡量指標。

1. 準確率。

```
1  m = metrics.Accuracy()
2  m.update_state(y_true, y_pred)
3
4  print(f'準確率:{m.result().numpy()}')
5  print(f'驗算={(tp+tn) / (tp+tn+fp+fn)}')
```

執行結果：0.625。

2. 計算精確率。

```
1  m = metrics.Precision()
2  m.update_state(y_true, y_pred)
3
4  print(f'精確率:{m.result().numpy()}')
5  print(f'驗算={(tp) / (tp+fp)}')
```

執行結果：0.75。

3. 計算召回率。

```
1  m = metrics.Recall()
2  m.update_state(y_true, y_pred)
3
4  print(f'召回率:{m.result().numpy()}')
5  print(f'驗算={(tp) / (tp+fn)}')
```

執行結果：0.6。

【範例 3】依資料檔 data/auc_data.csv 計算 AUC。

1. 讀取資料檔

```
1  # 讀取資料
2  import pandas as pd
3  df=pd.read_csv('./data/auc_data.csv')
4  df
```

執行結果：

	predict	actual
0	0.11	0
1	0.35	0
2	0.72	1
3	0.10	1
4	0.99	1
5	0.44	1
6	0.32	0
7	0.80	1
8	0.22	1
9	0.08	0
10	0.56	1

2. 以 Scikit-learn 函數計算 AUC

```
1  from sklearn.metrics import roc_curve, roc_auc_score, auc
2
3  # fpr：假陽率，tpr：真陽率, threshold：各種決策門檻
4  fpr, tpr, threshold = roc_curve(df['actual'], df['predict'])
5  print(f'假陽率={fpr}\n\n真陽率={tpr}\n\n決策門檻={threshold}')
```

執行結果：

```
假陽率=[0.          0.          0.          0.14285714 0.14285714 0.28571429
 0.28571429 0.57142857 0.57142857 0.71428571 0.71428571 1.          ]

真陽率=[0.          0.09090909 0.27272727 0.27272727 0.63636364 0.63636364
 0.81818182 0.81818182 0.90909091 0.90909091 1.          1.          ]

決策門檻=[1.99 0.99 0.8  0.73 0.56 0.48 0.42 0.32 0.22 0.11 0.1  0.03]
```

3. 繪製 AUC

```
1   # 繪圖
2   auc1 = auc(fpr, tpr)
3   ## Plot the result
4   plt.title('ROC/AUC')
5   plt.plot(fpr, tpr, color = 'orange', label = 'AUC = %0.2f' % auc1)
6   plt.legend(loc = 'lower right')
7   plt.plot([0, 1], [0, 1],'r--')
8   plt.xlim([0, 1])
9   plt.ylim([0, 1])
10  plt.ylabel('True Positive Rate')
11  plt.xlabel('False Positive Rate')
12  plt.show()
```

執行結果：

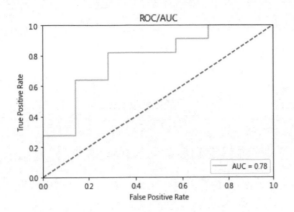

4. 以 TensorFlow 函數計算 AUC

```
1  m = metrics.AUC()
2  m.update_state(df['actual'], df['predict'])
3
4  print(f'AUC:{m.result().numpy()}')
```

執行結果：0.7792，與 Scikit-learn 的結果相去不遠。

4-8 超參數調校(Hyperparameter Tuning)

這一節來研究超參數(Hyperparameters)對效能的影響。在 4-1-3 只對單一變數進行調校，假如要同時調校多個超參數，有一些套件可以幫忙，包括 Keras Tuner、hyperopt、Ray Tune、Ax 等。

本節介紹 Keras Tuner 的用法。先安裝套件：pip install keras-tuner

【範例 1】超參數調校。

程式：04_04_keras_tuner_超參數調校.ipynb。

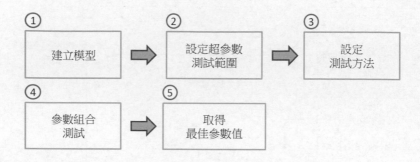

1. 首先要建立模型，並設定超參數測試的範圍：

- 學習率(learning rate) 測試選項：0.01, 0.001, 0.0001。
- 第一層 Dense 輸出神經元數：32、64、96、... 、512。

```
1   import kerastuner as kt
2
3   # 建立模型
4   def model_builder(hp):
5       # 學習率(learning rate)選項: 0.01, 0.001, or 0.0001
6       hp_learning_rate = hp.Choice('learning_rate', values = [0.01, 0.001, 0.0001])
7       # 第一層Dense輸出選項: 32、64、... 、512
8       hp_units = hp.Int('units', min_value = 32, max_value = 512, step = 32)
9
10      model = tf.keras.models.Sequential([
11        tf.keras.layers.Flatten(input_shape=(28, 28)),
12        tf.keras.layers.Dense(hp_units, activation='relu'),
13        tf.keras.layers.Dropout(0.1),
14        tf.keras.layers.Dense(10, activation='softmax')
15      ])
16
17      # 設定優化器(optimizer)、損失函數(loss)、效能衡量指標(metrics)的類別
18      model.compile(optimizer=tf.keras.optimizers.Adam(learning_rate = hp_learning_rate),
19                  loss='sparse_categorical_crossentropy',
20                  metrics=['accuracy'])
21
22      return model
```

2. 調校設定：呼叫 Hyperband()，設定下列參數。

- 目標函數(objective)：準確率。
- 最大執行週期(max_epochs)為 5。
- 執行週期數的遞減因子(factor) 為 3。
- 存檔目錄(directory)：my_dir。
- 專案名稱(project_name)：test1。

```
1   # 調校設定，Hyperband：針對所有參數組合進行測試
2   tuner = kt.Hyperband(model_builder,                    # 模型定義
3                        objective = 'val_accuracy',      # 目標函數
4                        max_epochs = 5,                  # 最大執行週期
5                        factor = 3,                      # 執行週期數的遞減因子
6                        directory = 'my_dir',            # 存檔目錄
7                        project_name = 'test1')          # 專案名稱
```

3. 執行參數調校：

- 設定 callbacks：每個參數組合測試完成後，清除輸出顯示。

- get_best_hyperparameters：取得最佳參數值。

```
1   # 參數調校
2   import IPython
3
4   # 每個參數組合測完後，清除顯示
5   class ClearTrainingOutput(tf.keras.callbacks.Callback):
6       def on_train_end(*args, **kwargs):
7           IPython.display.clear_output(wait = True)
8
9
10  # 調校執行
11  tuner.search(x_train_norm, y_train, epochs = 5,
12               validation_data = (x_test_norm, y_test),  # 驗證資料
13               callbacks = [ClearTrainingOutput()])       # 執行每個參數組合後回呼
14
15  # 顯示最佳參數值
16  best_hps = tuner.get_best_hyperparameters(num_trials = 1)[0]
17
18  print(f"最佳參數值\n第一層Dense輸出：{best_hps.get('units')}\n",
19        "學習率：{best_hps.get('learning_rate')}")
```

最佳參數組合為：

- 第一層 Dense 輸出：160。
- 學習率：0.001。

除此之外，Keras Tuner 還有很多的功能，包括：

1. 超參數測試範圍的設定，參閱 [18]。

- Boolean：真/假。
- Choice：多個設定選項。
- Int/Float：整數/浮點數的連續範圍。
- Fixed：測試所有參數(tune_new_entries=True)，除了目前的參數，也可依賴其他參數(parent_name)的設定。只有當其他參數值為特定值時，這個參數才會生效。

- conditional_scope：條件式，類似 Fixed，依賴其他參數，只有當其他參數值為特定值時，這個條件才會生效。

2. 測試方法(Tuners) 請參閱 [19]。
- Hyperband：測試所有組合。
- RandomSearch：若測試範圍過大，可隨機抽樣部份組合，加以測試。
- BayesianOptimization：搭配高斯過程(Gaussian process)，依照前次的測試結果，決定下次的測試內容。

3. Oracle：超參數調校的演算法，為測試方法(Tuners)的參數，可決定測試方法的下次測試組合，可參閱 [20]。

另外再加碼推薦，可搭配 Hiplot 視覺化套件，顯示每一種參數組合的設定值與損失/準確度，記得先安裝套件：

pip install hiplot

1. 解析 Keras Tuner 測試的日誌檔

```
1  # 解析 Keras Tuner 測試的日誌檔
2  import os
3  import json
4
5  vis_data = []
6  # 掃描目錄內每一個檔案
7  rootdir = 'my_dir/test1'
8  for subdirs, dirs, files in os.walk(rootdir):
9      for file in files:
10         if file.endswith("trial.json"):
11             with open(subdirs + '/' + file, 'r') as json_file:
12                 data = json_file.read()
13                 vis_data.append(json.loads(data))
```

2. 顯示參數組合與測試結果

```
1  # 顯示參數組合與測試結果
2  import hiplot as hip
3
4  # 建立字典，含參數組合與測試結果
5  data = [{'units': vis_data[idx]['hyperparameters']['values']['units'],
6          'learning_rate': vis_data[idx]['hyperparameters']['values']['learning_rate'],
7          'loss': vis_data[idx]['metrics']['metrics']['loss']['observations'][0]['value'],
8          'val_loss': vis_data[idx]['metrics']['metrics']['val_loss']['observations'][0]['value'],
9          'accuracy': vis_data[idx]['metrics']['metrics']['accuracy']['observations'][0]['value'],
10         'val_accuracy': vis_data[idx]['metrics']['metrics']['val_accuracy']['observations'][0]['value']}
11         for idx in range(len(vis_data))]
12
13 # 顯示
14 hip.Experiment.from_iterable(data).display()
```

執行結果：uid 為執行代碼，可以看出第一行(uid=7)有最高的準確率，獲
選為最佳參數組合。

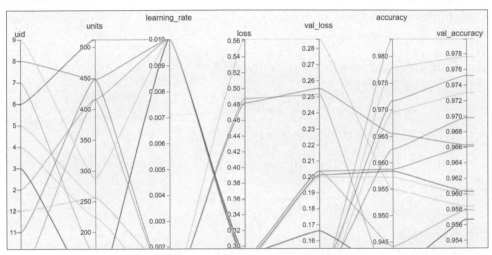

uid	from_uid	units	learning_rate	loss	val_loss	accuracy	val_a
■ 7	null	160	0.001	0.26826784014701843	0.13270235061645508	0.9832666516304016	0.9799000
■ 8	null	448	0.01	0.28484904766082764	0.20362061262130737	0.9587666392326355	0.9660999
■ 9	null	288	0.01	0.2784591615200043	0.20001891255378723	0.9572333097457886	0.9599000

Showing 11 to 13 of 13 entries   Previous 1 **2** Next

參數調校是深度學習中非常重要的步驟，因為深度學習是一個黑箱科學，
加上我們對於高維資料的聯合機率分配也是一無所知，唯有透過大量的實
驗，才能獲得較佳的模型。但困難的是，模型訓練的執行非常耗時，如何
透過各種方法或套件的協助，縮短調校時間，是工程師建構 AI 模型時須
思考如何改善的課題。

參考資料 (References)

[1] 維基百科 Activation Function 的介紹
     (https://en.wikipedia.org/wiki/Activation_function)
[2] TensorFlow 優化器的介紹

(https://www.tensorflow.org/api_docs/python/tf/keras/optimizers)

[3] Keras 優化器的介紹

(https://keras.io/api/optimizers/)

[4] TensorFlow 損失函數的介紹

(https://www.tensorflow.org/api_docs/python/tf/keras/optimizers)

[5] Keras 損失函數的介紹

(https://keras.io/api/losses/)

[6] TensorFlow 官網中效能衡量指標的介紹

(https://www.tensorflow.org/api_docs/python/tf/keras/metrics)

[7] TensorFlow 官網

(https://www.tensorflow.org)

[8] Keras 官網

(https://keras.io)

[9] TensorFlow 官網中 FashionMnist 的介紹

(https://www.tensorflow.org/datasets/catalog/fashion_mnist)

[10] Keras 官網中神經層的介紹

(https://keras.io/api/layers/)

[11] Keras 官網 Activation Function 說明

(https://keras.io/api/layers/activations/)

[12] Keras 官網 Activation Layers 說明

(https://keras.io/api/layers/activation_layers/)

[13] 維基百科關於鉸鏈損失函數的介紹

(https://zh.wikipedia.org/wiki/Hinge_loss)

[14] 《Understanding Nesterov Momentum (NAG)》, 2018

(https://dominikschmidt.xyz/nesterov-momentum/)

[15] Diederik P. Kingma、Jimmy Ba,《Adam: A Method for Stochastic Optimization》, 2014

(https://arxiv.org/abs/1412.6980)

[16] 深度學習於 NLP,《一文告訴你 Adam、AdamW、Amsgrad 區別和聯繫》, 2019

(https://zhuanlan.zhihu.com/p/39543160)

[17] Keras 官網中效能衡量指標的介紹

(https://keras.io/api/metrics/)

[18] Keras 官網中超參數測試範圍的設定

(https://keras-team.github.io/keras-tuner/documentation/hyperparameters/)

[19] Keras 官網效能調校(Tuners)的介紹(https://keras-team.github.io/keras-tuner/documentation/tuners/)

[20] Keras 官網中 Oracle 的介紹

(https://keras-team.github.io/keras-tuner/documentation/oracles/)

TensorFlow 其他常用指令

除了建構模型外，TensorFlow 還貼心地提供各種的工具和指令，方便在程式開發流程中使用，包括前置處理、模型存檔/載入/繪製、除錯等功能，現在我們就來認識這些功能吧。

5-1 特徵轉換

One-hot encoding：將類別變數轉為多個虛擬變數(Dummy variable)，每個虛擬變數只含真/假值(1/0)，避免被演算法誤認該變數類別有順序大小之分，例如顏色，紅/藍/綠會被轉換成三個變數：「是紅色嗎」、「是藍色嗎」、「是綠色嗎」，而非 1/2/3。

程式 05_01_特徵轉換.ipynb：

```
1  # One-hot encoding
2  # num_classes：類別個數，可不設定
3  tf.keras.utils.to_categorical([0, 1, 2, 3], num_classes=9)
```

執行結果如下，指定 9 種類別，即產生 9 個變數。

- num_classes：類別個數，此參數可以不設定，函數會從資料中統計出類別個數。

```
array([[1., 0., 0., 0., 0., 0., 0., 0., 0.],
       [0., 1., 0., 0., 0., 0., 0., 0., 0.],
       [0., 0., 1., 0., 0., 0., 0., 0., 0.],
       [0., 0., 0., 1., 0., 0., 0., 0., 0.]], dtype=float32)
```

修改 MNIST 手寫阿拉伯數字辨識程式：

```
1   mnist = tf.keras.datasets.mnist
2
3   # 載入 MNIST 手寫阿拉伯數字資料
4   (x_train, y_train),(x_test, y_test) = mnist.load_data()
5
6   # 特徵縮放，使用常態化(Normalization)，公式 = (x - min) / (max - min)
7   x_train_norm, x_test_norm = x_train / 255.0, x_test / 255.0
8
9   # One-hot encoding
10  y_train = tf.keras.utils.to_categorical(y_train)
11  y_test = tf.keras.utils.to_categorical(y_test)
12
13  # 建立模型
14  model = tf.keras.models.Sequential([
15      tf.keras.layers.Flatten(input_shape=(28, 28)),
16      tf.keras.layers.Dense(256, activation='relu'),
17      tf.keras.layers.Dropout(0.2),
18      tf.keras.layers.Dense(10, activation='softmax')
19  ])
20
21  # 設定優化器(optimizer)、損失函數(loss)、效能衡量指標(metrics)的類別
22  model.compile(optimizer='adam',
23                loss='categorical_crossentropy',
24                metrics=['accuracy'])
25
26  # 模型訓練
27  history = model.fit(x_train_norm, y_train, epochs=5, validation_split=0.2)
28
29  # 評分(Score Model)
30  score=model.evaluate(x_test_norm, y_test, verbose=0)
```

- 第 10、11 行將 y 作 One-hot encoding 轉換。
- 第 23 行損失函數使用 categorical_crossentropy，而非 sparse_categorical_crossentropy，因後者的 y 不需 One-hot encoding 轉換。

常態化(Normalization)：將所有資料標準化(Standardization)，使資料轉換為標準常態分配 N(0, 1)，類似 scikit-learn 的 StandardScaler()。

```
1   import numpy as np
2   import tensorflow as tf
3   from tensorflow.keras.layers.experimental import preprocessing
4
5   # 測試資料
6   data = np.array([[0.1, 0.2, 0.3], [0.8, 0.9, 1.0], [1.5, 1.6, 1.7],])
7   layer = preprocessing.Normalization()   # 常態化
8   layer.adapt(data)                 # 訓練
9   normalized_data = layer(data) # 轉換
10
11  # 顯示平均數、標準差
12  print(f"平均數: {normalized_data.numpy().mean():.2f}")
13  print(f"標準差: {normalized_data.numpy().std():.2f}")
```

執行結果：平均數: 0.00，標準差: 1.00

▍5-2 模型存檔與載入(Save and Load)

模型存檔，可以儲存下列資訊：

1. 模型結構與組態。
2. 權重，含偏差項(bias)。
3. 模型 compile 選項。
4. 優化器的狀態，這樣訓練即可由斷點處繼續執行。

目前支援兩種格式：

1. TensorFlow SavedModel 格式：官方建議採用這種，以目錄儲存，目錄下含多個檔案，各司其職。
2. Keras H5 格式：Keras 套件既有格式，以單一檔案儲存，注意，此種格式無法儲存自訂的神經層。

均使用 model.save(<file path>)指令，如果設定副檔名為.h5，則存檔成 Keras 既有格式。載入模型時呼叫 load_model(<file path>)指令。

【範例】模型存檔與載入。

程式 05_02_模型存檔與載入.ipynb。

```
1  model.save('my_model')
```

1. 執行結果如下，以整個目錄儲存模型資訊。

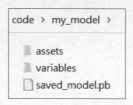

2. 載入模型，以變數名稱 model2 接收：

```
1  # 模型載入
2  model2 = tf.keras.models.load_model('my_model')
3
4  # 評分(Score Model)
5  score=model2.evaluate(x_test_norm, y_test, verbose=0)
6
7  for i, x in enumerate(score):
8      print(f'{model2.metrics_names[i]}: {score[i]:.4f}')
```

執行結果與之前存檔的模型相同。

可以使用 np.testing.assert_allclose 比較預測結果。

```
1  # 模型比較
2  import numpy as np
3
4  # 比較，若結果不同，會出現錯誤
5  np.testing.assert_allclose(
6      model.predict(x_test_norm), model2.predict(x_test_norm)
7  )
```

3. 可以只取得模型結構，不含權重，有兩種指令：

- get_config() / from_config()

```
1  # 取得模型結構
2  config = model.get_config()
3
4  # 載入模型結構
5  # Sequential model
6  new_model = tf.keras.Sequential.from_config(config)
7
8  # function API
9  # new_model = tf.keras.Model.from_config(config)
```

■ json

```
1  # 取得模型結構
2  json_config = model.to_json()
3
4  # 載入模型結構
5  new_model = tf.keras.models.model_from_json(json_config)
```

4. 只取得模型權重：

```
1  # 取得模型權重
2  weights = model.get_weights()
3  weights
```

5. 若有自訂神經層，需先註冊，才能從組態中還原模型，程式碼如下：

```
35  # Retrieve the config
36  config = model.get_config()
37
38  # Custom Layer 需註冊
39  custom_objects = {"CustomLayer": CustomLayer, "custom_activation": custom_activation}
40  with tf.keras.utils.custom_object_scope(custom_objects):
41      new_model = tf.keras.Model.from_config(config)
```

6. 載入模型權重，並不包含 compile() 選項，必須另外補填。

```
1  # 設定模型權重
2  new_model.set_weights(weights)
3
4  # 設定優化器(optimizer)、損失函數(loss)、效能衡量指標(metrics)的類別
5  new_model.compile(optimizer='adam',
6                    loss='sparse_categorical_crossentropy',
7                    metrics=['accuracy'])
8
9  # predict
10 score=new_model.evaluate(x_test_norm, y_test, verbose=0)
11 score
```

7. 模型權重存檔：

```
1  # 模型權重存檔
2  model.save_weights('my_h5_model.weight')
```

8. 載入模型權重檔：

```
1  # 載入模型權重檔
2  model.load_weights('my_h5_model.weight')
```

詳細相關資訊可參考「Keras Models API」[1]。

5-3 模型彙總與結構圖(Summary and Plotting)

如果設計一個複雜的模型，通常會希望有視覺化的呈現，讓我們一目瞭然，利於檢查結構是否 OK，TensorFlow 提供彙總資訊，以表格顯示，再提供模型結構圖，幫助我們釐清整體流程。

【範例】顯示模型彙總與繪製結構圖。

程式 05_03_模型彙總與結構圖.ipynb。

1. 顯示模型彙總：下列指令可取得彙總資訊表格，包含每一層的輸入/輸出/參數個數。

```
1  model.summary()
```

執行結果：

```
Model: "sequential"

Layer (type)                 Output Shape              Param #
=================================================================
flatten (Flatten)            (None, 784)               0

dense (Dense)                (None, 256)               200960

dropout (Dropout)            (None, 256)               0

dense_1 (Dense)              (None, 10)                2570
=================================================================
Total params: 203,530
Trainable params: 203,530
Non-trainable params: 0
```

2. 取得神經層資訊。

使用索引(index)取得神經層資訊。

```
1  # 以 index 取得神經層資訊
2  model.get_layer(index=0)
```

使用名稱取得神經層資訊。

```
1  # 以名稱取得神經層資訊
2  model.get_layer(name='dense_1')
```

取得神經層權重。

```
1  # 取得神經層權重
2  model.get_layer(name='dense').weights
```

繪製結構圖：要繪製模型結構，需先完成以下步驟，才能順利繪製圖形。

- 安裝 graphviz 軟體，網址為：https://www.graphviz.org/download，再把安裝目錄下的 bin 路徑加到環境變數 path 中。
- 安裝兩個 Python 套件：

  pip install graphviz

  pip install pydotplus

以下繪圖指令會先產生描述向量圖的文字檔(.dot)，再呼叫 graphviz 的 dot.exe，將.dot 檔案裡的內容轉化為影像檔案(.png)。

```
1  # 繪製結構圖
2  tf.keras.utils.plot_model(model)
```

加上不同的參數，可顯示各種的額外資訊，例如：

- show_shapes=True：顯示輸入/輸出的神經元個數。
- show_dtype=True：顯示輸入/輸出的資料型態。
- to_file="model.png"：同時存檔。

```
1  # 繪製結構圖
2  # show_shapes=True：可顯示輸入/輸出的神經元個數
3  # show_dtype=True：可顯示輸入/輸出的資料型態
4  # to_file：可同時存檔
5  tf.keras.utils.plot_model(model, show_shapes=True, show_dtype=True,
6                            to_file="model.png")
```

執行結果如下：

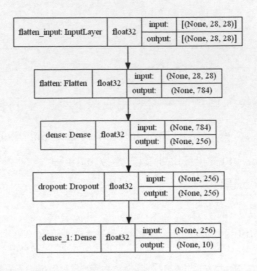

也可以產生 dot 格式：

```
1  # 產生 dot 格式及 png 檔
2  import pydotplus as pdp
3  from IPython.display import display, Image
4
5  # 產生 dot 格式
6  dot1 = tf.keras.utils.model_to_dot(model, show_shapes=True, show_dtype=True)
7  # 產生 png 檔
8  display(Image(dot1.create_png()))
```

5-4　回呼函數(Callbacks)

回呼函數(Callbacks)是在模型訓練過程中埋入要觸發的事件，在每一個週期執行之前與之後，都可以呼叫 Callback 函數，TensorFlow/Keras 提供許多類型的 Callbacks，功能如下：

1. 在訓練過程中記錄任何資訊。
2. 在每個查核點(Checkpoint)進行模型存檔。
3. 迫使訓練提前結束。
4. 結合 TensorBoard 視覺化工具，即時監看訓練過程。
5. 將訓練過程產生的資訊寫入 CSV 檔案。
6. 使用其他 PC 遠端監控訓練過程。

還有更多的其他功能，請參閱 Keras 官網的 Callback 介紹[2]。除了內建 Callback，也能夠自訂 Callback。

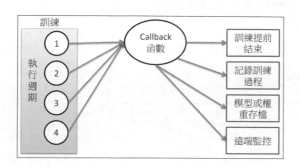

圖 5.1 回呼函數(Callbacks)

透過 Callback 可以完全解構模型訓練的過程，我們使用一些範例，說明各類型 Callback 的用法。

5-4-1 EarlyStopping Callbacks

EarlyStopping 是用來設定訓練提前結束的條件，我們可以設定較大的執行週期數，並搭配訓練提前結束的條件，效能一段時間內沒有持續改善時，就可提前結束訓練，這樣就能兼顧效能與訓練時間了。

【範例 1】實際操作 EarlyStopping Callback。
程式 05_04_Callback.ipynb。

1. 定義 Callback 函數為「要是連續三個執行週期 validation accuracy 沒改善就停止訓練」。

```
1  # validation loss 三個執行週期沒改善就停止訓練
2  my_callbacks = [
3      tf.keras.callbacks.EarlyStopping(patience=3, monitor = 'val_accuracy'),
4  ]
```

2. 訓練指令的 callback 參數設定為上述函數。

```
1  # 訓練 20 次，但實際只訓練 12次就停止了
2  history = model.fit(x_train_norm, y_train, epochs=20, validation_split=0.2,
3                      callbacks=my_callbacks)
```

3. 執行結果如下，預計訓練 20 次，但實際只訓練 12 次就停止了，注意，每次執行結果可能不同。

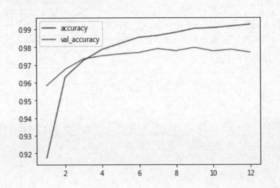

4. 效能指標也可以改為驗證的損失(val_loss)，只要連續 3 次沒改善停止訓練。

```
1  # validation loss 三個執行週期沒改善就停止訓練
2  my_callbacks = [
3      tf.keras.callbacks.EarlyStopping(patience=3, monitor = 'val_loss'),
4  ]
```

執行結果：

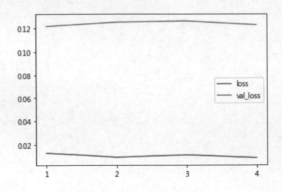

5-4-2 ModelCheckpoint Callbacks

若訓練過程過長，難免會發生訓練到一半就當掉的狀況，因此，我們可以利用 ModelCheckpoint Callback，在每一個檢查點(Checkpoint)存檔，當掉後再次執行時，就可以從中斷點延續，繼續訓練。

【**範例 2**】實際操作 ModelCheckpoint Callback。

程式 05_04_Callback.ipynb。

1. 定義 ModelCheckpoint callback。

```
1   # 定義 ModelCheckpoint callback
2   checkpoint_filepath = 'model.{epoch:02d}.h5' # 存檔名稱，可用 f-string 變數
3   model_checkpoint_callback = tf.keras.callbacks.ModelCheckpoint(
4       filepath=checkpoint_filepath, # 設定存檔名稱
5       save_weights_only=True,       # 只存權重
6       monitor='val_accuracy',       # 監看驗證資料的準確率
7       mode='max',                   # 設定save_best_only=True時，best是指 max or min
8       save_best_only=True)          # 只存最好的模型
9
10  EPOCHS = 3  # 訓練 3 次
11  model.fit(x_train_norm, y_train, epochs=EPOCHS, validation_split=0.2,
12            callbacks=[model_checkpoint_callback])
```

執行結果如下，最後的準確率等於 0.9859。

```
Epoch 1/3
1500/1500 [==============================] - 6s 4ms/step - loss: 0.0652 - accuracy: 0.9796 - val_loss: 0.0521 - val_accuracy:
0.9843
Epoch 2/3
1500/1500 [==============================] - 5s 3ms/step - loss: 0.0520 - accuracy: 0.9835 - val_loss: 0.0502 - val_accuracy:
0.9844
Epoch 3/3
1500/1500 [==============================] - 5s 3ms/step - loss: 0.0431 - accuracy: 0.9859 - val_loss: 0.0450 - val_accuracy:
0.9863
```

2. 再執行 3 個週期，準確率會接續上一次的結果，繼續改善。

```
1   # 再訓練 3 次，觀察 accuracy，會接續上一次，繼續改善 accuracy。
2   model.fit(x_train_norm, y_train, epochs=EPOCHS, validation_split=0.2,
3             callbacks=[model_checkpoint_callback])
```

執行結果如下，最後的準確率等於 0.9902。

```
Epoch 1/3
1500/1500 [==============================] - 5s 3ms/step - loss: 0.0378 - accuracy: 0.9876 - val_loss: 0.0559 - val_accuracy:
0.9842
Epoch 2/3
1500/1500 [==============================] - 5s 3ms/step - loss: 0.0316 - accuracy: 0.9892 - val_loss: 0.0580 - val_accuracy:
0.9827
Epoch 3/3
1500/1500 [==============================] - 5s 3ms/step - loss: 0.0300 - accuracy: 0.9902 - val_loss: 0.0548 - val_accuracy:
0.9837
```

5-4-3　TensorBoard Callbacks

TensorBoard 是 TensorFlow 所提供的視覺化診斷工具，功能非常強大，除了可以顯示訓練過程之外，也能夠顯示圖片、語音及文字訊息。將 TensorBoard 整合至 Callback 事件裡，就可以在訓練的過程中啟動 TensorBoard 網站，即時觀看訓練資訊。

【範例 3】實際操作 TensorBoard Callback。
程式 05_04_Callback.ipynb。

1. 定義觸發事件：

```
1  # 定義 tensorboard callback
2  tensorboard_callback = [tf.keras.callbacks.TensorBoard(log_dir='.\\logs')]
3
4  # 訓練 5 次
5  history = model.fit(x_train_norm, y_train, epochs=5, validation_split=0.2,
6                      callbacks=tensorboard_callback)
```

2. 啟動 TensorBoard 網站：

   tensorboard --logdir=.\logs

3. 使用瀏覽器輸入以下網址，即可觀看訓練資訊：

   http://localhost:6006/，畫面如下：

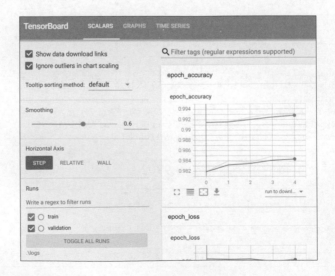

點選「Graph」頁籤，可以觀察模型的運算圖，顯示模型的運算順序。

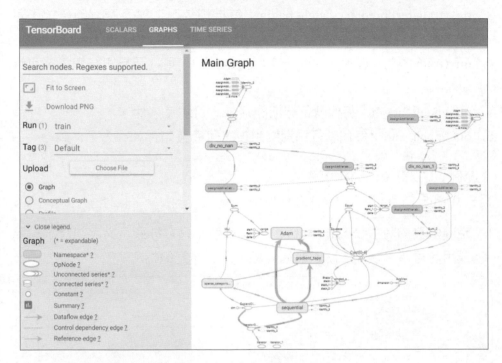

請注意，model.fit 的 Callback 參數值是一個 list，一次可以加入多個 Callback，在訓練時一併觸發，如下，同時設定提前結束訓練、檢查點及 TensorBoard。

```
1  # 可同時定義多個Callback事件
2  my_callbacks = [
3      tf.keras.callbacks.EarlyStopping(patience=3),
4      tf.keras.callbacks.ModelCheckpoint(filepath='model.{epoch:02d}.h5'),
5      tf.keras.callbacks.TensorBoard(log_dir='./logs'),
6  ]
7  model.fit(dataset, epochs=10, callbacks=my_callbacks)
```

5-4-4　自訂 Callback

如果內建的 Callback 不能滿足需求，也可以自訂 Callback，觸發時機可含 訓練、測試、預測階段的之前(Before)與之後(After)：

1. on_(train|test|predict)_begin：訓練、測試及預測開始前，可觸發事件。

2. on_(train|test|predict)_end：訓練、測試及預測**結束**後，可觸發事件。

3. on_(train|test|predict)_batch_begin：**每批**訓練、測試及預測**開始**前，可觸發事件。

4. on_(train|test|predict)_batch_end：**每批**訓練、測試及預測**結束**後，可觸發事件。

5. on_epoch_begin：**每個執行週期開始**前，可觸發事件。

6. on_epoch_end：**每個執行週期結束**後，可觸發事件。

【**範例 3**】自訂 Callback 實作。

程式 05_05_Custom_Callback.ipynb。

1. 定義觸發的時機及動作(Action)，以下動作只單純顯示文字訊息，實際運用時可取得當時的狀態或統計量寫入工作記錄檔。

```python
class CustomCallback(tf.keras.callbacks.Callback):
    def __init__(self):
        self.task_type=''
        self.epoch=0
        self.batch=0

    def on_train_begin(self, logs=None):
        self.task_type='訓練'
        print("訓練開始...")

    def on_train_end(self, logs=None):
        print("訓練結束.")

    def on_epoch_begin(self, epoch, logs=None):
        self.epoch=epoch
        print(f"{self.task_type}第 {epoch} 執行週期開始...")

    def on_epoch_end(self, epoch, logs=None):
        print(f"{self.task_type}第 {epoch} 執行週期結束.")

    def on_test_begin(self, logs=None):
        self.task_type='測試'
        print("測試開始...")

    def on_test_end(self, logs=None):
        print("測試結束.")

    def on_predict_begin(self, logs=None):
        self.task_type='預測'
        print("預測開始...")

    def on_predict_end(self, logs=None):
        print("預測結束.")
```

```
34
35      def on_train_batch_begin(self, batch, logs=None):
36          print(
37              f"訓練 第 {self.epoch} 執行週期, 第 {batch} 批次開始...")
38
39      def on_train_batch_end(self, batch, logs=None):
40          print(
41              f"訓練 第 {self.epoch} 執行週期, 第 {batch} 批次結束.")
42
43      def on_test_batch_begin(self, batch, logs=None):
44          print(
45              f"測試 第 {self.epoch} 執行週期, 第 {batch} 批次開始...")
46
47      def on_test_batch_end(self, batch, logs=None):
48          print(
49              f"測試 第 {self.epoch} 執行週期, 第 {batch} 批次結束.")
50
51      def on_predict_batch_begin(self, batch, logs=None):
52          print(
53              f"預測 第 {self.epoch} 執行週期, 第 {batch} 批次開始...")
54
55      def on_predict_batch_end(self, batch, logs=None):
56          print(
57              f"預測 第 {self.epoch} 執行週期, 第 {batch} 批次結束.")
```

2. 在訓練、測試、預測使用此 Callback，程式碼如下：

```
1   # 訓練
2   model.fit(
3       x_train_norm, y_train, epochs=5,
4       batch_size=256, verbose=0,
5       validation_split=0.2, callbacks=[CustomCallback()]
6   )
7
8   # 測試
9   model.evaluate(
10      x_test_norm, y_test, batch_size=128,
11      verbose=0, callbacks=[CustomCallback()]
12  )
13
14  # 預測
15  model.predict(
16      x_test_norm, batch_size=128,
17      callbacks=[CustomCallback()]
18  )
```

訓練顯示結果：   測試顯示結果：

```
訓練開始...
訓練第 0 執行週期開始...
訓練 第 0 執行週期, 第 0 批次開始...
訓練 第 0 執行週期, 第 0 批次結束.
訓練 第 0 執行週期, 第 1 批次開始...
訓練 第 0 執行週期, 第 1 批次結束.
訓練 第 0 執行週期, 第 2 批次開始...
訓練 第 0 執行週期, 第 2 批次結束.
訓練 第 0 執行週期, 第 3 批次開始...
訓練 第 0 執行週期, 第 3 批次結束.
訓練 第 0 執行週期, 第 4 批次開始...
訓練 第 0 執行週期, 第 4 批次結束.
訓練 第 0 執行週期, 第 5 批次開始...
訓練 第 0 執行週期, 第 5 批次結束.
訓練 第 0 執行週期, 第 6 批次開始...
訓練 第 0 執行週期, 第 6 批次結束.
訓練 第 0 執行週期, 第 7 批次開始...
訓練 第 0 執行週期, 第 7 批次結束.
訓練 第 0 執行週期, 第 8 批次開始...
訓練 第 0 執行週期, 第 8 批次結束.
訓練 第 0 執行週期, 第 9 批次開始...
訓練 第 0 執行週期, 第 9 批次結束.
```

```
測試開始...
測試 第 0 執行週期, 第 0 批次開始...
測試 第 0 執行週期, 第 0 批次結束.
測試 第 0 執行週期, 第 1 批次開始...
測試 第 0 執行週期, 第 1 批次結束.
測試 第 0 執行週期, 第 2 批次開始...
測試 第 0 執行週期, 第 2 批次結束.
測試 第 0 執行週期, 第 3 批次開始...
測試 第 0 執行週期, 第 3 批次結束.
測試 第 0 執行週期, 第 4 批次開始...
測試 第 0 執行週期, 第 4 批次結束.
測試 第 0 執行週期, 第 5 批次開始...
測試 第 0 執行週期, 第 5 批次結束.
測試 第 0 執行週期, 第 6 批次開始...
測試 第 0 執行週期, 第 6 批次結束.
測試 第 0 執行週期, 第 7 批次開始...
測試 第 0 執行週期, 第 7 批次結束.
測試 第 0 執行週期, 第 8 批次開始...
測試 第 0 執行週期, 第 8 批次結束.
```

5-4-5 自訂 Callback 應用

【範例 4】透過自訂的 callback，在每一批訓練結束時記錄損失，就可以在整個訓練過程結束後依據收集的資料繪製線圖。

程式 05_06_Custom_Callback_loss.ipynb。

1. 在每一批的訓練結束後記錄損失至 Pandas DataFrame 中。

```
1  class CustomCallback_2(tf.keras.callbacks.Callback):
2      ...
3      def on_train_batch_end(self, batch, logs=None):
4          # 新增資料至 df2 DataFrame
5          df2 = pd.DataFrame([[self.epoch, batch, logs["loss"]]],
6                          columns=['epoch', 'batch', 'metrics'])
7          self.df = self.df.append(df2, ignore_index=True)
```

2. 圖表顯示的結果很有意思，優化的過程中損失函數並不是一路遞減，而是起起伏伏，但整體趨勢向下遞減。

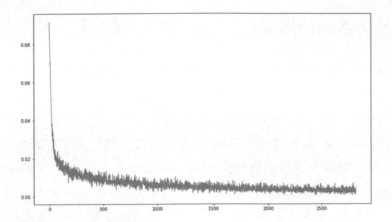

3. 依執行週期(epoch)作小計，取最小損失值，畫出線圖，如下。

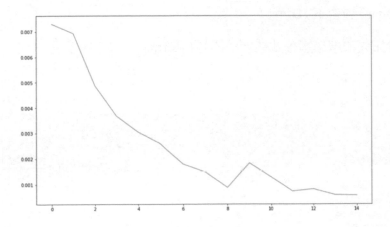

5-4-6 總結

藉由自訂 Callback，我們能夠更深入理解優化的過程。除了損失函數以外，要取得其他資訊也是可行的，比方可以在 Callback 中呼叫 self.model.get_weights() 來取得每一神經層的權重，另外，自訂 Callback 還有個好用的功能，就是用來除錯(Debug)，假如訓練出現錯誤時，像是 Nan 或優化無法收斂的情形，就可以利用 Callback 逐批檢查，更多 Callback 的用法，可參閱「Keras Callback API」[2]。

█ 5-5 TensorBoard

TensorBoard 是一種視覺化的診斷工具，功能非常強大，可以顯示模型結構、訓練過程，包括圖片、文字和音訊資料。在訓練的過程中啟動 TensorBoard，能夠即時觀看訓練過程。PyTorch 安裝時所包含的 TensorBoardX，其實與 TensorBoard 功能相似，這表示了 TensorFlow 和 PyTorch 雖然是相互競爭的死對頭，但 TensorBoard 仍然是 PyTorch 開發團隊也不得不承認的優秀工具。

5-5-1 TensorBoard 功能

TensorBoard 包含下列功能：

1. 追蹤損失和準確率等效能衡量指標(Metrics)，並以視覺化呈現。

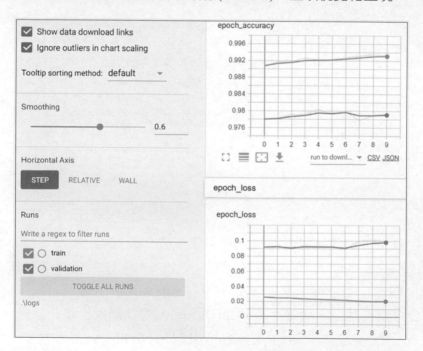

2. 顯示運算圖(Computational Graph)：包括張量運算(tensor operation)和神經層(layers)。

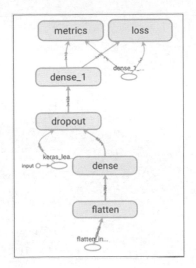

3. 直方圖(Histogram)：顯示訓練過程中的權重(weights)、偏差(bias)的機率分配。

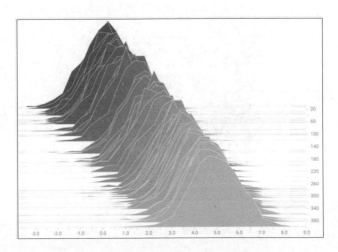

4. 詞嵌入(Word Embedding)展示：把詞嵌入向量降維，投影到三維空間來顯示。畫面右邊可輸入任意單字，例如 King，就會出現下圖，將與其相近的單字顯示出來，原理是透過詞向量(Word2Vec)將每個單字轉為向量，再利用 Cosine_Similarity 計算相似性，詳情會在後續章節介紹。

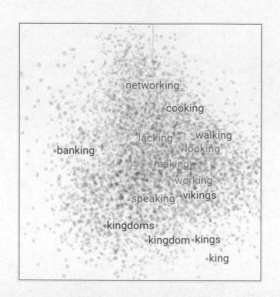

5. 顯示圖片、文字和音訊資料。

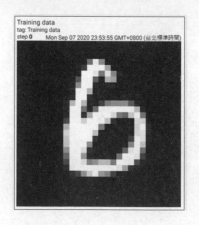

6. 剖析 TensorFlow 程式流程。

5-5-2　測試

之前在 model.fit()內指定 TensorBoard Callback，可寫入工作日誌檔(Log)，
內容類別是固定的，如果要自訂寫入的內容，可以直接在程式中寫入工作
日誌檔，但是不能使用 model.fit()，要改用自動微分 (tf.GradientTape) 的方
式訓練。

【**範例 5**】同樣拿 MNIST 辨識作測試，前面載入資料與建立模型程式碼的
流程不變，從 compile() 開始作一些調整，以下僅列出關鍵的程式碼，完
整的程式請參考 **05_07_TensorBoard.ipynb**。

1. 先設定優化器(optimizer)、損失函數(loss)、效能衡量指標(metrics)。

```
1  # 設定優化器(optimizer)、損失函數(loss)、效能衡量指標(metrics)的類別
2  loss_object = tf.keras.losses.SparseCategoricalCrossentropy()
3  optimizer = tf.keras.optimizers.Adam()
4
5  # Define 訓練及測試的效能衡量指標(Metrics)
6  train_loss = tf.keras.metrics.Mean('train_loss', dtype=tf.float32)
7  train_accuracy = tf.keras.metrics.SparseCategoricalAccuracy('train_accuracy')
8  test_loss = tf.keras.metrics.Mean('test_loss', dtype=tf.float32)
9  test_accuracy = tf.keras.metrics.SparseCategoricalAccuracy('test_accuracy')
```

2. 寫入效能衡量指標：使用自動微分(tf.GradientTape)的方式訓練模型。

```
1  def train_step(model, optimizer, x_train, y_train):
2      # 自動微分
3      with tf.GradientTape() as tape:
4          predictions = model(x_train, training=True)
5          loss = loss_object(y_train, predictions)
6      grads = tape.gradient(loss, model.trainable_variables)
7      optimizer.apply_gradients(
8          zip(grads, model.trainable_variables))
9
10     # 計算訓練的效能衡量指標
11     train_loss(loss)
12     train_accuracy(y_train, predictions)
13
14 def test_step(model, x_test, y_test):
15     # 預測
16     predictions = model(x_test)
17     # 計算損失
18     loss = loss_object(y_test, predictions)
19
20     # 計算測試的效能衡量指標
21     test_loss(loss)
22     test_accuracy(y_test, predictions)
```

3. 使用 tf.summary.create_file_writer 開啟 log 檔案。

```
10  train_summary_writer = tf.summary.create_file_writer(train_log_dir)
11  test_summary_writer = tf.summary.create_file_writer(test_log_dir)
```

4. 使用 tf.summary.scalar 寫入字串名稱(Name)及數值(Value)至 log 檔案。

```
6      with train_summary_writer.as_default():
7          tf.summary.scalar(
8              'loss', train_loss.result(), step=epoch)
9          tf.summary.scalar(
10             'accuracy', train_accuracy.result(), step=epoch)
```

5. 在終端機或 DOS 內執行以下指令，啟動 Tensorboard 網站：

tensorboard --logdir logs/gradient_tape

6. 也可以在 jupyter notebook 內啟動，分兩步驟：

先載入「TensorBoard notebook extension」擴充程式。

```
1  # 載入 TensorBoard notebook extension，
2  # 即可在 jupyter notebook 啟動 Tensorboard
3  %load_ext tensorboard
```

啟動 Tensorboard 網站。

```
1  # 啟動 Tensorboard
2  %tensorboard --logdir logs/gradient_tape
```

註：%為 jupyter notebook 的魔術方法前置符號，在終端機或 DOS 啟動，不需%。

7. 使用瀏覽器輸入以下網址，即可觀看訓練資訊：

http://localhost:6006/

5-5-3 寫入圖片

【範例 6】除了訓練過程的資訊，也可以隨時把資料寫入 Log，以下示範如何將圖片寫入工作日誌檔。

1. 設定工作日誌檔目錄，寫入圖像。

```
1   # 任意找一張圖片
2   img = x_train[0].numpy().reshape((-1, 28, 28, 1))
3   img.shape
4
5   # 指定 log 檔名
6   logdir = ".\\logs\\train_data\\" + datetime.datetime.now().strftime("%Y%m%d-%H%M%S")
7   # Creates a file writer for the log directory.
8   file_writer = tf.summary.create_file_writer(logdir)
9
10  # Using the file writer, log the reshaped image.
11  with file_writer.as_default():
12      # 將圖片寫入 log 檔
13      tf.summary.image("Training data", img, step=0)
```

2. 啟動 Tensorboard 網站：

   tensorboard --logdir logs/train_data

3. 使用瀏覽器觀看圖片(在 Images 頁籤)：

   http://localhost:6006/

4. 結果如下：

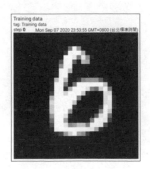

在後續章節介紹卷積神經網路(CNN)時，可以把卷積轉換後的圖片寫入，藉以瞭解訓練過程中圖片是如何被轉換的，有助於理解神經網路這個黑箱是如何辨識圖片的，這就是所謂的「可解釋的 AI」(Explainable AI, XAI)。

5-5-5 直方圖(Histogram)

要顯示訓練過程中的權重(weights)、偏差(bias)分佈，相當簡單，只要在定義 tensorboard callback 時加一個參數「histogram_freq=1」就完成了，意思是每一個執行週期繪製一張直方圖。

程式 05_04_Callback.ipynb。

```
1  # 定義 tensorboard callback
2  tensorboard_callback = [tf.keras.callbacks.TensorBoard(
3      log_dir='.\\logs', histogram_freq=1)]
```

啟動 Tensorboard 網站，點選「histograms」頁籤，顯示如下，前方的直方圖為訓練的最後結果。

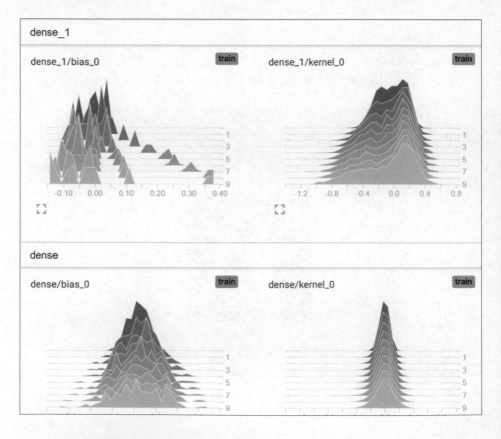

5-5-6　效能調校(Performance Tuning)

【範例 6】效能調校，找出最佳參數組合。

程式 05_08_TensorBoard_Tuning.ipynb。

1. TensorBoard 與 keras tuner 類似，首先設定多個調校的參數組合。

```
1  # 參數組合
2  from tensorboard.plugins.hparams import api as hp
3
4  HP_NUM_UNITS = hp.HParam('num_units', hp.Discrete([16, 32]))
5  HP_DROPOUT = hp.HParam('dropout', hp.RealInterval(0.1, 0.2))
6  HP_OPTIMIZER = hp.HParam('optimizer', hp.Discrete(['adam', 'sgd']))
```

2. 每一參數組合訓練一個模型。

```
1  # 依每一參數組合執行訓練
2  session_num = 0
3
4  for num_units in HP_NUM_UNITS.domain.values:
5      for dropout_rate in (
6          HP_DROPOUT.domain.min_value, HP_DROPOUT.domain.max_value):
7          for optimizer in HP_OPTIMIZER.domain.values:
8              hparams = {
9                      HP_NUM_UNITS: num_units,
10                     HP_DROPOUT: dropout_rate,
11                     HP_OPTIMIZER: optimizer,
12             }
13             run_name = "run-%d" % session_num
14             print('--- Starting trial: %s' % run_name)
15             print({h.name: hparams[h] for h in hparams})
16             run('logs/hparam_tuning/' + run_name, hparams)
17             session_num += 1
```

啟動 Tensorboard 網站，點選「hparams」頁籤，顯示如下，第 6 回合準確率最佳。

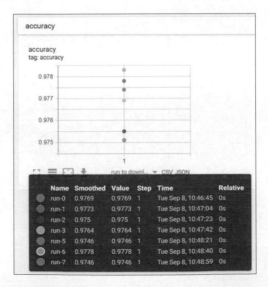

5-25

詳細資訊如下：

Trial ID	Show Metrics	dropout	num_units	optimizer	accuracy
3df0d7cf35bec5a...	☐	0.20000	32.000	sgd	0.97460
3ec2aed9e07589f...	☐	0.20000	32.000	adam	0.97780
53bf5bece9190fa...	☐	0.20000	16.000	adam	0.97500
5b97f3c2967245b...	☐	0.10000	16.000	adam	0.97690
6826c7fa3322d82...	☐	0.10000	32.000	adam	0.97390
7684dcc13358fd0...	☐	0.20000	16.000	sgd	0.97640
7b29a731e3daca7...	☐	0.10000	32.000	sgd	0.97460
ae235909ec4e4d9...	☐	0.10000	16.000	sgd	0.97730

從下圖的粗體線可以找到最佳參數：

- dropout rate=0.2
- 輸出神經元數(num_units)=32
- 優化器(optimizer)=adam
- 獲得最佳準確度 0.9775。

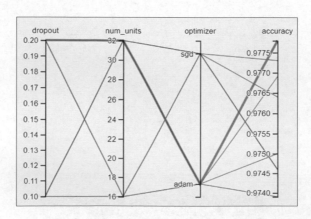

5-5-7　敏感度分析(What-If Tool, WIT)

敏感度分析能幫助我們更了解分類(classification)與迴歸(regression)模型，
它擁有許多超厲害的功能，包括：

1. 在下圖左邊的欄位中修改任一個觀察值,重新預測,即可觀察變動的影響。

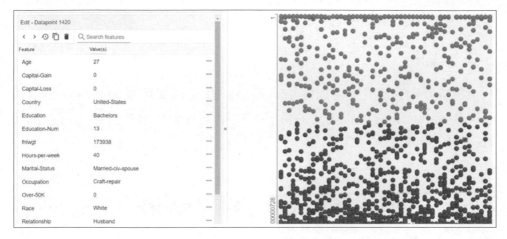

2. 點選「Partial dependence plots」,可以了解個別特徵對預測結果的影響。

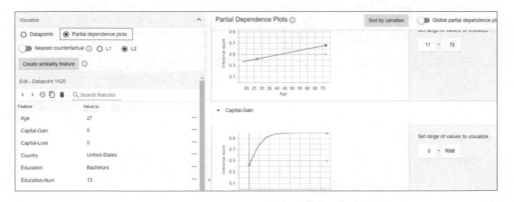

3. 切割訓練資料筆數:了解測試資料預測結果的敏感度分析。

Invoke What-If Tool for test data and the trained model

```
num_datapoints: 2000
```

```
tool_height_in_px: 1000
```

詳細操作說明可參考「A Walkthrough with UCI Census Data」[3]，「範例」[4]可在 Colaboratoy 環境中執行，這是一個二分類的模型，由於該範例不屬於深度學習模型，筆者就不多作說明了。

5-5-8 總結

TensorBoard 隨著時間增加的功能越來越多，都快可以另外寫成一本書了，以上我們只作了很簡單的實驗，如果需要更詳細的資訊，可以參閱 TensorBoard 官網的指南[5]。

5-6 模型佈署(Deploy)與 TensorFlow Serving

一般深度學習的模型安裝的選項如下：

1. 本地伺服器(Local Server)。
2. 雲端伺服器(Cloud Server)。
3. 邊緣運算(IoT Hub)：譬如要偵測全省的溫度，我們會在各縣市安裝上千個感測器，每個 IoT Hub 會負責多個感測器的信號接收、初步過濾和分析，分析完成後再將資料後送到資料中心。

呈現的方式可能是網頁、手機 App 或桌面程式，以下先就網頁開發作一說明。

5-6-1 自行開發網頁程式

若是自行開發網頁程式，並且安裝在本地伺服器的話，可以運用 Python 套件，例如 Django、Flask 或 Streamlit，快速建立網頁。其中以 Streamlit 最為簡單，不需要懂得 HTML/CSS/Javascript，只靠 Python 一招半式就可以搞定一個初階的網站，以下我們實際建立一個手寫阿拉伯數字的辨識網站。

1. 安裝 Streamlit 套件：

```
pip install streamlit
```

2. 執行 Python 程式，必須以 streamlit run 開頭，而非 python 執行，例如：

```
streamlit run 05_10_web.py
```

3. 網頁顯示後，拖曳 myDigits 目錄內的任一檔案至畫面中的上傳圖檔區域，就會顯示辨識結果，也可以使用小畫家等繪圖軟體書寫數字。

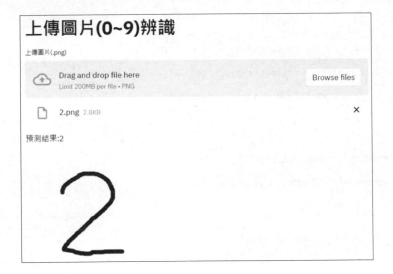

程式碼說明如下：

1. 載入相關套件。

```
1  # 載入套件
2  import streamlit as st
3  from skimage import io
4  from skimage.transform import resize
5  import numpy as np
6  import tensorflow as tf
```

2. 模型載入。

```
8  # 模型載入
9  model = tf.keras.models.load_model('./mnist_model.h5')
```

3. 上傳圖檔。

```
14  # 上傳圖檔
15  uploaded_file = st.file_uploader("上傳圖片(.png)", type="png")
```

檔案上傳後，執行下列工作：

- 第 18~22 行：把圖像縮小成寬高各為(28, 28)。
- 第 23 行：RGB 的白色為 255，但訓練資料 MNIST 的白色為 0，故需反轉顏色。
- 第 27 行：辨識上傳檔案。

```
16  if uploaded_file is not None:
17      # 讀取上傳圖檔
18      image1 = io.imread(uploaded_file, as_gray=True)
19      # 縮小圖形為(28, 28)
20      image_resized = resize(image1, (28, 28), anti_aliasing=True)
21      # 插入第一維，代表筆數
22      X1 = image_resized.reshape(1,28,28,1)
23      # 顏色反轉
24      X1 = np.abs(1-X1)
25
26      # 預測
27      predictions = model.predict_classes(X1)[0]
28      # 顯示預測結果
29      st.write(f'預測結果:{predictions}')
30      # 顯示上傳圖檔
31      st.image(image1)
```

完整程式請參閱 05_10_web.py。

5-6-2 TensorFlow Serving

另外也可以直接利用 TensorFlow Serving 架設網頁服務介面，它是一個高效能的服務系統，不需撰寫程式，就可提供 API，讓外界程式呼叫，支援 gPRC、REST API 兩種協定，依文件說明，目前只支援 Linux，因此，筆者以 Windows 內建的 Linux subsystem -- WSL 環境進行實驗。

先看一個簡單的例子，假使要佈署至雲端伺服器或 IoT Hub，考量到需要在短時間內佈署很多台，通常會選擇使用容器(Container)架構，建立虛擬機。我們以「TensorFlow Serving 官網」的案例[6]說明。

1. 安裝 Docker：有關 Docker 的安裝不在此說明，筆者以 Windows 內建的 WSL2 所整合的 Docker DeskTop 來說明整個程序。

2. 從 Docker Repository 下載一個 TensorFlow Serving 的 image：

```
docker pull tensorflow/serving
```

3. 使用 Git 指令複製 TensorFlow Serving 程式碼：

```
git clone https://github.com/tensorflow/serving
```

4. 指定模型：模型範例名稱為 half_plus_two，顧名思義，很簡單，就是將輸入除以 2，再加 2：

```
TESTDATA="$(pwd)/serving/tensorflow_serving/servables/tensorflow/te
stdata"
```

5. 啟動下載的 image：以下指令為同一行，提供 REST API：

```
docker run -t --rm -p 8501:8501 -v
"$TESTDATA/saved_model_half_plus_two_cpu:/models/half_plus_two" -e
MODEL_NAME=half_plus_two tensorflow/serving &
```

6. 測試資料預測：使用 curl 送出三筆資料分別為 1.0, 2.0, 5.0，進行預測，以下指令為同一行：

```
curl -d '{"instances": [1.0, 2.0, 5.0]}' -X POST
http://localhost:8501/v1/models/half_plus_two:predict
```

7. 傳回預測結果：{ "predictions": [2.5, 3.0, 4.5] }

8. 修改第 6 步驟的資料[10.0, 2.0, 5.0]，得到預測結果：

```
{ "predictions": [7.0, 3.0, 4.5] }
```

透過以上步驟，如法炮製，換上你的模型，程序是一樣的。

假設不使用 Docker，步驟如下：

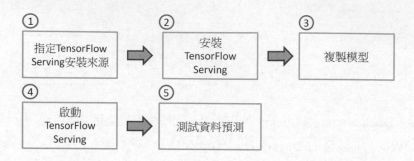

1. 首先指定 TensorFlow Serving 安裝來源，以下指令為同一行：

```
echo "deb [arch=amd64] http://storage.googleapis.com/tensorflow-
serving-apt stable tensorflow-model-server tensorflow-model-server-
universal" | sudo tee /etc/apt/sources.list.d/tensorflow-
serving.list && \
curl https://storage.googleapis.com/tensorflow-serving-
apt/tensorflow-serving.release.pub.gpg | sudo apt-key add -
```

2. 安裝 TensorFlow Serving：Ubuntu 作業系統使用 apt-get。

```
sudo apt-get update
sudo apt-get install tensorflow-model-server
```

3. 複製模型：將之前使用 SavedModel 存檔的目錄複製到/mnt/c/Users/
   mikec/，mikec 為筆者登入帳號的 Home 目錄，以下假設模型目錄為
   my_model。

4. 複製後需要在 my_model 下新增一個名為「1」的子目錄，代表版本
   別，將原有子目錄及檔案，搬移到「1」子目錄內，如下所示。

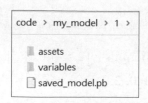

5. 啟動 TensorFlow Serving，模型名稱(model_name)可任意取名，以下取名為 MLP：

```
tensorflow_model_server --
model_base_path=/mnt/c/Users/mikec/my_model --model_name=MLP --
rest_api_port=8501
```

6. 測試資料預測：執行用戶端程式(05_11_tf_serving_client.py) 呼叫 TensorFlow Serving：

```
python 05_11_tf_serving_client.py
```

會傳回辨識的結果 4。

【範例 7】用戶端程式：05_11_tf_scrving_client.py：

```
1   import json
2   import numpy as np
3   import requests
4   from skimage import io
5   from skimage.transform import resize
6
7   uploaded_file='./myDigits/4.png'
8   image1 = io.imread(uploaded_file, as_gray=True)
9   # 縮小圖形為(28, 28)
10  image_resized = resize(image1, (28, 28), anti_aliasing=True)
11  # 插入第一維，代表筆數
12  X1 = image_resized.reshape(1,28,28,1)
13  # 顏色反轉
14  X1 = np.abs(1-X1)
15
16  # 將預測資料轉為 Json 格式
17  data = json.dumps({
18      "instances": X1.tolist()
19      })
20
21  # 呼叫 TensorFlow Serving API
22  headers = {"content-type": "application/json"}
23  json_response = requests.post(
24      'http://localhost:8501/v1/models/MLP:predict',
25      data=data, headers=headers)
26
27  # 解析預測結果
28  predictions = np.array(json.loads(json_response.text)['predictions'])
29  print(np.argmax(predictions, axis=-1))
```

程式説明：
- 第 7 行：指定要辨識的圖檔。

- 第 8~14 行：讀取圖檔，將像素轉為陣列。
- 第 17~19 行：將預測資料轉為 Json 格式。
- 第 21~25 行：呼叫 TensorFlow Serving API，送出圖形陣列。
- 第 28、29 行：接收 API 傳回的資料，解析預測結果。

由上述範例可以看出 Server 端完全不必撰寫程式，非常方便，詳細功能可詳閱「TensorFlow Serving 架構說明」[7]。

5-7 TensorFlow Dataset

TensorFlow Dataset 類似 Python Generator，可以視需要逐批讀取資料，不必一股腦把資料全部載入至記憶體，因為如果將龐大的資料量全部載入，記憶體可能就爆了。另外，它還有支援快取(Cache)、預取(Prefetch)、篩選(Filter)、轉換(Map)等功能，官網有許多範例都會使用到 Dataset，值得我們一探究竟。

5-7-1 產生 Dataset

建立 Dataset 有很多種方式：

1. from_tensor_slices()：自 List 或 NumPy ndarray 資料型態轉入。
2. from_tensors()：自 TensorFlow Tensor 資料型態轉入。
3. from_generator()：自 Python Generator 資料型態轉入。
4. TFRecordDataset()：自 TFRecord 資料型態轉入。
5. TextLineDataset()：自文字檔案轉入。

【範例 8】測試 Dataset 的相關操作。

程式：05_12_Dataset.ipynb。

1. 自 List 轉入 Dataset。

```
1  import tensorflow as tf
2
3  # 自 list 轉入
4  dataset = tf.data.Dataset.from_tensor_slices([8, 3, 0, 8, 2, 1])
```

2. 使用 for 迴圈即可自 Dataset 取出所有資料，必須以 numpy 函數轉換才能列印資料內容。

```
1  # 使用 for 迴圈可自 Dataset 取出所有資料
2  for elem in dataset:
3      print(elem.numpy())
```

3. 使用 iter()函數將 Dataset 轉成 Iterator，再使用 next 一次取一批資料。

```
1  # 轉成 iterator
2  it = iter(dataset)
3
4  # 一次取一筆
5  print(next(it).numpy())
6  print(next(it).numpy())
```

執行結果為前兩筆：8、3。

4. 依照維度小計(reduce)，如果資料維度是一維，即為總計。

```
1  # 依照維度小計(reduce)
2  import numpy as np
3
4  # 一維資料
5  ds = tf.data.Dataset.from_tensor_slices([1, 2, 3, 4, 5])
6
7  initial_state=0    # 起始值
8  print(ds.reduce(initial_state, lambda state, value: state + value).numpy())
```

執行結果：15。

5. 二維資料：按照行統計。

```
1  # 依照第一維度小計(reduce)
2  import numpy as np
3
4  # 二維資料
5  ds = tf.data.Dataset.from_tensor_slices(np.arange(1,11).reshape(2,5))
6
7  initial_state=0    # 起始值
8  print(ds.reduce(initial_state, lambda state, value: state + value).numpy())
```

執行結果：[7 9 11 13 15]。

6. 三維資料：依照第一維度小計(reduce)。

```
1  # 依照第一維度小計(reduce)
2  import numpy as np
3
4  # 三維資料
5  ds = tf.data.Dataset.from_tensor_slices(np.arange(1,13).reshape(2,2,3))
6
7  print('原始資料:\n', np.arange(1,13).reshape(2,2,3), '\n')
8
9  initial_state=0     # 起始值
10 print('計算結果:\n', ds.reduce(initial_state, lambda state, value: state + value).numpy())
```

執行結果：

```
原始資料:
 [[[ 1  2  3]
   [ 4  5  6]]

  [[ 7  8  9]
   [10 11 12]]]
計算結果:
 [[ 8 10 12]
  [14 16 18]]
```

7. map：以函數套用到 Dataset 內每個元素，下面程式碼將每個元素乘以
   2。

```
1  # 對每個元素應用函數(map)
2  import numpy as np
3
4  # 測試資料
5  ds = tf.data.Dataset.from_tensor_slices([1, 2, 3, 4, 5])
6
7  # 對每個元素應用函數(map)
8  ds = ds.map(lambda x: x * 2)
9
10 # 轉成 iterator，再顯示
11 print(list(ds.as_numpy_iterator()))
```

執行結果：[2, 4, 6, 8, 10]

8. 過濾(filter)：下面程式碼將偶數取出。

```
1  # 過濾(filter)
2  import numpy as np
3
4  # 測試資料
5  ds = tf.data.Dataset.from_tensor_slices([1, 2, 3, 4, 5])
6
7  # 對每個元素應用函數(map)
8  ds = ds.filter(lambda x: x % 2 == 0)
9
10 # 轉成 iterator，再顯示
11 print(list(ds.as_numpy_iterator()))
```

執行結果： [2, 4]

9. 複製(repeat)：有時候訓練資料過少，我們會希望複製訓練資料，來提高模型的準確度。

```
1  # 資料複製(repeat)
2  import numpy as np
3
4  # 測試資料
5  ds = tf.data.Dataset.from_tensor_slices([1, 2, 3, 4, 5])
6
7  # 重複 3 次
8  ds = ds.repeat(3)
9
10 # 轉成 iterator，再顯示
11 print(list(ds.as_numpy_iterator()))
```

執行結果： [1, 2, 3, 4, 5, 1, 2, 3, 4, 5, 1, 2, 3, 4, 5]

10. Dataset 分片(Shard)：將資料依固定間隔取樣，在分散式計算時，可利用此函數將資料分配給每一台工作站(Worker)進行運算。

```
1  # 分片(Shard)
2  import numpy as np
3
4  # 測試資料：0~10
5  ds = tf.data.Dataset.range(11)
6  print('原始資料:\n', list(ds.as_numpy_iterator()))
7
8  # 每 3 筆間隔取樣一筆，從第一筆開始
9  ds = ds.shard(num_shards=3, index=0)
10
11 # 轉成 iterator，再顯示
12 print('\n計算結果:\n', list(ds.as_numpy_iterator()))
```

執行結果：

```
原始資料：
 [0, 1, 2, 3, 4, 5, 6, 7, 8, 9, 10]

計算結果：
 [0, 3, 6, 9]
```

另外還有許多函數，比如 take、skip、unbatch、window、zip 等，請參閱 TensorFlow 官網關於 Dataset 的説明[8]。

11. 將 MNIST 資料轉入 Dataset。

```
 1  import tensorflow as tf
 2
 3  mnist = tf.keras.datasets.mnist
 4
 5  # 載入 MNIST 手寫阿拉伯數字資料
 6  (x_train, y_train),(x_test, y_test) = mnist.load_data()
 7
 8  # 特徵縮放，使用常態化(Normalization)，公式 = (x - min) / (max - min)
 9  x_train_norm, x_test_norm = x_train / 255.0, x_test / 255.0
10
11  # 轉為 Dataset，含 X/Y 資料
12  dataset = tf.data.Dataset.from_tensor_slices((x_train_norm, y_train))
13  print(dataset)
```

執行結果：會顯示資料型態及維度。

```
<TensorSliceDataset shapes: ((28, 28), ()), types: (tf.float64, tf.uint8)>
```

12. 逐批取得資料：

- shuffle(10000)：每次從 dataset 取出 10000 筆資料進行洗牌。
- batch(1000)：隨機抽出 1000 筆資料。

```
 1  # 每次隨機抽出 1000 筆
 2  # shuffle：每次從 60000 筆訓練資料取出 10000 筆洗牌，batch：隨機抽出 1000 筆
 3  train_dataset = dataset.shuffle(10000).batch(1000)
 4  i=0
 5  for (x_train, y_train) in train_dataset:
 6      if i == 0:
 7          print(x_train.shape)
 8          print(x_train[0])
 9
10      i+=1
11  print(i)
```

執行結果：顯示共 60 批資料，每批資料有 1000 筆。

13. 自隨機亂數產生 Dataset。

```
1  import tensorflow as tf
2
3  # 隨機亂數產生 Dataset
4  ds = tf.data.Dataset.from_tensor_slices(
5      tf.random.uniform([4, 10], minval=1, maxval=10, dtype=tf.int32))
6
7  # 轉成 iterator，再顯示
8  print(list(ds.as_numpy_iterator()))
```

執行結果：維度為(4, 10)，每個值介於(1, 10) 之間。

```
[array([1, 9, 1, 1, 6, 7, 5, 9, 8, 5]), array([3, 8, 1, 3, 9, 7, 1, 2, 3, 6]), array([4, 1, 5, 4, 1, 8, 5, 7, 7, 9]), array([1, 2, 7, 4, 4, 5, 2, 7, 3, 3])]
```

14. 從 TensorFlow Tensor 資料型態的變數轉入 Dataset。

```
1  import tensorflow as tf
2
3  # 稀疏矩陣
4  mat = tf.SparseTensor(indices=[[0, 0], [1, 2]], values=[1, 2],
5                        dense_shape=[3, 4])
6
7  # 轉入 Dataset
8  ds = tf.data.Dataset.from_tensors(mat)
9
10 # 使用迴圈自 Dataset 取出所有資料
11 for elem in ds:
12     print(tf.sparse.to_dense(elem).numpy())
```

執行結果：

```
[[1 0 0 0]
 [0 0 2 0]
 [0 0 0 0]]
```

5-7-2 圖像 Dataset

由於圖檔的尺寸通常都很大，不像 MNIST 的寬和高各只有(28, 28)，假使一次載入所有檔案至記憶體，恐怕會發生記憶體不足的狀況，因此，TensorFlow Dataset 針對影像和文字有做特殊處理，可以分批載入記憶體，同時提供資料增補(Data Augmentation)的功能，能夠在既有的圖像進行影像處理，產生更多的訓練資料，這些功能全都整合至 Dataset，可在訓練(fit)指令中指定資料來源為 Dataset，一氣呵成。

【範例 9】自 Python Generator 資料型態的變數轉入 Dataset，例如從網路取得壓縮檔，解壓縮後，進行資料增補(Data Augmentation)，作為訓練資料。資料增補是提高圖形辨識度非常有效的方法，利用影像處理的技巧，比如放大、縮小、偏移、旋轉、裁切等方式，產生各式的訓練資料，讓訓練資料更加多樣化，進而使模型有更高的辨識能力。

1. 從網路取得壓縮檔，解壓縮，並進行資料增補。

```
1  # 從網路取得壓縮檔，並解壓縮
2  flowers = tf.keras.utils.get_file(
3      'flower_photos',
4      'https://storage.googleapis.com/download.tensorflow.org/example_images/flower_photos.tgz',
5      untar=True)
6
7  # 定義參數
8  BATCH_SIZE = 32 # 批量
9  IMG_DIM = 224    # 影像寬度
10 NB_CLASSES = 5  # Label 類別數
11
12 # 資料增補，rescale：特徵縮放，rotation_range：自動增補旋轉20度內的圖片
13 img_gen = tf.keras.preprocessing.image.ImageDataGenerator(rescale=1./255, rotation_range=20)
```

2. 試取一批檔案，並顯示第一筆影像。

```
1  # 取一批檔案
2  images, labels = next(img_gen.flow_from_directory(flowers))
3
4  # 顯示第一筆影像
5  import matplotlib.pyplot as plt
6  plt.imshow(images[0])
7  plt.axis('off')
8  print('labels:', labels[0])
```

執行結果：

3. 定義 generator 的屬性，包括取出資料的邏輯，再將之轉為 Dataset。

```
1  # 定義 generator 的屬性：取出資料的邏輯
2  gen = img_gen.flow_from_directory(
3      flowers,
4      (IMG_DIM, IMG_DIM),
5      'rgb',
6      class_mode='categorical',
7      batch_size=BATCH_SIZE,
8      shuffle=False
9  )
10
11 # 轉入 Dataset
12 ds = tf.data.Dataset.from_generator(lambda: gen,
13     output_signature=(
14         tf.TensorSpec(shape=(BATCH_SIZE, IMG_DIM, IMG_DIM, 3)),
15         tf.TensorSpec(shape=(BATCH_SIZE, NB_CLASSES))
16                     )
17     )
```

4. 試取下一批資料。

```
1  # 取下一批資料
2  it = iter(ds)
3  images, label = next(it)
4  print(np.array(images).shape, np.array(label).shape)
```

執行結果：取出的影像及標記維度分別為 (32, 224, 224, 3) 、(32, 5)。

可在訓練指令中指定資料來源為產生的 Dataset，在下一章會有完整範例說明。

5-7-3 TFRecord 與 Dataset

TFRecord 是由 TensorFlow 團隊所開發，遵循 Google Protocol Buffer，希望實現跨平台、跨語言的資料結構(record-oriented binary format)，每筆記錄能夠儲存各種資料型態的欄位，類似 Json 格式，可序列化(serialization)為二進位的格式儲存。

在操作 TFRecord 時，需要藉由 tf.train.Example 將資料封裝成 protocol message，基本上 tf.train.Example 的格式為{"string": tf.train.Feature}，而 tf.train.Feature 可接受 BytesList、FloatList、Int64List 三種格式。BytesList 用於字串或二進位的資料，例如圖像、語音等。

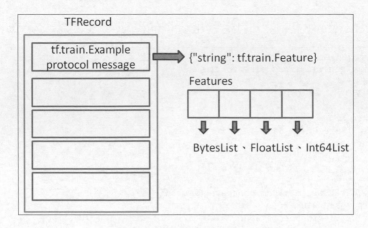

圖 5.2 TFRecord 結構

【範例 13】測試 TFRecord 相關操作。

程式：05_13_TFRecord.ipynb。

1. 定義 tf.train.Feature 轉換函數。

```
1  # 下列函數可轉換為 tf.train.Example 的 tf.train.Feature
2  def _bytes_feature(value):
3      """Returns a bytes_list from a string / byte."""
4      if isinstance(value, type(tf.constant(0))):
5          value = value.numpy()
6      return tf.train.Feature(bytes_list=tf.train.BytesList(value=[value]))
7
8  def _float_feature(value):
9      """Returns a float_list from a float / double."""
10     return tf.train.Feature(float_list=tf.train.FloatList(value=[value]))
11
12 def _int64_feature(value):
13     """Returns an int64_list from a bool / enum / int / uint."""
14     return tf.train.Feature(int64_list=tf.train.Int64List(value=[value]))
```

2. 簡單測試。

```
1  print(_bytes_feature(b'test_string'))
2  print(_bytes_feature(u'test_bytes'.encode('utf-8')))
3
4  print(_float_feature(np.exp(1)))
5
6  print(_int64_feature(True))
7  print(_int64_feature(1))
```

執行結果：

```
bytes_list {
  value: "test_string"
}

bytes_list {
  value: "test_bytes"
}

float_list {
  value: 2.7182817
}

int64_list {
  value: 1
}

int64_list {
  value: 1
}
```

3. 序列化(serialization)測試。

```
1  # 序列化(serialization)
2  feature = _float_feature(np.exp(1))
3  feature.SerializeToString()
```

執行結果：b'\x12\x06\n\x04T\xf8-@'，為二進位的格式，有經過壓縮。

4. 建立 tf.train.Example 訊息，含有 4 個 feature：0~3。

```
1   # 建立tf.train.Example訊息，含 4 個 feature
2
3   # The number of observations in the dataset.
4   n_observations = int(1e4)
5
6   # Boolean feature, encoded as False or True.
7   feature0 = np.random.choice([False, True], n_observations)
8
9   # Integer feature, random from 0 to 4.
10  feature1 = np.random.randint(0, 5, n_observations)
11
12  # String feature
13  strings = np.array([b'cat', b'dog', b'chicken', b'horse', b'goat'])
14  feature2 = strings[feature1]
15
16  # Float feature, from a standard normal distribution
17  feature3 = np.random.randn(n_observations)
```

5. 接下來要寫入 TFRecord 檔案，先定義 tf.train.Example 資料序列化函數。

```
1   # 序列化(serialization)
2   def serialize(feature0, feature1, feature2, feature3):
3       """
4       Creates a tf.train.Example message ready to be written to a file.
5       """
6       # Create a dictionary mapping the feature name to the tf.train.Example-compatible
7       # data type.
8       feature = {
9               'feature0': _int64_feature(feature0),
10              'feature1': _int64_feature(feature1),
11              'feature2': _bytes_feature(feature2),
12              'feature3': _float_feature(feature3),
13      }
14
15      # Create a Features message using tf.train.Example.
16
17      example_proto = tf.train.Example(features=tf.train.Features(feature=feature))
18      return example_proto.SerializeToString()
```

6. 將一筆記錄寫入 TFRecord 檔案。

```
1   # 將一筆記錄寫入 TFRecord 檔案
2   with tf.io.TFRecordWriter("test.tfrecords") as writer:
3       writer.write(serialized_example)
```

7. 讀取 TFRecord 檔案。

```
1   # 開啟 TFRecord 檔案
2   filenames = ["test.tfrecords"]
3   raw_dataset = tf.data.TFRecordDataset(filenames)
4
5   ## 取得序列化的資料
6   for raw_record in raw_dataset.take(10):
7       print(repr(raw_record))
```

執行結果：為二進位的格式。

```
<tf.Tensor: shape=(), dtype=string, numpy=b'\nR\n\x11\n\x08feature0\x12\x05\x1a\x03\n\x01\x00\n\x11\n\x08feature1\x12\x05\x1a\x03\n\x01\x04\n\x14\n\x08feature2\x12\x08\n\x06\n\x04goat\n\x14\n\x08feature3\x12\x08\x12\x06\n\x04[\xd3|?'>
```

8. 若要取得原始資料，先反序列化(Deserialize)，設定原始資料的欄位屬性，透過 parse_single_example()來進行反序列化。

```
1   # 設定原始資料的欄位屬性
2   feature_description = {
3           'feature0': tf.io.FixedLenFeature([], tf.int64, default_value=0),
4           'feature1': tf.io.FixedLenFeature([], tf.int64, default_value=0),
5           'feature2': tf.io.FixedLenFeature([], tf.string, default_value=''),
6           'feature3': tf.io.FixedLenFeature([], tf.float32, default_value=0.0),
7   }
8
9   # 將 tf.train.Example 訊息轉為 字典(dictionary)
10  def _parse_function(example_proto):
11      return tf.io.parse_single_example(example_proto, feature_description)
```

9. 取得每一個欄位值。

```
1   # 反序列化(Deserialize)
2   parsed_dataset = raw_dataset.map(_parse_function)
3
4   # 取得每一個欄位值
5   for parsed_record in parsed_dataset.take(10):
6       print(repr(parsed_record))
```

執行結果：

```
{'feature0': <tf.Tensor: shape=(), dtype=int64, numpy=0>, 'feature1': <tf.Tensor: shape=(), dtype=int64, numpy=4>, 'feature2':
<tf.Tensor: shape=(), dtype=string, numpy=b'goat'>, 'feature3': <tf.Tensor: shape=(), dtype=float32, numpy=0.9876>}
```

10. 從網路上取得官網的 TFRecord 檔案。

```
1   # 從網路上取的官網 TFRecord 檔案
2   file_path = "https://storage.googleapis.com/download.tensorflow.org/" +
3               "data/fsns-20160927/testdata/fsns-00000-of-00001"
4   fsns_test_file = tf.keras.utils.get_file("fsns.tfrec", file_path)
5
6   # 顯示存檔位置
7   fsns_test_file
```

執行結果：顯示預設存檔位置，其中 mikec 為使用者目錄：

C:\\Users\\mikec\\.keras\\datasets\\fsns.tfrec

11. 讀取 TFRecord 檔案。

```
1   # 讀取 TFRecord 檔案
2   dataset = tf.data.TFRecordDataset(filenames = [fsns_test_file])
3
4   # 取得下一筆資料
5   raw_example = next(iter(dataset))
6   parsed = tf.train.Example.FromString(raw_example.numpy())
7   parsed.features.feature['image/text']
```

執行結果,該欄位為一字串:

```
bytes_list {
  value: "Rue Perreyon"
}
```

5-7-4 TextLineDataset

文字檔也可以像二進位檔案一樣儲存在 Dataset 內,並且序列化後存檔。
同樣來示範個簡單的例子。

【範例 14】測試 TextLineDataset 相關操作。

程式:**05_14_TextLineDataset.ipynb**。

1. 讀取三個語料庫檔案,合併為一 TextLineDataset。

```
1   # 讀取三個檔案
2   directory_url = 'https://storage.googleapis.com/download.tensorflow.org/data/illiad/'
3   file_names = ['cowper.txt', 'derby.txt', 'butler.txt']
4
5   file_paths = [
6       tf.keras.utils.get_file(file_name, directory_url + file_name)
7       for file_name in file_names
8   ]
9
10  # 合併為一資料集
11  ds = tf.data.TextLineDataset(file_paths)
```

2. 讀取 5 筆資料。

```
1   # 讀取5筆資料
2   for line in ds.take(5):
3       print(line.numpy())
```

執行結果:

```
b"\xef\xbb\xbfAchilles sing, O Goddess! Peleus' son;"
b'His wrath pernicious, who ten thousand woes'
b"Caused to Achaia's host, sent many a soul"
b'Illustrious into Ades premature,'
b'And Heroes gave (so stood the will of Jove)'
```

3. 輪流 (interleave):每個檔案讀 3 筆資料即換下一個檔案讀取
   (cycle_length=3)。

```
1   # interleave：每個檔案輪流讀取一次
2   files_ds = tf.data.Dataset.from_tensor_slices(file_paths)
3   lines_ds = files_ds.interleave(tf.data.TextLineDataset, cycle_length=3)
4
5   # 各讀 3 筆，共 9 筆
6   for i, line in enumerate(lines_ds.take(9)):
7       if i % 3 == 0:
8           print()
9       print(line.numpy())
```

執行結果：

```
b"\xef\xbb\xbfAchilles sing, O Goddess! Peleus' son;"
b"\xef\xbb\xbfOf Peleus' son, Achilles, sing, O Muse,"
b'\xef\xbb\xbfSing, O goddess, the anger of Achilles son of Peleus, that brought'

b'His wrath pernicious, who ten thousand woes'
b'The vengeance, deep and deadly; whence to Greece'
b'countless ills upon the Achaeans. Many a brave soul did it send'

b"Caused to Achaia's host, sent many a soul"
b'Unnumbered ills arose; which many a soul'
b'hurrying down to Hades, and many a hero did it yield a prey to dogs and'
```

【範例 15】TextLineDataset 結合篩選(filter)函數。

1. 讀取鐵達尼文字檔案(.csv)，匯入至 TextLineDataset。

```
1   # 讀取鐵達尼文字檔案(.csv)，匯入至TextLineDataset
2   file_path = "https://storage.googleapis.com/tf-datasets/titanic/train.csv"
3   titanic_file = tf.keras.utils.get_file("train.csv", file_path)
4   titanic_lines = tf.data.TextLineDataset(titanic_file)
```

2. 篩選生存者的資料。

```
1   # 篩選生存者的資料
2   def survived(line):
3       return tf.not_equal(tf.strings.substr(line, 0, 1), "0")
4
5   # 篩選
6   survivors = titanic_lines.skip(1).filter(survived)
7
8   # 讀取10筆資料
9   for line in survivors.take(10):
10      print(line.numpy())
```

執行結果：

```
b'1,female,38.0,1,0,71.2833,First,C,Cherbourg,n'
b'1,female,26.0,0,0,7.925,Third,unknown,Southampton,y'
b'1,female,35.0,1,0,53.1,First,C,Southampton,n'
b'1,female,27.0,0,2,11.1333,Third,unknown,Southampton,n'
b'1,female,14.0,1,0,30.0708,Second,unknown,Cherbourg,n'
b'1,female,4.0,1,1,16.7,Third,G,Southampton,n'
b'1,male,28.0,0,0,13.0,Second,unknown,Southampton,y'
b'1,female,28.0,0,0,7.225,Third,unknown,Cherbourg,y'
b'1,male,28.0,0,0,35.5,First,A,Southampton,y'
b'1,female,38.0,1,5,31.3875,Third,unknown,Southampton,n'
```

【範例 16】TextLineDataset 結合 DataFrame。

1. 讀取鐵達尼文字檔案(.csv)。

```
1  import pandas as pd
2
3  df = pd.read_csv(titanic_file, index_col=None)
4  df.head()
```

2. 匯入 Dataset，讀取 1 筆資料。

```
1  # 匯入 Dataset
2  ds = tf.data.Dataset.from_tensor_slices(dict(df))
3
4  # 讀取1筆資料
5  for feature_batch in ds.take(1):
6      for key, value in feature_batch.items():
7          print(f"{key:20s}: {value}")
```

執行結果：

```
survived            : 0
sex                 : b'male'
age                 : 22.0
n_siblings_spouses  : 1
parch               : 0
fare                : 7.25
class               : b'Third'
deck                : b'unknown'
embark_town         : b'Southampton'
alone               : b'n'
```

5-7-5 Dataset 效能提升

使用 Dataset 時，可利用「預先讀取」(Prefetch)、快取(Cache)等指令，來提升資料讀取的效能。下面用時間軸的方式來展示 Prefetch 和 Cache 的用途。

1. prefetch：在訓練時只利用到 CPU/RAM，同時 TensorFlow 利用空檔先
   讀取下一批資料，並作轉換。

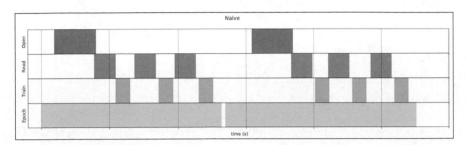

圖 5.3 不使用 prefetch 的話，開啟 Dataset、讀取資料、訓練這三個動作會依序進
行，拉長執行時間

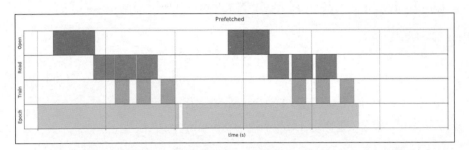

圖 5.4 使用 prefetch 的話，則會在訓練時，同時讀取下一批資料，故讀取資料和訓
練一起同步進行

2. cache：可將讀出的資料留在快取記憶體裡，之後可再重複使用。

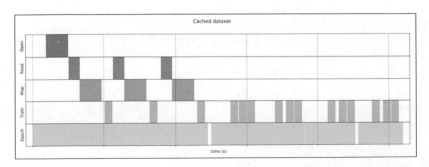

圖 5.5 使用 cache，能夠降低開啟 Dataset 讀取資料的次數，減少硬碟 IO

詳細的情形可參考「官網 Dataset 效能説明」[9]。

參考資料 (References)

[1] Keras 官網「Model saving & serialization API」
(https://keras.io/api/models/model_saving_apis/)

[2] Keras 官網「Callbacks API」
(https://keras.io/api/callbacks/)

[3] 「A Walkthrough with UCI Census Data」
(https://pair-code.github.io/what-if-tool/learn/tutorials/walkthrough/)

[4] 參考範例
(https://colab.research.google.com/github/pair-code/what-if-tool/blob/master/What_If_Tool_Notebook_Usage.ipynb)

[5] TensorFlow 官網的 TensorBoard 指南
(https://www.tensorflow.org/tensorboard/get_started)

[6] TensorFlow 官網中的「TensorFlow Serving with Docker」案例
(https://www.tensorflow.org/tfx/serving/docker)

[7] TensorFlow 官網中的 TensorFlow Serving 架構說明
(https://www.tensorflow.org/tfx/serving/architecture)

[8] TensorFlow 官網中關於 Dataset 的說明
(https://www.tensorflow.org/api_docs/python/tf/data/Dataset)

[9] TensorFlow 官網中關於 Dataset 效能的說明
(https://www.tensorflow.org/guide/data_performance)

卷積神經網路
(Convolutional Neural Network)

第三波人工智慧浪潮在自然使用者介面(Natural User Interface, NUI)有突破性的進展,包括影像(Image、Video)、語音(Voice)與文字(Text)的辨識/生成/分析,機器學會利用人類日常生活中所使用的溝通方式,與使用者的互動不僅更具親和力,也能對週遭的環境作出更合理、更有智慧的判斷與反應。將這種能力附加到產品上,可使產品的應用發展爆發無限可能,包括自駕車(Self-Driving)、無人機(Drone)、智慧家庭(Smart Home)、製造/服務機器人(Robot)、聊天機器人(ChatBot)等,不勝枚舉。

從這一章開始,我們逐一來探討影像(Image、Video)、語音(Voice)、文字(Text)的相關演算法。

▌6-1 卷積神經網路簡介

之前我們只用了十幾行程式即可辨識阿拉伯數字,令人相當興奮,但是,模型使用像素(Pixel)為特徵輸入,好像與人類辨識圖形的方式並不一致,我們應該不會逐點辨識圖形的內涵:

1. 手寫阿拉伯數字,通常都會集中在中央,故在中央像素的重要性應遠大於周邊的像素。
2. 像素之間應有所關聯,而非互相獨立,比如 1,為一垂直線。
3. 人類視覺應該不是逐個像素辨識,而是觀察數字的線條或輪廓。

因此，卷積神經網路(Convolutional Neural Network, CNN) 引進了卷積層
(Convolution Layer)，進行「特徵萃取」(Feature Extraction)，將像素轉換
為各種線條特徵，再交給 Dense 層辨識，這就是圖 1.7 機器學習流程的第
3 步驟 -- 特徵工程(Feature Engineering)。

卷積(Convolution)簡單說就是將圖形逐步抽樣化(Abstraction)，把不必要的
資訊刪除，例如色彩、背景等，下圖經過三層卷積後，有些圖依稀可辨識
出人臉的輪廓了，因此，模型就依據這些線條辨識出是人、車或其他動
物。

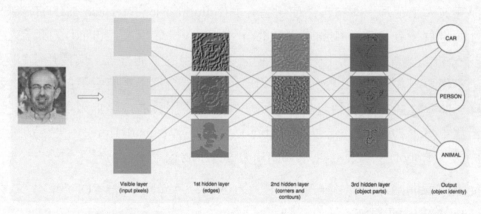

圖 6.1 卷積神經網路(Convolutional Neural Network, CNN) 的特徵萃取

卷積神經網路(Convolutional Neural Network)，以下簡稱 CNN，它的模型
結構如下：

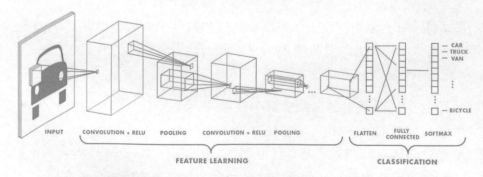

圖 6.2 卷積神經網路(Convolutional Neural Network, CNN) 的模型結構

1. 先輸入一張圖像，可以是彩色的，每個色彩通道(Channel)分別卷積再合併。

2. 圖像經過卷積層(Convolution Layer)運算，變成特徵圖(Feature Map)，卷積可以指定很多個，卷積矩陣不是固定的，而是由反向傳導推估出來的，與傳統的影像處理不同，卷積層後面通常會附加 ReLU Activation Function。

3. 卷積層後面會接一個池化層(Pooling)，作下採樣(Down Sampling)，以降低模型的參數個數，避免模型過於龐大。

4. 最後把特徵圖(Feature Map)壓扁成一維(Flatten)，交給 Dense 層辨識。

6-2 卷積(Convolution)

卷積是定義一個濾波器(Filter)或稱卷積核(Kernel)，對圖像進行「乘積和」運算，例如下圖所示，計算步驟如下：

1. 將輸入圖像依照濾波器裁切相同尺寸的部份圖像。
2. 裁切的圖像與濾波器相同的位置進行相乘。
3. 加總所有格的數值，即為輸出的第一格數值。
4. 逐步向右滑動視窗(如圖 6.4)，回到步驟 1，計算下一格的值。
5. 滑到最右邊後，再往下滑動視窗，繼續進行。

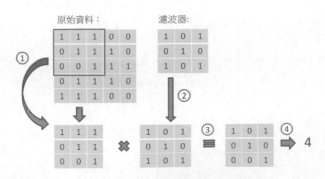

圖 6.3 卷積計算(1)

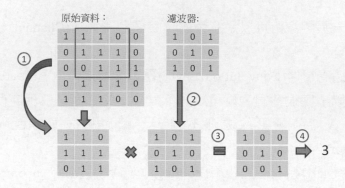

圖 6.4 卷積計算(2)

網路上有許多動畫或影片可以參考，例如「Convolutional Neural Networks—Simplified」[1]文中卷積計算的 GIF 動圖[2]。

【範例 1】使用程式計算卷積。

程式：**06_01_convolutions.ipynb**。

1. 準備資料及濾波器(Filter)。

```python
1  import numpy as np
2
3  # 測試資料
4  source_map = np.array(list('1110001110001110011001100')).astype(np.int)
5  source_map = source_map.reshape(5,5)
6  print('原始資料：')
7  print(source_map)
8
9  # 濾波器(Filter)
10 filter1 = np.array(list('101010101')).astype(np.int).reshape(3,3)
11 print('\n濾波器:')
12 print(filter1)
```

執行結果：

```
原始資料：
[[1 1 1 0 0]
 [0 1 1 1 0]
 [0 0 1 1 1]
 [0 0 1 1 0]
 [0 1 1 0 0]]

濾波器:
[[1 0 1]
 [0 1 0]
 [1 0 1]]
```

2. 計算卷積。

```
1  # 計算卷積
2  # 初始化計算結果的矩陣
3  width = height = source_map.shape[0] - filter1.shape[0] + 1
4  result = np.zeros((width, height))
5
6  # 計算每一格
7  for i in range(width):
8      for j in range(height):
9          value1 =source_map[i:i+filter1.shape[0], j:j+filter1.shape[1]] * filter1
10         result[i, j] = np.sum(value1)
11 print(result)
```

執行結果：

```
[4. 3. 4.]
[2. 4. 3.]
[2. 3. 4.]
```

3. 使用 SciPy 套件提供的卷積函數驗算，執行結果一致。

```
1  # 使用 scipy 計算卷積
2  import scipy
3
4  # convolve2d：二維卷積
5  scipy.signal.convolve2d(source_map, filter1, mode='valid')
```

卷積計算時，其實還有兩個參數：

1. 補零(Padding)：上面的卷積計算會使得圖像尺寸變小，因為，滑動視窗時，裁切的視窗會不足 2 個，即濾波器寬度減 1，因此 Padding 有兩個選項：

   - Padding='same'：在圖像周遭補上不足的列與行，使計算結果的矩陣尺寸不變(same)，與原始圖像尺寸相同。
   - Padding='valid'：不補零，計算後圖像尺寸變小。

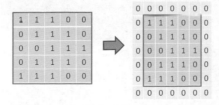

圖 6.5 Padding='same'，在圖像周遭補上不足的列與行

2. 滑動視窗的步數(Stride)：圖 6.4 是 Stride=1，圖 6.6 是 Stride=2，可減少要估算的參數個數。

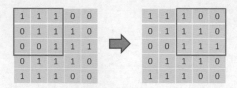

圖 6.6 Stride=2，一次滑動 2 格視窗

以上是二維的卷積(Conv2D)的運作，通常應用在圖像上。TensorFlow 還提供 Conv1D、Conv3D，其中 Conv1D 因只考慮上下文(Context Sensitive)，所以可應用於語音或文字方面，Conv3D 則可應用於立體的物件。還有 Conv2DTranspose 提供反卷積(Deconvolution)或稱上採樣(Up Sampling)的功能，反向由特徵圖重建圖像。卷積和反卷積兩者相結合，還可以組合成 AutoEncoder 模型，它是許多生成模型的基礎演算法，可以去除雜訊，生成乾淨的圖像。

6-3 各式卷積

雖然 CNN 會自動配置卷積的種類，不過我們還是來看看各式卷積的影像處理效果，進而加深大家對 CNN 的理解。

1. 首先定義一個卷積的影像轉換函數，如下：

```
1   # 卷積的影像轉換函數，padding='same'
2   from skimage.exposure import rescale_intensity
3
4   def convolve(image, kernel):
5       # 取得圖像與濾波器的寬高
6       (iH, iW) = image.shape[:2]
7       (kH, kW) = kernel.shape[:2]
8
9       # 計算 padding='same' 單邊所需的補零行數
10      pad = int((kW - 1) / 2)
11      image = cv2.copyMakeBorder(image, pad, pad, pad, pad, cv2.BORDER_REPLICATE)
12      output = np.zeros((iH, iW), dtype="float32")
13
14      # 卷積
```

```
15      for y in np.arange(pad, iH + pad):
16          for x in np.arange(pad, iW + pad):
17              roi = image[y - pad:y + pad + 1, x - pad:x + pad + 1]    # 裁切圖像
18              k = (roi * kernel).sum()                                 # 卷積計算
19              output[y - pad, x - pad] = k                            # 更新計算結果的矩陣
20
21      # 調整影像色彩深淺範圍至 (0, 255)
22      output = rescale_intensity(output, in_range=(0, 255))
23      output = (output * 255).astype("uint8")
24
25      return output         # 回傳結果影像
```

2. 需要安裝 Python OpenCV 套件，OpenCV 是一個影像處理的套件。

   pip install opencv-python

3. 將影像灰階化：skimage 全名為 scikit-image，也是一個影像處理的套件，功能較 OpenCV 簡易。

```
1  # pip install opencv-python
2  import skimage
3  import cv2
4
5  # 自 skimage 取得內建的圖像
6  image = skimage.data.chelsea()
7  cv2.imshow("original", image)
8
9  # 灰階化
10 gray = cv2.cvtColor(image, cv2.COLOR_BGR2GRAY)
11 cv2.imshow("gray", gray)
12
13 # 按 Enter 關閉視窗
14 cv2.waitKey(0)
15 cv2.destroyAllWindows()
```

執行結果：

原圖：   灰階化：

2. 模糊化(Blur)：濾波器設定為周圍點的平均，就可以讓圖像模糊化，一般用於消除紅眼現象或是雜訊。

```
1  # 小模糊 filter
2  smallBlur = np.ones((7, 7), dtype="float") * (1.0 / (7 * 7))
3
4  # 卷積
5  convoleOutput = convolve(gray, smallBlur)
6  opencvOutput = cv2.filter2D(gray, -1, smallBlur)
7  cv2.imshow("little Blur", convoleOutput)
8
9  # 大模糊
10 largeBlur = np.ones((21, 21), dtype="float") * (1.0 / (21 * 21))
11
12 # 卷積
13 convoleOutput = convolve(gray, largeBlur)
14 opencvOutput = cv2.filter2D(gray, -1, largeBlur)
15 cv2.imshow("large Blur", convoleOutput)
16
17 # 按 Enter 關閉視窗
18 cv2.waitKey(0)
19 cv2.destroyAllWindows()
```

小模糊：7x7 矩陣。   大模糊：21x21 矩陣，矩陣越大，影像越模糊。

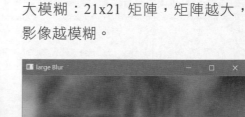

3. 銳化(sharpen)：可使圖像的對比更加明顯。

```
1  # sharpening filter
2  sharpen = np.array((
3      [0, -1, 0],
4      [-1, 5, -1],
5      [0, -1, 0]), dtype="int")
6
7  # 卷積
```

```
 8  convoleOutput = convolve(gray, sharpen)
 9  opencvOutput = cv2.filter2D(gray, -1, sharpen)
10  cv2.imshow("sharpen", convoleOutput)
11
12  # 按 Enter 關閉視窗
13  cv2.waitKey(0)
14  cv2.destroyAllWindows()
```

執行結果：卷積凸顯中間點，使圖像特徵越明顯。

4. Laplacian 邊緣偵測：可偵測圖像的輪廓。

```
 1  # Laplacian filter
 2  laplacian = np.array((
 3      [0, 1, 0],
 4      [1, -4, 1],
 5      [0, 1, 0]), dtype="int")
 6
 7  # 卷積
 8  convoleOutput = convolve(gray, laplacian)
 9  opencvOutput = cv2.filter2D(gray, -1, laplacian)
10  cv2.imshow("laplacian edge detection", convoleOutput)
11
12  # 按 Enter 關閉視窗
13  cv2.waitKey(0)
14  cv2.destroyAllWindows()
```

執行結果：卷積凸顯週邊，顯現圖像週邊線條。

5. Sobel X 軸邊緣偵測：沿著 X 軸偵測邊緣，故可偵測垂直線特徵。

```
1  # Sobel x-axis filter
2  sobelX = np.array((
3      [-1, 0, 1],
4      [-2, 0, 2],
5      [-1, 0, 1]), dtype="int")
6
7  # 卷積
8  convoleOutput = convolve(gray, sobelX)
9  opencvOutput = cv2.filter2D(gray, -1, sobelX)
10 cv2.imshow("x-axis edge detection", convoleOutput)
11
12 # 按 Enter 關閉視窗
13 cv2.waitKey(0)
14 cv2.destroyAllWindows()
```

執行結果：卷積行由小至大，顯現圖像垂直線條。

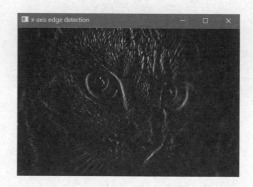

6. Sobel Y 軸邊緣偵測：沿著 Y 軸偵測邊緣，故可偵測水平線特徵。

```
1  # Sobel y-axis filter
2  sobelY = np.array((
3      [-1, -2, -1],
4      [0, 0, 0],
5      [1, 2, 1]), dtype="int")
6
7  # 卷積
8  convoleOutput = convolve(gray, sobelY)
9  opencvOutput = cv2.filter2D(gray, -1, sobelY)
10 cv2.imshow("y-axis edge detection", convoleOutput)
11
12 # 按 Enter 關閉視窗
13 cv2.waitKey(0)
14 cv2.destroyAllWindows()
```

執行結果：卷積列由小至大，顯現圖像水平線條。

6-4 池化層(Pooling Layer)

通常我們會設定每個卷積層的濾波器個數為 4 的倍數，因此總輸出等於(筆數 x W_out x H_out x 濾波器個數)，會使輸出尺寸變得很大，因此，我們必須透過池化層(Pooling Layer)進行採樣(Down Sampling)，只取滑動視窗的最大值或平均值，換句話說，就是將每個滑動視窗轉化為一個點，就能有效降低每一層輸入的尺寸了，同時也能保有每個視窗的特徵。我們來舉個例子說明會比較清楚。

以最大池化層(Max Pooling)為例：

1. 下圖左邊為原始圖像。
2. 假設濾波器尺寸為(2, 2)、Stride = 2。
3. 滑動視窗取(2, 2)，如下圖左上角的框，取最大值=6。
4. 接著再滑動 2 步，如圖 6.8，取最大值=8。

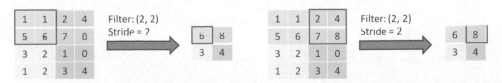

圖 6.7 最大池化層(Max Pooling)　　　　　圖 6.8 最大池化層 -- 滑動 2 步

6-5 CNN 模型實作

一般卷積會採用 3x3 或 5x5 的濾波器，尺寸越大，可以萃取越大的特徵，但相對的，較小的特徵就容易被忽略。而池化層通常會採用 2x2，stride=2 的濾波器，使用越大的尺寸，會使得參數個數減少很多，但萃取到的特徵也相對減少。

以下就先以 CNN 模型實作 MNIST 辨識。

【範例 1】將手寫阿拉伯數字辨識的模型改用 CNN。

程式：06_02_MNIST_CNN.ipynb。

1. 載入 MNIST 手寫阿拉伯數字資料，完全不需改變。

```
1  import tensorflow as tf
2  mnist = tf.keras.datasets.mnist
3
4  # 載入 MNIST 手寫阿拉伯數字資料
5  (x_train, y_train),(x_test, y_test) = mnist.load_data()
6
7
8  ## 步驟2：資料清理，此步驟無需進行
9
10 ## 步驟3：進行特徵工程，將特徵縮放成(0, 1)之間
11
12 # 特徵縮放，使用常態化(Normalization)，公式 = (x - min) / (max - min)
13 # 顏色範圍：0~255，所以，公式簡化為 x / 255
14 # 注意，顏色0為白色，與RGB顏色不同，(0,0,0) 為黑色。
15 x_train_norm, x_test_norm = x_train / 255.0, x_test / 255.0
```

2. 改用 CNN 模型：使用兩組 Conv2D/MaxPooling2D。

```
1  # 建立模型
2  from tensorflow.keras import layers
3  import numpy as np
4
5  input_shape=(28, 28, 1)
6  # 增加一維在最後面
7  x_train_norm = np.expand_dims(x_train_norm, -1)
8  x_test_norm = np.expand_dims(x_test_norm, -1)
9
10 # CNN 模型
11 model = tf.keras.Sequential(
12     [
13         tf.keras.Input(shape=input_shape),
14         layers.Conv2D(32, kernel_size=(3, 3), activation="relu"),
```

```
15          layers.MaxPooling2D(pool_size=(2, 2)),
16          layers.Conv2D(64, kernel_size=(3, 3), activation="relu"),
17          layers.MaxPooling2D(pool_size=(2, 2)),
18          layers.Flatten(),
19          layers.Dropout(0.5),
20          layers.Dense(10, activation="softmax"),
21      ]
22  )
```

CNN 的卷積層(Conv2D)的輸入多一個維度，代表色彩通道(Channel)，單色為 1，RGB 色系則設為 3。因此，輸入資料須增加一維，第 7、8 行程式即是使用 np.expand_dims 增加了一維在最後面。

3. 模型訓練，程式碼不需作任何改變。

```
1  # 設定優化器(optimizer)、損失函數(loss)、效能衡量指標(metrics)的類別
2  model.compile(optimizer='adam',
3               loss='sparse_categorical_crossentropy',
4               metrics=['accuracy'])
5
6  # 模型訓練
7  history = model.fit(x_train_norm, y_train, epochs=5, validation_split=0.2)
8
9  # 評分(Score Model)
10 score=model.evaluate(x_test_norm, y_test, verbose=0)
11
12 for i, x in enumerate(score):
13     print(f'{model.metrics_names[i]}: {score[i]:.4f}')
```

執行結果：準確率為 0.9892，較之前的模型略高。

注意事項：

1. 也是有模型採用連續兩個 Conv2D，再接一個 MaxPooling2D，並沒有硬性規定，可依照資料多寡與實驗，調校出最佳模型及最佳參數。

2. 再強調一次，CNN 不須指定要使用何種濾波器(Filter)，TensorFlow 會自動配置，且參數會在訓練過程中找到最佳參數值，我們只要指定濾波器的「個數」即可。

3. 可使用 model.summary()，觀察輸出維度及參數個數。
   卷積層輸出的寬度/高度公式如下：

$$W\_out = (W-F+2P)/S+1$$

其中縮寫表示：

- W_out：輸出的寬度
- W：原圖像的寬度
- F：濾波器(Filter)的寬度
- P：單邊補零的行數(Padding)
- S：滑動的步數(Stride)

```
Model: "sequential"

Layer (type)                    Output Shape              Param #
=================================================================
conv2d (Conv2D)                 (None, 26, 26, 32)        320

max_pooling2d (MaxPooling2D)    (None, 13, 13, 32)        0

conv2d_1 (Conv2D)               (None, 11, 11, 64)        18496

max_pooling2d_1 (MaxPooling2    (None, 5, 5, 64)          0

flatten (Flatten)               (None, 1600)              0

dropout (Dropout)               (None, 1600)              0

dense (Dense)                   (None, 10)                16010
=================================================================
Total params: 34,826
Trainable params: 34,826
Non-trainable params: 0
```

4. 依上述公式驗算第一層 Conv2D 輸出寬度(W_out)：

   W_out = floor((W−F+2P)/S+1) = (28 - 3 + 2*0)/1 + 1=26

5. 驗算第一層 Conv2D 輸出參數：

   Output Filter 數量 * (Filter 寬 * Filter 高 * Input Filter 數量 + 1) = 32 * (3 * 3 * 1 + 1) = 32 * 10 = 320。其中加 1 為迴歸線的偏差項(Bias)。

6. 第一層 MaxPooling2D 輸出寬度(W_out)：

   W_out = floor((W−F)/S+1) = (26 - 2) / 2 + 1 = 13 (無條件捨去)

7. 驗算第一層 Conv2D 輸出參數：

   Output Filter 數量 * (Filter 寬 * Filter 高 * Input Filter 數量 + 1) = 64 * (3 * 3 * 32 + 1) = 18496。

從卷積層運算觀察，CNN 模型有兩個特點：

1. 部分連接(Locally Connected or Sparse Connectivity)： Dense 每一層的
   神經元完全連接(Full Connected)至下一層的每個神經元，但卷積層的
   輸出神經元則只連接滑動視窗神經元，如下圖。想像一下，假設在手
   臂上拍打一下，手臂以外的神經元應該不會收到訊號，既然沒收到訊
   號，理所當然就不必往下一層傳送訊號了，所以，下一層的神經元只
   會收到上一層少數神經元的訊號，接收到的範圍稱之為「感知域」
   (Reception Field)。
   由於部分連接的關係，神經層中每條迴歸線的輸入特徵因而大幅減
   少，要估算的權重個數也就少了很多，於是模型即可大幅簡化。

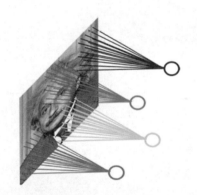

圖 6.9　部分連接 (Locally Connected)

2. 權重共享(Weight Sharing)：單一濾波器應用到滑動視窗時，卷積矩陣
   值都是一樣的，如下圖所示，基於這個假設，要估計的權重個數就減
   少許多，模型複雜度因而進一步簡化了。

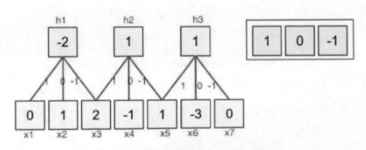

圖 6.10　權重共享(Weight Sharing)

所以，基於以上的兩個假設，CNN 模型訓練時間就不會過長。

另外為什麼 CNN 模型輸入資料要加入色彩通道(Channel)？是因為有些情況加入色彩，會比較容易辨識，比如獅子大部份是金黃色的，又或者偵測是否有戴口罩，只要圖像上有一塊白色的矩形，我們應該就能假定有戴口罩，當然目前口罩顏色已經是五花八門，需要更多的訓練資料，才能正確辨識。

以下我們使用 TensorFlow 內建的 Cifar 圖像[3]，比較單色與彩色的圖像辨識準確率。

【範例 2】單色的圖像辨識。

程式：**06_03_Cifar_gray_CNN.ipynb**。

1. 載入 Cifar10 資料。

```
1  import tensorflow as tf
2  cifar10 = tf.keras.datasets.cifar10
3
4  # 載入 cifar10 資料
5  (x_train, y_train),(x_test, y_test) = cifar10.load_data()
6
7  # 訓練/測試資料的 X/y 維度
8  print(x_train.shape, y_train.shape,x_test.shape, y_test.shape)
```

執行結果：訓練/測試資料各為 50,000 / 10,000 筆，圖像的寬和高均各為 32，為 RGB 色系。

```
(50000, 32, 32, 3) (50000, 1) (10000, 32, 32, 3) (10000, 1)
```

2. 轉成單色：使用 TensorFlow 內建的 rgb_to_grayscale 函數。

```
1  # 轉成單色：rgb_to_grayscale
2  x_train = tf.image.rgb_to_grayscale(x_train)
3  x_test = tf.image.rgb_to_grayscale(x_test)
4  print(x_train.shape, x_test.shape)
```

執行結果：最後一維為 1。

```
(50000, 32, 32, 1) (10000, 32, 32, 1)
```

3. 後續的程式碼與 MNIST 辨識相同。

執行結果：準確率只有 32%。

```
loss: 1.8816
accuracy: 0.3258
```

【範例 3】彩色的圖像辨識。

程式：06_04_Cifar_RGB_CNN.ipynb。

1. 載入 Cifar10 資料，與單色的圖像辨識相同，但不需轉換為單色。

2. 修改模型為 CNN，第 3 行 input_shape 為 3 維，最後一維是色彩通道。

```
1  # 建立模型
2  model = tf.keras.models.Sequential([
3      tf.keras.layers.Conv2D(32, (3, 3), activation='relu', input_shape=(32, 32, 3)),
4      tf.keras.layers.MaxPooling2D((2, 2)),
5      tf.keras.layers.Conv2D(64, (3, 3), activation='relu'),
6      tf.keras.layers.MaxPooling2D((2, 2)),
7      tf.keras.layers.Conv2D(64, (3, 3), activation='relu'),
8      tf.keras.layers.Flatten(),
9      tf.keras.layers.Dense(64, activation='relu'),
10     tf.keras.layers.Dense(10)
11 ])
```

執行結果：準確率提升為 70%，辨識效果顯著。

6-6 影像資料增補(Data Augmentation)

之前我們誇讚半天的辨識手寫阿拉伯數字程式，老實說有下列缺點：

1. 使用 MNIST 的測試資料，辨識率達 98%，但如果以在繪圖軟體裡使用滑鼠書寫的檔案測試，辨識率就差很多了。這是因為 MNIST 的訓練資料與滑鼠撰寫的樣式有所差異，MNIST 的資料應該是請受測者先寫在紙上，再掃描存檔的，所以圖像會有深淺不一的灰階和鋸齒狀，與我們直接使用滑鼠在繪圖軟體內書寫的情況不太一樣，所以，假使要實際應用，應該要自行收集訓練資料，準確率才會提升。

2. 若要自行收集資料，須找上萬個測試者，可能不太容易，又加上有些人書寫可能字體歪斜、偏一邊、或字體大小不同，影響預測準確度，我們可以藉由「資料增補」(Data Augmentation)的方法，自動產生各種變形的訓練資料，讓模型更強健(Robust)，可容忍這些缺點。

資料增補可將一張正常圖像，轉換成各式有缺陷的圖像，例如旋轉、偏移、拉近/拉遠、亮度等效果，再將這些資料當作訓練資料，訓練出來的模型，就較能辨識有缺陷的圖像。

TensorFlow/Keras 提供的資料增補函數 ImageDataGenerator 的參數很多元，包括：

1. width_shift_range：圖像寬度偏移的點(pixel)數或比例。
2. height_shift_range：圖像高度偏移的點(pixel)數或比例。
3. brightness_range：圖像亮度偏移的範圍。
4. shear_range：圖像順時鐘歪斜的範圍。
5. zoom_range：圖像拉近/拉遠的比例。
6. fill_mode：圖像填滿的方式，有四種方式 constant, nearest, reflect, wrap，詳見 Keras 官網。
7. horizontal_flip：圖像水平翻轉。
8. vertical_flip：圖像垂直翻轉。
9. rescale：特徵縮放。

詳細情形可參考 Keras 官網[4]。

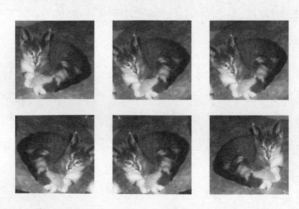

圖 6.11 左上角的原始圖像經過資料增補後，變成各種角度旋轉的圖像

【範例 1】MNIST 加上 Data Augmentation。

程式：06_05_Data_Augmentation_MNIST.ipynb。

1. 載入 MNIST 資料與模型定義，完全不需改變。
2. 訓練之前先進行資料增補。

```
1  # 參數設定
2  batch_size = 1000
3  epochs = 5
4
5  # 資料增補定義
6  datagen = tf.keras.preprocessing.image.ImageDataGenerator(
7          rescale=1./255,          # 特徵縮放
8          rotation_range=10,       # 旋轉 10 度
9          zoom_range=0.1,          # 拉遠/拉近 10%
10         width_shift_range=0.1,   # 寬度偏移  10%
11         height_shift_range=0.1)  # 高度偏移  10%
12
13 # 增補資料，進行模型訓練
14 datagen.fit(x_train)
15 history = model.fit(datagen.flow(x_train, y_train, batch_size=batch_size), epochs=epochs,
16         validation_data=datagen.flow(x_test, y_test, batch_size=batch_size), verbose=2,
17         steps_per_epoch=x_train.shape[0]//batch_size)
```

執行結果：

- 準確度並沒有提升，但沒關係，因為我們的目的是要看自行繪製的數字是否被正確辨識。
- 加入資料增補後，訓練時間拉長為兩倍多，以筆者的 PC 為例，由原本的 5 秒拉長至 12 秒。

3. 測試自行繪製的數字，原來的模型無法正確辨識筆者寫的 9，經過資料增補後，已經可以正確辨識了。

```
1  # 使用小畫家，繪製 0~9，實際測試看看
2  from skimage import io
3  from skimage.transform import resize
4  import numpy as np
5
6  # 讀取影像並轉為單色
7  uploaded_file = './myDigits/9.png'
8  image1 = io.imread(uploaded_file, as_gray=True)
9
10 # 縮為 (28, 28) 大小的影像
11 image_resized = resize(image1, (28, 28), anti_aliasing=True)
12 X1 = image_resized.reshape(1,28, 28, 1) #/ 255
13
14 # 反轉顏色，顏色0為白色，與 RGB 色碼不同，它的 0 為黑色
```

```
15  X1 = np.abs(1-X1)
16
17  # 預測
18  predictions = np.argmax(model.predict(X1), axis=-1)
19  print(predictions)
```

【範例 2】寵物資料集的處理。

程式：06_06_Data_Augmentation_Pets.ipynb。

之前都是使用 TensorFlow/Keras 內建的資料集，這個範例使用 Kaggle 所提供的資料集，它需要做前置處理，就是進行資料清理(Data Clean)，這會比較接近現實的狀況，但由於資料量較少，準確率較差，因此我們使用更複雜的 CNN 模型，再加上資料增補，以提升準確率。

Kaggle 為知名的 AI 競賽網站，也是一個很好的學習園地，這裡有很多佛心人士免費提供程式碼和資料集，各位讀者可以進去逛逛。網址為 https://www.kaggle.com/。

寵物資料集網址：

https://download.microsoft.com/download/3/E/1/3E1C3F21-ECDB-4869-8368-6DEBA77B919F/kagglecatsanddogs_3367a.zip。

原始程式來自 Keras 官網所提供的範例「Image classification from scratch」[5]，筆者拿來做了一些修改及註解。

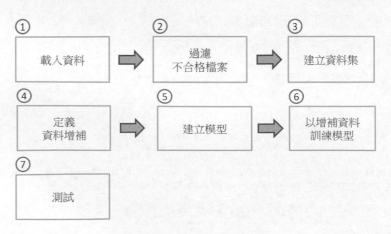

1. 從網路取得壓縮檔，並且解壓縮。

```python
1   # 從網路取得壓縮檔，並解壓縮
2   import os
3   import zipfile
4
5   # 壓縮檔 URL
6   zip_file_path = 'https://download.microsoft.com/download/3/E/1/'
7   zip_file_path += '3E1C3F21-ECDB-4869-8368-6DEBA77B919F/kagglecatsanddogs_3367a.zip'
8
9   # 存檔路徑
10  zip_file = os.path.join(os.getcwd(), 'CatAndDog.zip')
11
12  # 若壓縮檔案不存在，則下載檔案
13  if not os.path.exists(zip_file):
14      tf.keras.utils.get_file(
15          os.path.join(zip_file),
16          zip_file_path,
17          archive_format='auto'
18      )
19
20  # 若解壓縮目錄不存在，則解壓縮檔案至 unzip_path
21  unzip_path = os.path.join(os.getcwd(), 'CatAndDog')
22  if not os.path.exists(unzip_path):
23      with zipfile.ZipFile(zip_file, 'r') as zip_ref:
24          zip_ref.extractall(unzip_path)
```

2. 過濾不合格的檔案：掃描每一個檔案，若表頭不含"JFIF"，就不是圖檔，歸類為不合格的檔案，不納入訓練資料內。

```python
1   # 掃描每一個檔案，若表頭不含"JFIF"，即為不合格的檔案，不納入訓練資料內。
2   num_skipped = 0     # 記錄刪除的檔案個數
3   # 掃描目錄
4   for folder_name in ("Cat", "Dog"):
5       folder_path = os.path.join(unzip_path, "PetImages", folder_name)
6       for fname in os.listdir(folder_path):
7           fpath = os.path.join(folder_path, fname)
8           try:
9               fobj = open(fpath, "rb")
10              is_jfif = tf.compat.as_bytes("JFIF") in fobj.peek(10)
11          finally:
12              fobj.close()
13
14          if not is_jfif:
15              num_skipped += 1
16              # 刪除檔案
17              os.remove(fpath)
18
19  print(f"刪除 {num_skipped} 個檔案")
```

3. 以檔案目錄為基礎，建立訓練(Training)及驗證(Validation)資料集。

```
1  # image_dataset_from_directory : 讀取目錄中的檔案，存入 dataset
2  # image_dataset_from_directory : tf v2.3.0 才支援
3
4  image_size = (180, 180)   # 影像尺寸
5  batch_size = 32           # 批量
6
7  # 訓練資料集(Dataset)
8  train_ds = tf.keras.preprocessing.image_dataset_from_directory(
9      os.path.join(unzip_path, "PetImages"),
10     validation_split=0.2,
11     subset="training",
12     seed=1337,
13     image_size=image_size,
14     batch_size=batch_size,
15 )
16 # 驗證(Validation)資料集
17 val_ds = tf.keras.preprocessing.image_dataset_from_directory(
18     os.path.join(unzip_path, "PetImages"),
19     validation_split=0.2,
20     subset="validation",
21     seed=1337,
22     image_size=image_size,
23     batch_size=batch_size,
24 )
```

4. 定義資料增補(Data Augmentation)。

```
1  # RandomFlip("horizontal") : 水平翻轉
2  # RandomRotation(0.1) : 旋轉 0.1 比例
3  data_augmentation = keras.Sequential(
4      [
5          layers.experimental.preprocessing.RandomFlip("horizontal"),
6          layers.experimental.preprocessing.RandomRotation(0.1),
7      ]
8  )
```

5. 設定 prefetch：預先讀取訓練資料，以提升效能。

```
1  train_ds = train_ds.prefetch(buffer_size=32)
2  val_ds = val_ds.prefetch(buffer_size=32)
```

6. 建立模型：原作者使用較複雜的模型(類似 ResNet)，部份的神經層我們還沒說到，會在後續章節做講解。

```
1  # 定義模型
2  def make_model(input_shape, num_classes):
3      inputs = keras.Input(shape=input_shape)
4      # Image augmentation block
5      x = data_augmentation(inputs)
```

```
6
7      # 特徵縮放
8      x = layers.experimental.preprocessing.Rescaling(1.0 / 255)(x)
9      x = layers.Conv2D(32, 3, strides=2, padding="same")(x)
10     x = layers.BatchNormalization()(x)
11     x = layers.Activation("relu")(x)
12
13     x = layers.Conv2D(64, 3, padding="same")(x)
14     x = layers.BatchNormalization()(x)
15     x = layers.Activation("relu")(x)
16
17     previous_block_activation = x  # Set aside residual
18
19     for size in [128, 256, 512, 728]:
20         x = layers.Activation("relu")(x)
21         x = layers.SeparableConv2D(size, 3, padding="same")(x)
22         x = layers.BatchNormalization()(x)
23
24         x = layers.Activation("relu")(x)
25         x = layers.SeparableConv2D(size, 3, padding="same")(x)
26         x = layers.BatchNormalization()(x)
27
28         x = layers.MaxPooling2D(3, strides=2, padding="same")(x)
```

```
29
30         # Project residual
31         residual = layers.Conv2D(size, 1, strides=2, padding="same")(
32             previous_block_activation
33         )
34         x = layers.add([x, residual])  # Add back residual
35         previous_block_activation = x  # Set aside next residual
36
37     x = layers.SeparableConv2D(1024, 3, padding="same")(x)
38     x = layers.BatchNormalization()(x)
39     x = layers.Activation("relu")(x)
40
41     x = layers.GlobalAveragePooling2D()(x)
42     if num_classes == 2:
43         activation = "sigmoid"
44         units = 1
45     else:
46         activation = "softmax"
47         units = num_classes
48
49     x = layers.Dropout(0.5)(x)
50     outputs = layers.Dense(units, activation=activation)(x)
51     return keras.Model(inputs, outputs)
52
53 # 建立模型
54 model = make_model(input_shape=image_size + (3,), num_classes=2)
```

7. 訓練模型：因為標記只有兩種，故使用 binary_crossentropy 損失函數。
   fit()可直接使用 Dataset。

```
1   epochs = 5
2
3   # 設定優化器(optimizer)、損失函數(loss)、效能衡量指標(metrics)的類別
4   model.compile(
5       optimizer=keras.optimizers.Adam(1e-3),
6       loss="binary_crossentropy",
7       metrics=["accuracy"],
8   )
9
10  # 模型訓練
11  model.fit(
12      train_ds, epochs=epochs, validation_data=val_ds
13  )
```

執行結果：

- 訓練模型需時甚久，可將程式改在 Google Colaboratory 雲端環境執行，記得要設定使用 GPU 或 TPU。
- 訓練 5 epochs 準確率約 76%。
- 依據原作者實驗，若訓練 50 epochs，驗證準確率可達 96%。

8. 從目錄中任選一個檔案進行測試，建議從網路上下載檔案來測試比較好，但由於涉及圖檔版權，就請讀者自行修改第 3 行檔案路徑了。

```
1   # 任取一筆資料測試
2   img = keras.preprocessing.image.load_img(
3       os.path.join(unzip_path, "PetImages/Cat/18.jpg"), target_size=image_size
4   )
5   img_array = keras.preprocessing.image.img_to_array(img)  # 將影像轉為陣列
6   img_array = tf.expand_dims(img_array, 0)  # 增加一維在最前面，代表一筆資料
7
8   predictions = model.predict(img_array)
9   score = predictions[0][0]
10  print(f"是貓的機率= {(100 * score):.2f}")
```

執行結果：是貓的機率= 97.27%。

除了 TensorFlow/Keras 提供的資料增補功能之外，還有其他的函數庫，提供更多的資料增補效果，比方 Albumentations[6]，包含的類型多達 70 種，很多都是 TensorFlow/Keras 所沒有的效果，例如下圖的顏色資料增補：

6-7 可解釋的 AI(eXplainable AI, XAI)

雖然前文有說過深度學習是黑箱科學，但是，科學家們依然試圖解釋模型是如何辨識的，這方面的研究領域統稱為「可解釋的 AI」(eXplainable AI, XAI)，研究目的如下：

1. 確認模型辨識的結果是合理的：深度學習永遠不會跟你說錯，「垃圾進、垃圾出」(Garbage In, Garbage Out)，確認模型推估的合理性是相當重要的。

2. 改良演算法：唯有知其所以然，才能有較大的進步，光是靠參數的調校，只能有微幅的改善。目前機器學習還只能從資料中學習到知識

(Knowledge Discovery from Data, KDD)，要進階到機器能具有智慧(Wisdom)及感知(Feeling)能力，實現真正的人工智慧，勢必要有更突破性的發展。

目前 XAI 用視覺化的方式呈現特徵對模型的影響力，例如：

1. 使用卷積層萃取圖像的線條特徵，我們可以觀察到轉換後的結果嗎？
2. 甚至更進一步，我們可以知道哪些線條對辨識最有幫助嗎？

接下來我們以兩個實例展示相關的作法。

【範例 1】重建卷積層處理後的影像：觀察線條特徵。透過多次的卷積層/池化層處理，觀察圖像會有何種變化。

程式 **06_07_CNN_Visualization.ipynb**，此程式修改自「Machine Learning Mastery」中的部落文[7]。

1. 載入套件。

```
1  # 載入套件
2  import tensorflow as tf
3  from tensorflow.keras.applications.vgg16 import VGG16
4  import matplotlib.pyplot as plt
5  import numpy as np
```

2. 載入 VGG16 模型：VGG16 為知名的影像辨識模型，TensorFlow 當然沒有錯過內建此模型，包含已訓練好的模型參數，後續章節會介紹到此類預先訓練好的模型用法。

```
1  # 載入 VGG16 模型
2  model = VGG16()
3  model.summary()
```

執行結果：包括 16 層卷積/池化層。

```
Model: "vgg16"

Layer (type)                  Output Shape              Param #
=================================================================
input_1 (InputLayer)          [(None, 224, 224, 3)]     0

block1_conv1 (Conv2D)         (None, 224, 224, 64)      1792

block1_conv2 (Conv2D)         (None, 224, 224, 64)      36928

block1_pool (MaxPooling2D)    (None, 112, 112, 64)      0

block2_conv1 (Conv2D)         (None, 112, 112, 128)     73856

block2_conv2 (Conv2D)         (None, 112, 112, 128)     147584

block2_pool (MaxPooling2D)    (None, 56, 56, 128)       0

block3_conv1 (Conv2D)         (None, 56, 56, 256)       295168

block3_conv2 (Conv2D)         (None, 56, 56, 256)       590080

block3_conv3 (Conv2D)         (None, 56, 56, 256)       590080
```

3. 定義視覺化濾波器的函數。

```
1  # 視覺化特定層的特徵圖(Feature Map)
2  def Visualize(layer_no=1, n_filters=6):
3      # 取得權重(weight)
4      filters, biases = model.layers[layer_no].get_weights()
5      # 常態化(Normalization)
6      f_min, f_max = filters.min(), filters.max()
7      filters = (filters - f_min) / (f_max - f_min)
8
9      # 繪製特徵圖
10     ix = 1
11     for i in range(n_filters):
12         f = filters[:, :, :, i]    # 取得每一個特徵圖
13         for j in range(3):         # 每列 3 張圖
14             ax = plt.subplot(n_filters, 3, ix)  # 指定子視窗
15             ax.set_xticks([])      # 無X軸刻度
16             ax.set_yticks([])      # 無Y軸刻度
17             plt.imshow(f[:, :, j], cmap='gray') # 以灰階繪圖
18             ix += 1
19     plt.show()
```

4. 視覺化第一層的濾波器。

```
1  Visualize(1)
```

執行結果：可以看出每個濾波器均不相同，表示做了不同的影像處理。

5. 視覺化第 15 層的濾波器。

```
1  Visualize(15)
```

執行結果：與上一張圖相對照，可以看出與第一層不同，表示又做了不同的影像處理。

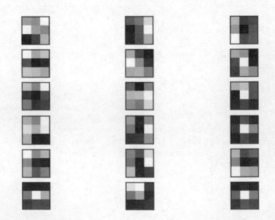

6. 重建第一個卷積層的輸出圖像：以鳥的圖片為例，先進行卷積，接著再重建圖像。

```
1   # 設定第一個卷積層的輸出為模型輸出
2   model2 = tf.keras.models.Model(inputs=model.inputs, outputs=model.layers[1].output)
3
4   # 載入測試的圖像
5   img = tf.keras.preprocessing.image.load_img('./images/bird.jpg', target_size=(224, 224))
6   img = tf.keras.preprocessing.image.img_to_array(img)      # 圖像轉為陣列
7   img = np.expand_dims(img, axis=0)                          # 加一維作為筆數
8   img = tf.keras.applications.vgg16.preprocess_input(img) # 前置處理(常態化)
9
10  # 預測
11  feature_maps = model2.predict(img)
12
13  # 將結果以 8x8 視窗顯示
14  square = 8
15  ix = 1
16  plt.figure(figsize=(12,8))
17  for _ in range(square):
18      for _ in range(square):
19          ax = plt.subplot(square, square, ix)
20          ax.set_xticks([])
21          ax.set_yticks([])
22          plt.imshow(feature_maps[0, :, :, ix-1], cmap='gray')
23          ix += 1
24  plt.show()
```

執行結果：可以看見第一層影像處理結果，有的濾波器可以抓到線條，有的則是漆黑一片。

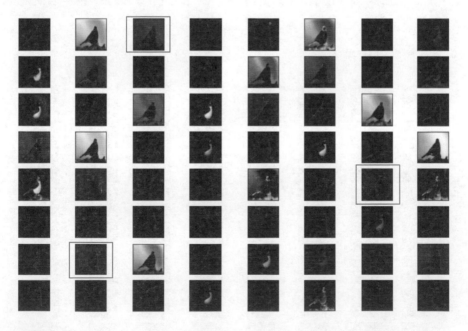

7. 重建 2, 5, 9, 13, 17 多層卷積層的輸出圖像。

```
1  # 取得 2, 5, 9, 13, 17 卷積層輸出
2  ixs = [2, 5, 9, 13, 17]
3  outputs = [model.layers[i].output for i in ixs]
4  model2 = tf.keras.models.Model(inputs=model.inputs, outputs=outputs)
5
6  # 載入測試的圖像
7  img = tf.keras.preprocessing.image.load_img('./images/bird.jpg', target_size=(224, 224))
8  img = tf.keras.preprocessing.image.img_to_array(img)      # 圖像轉為陣列
9  img = np.expand_dims(img, axis=0)                         # 加一維作為筆數
10 img = tf.keras.applications.vgg16.preprocess_input(img) # 前置處理(常態化)
11
12 # 預測
13 feature_maps = model2.predict(img)
14
15 # 將結果以 8x8 視窗顯示
16 square = 8
17 for fmap in feature_maps:
18     ix = 1
19     plt.figure(figsize=(12,8))
20     for _ in range(square):
21         for _ in range(square):
22             ax = plt.subplot(square, square, ix)
23             ax.set_xticks([])
24             ax.set_yticks([])
25             plt.imshow(fmap[0, :, :, ix-1], cmap='gray')
26             ix += 1
27     plt.show()
```

執行結果：在第 9 層影像處理結果中，還能夠明顯看到線條，但到第 17 層影像處理結果，已經是抽象到認不出來是鳥的地步了。

第 9 層影像處理結果：

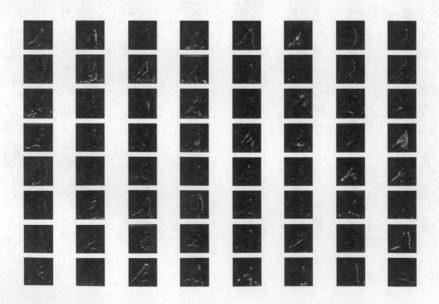

第 17 層影像處理結果：

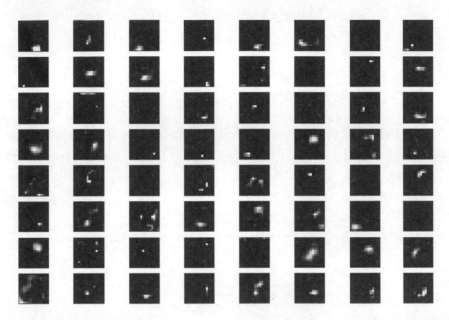

從以上的實驗，可以很清楚看到 CNN 的處理過程，我們雖然不明白辨識的邏輯，但是至少能夠觀察到整個模型處理的過程。

【範例 2】使用 Shap 套件，觀察圖像的哪些位置對辨識最有幫助。

SHAP (SHapley Additive exPlanations)套件是由 Scott Lundberg 及 Su-In Lee 所開發的，提供 Shapley value 的計算，並具有視覺化的介面，目標希望能解釋各種機器學習模型。套件使用說明可參考這個網址[8]，以下僅說明神經網路的應用。

Shapley value 是由多人賽局理論(Game Theory)而發展出來的，原本是用來分配利益給團隊中的每個人時所使用的分配函數，沿用到了機器學習的領域，則被應用在特徵對預測結果的個別影響力評估。詳細的介紹可參考維基百科[9]。

Shap 套件安裝：pip install shap。
程式 **06_08_Shap_MNIST.ipynb**。

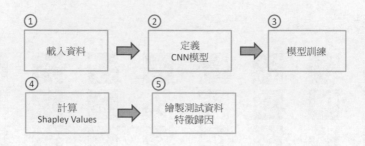

1. 載入 MNIST 資料集：請注意，目前 TensorFlow 2.x 版執行 shap 有 bug，所以務必使用 tf.compat.v1.disable_v2_behavior()，切換回 TensorFlow 1.x 版。

```
1  import tensorflow as tf
2
3  # 目前 tensorflow 2.x 版執行 shap 有 bug
4  tf.compat.v1.disable_v2_behavior()
5
6  # 載入 MNIST 手寫阿拉伯數字資料
7  mnist = tf.keras.datasets.mnist
8  (x_train, y_train),(x_test, y_test) = mnist.load_data()
```

2. 定義 CNN 模型：與前面模型相同，也可使用其他模型做測試。

```
1   # 建立模型
2   from tensorflow.keras import layers
3   import numpy as np
4
5   # 增加一維在最後面
6   x_train = np.expand_dims(x_train, -1)
7   x_test = np.expand_dims(x_test, -1)
8   x_train_norm, x_test_norm = x_train / 255.0, x_test / 255.0
9
10  # CNN 模型
11  input_shape=(28, 28, 1)
12  model = tf.keras.Sequential(
13      [
14          tf.keras.Input(shape=input_shape),
15          layers.Conv2D(32, kernel_size=(3, 3), activation="relu"),
16          layers.MaxPooling2D(pool_size=(2, 2)),
17          layers.Conv2D(64, kernel_size=(3, 3), activation="relu"),
18          layers.MaxPooling2D(pool_size=(2, 2)),
19          layers.Flatten(),
20          layers.Dropout(0.5),
21          layers.Dense(10, activation="softmax"),
22      ]
23  )
```

```
24
25  # 設定優化器(optimizer)、損失函數(loss)、效能衡量指標(metrics)的類別
26  model.compile(optimizer='adam',
27                loss='sparse_categorical_crossentropy',
28                metrics=['accuracy'])
```

3. 模型訓練：與前面相同。

```
1  # 模型訓練
2  history = model.fit(x_train_norm, y_train, epochs=5, validation_split=0.2)
3
4  # 評分(Score Model)
5  score=model.evaluate(x_test_norm, y_test, verbose=0)
6
7  for i, x in enumerate(score):
8      print(f'{model.metrics_names[i]}: {score[i]:.4f}')
```

4. Shapley Values 計算：測試第 1 筆資料。

```
1  import shap
2  import numpy as np
3
4  # 計算 Shap value 的 base
5  # 目前 tensorflow 2.x 版執行 shap 有 bug
6  # background = x_train[np.random.choice(x_train_norm.shape[0], 100, replace=False)]
7  # e = shap.DeepExplainer(model, background)        # shap values 不明顯
8  e = shap.DeepExplainer(model, x_train_norm[:100])
9
10 # 測試第 1 筆
11 shap_values = e.shap_values(x_test_norm[:1])
12 shap_values
```

執行結果：會顯示圖像中每一個像素的歸因，每個像素一共有 10 個數值，每個數值代表辨識為 0~9 的貢獻率。可使用下列指令觀察執行結果的維度(10, 1, 28, 28, 1)。

```
1  np.array(shap_values).shape
```

5. 繪製 5 筆測試資料的特徵歸因：紅色的區塊(請參看程式)代表貢獻率較大的區域。

```
1  # 繪製特徵的歸因(feature attribution)
2  shap.image_plot(shap_values, -x_test_norm[:5])
```

執行結果：每一列第一個數字為真實的標記，後面為預測每個數字貢獻率較大的區域。

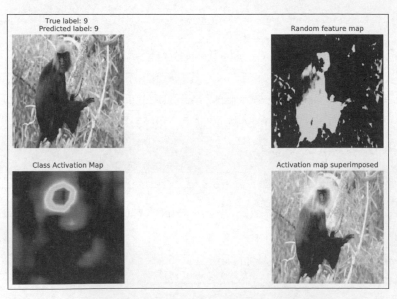

從 Shap 套件的功能,我們很容易判斷出中央位置是辨識的重點區域,這與我們認知是一致的。另一個名為 LIME[10]的套件,與 Shap 套件齊名,讀者如果對這領域有興趣,可以由此深入研究,筆者就偷懶一下嘍。

還有一篇論文[11]提出了 Class Activation Mapping 概念,可以描繪辨識的熱區,如下圖所示。Kaggle 也有一篇超讚的實作[12],值得大家好好欣賞一番。

圖 6.12 左上角的圖像為原圖,左下角的圖像顯示了辨識熱區,即猴子的頭和頸部都是辨識的主要關鍵區域

透過以上視覺化的輔助，不只可以幫助我們更瞭解 CNN 模型的運作，也能夠讓我們在收集資料時，有較明確的方向知道重點應該要放在哪裡，當然，如果未來能有更創新的想法，來改良演算法，那就可以開香檳慶祝了。

參考資料 (References)

[1] Prateek Karkare, 《Convolutional Neural Networks—Simplified》, 2019
(https://medium.com/x8-the-ai-community/cnn-9c5e63703c3f)

[2] 《Convolutional Neural Networks—Simplified》文中卷積計算的 GIF 動圖
(https://miro.medium.com/max/963/1*wpbLgTW_lopZ6JtDqVByuA.gif)

[3] TensorFlow 內建的 Cifar 圖像
(https://www.tensorflow.org/datasets/catalog/cifar10?hl=zh-tw)

[4] TensorFlow/Keras 提供的資料增補函數 ImageDataGenerator 參數
(https://keras.io/api/preprocessing/image/#imagedatagenerator-class)

[5] Keras 官網提供的範例
(https://keras.io/examples/vision/image_classification_from_scratch/)

[6] Albumentations
(https://github.com/albumentations-team/albumentations)

[7] Jason Brownlee, 《How to Visualize Filters and Feature Maps in Convolutional Neural Networks》, 2019
(https://machinelearningmastery.com/how-to-visualize-filters-and-feature-maps-in-convolutional-neural-networks/)

[8] SHAP 套件的安裝與介紹說明
(https://github.com/slundberg/shap)

[9] 維基百科中關於 Shapley value 的介紹
(https://en.wikipedia.org/wiki/Shapley_value)

[10] LIME 套件的安裝與介紹說明

(https://github.com/marcotcr/lime)

[11] Bolei Zhou, Aditya Khosla, Agata Lapedriza et al,《Learning Deep Features for Discriminative Localization》, 2015

(https://arxiv.org/pdf/1512.04150.pdf)

[12] Kaggle 中介紹的實作

(https://www.kaggle.com/aakashnain/what-does-a-cnn-see)

預先訓練的模型(Pre-trained Model)

透過 CNN 模型和資料增補的強化，我們已經能夠建立準確度還不錯的模型，然而，與近幾午影像辨識競賽中的冠、亞軍模型相比較，只能算是小巫見大巫了，冠、亞軍模型的神經層數量有些高達 100 多層，若要自行訓練這些模型就需要花上幾天甚至幾個星期的時間，難道縮短訓練時間的辦法，只剩購置企業級伺服器這個選項嗎？

幸好 TensorFlow/Keras、PyTorch 等深度學習框架早已為我們這些中小企業設想好了，套件提供事先訓練好的模型，我們可以直接套用，也可以只採用部份模型，再接上自訂的神經層，進行其他物件的辨識，這些預先訓練好的模型就稱為「Pre-trained Model」或「Keras Applications」。

7-1 預先訓練模型的簡介

在 ImageNet 歷年舉辦的競賽(ILSVRC)當中，近幾年產生的冠亞軍，大都是 CNN 模型的變型，整個演進過程非常精彩，簡述如下：

1. 2012 年冠軍 AlexNet 一舉將錯誤率減少 10%以上，且首度導入 Dropout層。

2. 2014 年亞軍 VGGNet 承襲 AlexNet 思路，建立更多層的模型，VGG 16/19 分別包括 16 及 19 層卷積層及池化層。

3. 2014 年圖像分類冠軍 GoogNet & Inception 同時導入多種不同尺寸的

Kernel，讓 系 統 決 定 最 佳 Kernel 尺 寸。Inception 引 入 Batch Normalization 等觀念，參見「Batch Normalization: Accelerating Deep Network Training by Reducing Internal Covariate Shift」[1]。

4. 2015 年冠軍 ResNets 發現到 20 層以上的模型其前面幾層會發生退化 (degradation)的狀況，因而提出以「殘差」(Residual)方法來解決問題，參見「Deep Residual Learning for Image Recognition」[2]。

Keras 收錄許多預先訓練的模型，稱為 Keras Applications[3]，隨著版本的更新，提供的模型愈來愈多，目前(2021 年) 包括：

Model	Size	Top-1 Accuracy	Top-5 Accuracy	Parameters	Depth
Xception	88 MB	0.790	0.945	22,910,480	126
VGG16	528 MB	0.713	0.901	138,357,544	23
VGG19	549 MB	0.713	0.900	143,667,240	26
ResNet50	98 MB	0.749	0.921	25,636,712	-
ResNet101	171 MB	0.764	0.928	44,707,176	-
ResNet152	232 MB	0.766	0.931	60,419,944	-
ResNet50V2	98 MB	0.760	0.930	25,613,800	-
ResNet101V2	171 MB	0.772	0.938	44,675,560	-
ResNet152V2	232 MB	0.780	0.942	60,380,648	-
InceptionV3	92 MB	0.779	0.937	23,851,784	159
InceptionResNetV2	215 MB	0.803	0.953	55,873,736	572
MobileNet	16 MB	0.704	0.895	4,253,864	88
MobileNetV2	14 MB	0.713	0.901	3,538,984	88
DenseNet121	33 MB	0.750	0.923	8,062,504	121
DenseNet169	57 MB	0.762	0.932	14,307,880	169
DenseNet201	80 MB	0.773	0.936	20,242,984	201
NASNetMobile	23 MB	0.744	0.919	5,326,716	-
NASNetLarge	343 MB	0.825	0.960	88,949,818	-
EfficientNetB0	29 MB	-	-	5,330,571	-
EfficientNetB1	31 MB	-	-	7,856,239	-
EfficientNetB2	36 MB	-	-	9,177,569	-
EfficientNetB3	48 MB	-	-	12,320,535	-
EfficientNetB4	75 MB	-	-	19,466,823	-
EfficientNetB5	118 MB	-	-	30,562,527	-
EfficientNetB6	166 MB	-	-	43,265,143	-
EfficientNetB7	256 MB	-	-	66,658,687	-

圖 7.1 Keras 提供的預先訓練模型(Pre-trained Model)

上述表格的欄位説明如下：

1. Size：模型檔案大小。
2. Top-1 Accuracy：預測一次就正確的準確率。
3. Top-5 Accuracy：預測五次中有一次正確的準確率。
4. Parameters：模型參數(權重、偏差)的數目。
5. Depth：模型層數。

Keras 研發團隊將這些模型先進行訓練與參數調校，並且存檔，使用者就不用自行訓練，直接套用即可，故稱為預先訓練的模型(Pre-trained Model)。

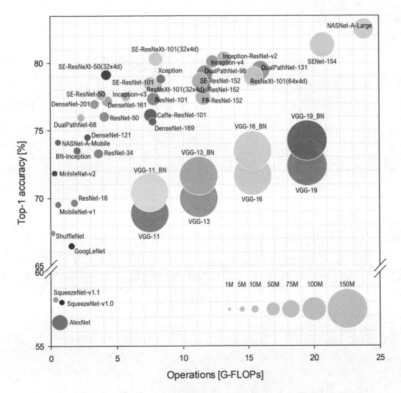

圖 7.2　預先訓練模型的準確率與計算速度之比較，圖形來源：
How to Choose the Best Keras Pre-Trained Model for Image Classification [4]

這些預先訓練的模型主要應用在圖像辨識，各模型結構的複雜度和準確率有所差異，圖 7.2 是各模型的比較，這裡提供各位一個簡單的選用原則，如果是注重準確率，可選擇準確率較高的模型，例如 ResNet 152，反之，如果要佈署在手機上，就可考慮使用檔案較小的模型，例如 MobileNet。

這些模型使用 ImageNet 100 萬張圖片作為訓練資料集，內含 1,000 種類別，詳情請參考 yrevar GitHub[5]，幾乎涵蓋了日常生活中會看到的物件類別，例如動物、植物、交通工具等，所以如果要辨識的物件屬於這 1000種，就可以直接套用模型，反之，如果要辨識這 1000 種以外的物件，就需要接上自訂的輸入層及辨識層(Dense)，只利用預先訓練模型的中間層萃取特徵。

因此應用這些預先訓練的模型，有三種方式：

1. 採用完整的模型，可辨識 ImageNet 所提供 1000 種物件。
2. 採用部分的模型，只萃取特徵，不作辨識。
3. 採用部分的模型，並接上自訂的輸入層和辨識層(Dense)，即可辨識這 1000 種以外的物件。

以下我們就依照這三種方式各實作一次。

7-2 採用完整的模型

預先訓練的模型的第一種用法，是採用完整的模型來辨識 1000 種物件，直接了當。

【範例 1】使用 VGG16 模型進行物件的辨識。

程式：**07_01_Keras_applications.ipynb**。

1. 載入套件：

```
1  import tensorflow as tf
2  from tensorflow.keras.applications.vgg16 import VGG16
3  from tensorflow.keras.preprocessing import image
4  from tensorflow.keras.applications.vgg16 import preprocess_input
5  from tensorflow.keras.applications.vgg16 import decode_predictions
6  import numpy as np
```

2. 載入 VGG16 模型：顯示和繪製模型結構。

```
1  model = VGG16(weights='imagenet')
2  print(model.summary())
3
4  # 繪製模型結構
5  tf.keras.utils.plot_model(model, to_file='vgg16.png')
```

執行 VGG16()時，系統會先下載模型檔案至使用者 Home 目錄下的/.keras/
models/。

■ Linux/Mac：~/.keras/models/

■ Windows：%HomePath%/.keras/models/

檔案名稱為 vgg16_weights_tf_dim_ordering_tf_kernels.h5。

參數 weight 有三種選項：

■ None：表示此模型還未經訓練，只有模型結構。

■ imagenet：已使用 ImageNet 圖片完成訓練，載入該模型權重。

■ 檔案路徑：使用自訂的權重檔。

include_top：

■ True：預設值，表示採用完整的模型。

■ False：不包含最上面的三層，一層是 Flatten、另外兩層則是 Dense。
   注意，最上面是指最後面的神經層，檔案名稱會包括 notop，為
   vgg16_weights_tf_dim_ordering_tf_kernels_notop.h5。

執行結果：VGG 16 使用多組的卷積/池化層，共有 16 層的卷積/池化層
（下頁左圖）。

模型結構圖：為單純的順序型模型，後三層為 Dense（下頁右圖）。

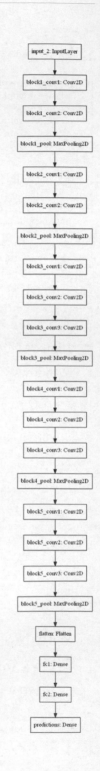

```
Model: "vgg16"

Layer (type)                 Output Shape              Param #
=================================================================
input_2 (InputLayer)         [(None, 224, 224, 3)]     0

block1_conv1 (Conv2D)        (None, 224, 224, 64)      1792

block1_conv2 (Conv2D)        (None, 224, 224, 64)      36928

block1_pool (MaxPooling2D)   (None, 112, 112, 64)      0

block2_conv1 (Conv2D)        (None, 112, 112, 128)     73856

block2_conv2 (Conv2D)        (None, 112, 112, 128)     147584

block2_pool (MaxPooling2D)   (None, 56, 56, 128)       0

block3_conv1 (Conv2D)        (None, 56, 56, 256)       295168

block3_conv2 (Conv2D)        (None, 56, 56, 256)       590080

block3_conv3 (Conv2D)        (None, 56, 56, 256)       590080

block3_pool (MaxPooling2D)   (None, 28, 28, 256)       0

block4_conv1 (Conv2D)        (None, 28, 28, 512)       1180160

block4_conv2 (Conv2D)        (None, 28, 28, 512)       2359808

block4_conv3 (Conv2D)        (None, 28, 28, 512)       2359808

block4_pool (MaxPooling2D)   (None, 14, 14, 512)       0

block5_conv1 (Conv2D)        (None, 14, 14, 512)       2359808

block5_conv2 (Conv2D)        (None, 14, 14, 512)       2359808

block5_conv3 (Conv2D)        (None, 14, 14, 512)       2359808

block5_pool (MaxPooling2D)   (None, 7, 7, 512)         0

flatten (Flatten)            (None, 25088)             0

fc1 (Dense)                  (None, 4096)              102764544

fc2 (Dense)                  (None, 4096)              16781312

predictions (Dense)          (None, 1000)              4097000
=================================================================
```

3. 任選一張圖片，比如大象的側面照，進行模型預測。

```
1   # 任選一張圖片，例如大象側面照
2   img_path = './images/elephant.jpg'
3   # 載入圖檔，並縮放寬高為 (224, 224)
4   img = image.load_img(img_path, target_size=(224, 224))
5
6   # 加一維，變成 (1, 224, 224)
7   x = image.img_to_array(img)
8   x = np.expand_dims(x, axis=0)
9   x = preprocess_input(x)
10
11  # 預測
12  preds = model.predict(x)
13  # decode_predictions : 取得前 3 名的物件，每個物件屬性包括 ( 類別代碼,  名稱,  機率)
14  print('Predicted:', decode_predictions(preds, top=3)[0])
```

執行結果：前三名的結果分別是印度象、非洲象、圖斯克象。

```
[('n02504013', 'Indian_elephant', 0.71942127), ('n02504458',
'African_elephant', 0.24141161), ('n01871265', 'tusker',
0.03627622)]
```

4. 再換一張圖片，比方大象的正面照，進行模型預測。

```
1   # 任選一張圖片，例如大象正面照
2   img_path = './images/elephant2.jpg'
3   # 載入圖檔，並縮放寬高為 (224, 224)
4   img = image.load_img(img_path, target_size=(224, 224))
5   # 加一維，變成 (1, 224, 224)
6   x = image.img_to_array(img)
7   x = np.expand_dims(x, axis=0)
8   x = preprocess_input(x)
9
10  # 預測
11  preds = model.predict(x)
12  # decode_predictions : 取得前 3 名的物件，每個物件屬性包括 ( 類別代碼,  名稱,  機率)
13  print('Predicted:', decode_predictions(preds, top=3)[0])
```

執行結果：前三名的結果分別是圖斯克象、非洲象、印度象。

```
[('n01871265', 'tusker', 0.6267539), ('n02504458',
'African_elephant', 0.3303416), ('n02504013', 'Indian_elephant',
0.04290244)]
```

不論正面或是側面，都可以正確辨識，也不用另外去背。

5. 改用並載入 ResNet 50 模型，使用其他模型亦可。

```
1  from tensorflow.keras.applications.resnet50 import ResNet50
2  from tensorflow.keras.preprocessing import image
3  from tensorflow.keras.applications.resnet50 import preprocess_input
4  from tensorflow.keras.applications.resnet50 import decode_predictions
5  import numpy as np
6
7  # 預先訓練好的模型 -- ResNet50
8  model = ResNet50(weights='imagenet')
```

6. 任選一張圖片，例如老虎的大頭照，進行模型預測。

```
1  # 任意一張圖片，例如老虎大頭照
2  img_path = './images/tiger3.jpg'
3  # 載入圖檔，並縮放寬高為 (224, 224)
4  img = image.load_img(img_path, target_size=(224, 224))
5
6  # 加一維，變成 (1, 224, 224)
7  x = image.img_to_array(img)
8  x = np.expand_dims(x, axis=0)
9  x = preprocess_input(x)
10
11 # 預測
12 preds = model.predict(x)
13 # decode the results into a list of tuples (class, description, probability)
14 # (one such list for each sample in the batch)
15 print('Predicted:', decode_predictions(preds, top=3)[0])
```

執行結果：前三名的結果分別是老虎、虎貓、美洲虎。

```
[('n02129604', 'tiger', 0.8657895), ('n02123159', 'tiger_cat',
0.13371062), ('n02128925', 'jaguar', 0.00046872292)]
```

可以改用 tiger1.jpg、tiger2.jpg 再嘗試看看，結果應該相去不遠。

7-3 採用部分模型

預先訓練的模型的第二種用法，是採用部分模型，只萃取特徵，不作辨識。例如，一個 3D 模型的網站，提供模型搜尋功能，首先使用者上傳要搜尋的圖檔，網站即時比對出相似的圖檔，顯示在網頁上讓使用者勾選下載，操作請參考 Sketchfab 網站(https://sketchfab.com/)，類似的功能應可適用到許多場域，譬如比對嫌疑犯、商品推薦等。

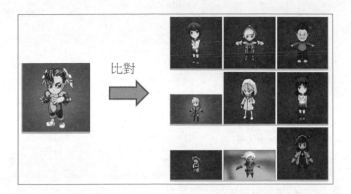

圖 7.3　3D 模型搜尋，資料來源：Using Keras' Pretrained Neural Networks for
Visual Similarity Recommendations[6]

【範例 2】使用 VGG16 模型進行物件的辨識。

程式：07_02_圖像相似度比較.ipynb。

1. 載入套件。

```
1  from tensorflow.keras.applications.vgg16 import VGG16
2  from tensorflow.keras.preprocessing import image
3  from tensorflow.keras.applications.vgg16 import preprocess_input
4  import numpy as np
```

2. 載入 VGG 16 模型：include_top=False 表示不包含最上面的三層(辨識
   層)。

```
1  # 載入VGG 16 模型, 不含最上面的三層(辨識層)
2  model = VGG16(weights='imagenet', include_top=False)
3  model.summary()
```

執行結果：模型不包含 Dense。

```
Model: "vgg16"

Layer (type)                Output Shape              Param #
=================================================================
input_1 (InputLayer)        [(None, None, None, 3)]   0

block1_conv1 (Conv2D)       (None, None, None, 64)    1792

block1_conv2 (Conv2D)       (None, None, None, 64)    36928

block1_pool (MaxPooling2D)  (None, None, None, 64)    0

block2_conv1 (Conv2D)       (None, None, None, 128)   73856

block2_conv2 (Conv2D)       (None, None, None, 128)   147584

block2_pool (MaxPooling2D)  (None, None, None, 128)   0

block3_conv1 (Conv2D)       (None, None, None, 256)   295168

block3_conv2 (Conv2D)       (None, None, None, 256)   590080

block3_conv3 (Conv2D)       (None, None, None, 256)   590080

block3_pool (MaxPooling2D)  (None, None, None, 256)   0

block4_conv1 (Conv2D)       (None, None, None, 512)   1180160

block4_conv2 (Conv2D)       (None, None, None, 512)   2359808

block4_conv3 (Conv2D)       (None, None, None, 512)   2359808

block4_pool (MaxPooling2D)  (None, None, None, 512)   0

block5_conv1 (Conv2D)       (None, None, None, 512)   2359808

block5_conv2 (Conv2D)       (None, None, None, 512)   2359808

block5_conv3 (Conv2D)       (None, None, None, 512)   2359808

block5_pool (MaxPooling2D)  (None, None, None, 512)   0
=================================================================
```

3. 萃取特徵：任選一張圖片，例如大象的側面照，取得圖檔的特徵向量。

```
1   # 任選一張圖片，例如大象側面照，取得圖檔的特徵向量
2   img_path = './images_test/elephant.jpg'
3
4   # 載入圖檔，並縮放寬高為 (224, 224)
5   img = image.load_img(img_path, target_size=(224, 224))
6
7   # 加一維，變成 (1, 224, 224)
8   x = image.img_to_array(img)
9   x = np.expand_dims(x, axis=0)
10  x = preprocess_input(x)
11
12  # 取得圖檔的特徵向量
13  features = model.predict(x)
14  print(features[0])
```

執行結果：得到圖檔的特徵向量如下。

```
[[[ 0.          0.          41.877056  ...  0.          0.          0.        ]
  [ 0.          0.          0.         ...  0.          0.          0.        ]
  [ 1.0921738   0.          22.865002  ...  0.          0.          0.        ]
  ...
  [ 0.          0.          0.         ...  0.          0.          0.        ]
  [ 0.          0.          0.         ...  0.          0.          0.        ]
  [ 0.          0.          0.         ...  0.          0.          0.        ]]

 [[ 0.          0.          36.385143  ...  0.          0.          3.2606328]
  [ 0.          0.          80.49929   ...  8.425463    0.          0.        ]
  [ 0.          0.          48.48268   ...  0.          0.          0.        ]
  ...
  [ 0.          0.          0.         ...  4.342996    0.          0.        ]
  [ 0.          0.          0.         ...  0.          0.          0.        ]
  [ 0.          0.          0.         ...  0.          0.          0.        ]]
```

4. 相似度比較：使用 cosine_similarity 比較特徵向量。

5. 先取得 images_test 目錄下所有.jpg 檔案名稱。

```
1  from os import listdir
2  from os.path import isfile, join
3
4  # 取得 images_test 目錄下所有 .jpg 檔案名稱
5  img_path = './images_test/'
6  image_files = np.array([f for f in listdir(img_path)
7          if isfile(join(img_path, f)) and f[-3:] == 'jpg'])
8  image_files
```

執行結果：

```
array(['bird.jpg', 'bird2.jpg', 'deer.jpg', 'elephant.jpg',
       'elephant2.jpg', 'lion1.jpg', 'lion2.jpg', 'panda1.jpg',
       'panda2.jpg', 'panda3.jpg', 'tiger1.jpg', 'tiger2.jpg',
       'tiger3.jpg'], dtype='<U13')
```

6. 取得 images_test 目錄下所有.jpg 檔案的像素。

```
1  import numpy as np
2
3  # 合併所有圖檔的像素
4  X = np.array([])
5  for f in image_files:
6      image_file = join(img_path, f)
7      # 載入圖檔，並縮放寬高為 (224, 224)
8      img = image.load_img(image_file, target_size=(224, 224))
9      img2 = image.img_to_array(img)
10     img2 = np.expand_dims(img2, axis=0)
11     if len(X.shape) == 1:
12         X = img2
13     else:
14         X = np.concatenate((X, img2), axis=0)
15
16 X = preprocess_input(X)
```

7. 取得所有圖檔的特徵向量。

```
1  # 取得所有圖檔的特徵向量
2  features = model.predict(X)
3
4  features.shape, X.shape
```

執行結果： 輸出與輸入的維度比較。

```
((13, 7, 7, 512), (13, 224, 224, 3))
```

8. 使用 cosine_similarity 函數比較特徵向量相似度。Cosine Similarity 計算兩個向量的夾角，如下圖，判斷兩個向量的方向是否近似，Cosine 介於(-1, 1) 之間，越接近 1，表示方向越相近。

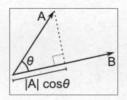

圖 7.4 夾角與 Cosine 函數

```
1  # 使用 cosine_similarity 比較特徵向量
2  from sklearn.metrics.pairwise import cosine_similarity
3
4
5  # 比較 Tiger2.jpg 與其他圖檔特徵向量
6  no=-2
7  print(image_files[no])
8
9  # 轉為二維向量，類似扁平層(Flatten)
10 features2 = features.reshape((features.shape[0], -1))
11
12 # 排除 Tiger2.jpg 的其他圖檔特徵向量
13 other_features = np.concatenate((features2[:no], features2[no+1:]))
14
15 # 使用 cosine_similarity 計算 Cosine 函數
16 similar_list = cosine_similarity(features2[no:no+1], other_features,
17                                  dense_output=False)
18
19 # 顯示相似度，由大排到小
20 print(np.sort(similar_list[0])[::-1])
21
22 # 依相似度，由大排到小，顯示檔名
23 image_files2 = np.delete(image_files, no)
24 image_files2[np.argsort(similar_list[0])[::-1]]
```

執行結果：與 tiger2.jpg 比較的相似度。

```
[0.35117537 0.26661643 0.19401284 0.19142228 0.1704499  0.14298241
 0.10661671 0.10612212 0.09741708 0.09370482 0.08440351 0.08097083]
```

對應的檔名：

```
['tiger1.jpg', 'tiger3.jpg', 'lion1.jpg', 'elephant.jpg',
 'elephant2.jpg', 'lion2.jpg', 'panda2.jpg', 'panda3.jpg',
 'bird.jpg', 'panda1.jpg', 'bird2.jpg', 'deer.jpg'], dtype='<U13')
```

觀察比對的結果，如預期一樣是正確的。利用這種方式，不只能夠比較 ImageNet 1000 類中的物件，也可以比較其他的物件，比如 3D 模型圖檔，讀者可以在網路上自行下載一些圖檔測試看看嘍。

▌ 7-4 轉移學習(Transfer Learning)

預先訓練的模型的第三種用法，是採用部分的模型，再加上自訂的輸入層和辨識層(Dense)，如此就能夠不受限於模型原先辨識的物件，也就是所謂的「轉移學習」(Transfer Learning)或者翻譯為「遷移學習」。其實不使用預先訓練的模型，直接建構 CNN 模型，也是可以辨識出任何物件的，然而為什麼要使用預先訓練的模型呢？原因歸納如下：

1. 使用大量高品質的資料(ImageNet 為普林斯頓大學與史丹福大學所主導的專案，有名校掛保證! ☺)，又加上設計較複雜的模型結構，例如 ResNet 高達有 150 層，準確率因此大大提高。
2. 使用較少的訓練資料：因為模型前半段已經訓練好了。
3. 訓練速度比較快：只需要重新訓練自訂的辨識層(Dense)即可。

一般的轉移學習分為兩階段：

1. 建立預先訓練的模型(Pre-trained Model)：包括之前的 Keras Applications，以及後面章節會談到的自然語言模型 -- Transformer 模型，它包含目前最夯的 BERT，利用大量的訓練資料和複雜的模型結構，取得通用性的圖像與自然語言特徵向量。

2. 微調(Fine Tuning)：依照特定應用領域的需求，個別建模並訓練，例如本節所述，利用預先訓練模型的前半段，再加入自訂的神經層，進行特殊類別的辨識。

【範例 3】使用 ResNet152V2 模型，辨識花朵資料集，程式源自 TensorFlow 官網所提供的範例「Load images」[7]，筆者進行一些修改和註解。

程式 07_03_Flower_ResNet.ipynb。

1. 載入套件：引進 ResNet152V2 模型。

```
1  import tensorflow as tf
2  from tensorflow.keras.applications.resnet_v2 import ResNet152V2
3  from tensorflow.keras.preprocessing import image
4  from tensorflow.keras.applications.resnet_v2 import preprocess_input
5  from tensorflow.keras.applications.resnet_v2 import decode_predictions
6  from tensorflow.keras.layers import Dense, GlobalAveragePooling2D
7  from tensorflow.keras.models import Model
8  import numpy as np
```

2. 載入 Flower 資料集。

```
1  # 資料集來源 : https://www.tensorflow.org/tutorials/load_data/images
2
3  # 參數設定
4  batch_size = 64
5  img_height = 224
6  img_width = 224
7  data_dir = './flower_photos/'
8
9  # 載入 Flower 訓練資料
10 train_ds = tf.keras.preprocessing.image_dataset_from_directory(
11   data_dir,
12   validation_split=0.2,
13   subset="training",
14   seed=123,
```

```
15    image_size=(img_height, img_width),
16    batch_size=batch_size)
17
18  # 載入 Flower 驗證資料
19  val_ds = tf.keras.preprocessing.image_dataset_from_directory(
20    data_dir,
21    validation_split=0.2,
22    subset="validation",
23    seed=123,
24    image_size=(img_height, img_width),
25    batch_size=batch_size)
```

執行結果：共 3670 個檔案、5 種類別(class)，其中 2936 個檔案作為訓練之
用，734 個檔案作為驗證。

3. 進行特徵工程，將特徵縮放成(0, 1)之間。

```
1  from tensorflow.keras import layers
2
3  normalization_layer = tf.keras.layers.experimental.preprocessing.Rescaling(1./255)
4  normalized_ds = train_ds.map(lambda x, y: (normalization_layer(x), y))
5  normalized_val_ds = val_ds.map(lambda x, y: (normalization_layer(x), y))
```

4. 顯示 ResNet152V2 完整的模型結構。

```
1  base_model = ResNet152V2(weights='imagenet')
2  print(base_model.summary())
```

執行結果：

- 共有 152 層卷積/池化層，再加上其他類型的神經層，總共有 566 層。
- 輸入層(InputLayer)維度為 (224, 224, 3)。

```
Model: "resnet152v2"

Layer (type)                 Output Shape
=================================================
input_5 (InputLayer)         [(None, 224, 224, 3)
```

- 最後兩層為 GlobalAveragePooling、Dense，若加上 include_top=False，
  這兩層則會被移除。

```
post_bn (BatchNormalization)    (None, 7, 7, 2048)

post_relu (Activation)          (None, 7, 7, 2048)

avg_pool (GlobalAveragePooling2 (None, 2048)

predictions (Dense)             (None, 1000)
=================================================
```

5. 建立模型結構：使用 Function API 加上自訂的辨識層(GlobalAverage Pooling、Dense)，再指定 Model 的輸入/輸出。

```
1  # 預先訓練好的模型 -- ResNet152V2
2  base_model = ResNet152V2(weights='imagenet', include_top=False)
3  print(base_model.summary())
4
5  # 加上自訂的辨識層(Dense)
6  x = base_model.output
7  x = GlobalAveragePooling2D()(x)
8  predictions = Dense(10, activation='softmax')(x)
9
10 # 指定自訂的輸入層及辨識層(Dense)
11 model = Model(inputs=base_model.input, outputs=predictions)
12
13 # 模型前段不需訓練了
14 for layer in base_model.layers:
15     layer.trainable = False
16
17 # 設定優化器(optimizer)、損失函數(loss)、效能衡量指標(metrics)的類別
18 model.compile(optimizer='rmsprop', loss='sparse_categorical_crossentropy',
19               metrics=['accuracy'])
```

6. 模型訓練：設定快取(cache)、prefetch，提升訓練效率。

```
1  # 設定快取(cache)、prefetch，以增進訓練效率
2  AUTOTUNE = tf.data.AUTOTUNE
3  normalized_ds = normalized_ds.cache().prefetch(buffer_size=AUTOTUNE)
4  normalized_val_ds = normalized_val_ds.cache().prefetch(buffer_size=AUTOTUNE)
5
6  # 模型訓練
7  history = model.fit(normalized_ds, validation_data = normalized_val_ds, epochs=5)
```

執行結果：

- 訓練準確率：93.33%。

- 驗證準確率：87.74%。

7. 繪製訓練過程的準確率/損失函數。

```
1  # 對訓練過程的準確率繪圖
2  plt.figure(figsize=(8, 6))
3  plt.plot(history.history['accuracy'], 'r', label='訓練準確率')
4  plt.plot(history.history['val_accuracy'], 'g', label='驗證準確率')
5  plt.xlabel('Epoch')
6  plt.ylabel('準確率')
7  plt.legend()
```

執行結果：下方圖表可見，隨著訓練週期的增長，驗證準確率並沒有提

高，這是因為預先訓練的模型已將大部份的神經層訓練過了。

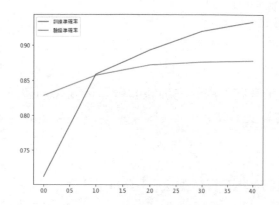

8. 顯示辨識的類別。

```
1  # 辨識的類別
2  class_names = train_ds.class_names
3  print(class_names)
```

執行結果：['daisy', 'dandelion', 'roses', 'sunflowers', 'tulips']。

9. 預測：任選一張圖片，譬如玫瑰花，預測結果還蠻正確的。

```
1  # 任選一張圖片，例如玫瑰
2  img_path = './images_test/rose.png'
3  # 載入圖檔，並縮放寬高為 (224, 224)
4  img = image.load_img(img_path, target_size=(224, 224))
5
6  # 加一維，變成 (1, 224, 224, 3)
7  x = image.img_to_array(img)
8  x = np.expand_dims(x, axis=0)
9  x = preprocess_input(x)
10
11  # 預測
12  preds = model.predict(x)
13
14  # 顯示預測結果
15  y_pred = [round(i * 100, 2) for i in preds[0]]
16  print(f'預測機率(%)：{y_pred}')
17  print(f'預測類別：{class_names[np.argmax(preds)]}')
```

執行結果：

```
預測機率(%)：[0.03, 0.0, 99.78, 0.04, 0.15, 0.0, 0.0, 0.0, 0.0, 0.0]
預測類別：roses
```

再任選一張圖片,例如雛菊,執行結果也是 OK。

注意,筆者一開始使用 cifar 10 內建資料集,它的圖片寬高只有(28, 28),而 ResNet 的訓練模型的輸入維度則為(224, 224),雖然還是可以訓練,因為 ResNet 會自動將 cifar 10 資料放大,但也因此造成圖像模糊,模型辨識能力變差。提醒讀者在應用預先訓練的模型時要特別留意,大部份模型的輸入維度都是(224, 224)以上。

7-5 Batch Normalization 說明

上一節我們使用複雜的 ResNet152V2 模型,其中內含許多的 Batch Normalization 神經層,它在神經網路的反向傳導時可消除梯度消失(Gradient Vanishing)或梯度爆炸(Gradient Exploding)現象,所以,我們花點時間研究其原理與應用時機。

當神經網路包含很多神經層時,經常會在其中放置一些 Batch Normalization 層,顧名思義,它的用途應該是特徵縮放,然而,究竟內部是如何運作的?有哪些好處?運用的時機?擺放的位置?

Sergey Ioffe 與 Christian Szegedy 在 2015 年首次提出 Batch Normalization,論文標題為「Batch Normalization: Accelerating Deep Network Training by Reducing Internal Covariate Shift」[8]。簡單來説,Batch Normalization 即為特徵縮放,將前一層的輸出標準化後,再轉至下一層,標準化公式如下:

$$\frac{x - \mu}{\delta}$$

標準化的好處就是讓收斂速度快一點,假如沒有標準化的話,模型通常會針對梯度較大的變數先優化,進而造成收斂路線曲折前進,如下圖,左圖是特徵未標準化的優化路徑,右圖則是標準化後的優化路徑。

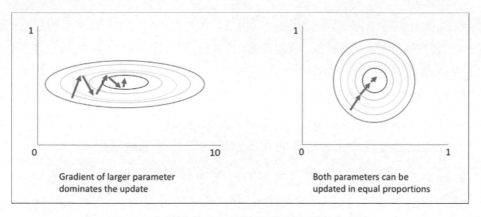

圖 7.5 未標準化 vs. 標準化優化過程的示意圖

圖片來源：Why Batch Normalization Matters? [9]

Batch Normalization 另外再引進兩個變數 γ、β，分別控制規模縮放(Scale)和偏移(Shift)。

$$
\begin{aligned}
&\textbf{Input: } \text{Values of } x \text{ over a mini-batch: } \mathcal{B} = \{x_{1...m}\}; \\
&\qquad\quad \text{Parameters to be learned: } \gamma, \beta \\
&\textbf{Output: } \{y_i = BN_{\gamma,\beta}(x_i)\} \\[6pt]
&\mu_{\mathcal{B}} \leftarrow \frac{1}{m} \sum_{i=1}^{m} x_i \qquad\qquad\qquad \text{// mini-batch mean} \\
&\sigma_{\mathcal{B}}^2 \leftarrow \frac{1}{m} \sum_{i=1}^{m} (x_i - \mu_{\mathcal{B}})^2 \qquad\;\; \text{// mini-batch variance} \\
&\widehat{x}_i \leftarrow \frac{x_i - \mu_{\mathcal{B}}}{\sqrt{\sigma_{\mathcal{B}}^2 + \epsilon}} \qquad\qquad\quad\;\; \text{// normalize} \\
&y_i \leftarrow \gamma \widehat{x}_i + \beta \equiv BN_{\gamma,\beta}(x_i) \qquad \text{// scale and shift}
\end{aligned}
$$

圖 7.6 Batch Normalization 公式，圖片來源：Why Batch Normalization Matters? [9]

補充說明：

1. 標準化是在訓練時「逐批」處理的，而非同時所有資料一起標準化，通常加在 Activation Function 之前。

2. ε是為了避免分母為 0 而加上的一個微小正數。

3. γ、β 值是由訓練過程中計算出來的，並不是事先設定好的。

假設我們要建立小狗的辨識模型，卻收集黃狗的圖片進行訓練，模型完成後，拿花狗的圖片來辨識，效果想當然會變差，要改善的話則必須重新收集資料再訓練一次，這種現象就稱為「Covariate Shift」，正式的定義是「假設我們要使用 X 預測 Y 時，當 X 的分配隨著時間有所變化時，模型就會逐漸失效」。股價預測也是類似的情形，當股價長期趨勢上漲時，原來的模型就慢慢失準了，除非納入最新的資料重新訓練模型。

由於神經網路的權重會隨著反向傳導不斷更新，每一層的輸出都會受到上一層的輸出影響，這是一種迴歸的關係，隨著神經層越多，整個神經網路的輸出有可能會逐漸偏移，此種現象稱之為「Internal Covariate Shift」。

而 Batch Normalization 就可以矯正「Internal Covariate Shift」現象，它在輸出至下一層的神經層時，每批資料都會先被標準化，這使得輸入資料的分布全屬於 N(0, 1) 的標準常態分配，因此，不管有多少層神經層，都不用擔心發生輸出逐漸偏移的問題。

至於什麼是梯度消失和梯度爆炸？這是由於 CNN 模型共享權值(Shared Weights)的關係，使得梯度逐漸消失或爆炸，原因如下，相同的 W 值若是經過很多層：

- 如果 W<1 ➜ 模型前幾層的 n 愈大，W^n 會趨近於 0，則影響力逐漸消失，即梯度消失(Gradient Vanishing)。
- 如果 W>1 ➜ 模型前幾層的 n 愈大，W^n 會趨近於∞，則造成模型優化無法收斂，即梯度爆炸(Gradient Explosion)。

只要經過 Batch Normalization，將每一批標準化後，梯度都會重新計算，這樣就不會有梯度消失和梯度爆炸的狀況發生了。除此之外，根據原作者的說法，Batch Normalization 還有以下優點：

- 優化收斂速度快(Train faster)。
- 可使用較大的學習率(Use higher learning rates)，加速訓練過程。
- 權重初始化較容易(Parameter initialization is easier)。

- 不使用 Batch Normalization 時，Activation function 容易在訓練過程中消失或提早停止學習，但如果經過 Batch Normalization 則又會再復活 (Makes activation functions viable by regulating the inputs to them)。

- 準確率全面性提升(Better results overall)。

- 類似 Dropout 的效果，可防止過度擬合(It adds noise which reduces overfitting with a regularization effect)，所以，當使用 Batch Normalization 時，就**不需要加 Dropout 層了**，為避免效果加乘過強，反而造成低度擬合(Underfitting)。

有一篇文章「On The Perils of Batch Norm」[10]做了一個很有趣的實驗，使用兩個資料集模擬「Internal Covariate Shift」現象，一個是 MNIST 資料集，背景是單純白色，另一個則是 SVHN 資料集，有複雜的背景，實驗過程如下：

1. 首先合併兩個資料集來訓練第一種模型，如下圖所示：

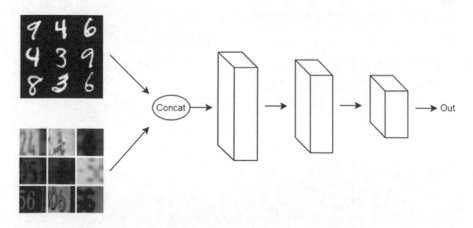

圖 7.7　合併兩個資料集來訓練一個模型

2. 再使用兩個資料集各自分別訓練模型,但共享權值,為第二種模型,
   如下圖:

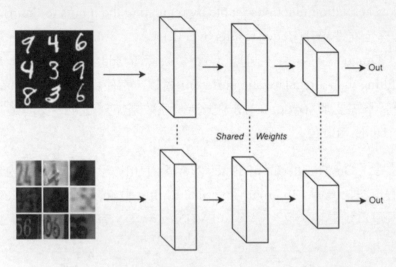

圖 7.8 使用兩個資料集個別訓練模型,但共享權值

兩種模型都有插入 Batch Normalization,比較結果,前者即單一模型準確
度較高,因為 Batch Normalization 可以矯正「Internal Covariate Shift」現
象。後者則由於資料集內容的不同,兩個模型共享權值本來就不合理。

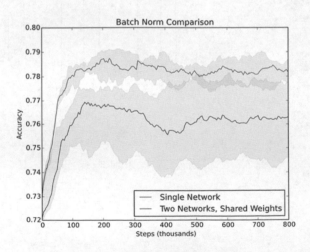

圖 7.9 兩種模型準確率比較

3. 第三種模型：使用兩個資料集訓練兩個模型，個別作 Batch Normalization，但不共享權值。比較結果，第三種模型效果最好。

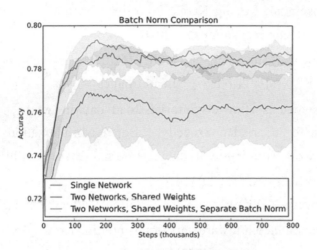

圖 7.10 三種模型準確率的比較

參考資料 (References)

[1] Sergey Ioffe、Christian Szegedy,《Batch Normalization: Accelerating Deep Network Training by Reducing Internal Covariate Shift》, 2015 (http://proceedings.mlr.press/v37/ioffe15.pdf)

[2] Kaiming He、Xiangyu Zhang、Shaoqing Ren、Jian Sun,《Deep Residual Learning for Image Recognition》, 2015 (https://arxiv.org/abs/1512.03385)

[3] Keras 官網關於 Keras Applications 的介紹 (https://keras.io/api/applications/)

[4] Marie Stephen Leo,《How to Choose the Best Keras Pre-Trained Model for Image Classification》, 2020 (https://towardsdatascience.com/how-to-choose-the-best-keras-pre-trained-model-for-image-classification-b850ca4428d4)

[5] yrevar GitHub

(https://gist.github.com/yrevar/942d3a0ac09ec9e5eb3a)

[6] Ethan Rosenthal,《Using Keras' Pretrained Neural Networks for Visual Similarity Recommendations》, 2016

(https://www.ethanrosenthal.com/2016/12/05/recasketch-keras/)

[7] TensorFlow 官網提供的範例「Load images」

(https://www.tensorflow.org/tutorials/load_data/images)

[8] Sergey Ioffe、Christian Szegedy,《Batch Normalization: Accelerating Deep Network Training by Reducing Internal Covariate Shift》, 2015

(https://arxiv.org/pdf/1502.03167.pdf)

[9] Aman Sawarn,《Why Batch Normalization Matters?》, 2020

(https://medium.com/towards-artificial-intelligence/why-batch-normalization-matters-4a6d753ba309)

[10] alexirpan,《On The Perils of Batch Norm》, 2017

(https://www.alexirpan.com/2017/04/26/perils-batch-norm.html)

| 第三篇 | 進階的影像應用

恭喜各位勇士們通過卷積神經網路(CNN)關卡，越過一座高山，本篇就來好好秀一下努力的成果，展現 CNN 在各領域應用上有哪些厲害的功能吧！

本篇的菜色超澎湃，包括下列主題：

- 物件偵測(Object Detection)。
- 語義分割(Semantic Segmentation)。
- 人臉辨識(Facial Recognition)。
- 風格轉換(Style Transfer)。
- 光學文字辨識(Optical Character Recognition, OCR)。

物件偵測(Object Detection)

前面介紹的圖像辨識模型,一張圖片中僅含有一個物件,接下來要登場的物件偵測可以同時偵測多個物件,並且標示出物件的位置。但是標示位置有什麼用處呢?現今最熱門的物件偵測演算法 YOLO,發明人 Joseph Redmon 提出一張有趣的照片:

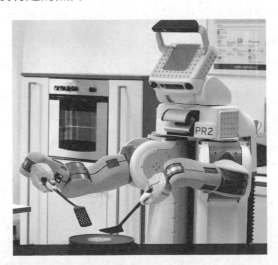

圖 8.1 機器人煎餅

圖片來源:Real-Time Grasp Detection Using Convolutional Neural Networks[1]

機器人要能完成煎餅的任務,它必須知道煎餅的所在位置,才能夠將餅翻面,如果有兩張以上的餅,還需知道要翻哪一張。不只機器人工作時需要電腦視覺,其他領域也會用到物件偵測,譬如:

1. 自駕車(Self-driving Car)：需要即時掌握前方路況及閃避障礙物。
2. 智慧交通：車輛偵測，利用一輛車在兩個時間點的位置，計算車速，進而可以推算道路壅塞的狀況，也可以用來偵測違規車輛。
3. 玩具、無人機、飛彈等都可以作類似的應用。
4. 異常偵測(Anomaly Detection)：可以在生產線上架設攝影機，即時偵測異常的瑕疵，像是印刷電路板、產品外觀等。
5. 無人商店的購物籃掃描，自動結帳。

8-1 圖像辨識模型的發展

綜觀歷年 ImageNet ILSVRC 挑戰賽(Large Scale Visual Recognition Challenge)的競賽題目，從 2011 年的影像分類(Classification)與定位(Classification with Localization) [2]，到 2017 年，題目擴展至物體定位(Object Localization)、物體偵測(Object Detection)、影片物體偵測(Object Detection from Video) [3]。我們可從中觀察到圖像辨識模型的發展史，了解到整個技術的演進。目前圖像辨識大概分為下列四大類型，如下圖所示：

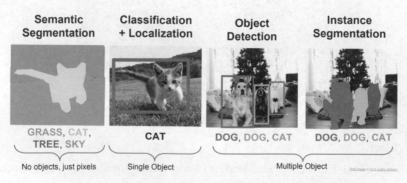

圖 8.2 物件偵測類型，圖片來源：Detection and Segmentation[4]

1. 語義分割(Semantic Segmentation)：按照物件類別來劃分像素區域，但不區分實例 (Instance)。拿上圖的第 4 張照片為例，照片中有 2 隻狗，都使用同一種顏色表達，即是語義分割， 2 隻狗使用不同顏色來表示，區分實例，則稱為實例分割。

2. 定位(Classification + Localization)：標記單一物件(Single Object)的類別與所在的位置。

3. 物件偵測(Object Detection)：標註多個物件(Multiple Object)的類別與所在的位置。

4. 實例分割(Instance Segmentation)：標記實例(Instance)，同一類的物件可以區分，並標示個別的位置，尤其是物件之間有重疊時。

接下來就逐一介紹上述四類演算法，並說明如何利用 TensorFlow 實作。

8-2 滑動視窗(Sliding Window)

物件偵測要能夠同時辨識物件的類別與位置，如果拆開來看就是兩項任務(Task)：

1. 分類(Classification)：辨識物件的類別。

2. 迴歸(Regression)：找到物件的位置，包括物件左上角的座標和寬度/高度。

最原始的演算法是採用滑動視窗(Sliding Window)，與前面介紹的卷積作法相似，步驟如下：

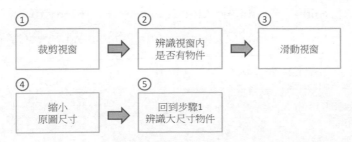

1. 設定某一尺寸的視窗，比如寬高各為 128 像素，由原圖左上角起裁剪成視窗大小。

2. 辨識視窗內是否有物件存在。

3. 滑動視窗,再次裁剪,並回到步驟 2,直到全圖掃描完為止。

4. 縮小原圖尺寸後,再重新回到步驟 1,尋找更大尺寸的物件。

這種將原圖縮小成各種尺寸的方式稱為「影像金字塔」(Image Pyramid) ,詳情請參閱「Image Pyramids with Python and OpenCV」一文[5],如下圖所示:

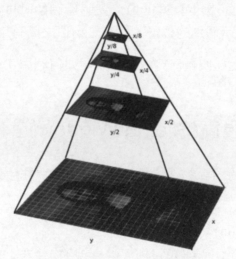

圖 8.3 影像金字塔(Image Pyramid),最下層為原圖,往上逐步縮小原圖尺寸,圖片來源:IIPImage[6]

【範例 1】對圖片滑動視窗並作影像金字塔(Image Pyramid)。

程式:**08_01_Sliding_Window_And_Image_Pyramid.ipynb**,程式修改自「Sliding Windows for Object Detection with Python and OpenCV」[7]。

1. 載入套件,需先安裝 OpenCV、imutils。

```
1  # 載入套件
2  import cv2
3  import time
4  import imutils
```

2. 定義影像金字塔操作函數:逐步縮小原圖尺寸,以便找到較大尺寸的物件。

```
1   # 影像金字塔操作
2   # image：原圖，scale：每次縮小倍數，minSize：最小尺寸
3   def pyramid(image, scale=1.5, minSize=(30, 30)):
4       # 第一次傳回原圖
5       yield image
6
7       while True:
8           # 計算縮小後的尺寸
9           w = int(image.shape[1] / scale)
10          # 縮小
11          image = imutils.resize(image, width=w)
12          # 直到最小尺寸為止
13          if image.shape[0] < minSize[1] or image.shape[1] < minSize[0]:
14              break
15          # 傳回縮小後的圖像
16          yield image
```

3. 定義滑動視窗函數。

```
1   # 滑動視窗
2   def sliding_window(image, stepSize, windowSize):
3       for y in range(0, image.shape[0], stepSize):      # 向下滑動 stepSize 格
4           for x in range(0, image.shape[1], stepSize):  # 向右滑動 stepSize 格
5               # 傳回裁剪後的視窗
6               yield (x, y, image[y:y + windowSize[1], x:x + windowSize[0]])
```

4. 測試。

```
1   # 讀取一個圖檔
2   image = cv2.imread('./lena.jpg')
3
4   # 視窗尺寸
5   (winW, winH) = (128, 128)
6
7   # 取得影像金字塔各種尺寸
8   for resized in pyramid(image, scale=1.5):
9       # 滑動視窗
10      for (x, y, window) in sliding_window(resized, stepSize=32,
11                                    windowSize=(winW, winH)):
12          # 視窗尺寸不合即放棄，滑動至邊緣時，尺寸過小
13          if window.shape[0] != winH or window.shape[1] != winW:
14              continue
15          # 標示滑動的視窗
16          clone = resized.copy()
17          cv2.rectangle(clone, (x, y), (x + winW, y + winH), (0, 255, 0), 2)
18          cv2.imshow("Window", clone)
19          cv2.waitKey(1)
20          # 暫停
21          time.sleep(0.025)
22
23  # 結束時關閉視窗
24  cv2.destroyAllWindows()
```

執行結果：

全程的執行結果可參閱影片檔：

video\Sliding_Window_And_Image_Pyramid.mp4。

8-3 方向梯度直方圖(HOG)

再看另一種作法「方向梯度直方圖」(Histogram of oriented gradient, HOG)，是抓取圖像輪廓線條的演算法，先將圖片切成很多個區域(Cell)，從每個區域中找出方向梯度，並把它描繪出來，就形成了物件的輪廓，與其它邊緣萃取的演算法比起來，它對環境的變化，例如光線，有較強(Robust)的辨識能力。有關 HOG 演算法的詳細處理方法可參閱「方向梯度直方圖（HOG）」[8]。

圖 8.4 HOG 處理：左圖為原圖，右圖為 HOG 處理過後的圖

根據「Histogram of Oriented Gradients and Object Detection」[9]一文的介紹，結合了 HOG 的物件偵測，流程如下：

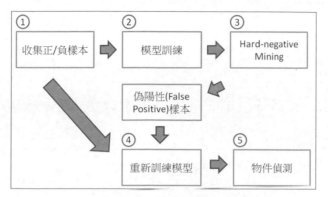

圖 8.5 結合 **HOG** 的物件偵測之流程圖

1. 收集正樣本(Positive set)：集結目標物件的各式圖像樣本，包括不同視角、尺寸、背景的圖像。

2. 收集負樣本 (Negative set)：集結無目標物件的各式圖像樣本，若有找到相近的物件則更好，可增加辨識準確度。

3. 使用以上正/負樣本與分類演算法訓練二分類模型，判斷是否包含目標物件，一般使用「支援向量機」(SVM)演算法。

4. Hard-negative Mining：掃描負樣本，使用滑動視窗的技巧，將每個視窗餵入模型來預測，如果有偵測到目標物件，即是偽陽性(False Positive)，接著將這些圖像加到訓練資料集中重新進行訓練，這個步驟可以重複很多次，能夠有效地提高模型準確率，類似 Boosting 整體學習的演算法。

5. 使用最後的模型進行物件偵測：將目標物件的圖像使用滑動視窗與影像金字塔技巧，餵入模型進行辨識，找出合格的視窗。

6. 篩選合格的視窗：使用 Non-Maximum Suppression (NMS)演算法，剔除多餘重疊的視窗。

【範例 2】使用 HOG、滑動視窗及 SVM 進行物件偵測。

程式：08_02_HOG-Face-Detection.ipynb，修改自 Scikit-Image 的範例。

1. 載入套件：本例使用 scikit-image 套件，OpenCV 也有支援類似的函數。

```
1  # Scikit-Image 的範例
2  # 載入套件
3  import numpy as np
4  import matplotlib.pyplot as plt
5  from skimage.feature import hog
6  from skimage import data, exposure
```

2. HOG 測試：使用 Scikit-Image 內建的女太空人圖像來測試 HOG 的效果。

```
1  # 測試圖片
2  image = data.astronaut()
3
4  # 取得圖片的 hog
5  fd, hog_image = hog(image, orientations=8, pixels_per_cell=(16, 16),
6                      cells_per_block=(1, 1), visualize=True, multichannel=True)
7
8  # 原圖與 hog圖比較
9  fig, (ax1, ax2) = plt.subplots(1, 2, figsize=(12, 6), sharex=True, sharey=True)
10
11 ax1.axis('off')
12 ax1.imshow(image, cmap=plt.cm.gray)
13 ax1.set_title('Input image')
14
15 # 調整對比，讓顯示比較清楚
16 hog_image_rescaled = exposure.rescale_intensity(hog_image, in_range=(0, 10))
17
18 ax2.axis('off')
19 ax2.imshow(hog_image_rescaled, cmap=plt.cm.gray)
20 ax2.set_title('Histogram of Oriented Gradients')
21 plt.show()
```

執行結果：原圖與 HOG 處理過後的圖比較。

Input image　　　　　　　Histogram of Oriented Gradients

3. 收集正樣本(positive set)：使用 scikit-learn 內建的人臉資料集作為正樣本，共有 13233 筆。

```
1  # 收集正樣本 (positive set)
2  # 使用 scikit-Learn 的人臉資料集
3  from sklearn.datasets import fetch_lfw_people
4  faces = fetch_lfw_people()
5  positive_patches = faces.images
6  positive_patches.shape
```

4. 觀察正樣本中部份的圖片。

```
1  # 顯示正樣本部份圖片
2  fig, ax = plt.subplots(4,6)
3  for i, axi in enumerate(ax.flat):
4      axi.imshow(positive_patches[500 * i], cmap='gray')
5      axi.axis('off')
```

執行結果：每張圖片寬高為(62, 47)。

5. 收集負樣本(negative set)：使用 Scikit-Image 內建的資料集，共有 9 筆。

```
1  # 收集負樣本 (negative set)
2  # 使用 Scikit-Image 的非人臉資料
3  from skimage import data, transform, color
4
5  imgs_to_use = ['hubble_deep_field', 'text', 'coins', 'moon',
6                 'page', 'clock','coffee','chelsea','horse']
7  images = [color.rgb2gray(getattr(data, name)())
8            for name in imgs_to_use]
9  len(images)
```

6. 增加負樣本筆數：將負樣本轉換為不同的尺寸，也可以使用資料增補
   技術。

```
1   # 將負樣本轉換為不同的尺寸
2   from sklearn.feature_extraction.image import PatchExtractor
3
4   # 轉換為不同的尺寸
5   def extract_patches(img, N, scale=1.0, patch_size=positive_patches[0].shape):
6       extracted_patch_size = tuple((scale * np.array(patch_size)).astype(int))
7       # PatchExtractor：產生不同尺寸的圖像
8       extractor = PatchExtractor(patch_size=extracted_patch_size,
9                                  max_patches=N, random_state=0)
10      patches = extractor.transform(img[np.newaxis])
11      if scale != 1:
12          patches = np.array([transform.resize(patch, patch_size)
13                              for patch in patches])
14      return patches
15
16  # 產生 27000 筆圖像
17  negative_patches = np.vstack([extract_patches(im, 1000, scale)
18                               for im in images for scale in [0.5, 1.0, 2.0]])
19  negative_patches.shape
```

執行結果：產生 27000 筆圖像。

7. 觀察負樣本中部份的圖片。

```
1   # 顯示部份負樣本
2   fig, ax = plt.subplots(4,6)
3   for i, axi in enumerate(ax.flat):
4       axi.imshow(negative_patches[600 * i], cmap='gray')
5       axi.axis('off')
```

執行結果：

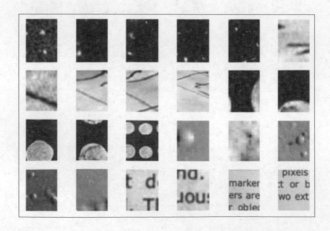

8. 合併正樣本與負樣本。

```
1   # 合併正樣本與負樣本
2   from skimage import feature    # To use skimage.feature.hog()
3   from itertools import chain
4
5   X_train = np.array([feature.hog(im)
6                       for im in chain(positive_patches,
7                                       negative_patches)])
8   y_train = np.zeros(X_train.shape[0])
9   y_train[:positive_patches.shape[0]] = 1
```

9. 使用 SVM 進行二分類的訓練：使用 GridSearchCV 尋求最佳參數值。

```
1   # 使用 SVM 作二分類的訓練
2   from sklearn.svm import LinearSVC
3   from sklearn.model_selection import GridSearchCV
4
5   # C為矯正過度擬合強度的倒數，使用 GridSearchCV 尋找最佳參數值
6   grid = GridSearchCV(LinearSVC(dual=False), {'C': [1.0, 2.0, 4.0, 8.0]},cv=3)
7   grid.fit(X_train, y_train)
8   grid.best_score_
```

執行結果：最佳模型準確率為 98.77%。

10. 取得最佳參數值。

```
1   # C 最佳參數值
2   grid.best_params_
```

11. 依最佳參數值再訓練一次，取得最終模型。

```
1   # 依最佳參數值再訓練一次
2   model = grid.best_estimator_
3   model.fit(X_train, y_train)
```

12. 新圖像測試：需先轉為灰階圖像。

```
1   # 取新圖像測試
2   test_img = data.astronaut()
3   test_img = color.rgb2gray(test_img)
4   test_img = transform.rescale(test_img, 0.5)
5   test_img = test_img[:120, 60:160]
6
7
8   plt.imshow(test_img, cmap='gray')
9   plt.axis('off');
```

執行結果:

13. 定義滑動視窗函數。

```
1   # 滑動視窗函數
2   def sliding_window(img, patch_size=positive_patches[0].shape,
3                      istep=2, jstep=2, scale=1.0):
4       Ni, Nj = (int(scale * s) for s in patch_size)
5       for i in range(0, img.shape[0] - Ni, istep):
6           for j in range(0, img.shape[1] - Ni, jstep):
7               patch = img[i:i + Ni, j:j + Nj]
8               if scale != 1:
9                   patch = transform.resize(patch, patch_size)
10              yield (i, j), patch
```

14. 計算 Hog：使用滑動視窗來計算每一滑動視窗的 Hog，餵入模型辨識。

```
1   # 使用滑動視窗計算每一視窗的 Hog
2   indices, patches = zip(*sliding_window(test_img))
3   patches_hog = np.array([feature.hog(patch) for patch in patches])
4
5   # 辨識每一視窗
6   labels = model.predict(patches_hog)
7   labels.sum() # 偵測到的總數
```

執行結果: 共有 55 個合格視窗。

15. 顯示這 55 個合格視窗。

```
1   # 將每一個偵測到的視窗顯示出來
2   fig, ax = plt.subplots()
3   ax.imshow(test_img, cmap='gray')
4   ax.axis('off')
5
6   # 取得左上角座標
7   Ni, Nj = positive_patches[0].shape
8   indices = np.array(indices)
9
10  # 顯示
11  for i, j in indices[labels == 1]:
12      ax.add_patch(plt.Rectangle((j, i), Nj, Ni, edgecolor='red',
13                                 alpha=0.3, lw=2, facecolor='none'))
```

執行結果：

16. 篩選合格視窗：使用 Non-Maximum Suppression(NMS)演算法，剔除多餘的視窗。以下採用「Non-Maximum Suppression for Object Detection in Python」[10]一文的程式碼。

定義 NMS 演算法函數：這是由 Pedro Felipe Felzenszwalb 等學者發明的演算法，執行速度較慢，Tomasz Malisiewicz [11]因而提出改善的演算法。函數的重疊比例門檻(overlapThresh)參數一般設為 0.3~0.5 之間。

```python
1   # Non-Maximum Suppression演算法 by Felzenszwalb et al.
2   # boxes：所有候選的視窗，overlapThresh：視窗重疊的比例門檻
3   def non_max_suppression_slow(boxes, overlapThresh=0.5):
4       if len(boxes) == 0:
5           return []
6
7       pick = []          # 儲存篩選的結果
8       x1 = boxes[:,0]    # 取得候選的視窗的左/上/右/下 座標
9       y1 = boxes[:,1]
10      x2 = boxes[:,2]
11      y2 = boxes[:,3]
12
13      # 計算候選視窗的面積
14      area = (x2 - x1 + 1) * (y2 - y1 + 1)
15      idxs = np.argsort(y2)     # 依視窗的底Y座標排序
```

```python
17      # 比對重疊比例
18      while len(idxs) > 0:
19          # 最後一筆
20          last = len(idxs) - 1
21          i = idxs[last]
22          pick.append(i)
23          suppress = [last]
24
25          # 比對最後一筆與其他視窗重疊的比例
26          for pos in range(0, last):
```

```
27              j = idxs[pos]
28
29              # 取得所有視窗的涵蓋範圍
30              xx1 = max(x1[i], x1[j])
31              yy1 = max(y1[i], y1[j])
32              xx2 = min(x2[i], x2[j])
33              yy2 = min(y2[i], y2[j])
34              w = max(0, xx2 - xx1 + 1)
35              h = max(0, yy2 - yy1 + 1)
36
37              # 計算重疊比例
38              overlap = float(w * h) / area[j]
39
40              # 如果大於門檻值，則儲存起來
41              if overlap > overlapThresh:
42                  suppress.append(pos)
43
44          # 刪除合格的視窗，繼續比對
45          idxs = np.delete(idxs, suppress)
46
47      # 傳回合格的視窗
48      return boxes[pick]
```

17. 呼叫 non_max_suppression_slow 函數，剔除多餘的視窗。

```
1  # 使用 Non-Maximum Suppression演算法，剔除多餘的視窗。
2  candidate_boxes = []
3  for i, j in indices[labels == 1]:
4      candidate_boxes.append([j, i, Nj, Ni])
5  final_boxes = non_max_suppression_slow(np.array(candidate_boxes).reshape(-1, 4))
6
7  # 將每一個合格的視窗顯示出來
8  fig, ax = plt.subplots()
9  ax.imshow(test_img, cmap='gray')
10 ax.axis('off')
11
12 # 顯示
13 for i, j, Ni, Nj in final_boxes:
14     ax.add_patch(plt.Rectangle((i, j), Ni, Nj, edgecolor='red',
15                                 alpha=0.3, lw=2, facecolor='none'))
```

執行結果： 得到兩個合格視窗。

以上範例的過程中省略了一些細節，譬如 Hard-negative mining、影像金字塔，這個例子無法偵測多個不同實體(Instance)與不同尺寸的物件。所以我們再來看一個範例，可使用任何 CNN 模型結合影像金字塔，進行多物件、多實體的偵測。

【**範例 2**】使用 ResNet50 進行物件偵測，並標示出位置。

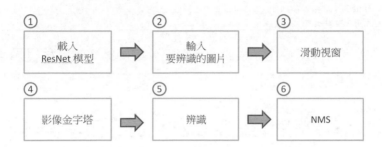

程式：**08_03_Object_Detection.ipynb**。

1. 載入套件：需要額外安裝 imutils 套件，它是一個簡單的影像處理套件。

```
1  # 載入套件，需額外安裝 imutils 套件
2  from tensorflow.keras.applications import ResNet50
3  from tensorflow.keras.applications.resnet import preprocess_input
4  from tensorflow.keras.preprocessing.image import img_to_array
5  from tensorflow.keras.applications import imagenet_utils
6  from imutils.object_detection import non_max_suppression
7  import numpy as np
8  import imutils
9  import time
10 import cv2
```

2. 參數設定：此範例是辨識三隻斑馬同時存在的圖像，另外也可以辨識騎自行車的圖像(bike.jpg)。

```
1  # 參數設定
2  image_path = './images_test/zebra.jpg'  # 要辨識的圖檔
3  WIDTH = 600                # 圖像縮放為 (600, 600)
4  PYR_SCALE = 1.5            # 影像金字塔縮放比例
5  WIN_STEP = 16             # 視窗滑動步數
6  ROI_SIZE = (250, 250)     # 視窗大小
7  INPUT_SIZE = (224, 224)  # CNN的輸入尺寸
```

3. 載入 ResNet50 模型。

```
1  # 載入 ResNet50 模型
2  model = ResNet50(weights="imagenet", include_top=True)
```

4. 讀取要辨識的圖片。

```
1  # 讀取要辨識的圖片
2  orig = cv2.imread(image_path)
3  orig = imutils.resize(orig, width=WIDTH)
4  (H, W) = orig.shape[:2]
```

5. 定義滑動視窗和影像金字塔函數，這部分與範例 2 的流程相同。

```
1  # 定義滑動視窗與影像金字塔函數
2
3  # 滑動視窗
4  def sliding_window(image, step, ws):
5      for y in range(0, image.shape[0] - ws[1], step):      # 向下滑動 stepSize 格
6          for x in range(0, image.shape[1] - ws[0], step): # 向右滑動 stepSize 格
7              # 傳回裁剪後的視窗
8              yield (x, y, image[y:y + ws[1], x:x + ws[0]])
9
10 # 影像金字塔操作
11 # image：原圖，scale：每次縮小倍數，minSize：最小尺寸
12 def image_pyramid(image, scale=1.5, minSize=(224, 224)):
13     # 第一次傳回原圖
14     yield image
15
16     # keep looping over the image pyramid
17     while True:
18         # 計算縮小後的尺寸
19         w = int(image.shape[1] / scale)
20         image = imutils.resize(image, width=w)
21
22         # 直到最小尺寸為止
23         if image.shape[0] < minSize[1] or image.shape[1] < minSize[0]:
24             break
25
26         # 傳回縮小後的圖像
27         yield image
```

6. 產生影像金字塔，並逐一進行視窗辨識。

```
1  # 輸出候選框
2  rois = []       # 候選框
3  locs = []       # 位置
4  SHOW_BOX = False  # 是否顯示要找的框
5
6  # 產生影像金字塔
7  pyramid = image_pyramid(orig, scale=PYR_SCALE, minSize=ROI_SIZE)
8  # 逐一視窗辨識
```

```
 9  for image in pyramid:
10      # 框與原圖的比例
11      scale = W / float(image.shape[1])
12
13      # 滑動視窗
14      for (x, y, roiOrig) in sliding_window(image, WIN_STEP, ROI_SIZE):
15          # 取得候選框
16          x = int(x * scale)
17          y = int(y * scale)
18          w = int(ROI_SIZE[0] * scale)
19          h = int(ROI_SIZE[1] * scale)
20
21          # 縮放圖形以符合模型輸入規格
22          roi = cv2.resize(roiOrig, INPUT_SIZE)
23          roi = img_to_array(roi)
24          roi = preprocess_input(roi)
25
```

```
26          # 加入輸出變數中
27          rois.append(roi)
28          locs.append((x, y, x + w, y + h))
29
30          # 是否顯示要找的框
31          if SHOW_BOX:
32              clone = orig.copy()
33              cv2.rectangle(clone, (x, y), (x + w, y + h),
34                  (0, 255, 0), 2)
35
36              # 顯示正在找的框
37              cv2.imshow("Visualization", clone)
38              cv2.imshow("ROI", roiOrig)
39              cv2.waitKey(0)
40
41  cv2.destroyAllWindows()
```

7. 預測：辨識機率必須大於設定值，並進行 NMS。

```
 1  # 預測
 2  MIN_CONFIDENCE = 0.9  # 辨識機率門檻值
 3
 4  rois = np.array(rois, dtype="float32")
 5  preds = model.predict(rois)
 6  preds = imagenet_utils.decode_predictions(preds, top=1)
 7  labels = {}
 8
 9  # 檢查預測結果，機率須大於設定值
10  for (i, p) in enumerate(preds):
11      # grab the prediction information for the current ROI
12      (imagenetID, label, prob) = p[0]
13
14      # 機率大於設定值，則放入候選名單
15      if prob >= MIN_CONFIDENCE:
16          # 放入候選名單
17          box = locs[i]
18          L = labels.get(label, [])
19          L.append((box, prob))
20          labels[label] = L
```

```
22  # 掃描每一個類別
23  for label in labels.keys():
24      # 複製原圖
25      clone = orig.copy()
26
27      # 畫框
28      for (box, prob) in labels[label]:
29          (startX, startY, endX, endY) = box
30          cv2.rectangle(clone, (startX, startY), (endX, endY),
31              (0, 255, 0), 2)
32
33      # 顯示 NMS(non-maxima suppression) 前的框
34      cv2.imshow("Before NMS", clone)
35      clone = orig.copy()
36
37      # NMS
38      boxes = np.array([p[0] for p in labels[label]])
39      proba = np.array([p[1] for p in labels[label]])
40      boxes = non_max_suppression(boxes, proba)
41
42      for (startX, startY, endX, endY) in boxes:
43          # 畫框及類別
44          cv2.rectangle(clone, (startX, startY), (endX, endY), (0, 255, 0), 2)
45          y = startY - 10 if startY - 10 > 10 else startY + 10
46          cv2.putText(clone, label, (startX, y),
47              cv2.FONT_HERSHEY_SIMPLEX, 0.45, (0, 255, 0), 2)
48
49      # 顯示
50      cv2.imshow("After NMS", clone)
51      cv2.waitKey(0)
52
53  cv2.destroyAllWindows()     # 關閉所有視窗
```

執行結果：因為圖中的斑馬有重疊，所以少抓到一匹馬。

試試看騎自行車的圖像，images_test/bike.jpg，結果也只抓到兩輛，後續我們會運用其他演算法來改善這個缺點。

由於物件偵測的應用範圍廣大，因此有許多學者前仆後繼地提出各種改良的演算法，試圖提高準確率並加快辨識速度，接下來我們就沿著前輩們的研究軌跡，逐步深入探討。

8-4 R-CNN 物件偵測

滑動視窗並結合 HOG 的演算法雖然很好用，但是它還是有以下的缺點：

1. 滑動視窗加上影像金字塔，需要檢查的視窗個數太多了，耗時過久。
2. 一個 SVM 分類器只能偵測一個物件。
3. 通用性的 CNN 模型辨識不並準確，尤其是重疊的物件。

因此，從 2014 年開始，每年都有改良的演算法出現，如下圖：

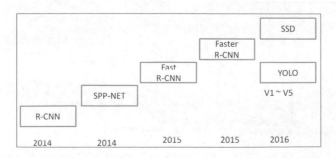

圖 8.6 物件偵測演算法的發展過程

第一個神經網路的演算法 Regions with CNN，以下簡稱 R-CNN，於 2014
年由 Ross B. Girshick 提出，論文標題為「Rich feature hierarchies for
accurate object detection and semantic segmentation」[12]。

架構如下：

1. 讀取要辨識的圖片。
2. 使用區域推薦(Region Proposal)演算法，找到 2000 個候選視窗。
3. 使用 CNN 萃取特徵。
4. 使用 SVM 辨識。

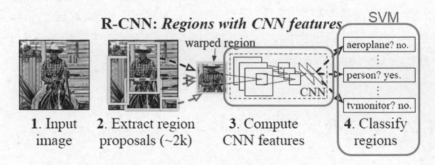

圖 8.7 R-CNN 架構，圖片來源：Rich feature hierarchies for accurate object
detection and semantic segmentation

更詳細的架構如下：

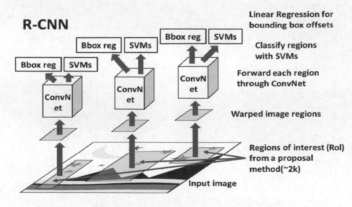

圖 8.8 另一視角的 R-CNN 架構

程式處理流程如下：

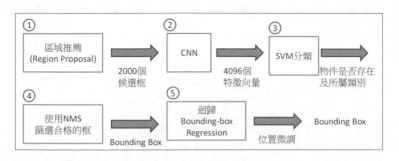

圖 8.9 R-CNN 處理流程

1. 區域推薦(Region Proposal)：用途為改善滑動視窗的過程檢查過多視窗的問題，使用區域推薦演算法，只找出 2000 個候選框(Bounding Box)輸入到模型。

   區域推薦(Region Proposal)也有多種演算法，R-CNN 所採用的是 Selective Search，它會依據顏色(color)、紋理(texture)、規模(Scale)、空間關係(Enclosure)來進行合併，接著再選取 2000 個最有可能包含物件的區域，稱之為候選框(Bounding Box)。

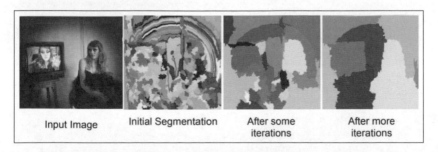

圖 8.10 區域推薦(Region Proposal)：最左邊的圖為原圖，將顏色、紋理、規模、空間關係相近的區域合併，最後變成最右邊圖的區域。

2. 特徵萃取(Feature Extractor)：將 2000 個候選框使用影像變形轉換 (Image Warping)，轉成固定尺寸 227 x 227 的圖像，餵入 CNN 進行特徵萃取，如果採用 AlexNet 的話，則每個候選框轉換成 4096 個特徵向量。

3. SVM 分類器：比對特徵向量，偵測物件是否存在與所屬的類別，注意，一種類使用一個二分類 SVM。

4. 使用 Non-Maximum Suppression (NMS)篩選合格的框：選取可信度較高的候選框為基準，計算與基準框的 IoU(Intersection-over Union)，高 IoU 值表示高度重疊，就可以把它們過濾掉，類似上一節的作法。

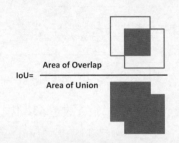

$$IoU= \frac{\text{Area of Overlap}}{\text{Area of Union}}$$

圖 8.11 IoU：分母為與目標框聯集的面積，分子為與目標框交集的面積

5. 位置微調：利用迴歸(Bounding-box Regression) 微調候選框的位置。利用迴歸計算候選框的四個變數：中心點(P_x, P_y)與寬高(P_w, P_h)，其微調公式如下，G 為預估值。推論過程有點複雜，詳情可參考原文附錄 C。

$$\hat{G}_x = P_w d_x(P) + P_x \qquad (1)$$
$$\hat{G}_y = P_h d_y(P) + P_y \qquad (2)$$
$$\hat{G}_w = P_w \exp(d_w(P)) \qquad (3)$$
$$\hat{G}_h = P_h \exp(d_h(P)). \qquad (4)$$

損失函數如下，採用 Ridge Regression，以最小平方法估算出來的權重：

$$\mathbf{w}_\star = \underset{\hat{\mathbf{w}}_\star}{\operatorname{argmin}} \sum_i^N (t_\star^i - \hat{\mathbf{w}}_\star^{\mathsf{T}} \phi_5(P^i))^2 + \lambda \|\hat{\mathbf{w}}_\star\|^2. \qquad (5)$$

微調後的目標值 t_*：

$$t_x = (G_x - P_x)/P_w \qquad (6)$$
$$t_y = (G_y - P_y)/P_h \qquad (7)$$
$$t_w = \log(G_w/P_w) \qquad (8)$$
$$t_h = \log(G_h/P_h). \qquad (9)$$

整個 R-CNN 處理流程涉及相當多的演算法，包括：

1. 區域推薦(Region Proposal)：Selective Search。
2. 特徵萃取(Feature Extractor)：AlexNet，也可採取 VGG 或者其他 CNN 模型。
3. SVM 分類器。
4. Non-Maximum Suppression (NMS)。
5. Bounding-box Regression)。

【範例 4】使用 R-CNN 偵測空照圖中的飛機。

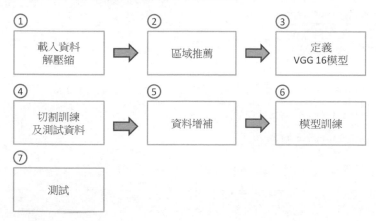

程式：08_04_RCNN.ipynb。

1. 需安裝 OpenCV 擴展版：先解除安裝 OpenCV，再安裝擴展版，一般版與擴展版只能擇其一。

```
pip uninstall opencv-contrib-python opencv-python
pip install opencv-contrib-python
```

2. 解壓縮圖像訓練資料。

```
1  import zipfile
2  import os
3
4  # 圖像訓練資料
5  path_to_zip_file = './images_Object_Detection/Images.zip'
6  directory_to_extract_to = './images_Object_Detection/'
7
```

```
 8    # 檢查目錄是否存在
 9    if not os.path.isdir(directory_to_extract_to):
10        # 解壓縮
11        with zipfile.ZipFile(path_to_zip_file, 'r') as zip_ref:
12            zip_ref.extractall(directory_to_extract_to)
```

3. 解壓縮標註訓練資料。

```
1    # 標註訓練資料
2    path_to_zip_file = './images_Object_Detection/Airplanes_Annotations.zip'
3    directory_to_extract_to = './images_Object_Detection/'
4
5    # 檢查目錄是否存在
6    if not os.path.isdir(directory_to_extract_to):
7        # 解壓縮
8        with zipfile.ZipFile(path_to_zip_file, 'r') as zip_ref:
9            zip_ref.extractall(directory_to_extract_to)
```

4. 載入套件。

```
1    # 載入套件
2    import os,cv2
3    import pandas as pd
4    import matplotlib.pyplot as plt
5    import numpy as np
6    import tensorflow as tf
```

5. 顯示 1 張圖像訓練資料並包含標註。

```
 1    # 設定圖像及標註目錄
 2    path = "./images_Object_Detection/Images"
 3    annot = "./images_Object_Detection/Airplanes_Annotations"
 4
 5    # 顯示1張圖像訓練資料含標註
 6    for e,i in enumerate(os.listdir(annot)):
 7        if e < 10:
 8            # 讀取圖像
 9            filename = i.split(".")[0]+".jpg"
10            print(filename)
11            img = cv2.imread(os.path.join(path,filename))
12            df = pd.read_csv(os.path.join(annot,i))
13            plt.axis('off')
14            plt.imshow(img)
15            for row in df.iterrows():
16                x1 = int(row[1][0].split(" ")[0])
17                y1 = int(row[1][0].split(" ")[1])
18                x2 = int(row[1][0].split(" ")[2])
19                y2 = int(row[1][0].split(" ")[3])
20                cv2.rectangle(img,(x1,y1),(x2,y2),(255,0,0), 2)
21            plt.figure()
22            plt.axis('off')
23            plt.imshow(img)
24            break
```

執行結果：

6. 區域推薦(Region Proposal)：使用 Selective Search 演算法，OpenCV 擴展版提供現成的函數 createSelectiveSearchSegmentation()，假如要自行開發，可以參照「R-CNN 學習筆記, LaptrinhX」[13]一文的內容。

```
1  # 區域推薦(Region Proposal)：Selective Search
2  # 讀取圖像
3  im = cv2.imread(os.path.join(path,"42850.jpg"))
4
5  # Selective Search
6  cv2.setUseOptimized(True);
7  ss = cv2.ximgproc.segmentation.createSelectiveSearchSegmentation()
8  ss.setBaseImage(im)
9  ss.switchToSelectiveSearchFast()
10 rects = ss.process()
11
12 # 輸出
13 imOut = im.copy()
14 for i, rect in (enumerate(rects)):
15     x, y, w, h = rect
16 #     print(x,y,w,h)
17 #     imOut = imOut[x:x+w,y:y+h]
18     cv2.rectangle(imOut, (x, y), (x+w, y+h), (0, 255, 0), 1, cv2.LINE_AA)
19
20 plt.imshow(imOut)
```

執行結果：會依顏色、紋理等萃取出 2000 個候選框(綠色)。

7. 定義 IoU 計算函數：計算兩個框的 IoU。

```
1   # 定義 IoU 計算函數
2   def get_iou(bb1, bb2):
3       assert bb1['x1'] < bb1['x2']
4       assert bb1['y1'] < bb1['y2']
5       assert bb2['x1'] < bb2['x2']
6       assert bb2['y1'] < bb2['y2']
7
8       x_left = max(bb1['x1'], bb2['x1'])
9       y_top = max(bb1['y1'], bb2['y1'])
10      x_right = min(bb1['x2'], bb2['x2'])
11      y_bottom = min(bb1['y2'], bb2['y2'])
12
13      if x_right < x_left or y_bottom < y_top:
14          return 0.0
15
16      intersection_area = (x_right - x_left) * (y_bottom - y_top)
17
18      bb1_area = (bb1['x2'] - bb1['x1']) * (bb1['y2'] - bb1['y1'])
19      bb2_area = (bb2['x2'] - bb2['x1']) * (bb2['y2'] - bb2['y1'])
20
21      iou = intersection_area / float(bb1_area + bb2_area - intersection_area)
22      assert iou >= 0.0
23      assert iou <= 1.0
24      return iou
```

8. 篩選訓練資料：找出檔名為 airplane 開頭的檔案，並使用區域推薦將每個檔案各取出 2000 個候選框。要留意的是，每個檔案必須包含 30 個以上的正樣本(IoU>70%)與 30 個以上的負樣本(IoU<30%)，才能被列為訓練資料。

```
1   # 篩選訓練資料
2
3   # 儲存正樣本及負樣本的候選框
4   train_images=[]
5   train_labels=[]
6
7   # 掃描每一個標註
8   for e,i in enumerate(os.listdir(annot)):
9       try:
10          # 取得飛機的圖像
11          if i.startswith("airplane"):
12              filename = i.split(".")[0]+".jpg"
13              print(e,filename)
14
15              # 讀取標註檔案
16              image = cv2.imread(os.path.join(path,filename))
17              df = pd.read_csv(os.path.join(annot,i))
18
19              # 取得所有標註的座標
```

```
20              gtvalues=[]
21              for row in df.iterrows():
22                  x1 = int(row[1][0].split(" ")[0])
23                  y1 = int(row[1][0].split(" ")[1])
24                  x2 = int(row[1][0].split(" ")[2])
25                  y2 = int(row[1][0].split(" ")[3])
26                  gtvalues.append({"x1":x1,"x2":x2,"y1":y1,"y2":y2})
```

```
28              # 區域推薦
29              ss.setBaseImage(image)
30              ss.switchToSelectiveSearchFast()
31              ssresults = ss.process()
32              imout = image.copy()
33
34              # 初始化
35              counter = 0          # 正樣本筆數
36              falsecounter = 0     # 負樣本筆數
37              flag = 0             # 1:正負樣本筆數均 >= 30
38              fflag = 0            # 1:正樣本筆數 >= 30
39              bflag = 0            # 1:負樣本筆數 >= 30
40
41              # 掃描每一個候選框
42              for e,result in enumerate(ssresults):
43                  if e < 2000 and flag == 0:
44                      for gtval in gtvalues:
45                          x,y,w,h = result
46                          # 比較區域推廣區域與標註的 IoU
47                          iou = get_iou(gtval,{"x1":x,"x2":x+w,"y1":y,"y2":y+h})
48
49                          # 收集30筆正樣本
50                          if counter < 30:
51                              if iou > 0.70:
52                                  timage = imout[y:y+h,x:x+w]
53                                  resized = cv2.resize(timage, (224,224),
54                                              interpolation = cv2.INTER_AREA)
```

```
55                                  train_images.append(resized)
56                                  train_labels.append(1)
57                                  counter += 1
58                          else :
59                              fflag =1
60
61                          # 收集30筆負樣本
62                          if falsecounter <30:
63                              if iou < 0.3:
64                                  timage = imout[y:y+h,x:x+w]
65                                  resized = cv2.resize(timage, (224,224),
66                                              interpolation = cv2.INTER AREA)
67                                  train_images.append(resized)
68                                  train_labels.append(0)
69                                  falsecounter += 1
70                          else :
71                              bflag = 1
72
73                      # 超過30筆正樣本及負樣本，表有物件在框裡面
74                      if fflag == 1 and bflag == 1:
75                          print("inside")
```

```
76                              flag = 1
77      except Exception as e:
78          print(e)
79          print("error in "+filename)
80          continue
```

9. 定義模型：使用 VGG 16，加上自訂的神經層。

```
1  from tensorflow.keras.layers import Dense
2  from tensorflow.keras import Model
3  from tensorflow.keras import optimizers
4  from tensorflow.keras.preprocessing.image import ImageDataGenerator
5  from tensorflow.keras.applications.vgg16 import VGG16
6
7  vggmodel = VGG16(weights='imagenet', include_top=True)
8
9  # VGG16 前端的神經層不重作訓練
10 for layers in (vggmodel.layers)[:15]:
11     print(layers)
12     layers.trainable = False
13
14 # 接自訂神經層作辨識
15 X= vggmodel.layers[-2].output
16 predictions = Dense(2, activation="softmax")(X)
17 model_final = Model(inputs = vggmodel.input, outputs = predictions)
18
19 # 訂定損失函數、優化器、效能衡量指標
20 from tensorflow.keras.optimizers import Adam
21 opt = Adam(lr=0.0001)
22 model_final.compile(loss = tf.keras.losses.categorical_crossentropy,
23                     optimizer = opt, metrics=["accuracy"])
24 model_final.summary()
```

10. 定義轉換函數：將標記 Y 轉為二個變數。

```
1  # 定義函數，將標記 Y 轉為二個變數，
2  from sklearn.preprocessing import LabelBinarizer
3
4  class MyLabelBinarizer(LabelBinarizer):
5      def transform(self, y):
6          Y = super().transform(y)
7          if self.y_type_ == 'binary':
8              return np.hstack((Y, 1-Y))
9          else:
10             return Y
11     def inverse_transform(self, Y, threshold=None):
12         if self.y_type_ == 'binary':
13             return super().inverse_transform(Y[:, 0], threshold)
14         else:
15             return super().inverse_transform(Y, threshold)
```

11. 前置處理及訓練資料/測試資料分割。

```
1  # 資料前置處理，切割訓練及測試資料
2  from sklearn.model_selection import train_test_split
3
4  # 筆者 PC 記憶體不足，只取 10000
5  X_new = np.array(train_images[:10000])
6  y_new = np.array(train_labels[:10000])
7
8  # 標記 Y 轉為二個變數，
9  lenc = MyLabelBinarizer()
10 Y =  lenc.fit_transform(y_new)
11
12 # 切割訓練及測試資料
13 X_train, X_test , y_train, y_test = train_test_split(X_new, Y, test_size=0.10)
14 print(X_train.shape,X_test.shape,y_train.shape,y_test.shape)
```

12. 進行資料增補(Data Augmentation)，以提高模型準確率，因為飛機停放
    的方向可能會有偏斜。

```
1  # 資料增補(Data Augmentation)
2  trdata = ImageDataGenerator(horizontal_flip=True,
3                              vertical_flip=True, rotation_range=90)
4  traindata = trdata.flow(x=X_train, y=y_train)
5  tsdata = ImageDataGenerator(horizontal_flip=True,
6                              vertical_flip=True, rotation_range=90)
7  testdata = tsdata.flow(x=X_test, y=y_test)
```

13. 模型訓練：原作者訓練週期(epoch)達 1000 次之多，故設定檢查點與提
    前結束的 Callback，以縮短訓練時間。然而筆者只測試 20 epochs，未
    使用到檢查點與提前結束的 Callback。

```
1  # 模型訓練
2  from tensorflow.keras.callbacks import ModelCheckpoint, EarlyStopping
3  # 定義模型存檔及提早結束的 Callback
4  checkpoint = ModelCheckpoint("ieeercnn_vgg16_1.h5", monitor='val_loss',
5                              verbose=1, save_best_only=True,
6                              save_weights_only=False, mode='auto', period=1)
7  early = EarlyStopping(monitor='val_loss', min_delta=0, patience=100,
8                      verbose=1, mode='auto')
9
10 # 模型訓練，節省時間，只訓練 20 epochs，正式專案還是要訓練較多次
11 # hist = model_final.fit_generator(generator= traindata, steps_per_epoch= 10,
12 #        epochs= 1000, validation_data= testdata, validation_steps=2,
13 #        callbacks=[checkpoint,early])
14 hist = model_final.fit_generator(generator= traindata, steps_per_epoch= 10,
15         epochs= 20, validation_data= testdata, validation_steps=2,
16         callbacks=[checkpoint,early])
```

14. 繪製模型訓練過程的準確率。

```
1  # 繪製模型訓練過程的準確率
2  import matplotlib.pyplot as plt
3  plt.plot(hist.history['accuracy'])
4  plt.plot(hist.history['val_accuracy'])
5  plt.ylabel("Accuracy")
6  plt.xlabel("Epoch")
7  plt.legend(["Accuracy","Validation Accuracy"])
8  plt.show()
```

執行結果：準確率並未穩定上升，這表示訓練週期不足，由於筆者只著重在演算法的研究，所以沒有繼續訓練下去，如果用於正式專案務必多訓練幾個週期比較妥當。

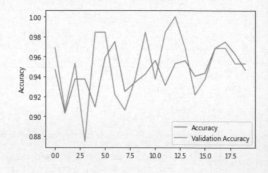

15. 任選一張圖片測試。

```
1   # 任選一張圖片測試
2   im = X_test[100]
3   plt.imshow(im)
4   img = np.expand_dims(im, axis=0)
5   out= model_final.predict(img)
6
7   # 顯示預測結果
8   if out[0][0] > out[0][1]:
9       print("有飛機")
10  else:
11      print("沒有飛機")
```

執行結果：圖片有偵測到飛機。

有飛機

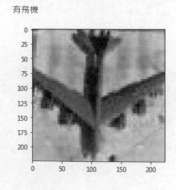

16. 測試所有檔名為 4 開頭的檔案。

```
1   # 測試所有檔名為 4 開頭的檔案
2   z=0
3   for e,i in enumerate(os.listdir(path)):
4       if i.startswith("4"):
5           z += 1
6           img = cv2.imread(os.path.join(path,i))
7           # 區域推薦
8           ss.setBaseImage(img)
9           ss.switchToSelectiveSearchFast()
10          ssresults = ss.process()
11          imout = img.copy()
12
13          # 物件偵測
14          for e,result in enumerate(ssresults):
15              if e < 2000:
16                  x,y,w,h = result
17                  timage = imout[y:y+h,x:x+w]
18                  resized = cv2.resize(timage, (224,224), interpolation = cv2.INTER_AREA)
19                  img = np.expand_dims(resized, axis=0)
20                  out= model_final.predict(img)
21
22                  # 機率 > 0.65 才算偵測到飛機
23                  if out[0][0] > 0.65:
24                      cv2.rectangle(imout, (x, y), (x+w, y+h), (0, 255, 0), 1, cv2.LINE_AA)
25          plt.figure()
26          plt.imshow(imout)
```

執行結果：下面這張圖片有偵測到飛機，但其他部分的圖片並沒有正確偵測到。

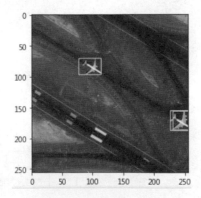

筆者並沒有找到原發明人 Ross B. Girshick 的程式碼，故以上的範例並未使用 Non-Maximum Suppression (NMS)、Bounding-box Regression，應該是作者之後又提出更好的演算法 Faster R-CNN，所以 R-CNN 程式碼就被取代了，後面我們會針對 Faster R-CNN 再進行測試。

R-CNN 依然不盡理想的原因如下:

1. 每張圖經由區域推薦處理過後,各會產生出 2000 個候選框,然後每個框都需經過辨識,執行時間還是過長,而且區域推薦也不具備自我學習能力。

2. 接著再透過 CNN 模型萃取 4096 個特徵向量,合計有 2000 x 4096 = 8,192,000 個特徵向量,記憶體消耗也很大。

3. 每筆資料都要經過 CNN、SVM、迴歸三個模型的訓練與預測,過於複雜。

總體而論,物件偵測不只追求高準確率,更要求能夠即時偵測,像是自駕車,總不能等撞到障礙物後才偵測到,那就悲劇了。原作者雖然以 Caffe(C++)開發 R-CNN,企圖縮短偵測時間,但仍需要 40 多秒才能偵測一張圖像,因此引發一波演算法的改良浪潮,參閱圖 8.6。

接下來,我們就來介紹各個改良演算法的發想。

8-5 R-CNN 改良

首先 Kaiming He 等學者提出 SPP-Net(Spatial Pyramid Pooling in Deep Convolutional Networks for Visual Recognition)演算法,針對 R-CNN 把每個候選框都視為單一圖像並需經過辨識的缺點,進行改良,作法如下:

1. R-CNN 一個尺寸候選框就占用掉一個 CNN 模型,而 SPP-Net 則是一張圖的全部候選框都只用一個 CNN。作者所提出的「Spatial pyramid pooling」(SPP),概念是不管圖像尺寸大小,它都能產生一個固定長度的輸出,因此,作者在最後一個卷積層上增加了一個 SPP 層,這樣就能接上一個可輸入固定大小維度的 Dense。

2. 之後的流程與 R-CNN 一樣。

R-CNN 與 SPP-Net 的模型結構比較，示意圖如下：

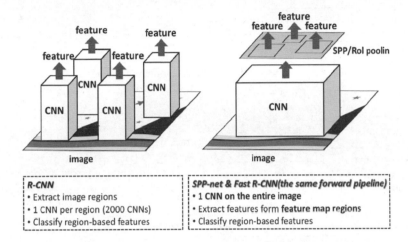

R-CNN
* Extract image regions
* 1 CNN per region (2000 CNNs)
* Classify region-based features

SPP-net & Fast R-CNN(the same forward pipeline)
* 1 CNN on the entire image
* Extract features form **feature map regions**
* Classify region-based features

圖 8.12 R-CNN 與 SPP-Net 模型結構比較

SPP 還是有缺點：

1. 雖然解決了 CNN 計算過多的狀況，但沒有處理分類(SVM)與迴歸過慢的問題。
2. 特徵向量太占記憶體空間。

詳細處理流程可參閱「Spatial Pyramid Pooling in Deep Convolutional Networks for Visual Recognition」[14]一文，中文說明可參閱「SPP-Net 論文詳解」[15]的內容。

接著 Ross B. Girshick 接續提出 Fast R-CNN、Faster R-CNN 等演算法。

Fast R-CNN 作法：

1. 將原始圖像直接經由 CNN 轉成特徵向量，不用再藉由個別候選框轉換。
2. 透過候選框與原始圖像的對照關係，換算出每個候選框的特徵向量。
3. 之後的流程與 R-CNN 一樣。

Fast R-CNN 模型結構如下：

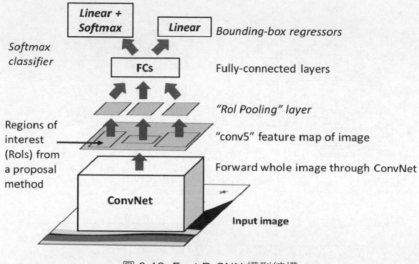

圖 8.12 Fast R-CNN 模型結構

優點：

1. CNN 模型只需訓練原圖就好，不用訓練 2000 個候選框。
2. 透過 ROI pooling 得到固定尺寸的特徵後，只要連接一個 Dense 進行分類即可。

缺點：

1. 用區域推薦演算法找 2000 張候選框，耗時太久。

Ross B. Girshick 決定放棄使用 selective search，引進 RPN (Region Proposal Network) 神經層，開發 Faster R-CNN 模型，在訓練的階段挑選 9 個尺寸的框，而這 9 個框稱為 Anchor Box，然後再利用滑動視窗找出要比對的候選框。

Faster R-CNN 模型結構如下：

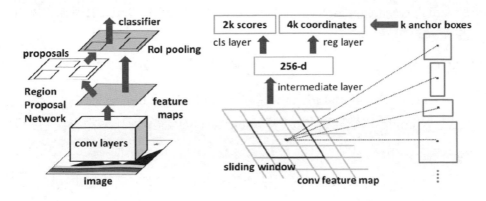

圖 8.13 Faster R-CNN 模型結構

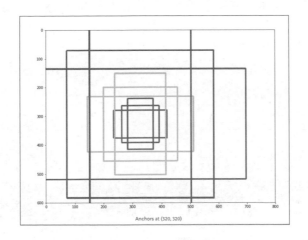

圖 8.14 Anchor Box

雖然 Ross B. Girshick 在 GitHub 放上 Faster R-CNN 程式碼[16]，但安裝不僅超複雜，執行環境的要求也很高(Caffe/C++)，所以建議大家直接使用 Detectron 套件，目前已開發至第二版(Detectron 2)，它使用 PyTorch 框架，只能安裝在 Linux/Mac 環境，Windows 使用者可以在 Google Colaboratory 上進行測試。

詳細說明請參閱「Getting Started with Detectron2」：

https://detectron2.readthedocs.io/en/latest/tutorials/getting_started.html

開啟範例檔：

https://colab.research.google.com/drive/16jcaJoc6bCFAQ96jDe2HwtXj7BMD_-m5。

【範例 5】使用 Detectron2 套件進行物件偵測。

程式：Detectron2 Tutorial.ipynb，需在 Google Colaboratory 上執行，請上傳程式至 Google 雲端硬碟，接著再 double click 檔案即可。記得要在選單「執行階段」選取 GPU。

1. 確認 PyTorch、gcc 安裝 OK，且 PyTorch 版本須為 1.7 或以上。
2. 安裝 Detectron2 套件。
3. 自 Model Zoo 下載 Detectron2 預先訓練的模型。
4. 預測。

```
cfg = get_cfg()
# add project-specific config (e.g., TensorMask) here if you're not running a model in detectron2's core library
cfg.merge_from_file(model_zoo.get_config_file("COCO-InstanceSegmentation/mask_rcnn_R_50_FPN_3x.yaml"))
cfg.MODEL.ROI_HEADS.SCORE_THRESH_TEST = 0.5  # set threshold for this model
# Find a model from detectron2's model zoo. You can use the https://dl.fbaipublicfiles... url as well
cfg.MODEL.WEIGHTS = model_zoo.get_checkpoint_url("COCO-InstanceSegmentation/mask_rcnn_R_50_FPN_3x.yaml")
predictor = DefaultPredictor(cfg)
outputs = predictor(im)
```

5. 顯示物件偵測結果。

執行結果：效果超好，非常厲害，就連背景中旁觀的人群都可以被正確偵測。

6. 上傳之前用 ResNet50 偵測結果失敗的斑馬照片，測試看看。

```
# Upload the results
from google.colab import files
files.upload()
```

7. 讀取檔案，進行物件偵測。

```
# 讀取檔案，進行物件偵測
im = cv2.imread("./zebra.jpg")
cv2_imshow(im)
predictor = DefaultPredictor(cfg)
outputs = predictor(im)
outputs
```

執行結果，三個框：

```
[ 46.5412,  94.6141, 234.9006, 258.9107],
[180.8245,  86.8508, 418.6142, 261.7740],
[342.8438, 103.8605, 563.8304, 266.2300]
```

和三個信賴度：[0.9992, 0.9986, 0.9983]，機率都相當高。

8. 顯示物件偵測結果。

```
v = Visualizer(im[:, :, ::-1], MetadataCatalog.get(cfg.DATASETS.TRAIN[0]), scale=1.2)
out = v.draw_instance_predictions(outputs["instances"].to("cpu"))
cv2_imshow(out.get_image()[:, :, ::-1])
```

執行結果：

這個套件真的超強，除了成功抓到所有物件之外，更是已經做到了實例分割(Instance Segmentation)，掃描到的物件不僅有框(Bounding Box)，還有準確的遮罩(Mask)。

檔案後面還示範了以下功能：

1. 使用自訂的資料集，偵測自己有興趣的物件。在 Google Colaboratory 上訓練只需幾分鐘的時間就可以完成。
2. 人體骨架的偵測。
3. 全景影片的物件偵測(筆者測試時有出現錯誤)。

8-5 YOLO 演算法簡介

由於 R-CNN 屬兩階段(Two Stage)的演算法，第一階段先利用區域推薦找出候選框，第二階段才是進行物件偵測，所以在偵測速度上始終是一個瓶頸，難以滿足即時偵測的要求，後來有學者提出了一階段(Single Shot)的演算法，主要區分為兩類：YOLO 及 SSD。

R-CNN 經過一連串的改良後，物件偵測的速度比較，如下表，最新版速度比原版增快了 250 倍。

	R-CNN	Fast R-CNN	Faster R-CNN
Test Time per Image	50 Seconds	2 Seconds	0.2 Seconds
Speed Up	1x	25x	250x

圖 8.15 R-CNN 各演算法之物件偵測的速度

看似很好了，然而 YOLO 發明人 Joseph Redmon 在 2016 年的 CVPR 研討會(You Only Look Once: Unified, Real-Time Object Detection)上，有兩張投影片非常有意思，一輛轎車平均車身長約 8 英呎(Feet)，假如使用 Faster R-CNN 偵測下一個路況的話，車子早已行駛了 12 英呎，也就是車子又開了 1 又 1/2 個車身的距離，相對的，如果使用 YOLO 偵測下一個路況，車子則只行駛了 2 英呎，即 1/4 個車身的距離，安全性是否會提高許多？相信答案已不言而喻，非常有說服力。

	Pascal 2007 mAP	Speed	
DPM v5	33.7	.07 FPS	14 s/img
R-CNN	66.0	.05 FPS	20 s/img
Fast R-CNN	70.0	.5 FPS	2 s/img
Faster R-CNN	73.2	7 FPS	140 ms/img

8 feet

12 feet

圖 8.16 Faster R-CNN 演算法物件偵測的速度

	Pascal 2007 mAP	Speed	
DPM v5	33.7	.07 FPS	14 s/img
R-CNN	66.0	.05 FPS	20 s/img
Fast R-CNN	70.0	.5 FPS	2 s/img
Faster R-CNN	73.2	7 FPS	140 ms/img
YOLO	63.4	45 FPS	22 ms/img

2 feet

圖 8.17 YOLO 演算法物件偵測的速度

YOLO(You Only Look Once)是現在最夯的物件偵測演算法，於 2016 年由 Joseph Redmon 提出，他本人開發至第三版，但因某些因素離開此研究領域，其他學者繼續接手，直至 2020 年已開發到第五版了。

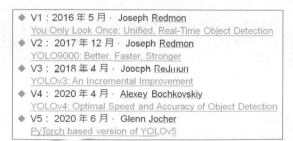

◆ V1：2016 年 5 月，Joseph Redmon
You Only Look Once: Unified, Real-Time Object Detection
◆ V2：2017 年 12 月，Joseph Redmon
YOLO9000: Better, Faster, Stronger
◆ V3：2018 年 4 月，Joocph Redmon
YOLOv3: An Incremental Improvement
◆ V4：2020 年 4 月，Alexey Bochkovskiy
YOLOv4: Optimal Speed and Accuracy of Object Detection
◆ V5：2020 年 6 月，Glenn Jocher
PyTorch based version of YOLOv5

圖 8.18 YOLO 版本演進

YOLO 各版本的平均準確度(mAP)與速度的比較，如下圖所示：

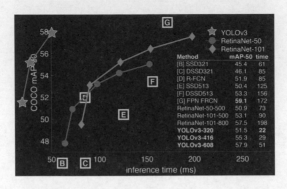

圖 8.19 YOLO 版本 v1~v3 的比較，圖片來源：YOLO 官網[17]

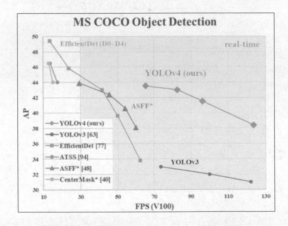

圖 8.20 YOLO 版本 v4、v3 的比較

圖片來源：YOLOv4: Optimal Speed and Accuracy of Object Detection[18]

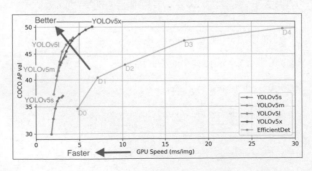

圖 8.21 YOLO 版本 v5 的各模型比較，圖片來源：YOLO5 GitHub[19]

YOLO 的快速，部分是犧牲準確率所換來的，它作法如下：

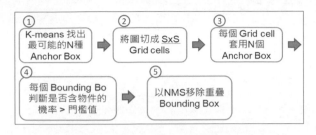

圖 8.22　YOLO 的處理流程

1. 放棄區域推薦，以集群演算法 K-Means，從訓練資料中找出最常見的 N 種尺寸的 Anchor Box。

2. 直接將圖像劃分成(s, s)個網格(Grid)：每個網格只檢查多種不同尺寸的 Anchor Box 是否含有物件而已。

3. 輸入 CNN 模型，計算每個 Anchor Box 含有物件的機率。

4. 同時計算每一個網格可能含有各種物件的機率，假設每一網格最多只含一個物件。

5. 合併步驟 3、4 的資訊，並找出合格的候選框。

6. 以 NMS 移除重疊 Bounding Box。

觀察下面示意圖，有助於 YOLO 的理解。

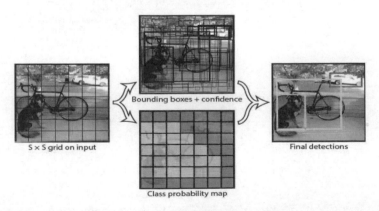

圖 8.23　YOLO 處理流程的示意圖

圖片來源：You Only Look Once: Unified, Real-Time Object Detection[20]

YOLO 為求速度快，程式碼採用 C/CUDA 開發，稱為 Darknet 架構，本書不剖析原始程式碼，只聚焦下列重點：

1. 環境建置。
2. 範例應用。
3. 自訂資料集。

8-6 YOLO 環境建置

本書以 YOLO v4 為例來示範環境建置的步驟，其他版本也差不多。官網同時提供 Linux 與 Windows 作業系統下的建置程序，在 Linux 上用 GCC 編譯器建置比較簡單，不過筆者習慣使用 Windows 作業系統，因此，本文主要是介紹如何在 Windows 下建置 Darknet。

先介紹比較簡單的方式，再聊聊官網所建議的方式，因為大神 Joseph Redmon 已經不玩了，後續有很多學者投入開發，所以 YOLO v4 在 GitHub 上百花齊放有非常多的版本，筆者就以 Alexey Bochkovskiy 的版本來說明建置步驟。

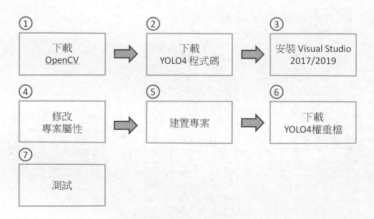

1. 下載 OpenCV：自 OpenCV 官網（https://opencv.org/releases/）下載 OpenCV Windows 版，要注意是 C 的版本，不是 OpenCV-Python 喔。

2. 解壓縮至 c:\ 或 d:\，以下假設安裝在 d:\。

3. 下載 YOLO4 程式碼：自 https://github.com/AlexeyAB/darknet 下載程式碼，並解壓縮，以下假設安裝在 D:\darknet-master。

4. 安裝 Visual Studio 2017 或 2019，以 Visual Studio 開啟 D:\darknet-master\build\darknet\darknet.sln 檔案，就會出現升級視窗，點選「確定」。提醒一下，若無 NVidia 獨立顯卡，請改成開啟 darknet_no_gpu.sln。

5. 將專案組態(Configuration)改為 x64 (64 位元)。

6. 修改專案屬性，編修「VC++ Directories」>「Include Directories」，加上：

```
D:\opencv\build\include
D:\opencv\build\include\opencv2
```

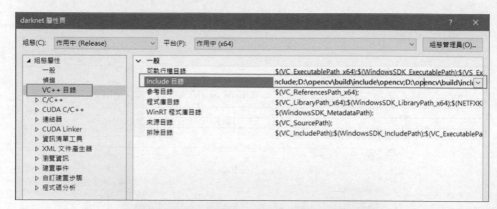

7. 編修「連結器」>「輸入」>「其他相依性」，加上：

```
D:\openCV\build\x64\vc15\lib\opencv_world430.lib
```

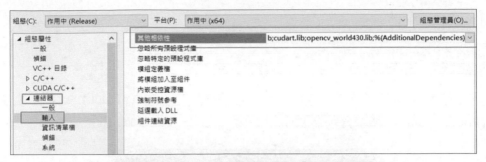

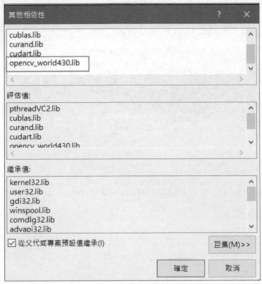

8. 在 darknet 專案上按滑鼠右鍵,選「重建」,若出現「建置成功」,即表示大功告成了,執行檔放在 D:\darknet-master\build\darknet\x64 目錄下。

9. 複製 D:\openCV\build\x64\vc15\lib\opencv_world430.lib 至 D:\darknet-master\build\darknet\x64 目錄下。留意若是使用 VS 2017,目錄應改為 vc14。

10. 依照 https://github.com/AlexeyAB/darknet 指示,從 https://github.com/AlexeyAB/darknet/releases/download/darknet_yolo_v3_optimal/yolov4.weights 下載 yolov4.weights,放入 D:\darknet-master\build\darknet\x64\weights 目錄中,如果不存在,可建立此目錄。

11. 執行下列指令進行測試：

```
darknet.exe
detect .\cfg\yolov4.cfg .\weights\yolov4.weights .\data\dog.jpg
```

執行結果：可以偵測到自行車(bike)、狗(dog)、貨車(Truck)、盆栽植物
(Potted plant)，機率分別為 92%、98%、92%、33%。

12. 另外目錄下還有許多.cmd 指令檔可以測試。

13. 若要使用 Python 直接呼叫 Darknet API，需建置 yolo_cpp_dll.sln，專案
    屬性不需任何修改。

14. 複製必要的函數庫至目前的目錄(D:\darknet-master\build\darknet\x64)
    下：

```
D:\darknet-master\3rdparty\pthreads\bin\pthreadGC2.dll、
pthreadVC2.dll。
D:\openCV\build\bin\*.dll。
D:\openCV\build\x64\vc15\lib\*.lib。
```

15. 使用 python 呼叫 yolo_cpp_dll.dll，執行下列指令來測試：

```
python darknet.py
```

darknet.py 內含 performBatchDetect 函數，可一次測試多個檔案。

執行結果會出現以下錯誤：

```
Traceback (most recent call last):
  File "darknet.py", line 211, in <module>
    lib = CDLL(winGPUdll, RTLD_GLOBAL)
  File "C:\Anaconda3\lib\ctypes\__init__.py", line 373, in __init__
    self._handle = _dlopen(self._name, mode)
FileNotFoundError: Could not find module 'yolo_cpp_dll.dll' (or one of its dependencies)
```

筆者之前曾經正確執行過，應該是欠缺某些 yolo_cpp_dll.dll 依賴的檔案所導致。後續會在下一節介紹 TensorFlow 使用 YOLO 權重檔，來排除此錯誤。

16. 使用 C++呼叫 yolo_cpp_dll.dll，可先編譯 yolo_console_dll.sln，然後再執行下列指令進行測試：

```
yolo_console_dll.exe data/coco.names cfg/yolov4.cfg weights/
yolov4.weights dog.jpg
```

執行結果如下，包括物件名稱/Id/框座標和寬高/機率：

```
bicycle - obj_id = 1,  x = 114, y = 127, w = 458, h = 298, prob = 0.923
dog - obj_id = 16,  x = 128, y = 225, w = 184, h = 316, prob = 0.979
car - obj_id = 2,  x = 468, y = 76, w = 211, h = 92, prob = 0.229
truck - obj_id = 7,  x = 463, y = 76, w = 220, h = 93, prob = 0.923
```

注意! 假如使用 VS 2019，須更改下列事項：

1. 修改 darknet.vcxproj、yolo_cpp_dll.vcxproj，將所有的 CUDA 10.0 改為對應的版本，因為 v10.0 只支援 VS 2015、VS 2017。

2. 複製 C:\Program Files\NVIDIA GPU Computing Toolkit\CUDA\v10.1\extras\visual_studio_integration\MSBuildExtensions*.* 至 「C:\Program Files (x86)\Microsoft Visual Studio\2019\Community\MSBuild\Microsoft\VC\v160\BuildCustomizations」目錄內。

3. 修改專案屬性 > C/C++ > 命令列，在其他選項加 /FS，點選「套用」鈕，清除舊專案後，再建置新的專案。

4. 若出現「dropout_layer_kernels.cu error code 2」的錯誤，則在「工具>選項>專案和方案>建置並執行」中的「平時專案組見的最大數目」項目修改為 1。

另外，官網介紹兩種 Windows 版的建置方法：

1. CMake：官網比較建議的方式。

2. vcpkg：程序相對複雜。

第一種方法雖然很順利地建置成功，但卻在測試時發生了以下錯誤：

```
Done! Loaded 162 layers from weights-file
CUDA status Error: file: F:/darknet-master/src/blas_kernels.cu : add_bias_gpu()
09:29:26

 CUDA Error: invalid device function

CUDA Error: invalid device function: Invalid argument
```

第二種方法較花時間,因為它會下載許多軟體,包括 NVidia SDK、ffmpeg 等原始程式碼,需重新建置,所以要耐心等候。

1. 前置作業須先安裝下列軟體:

- CMake (https://cmake.org/download/)

- VS 2017 或 2019:須安裝 VC toolset、English language pack 元件。

- NVidia CUDA SDK:CUDA 版本 > 10.0,cuDNN 版本 > 7.0。

- OpenCV:版本須 > 2.4。

- Git for Windows (https://git-scm.com/download/win)。

2. 複製 C:\Program Files\NVIDIA GPU Computing Toolkit\CUDA\v10.1\extras\visual_studio_integration\MSBuildExtensions*.* 至「C:\Program Files (x86)\Microsoft Visual Studio\2019\Community\MSBuild\Microsoft\VC\v160\BuildCustomizations」目錄內。

3. 建立一個新目錄,在「開始」按滑鼠右鍵,開啟「Windows Powershell」,執行以下指令:

```
cd <新目錄>
git clone https://github.com/microsoft/vcpkggit clone
https://github.com/microsoft/vcpkg
cd vcpkg
$env:VCPKG_ROOT=$PWD
.\bootstrap-vcpkg.bat
.\vcpkg install darknet[opencv-base,cuda,cudnn]:x64-windows
```

執行約需要 20 分鐘。

```
cd ..
git clone https://github.com/AlexeyAB/darknet
cd darknet
powershell -ExecutionPolicy Bypass -File .\build.ps1
```

執行約需要 10 分鐘。

大功告成後，執行檔會放在 darknet\build_win_release 目錄內，自 https://github.com/AlexeyAB/darknet/releases/download/darknet_yolo_v3_opt imal/yolov4.weights 下載 yolov4.weights，放入 build_win_release\weights 目錄，執行

```
darknet.exe
detect ..\cfg\yolov4.cfg .\weights\yolov4.weights ..\data\dog.jpg
```

執行結果：可以偵測到自行車(bike)、狗(dog)、貨車(Truck)、盆栽植物 (Potted plant)，機率分別為 92%、98%、92%、33%。

YOLO 特別強調速度，因此，要實際應用於專案，應採用 C/C++建置辨識 的模組，可依照 darknet-master\build\darknet\yolo_console_dll.sln 方案來進 行修改。

8-7 以 TensorFlow 實作 YOLO 模型

要使用 C/C++來開發程式，大部份的人可能都面有難色，所以網路上有許多程式碼，讓 TensorFlow/Keras 也可以使用 YOLO 模型，主要的方式有兩種：

1. 將 YOLO 權重檔轉為 TensorFlow/Keras 格式檔(.h5 或 SaveModel)。
2. 直接使用 YOLO 權重檔。

「How to Perform Object Detection With YOLOv3 in Keras」[21] 一文説明將 YOLO 權重檔轉為 Keras 格式檔(.h5)的步驟，很簡單，用 Keras 重建 YOLO 模型，並載入權重檔，變成完成訓練的模型，之後再呼叫 Save()即可。

【範例 6】將 YOLO 權重檔轉為 Keras 格式檔(.h5)。

程式：08_05_YOLO_Keras_Conversion.ipynb。

1. 載入相關套件。

```
1   # 載入套件
2   import struct
3   import numpy as np
4   from tensorflow.keras.layers import Conv2D
5   from tensorflow.keras.layers import Input
6   from tensorflow.keras.layers import BatchNormalization
7   from tensorflow.keras.layers import LeakyReLU
8   from tensorflow.keras.layers import ZeroPadding2D
9   from tensorflow.keras.layers import UpSampling2D
10  from tensorflow.keras.layers import add, concatenate
11  from tensorflow.keras.models import Model
```

2. 定義模型的卷積層。

```
1   # 定義建立卷積層的函數
2   def _conv_block(inp, convs, skip=True):
3       x = inp
4       count = 0
5       for conv in convs:
6           if count == (len(convs) - 2) and skip:
7               skip_connection = x
8           count += 1
9           # darknet 特殊的 Padding 設計，只補零左/上邊
10          if conv['stride'] > 1: x = ZeroPadding2D(((1,0),(1,0)))(x)
11          x = Conv2D(conv['filter'],
12                     conv['kernel'],
13                     strides=conv['stride'],
14                     padding='valid' if conv['stride'] > 1 else 'same',
15                     name='conv_' + str(conv['layer_idx']),
16                     use_bias=False if conv['bnorm'] else True)(x)
17          # 加 BatchNormalization 層
18          if conv['bnorm']: x = BatchNormalization(epsilon=0.001,
19                                  name='bnorm_' + str(conv['layer_idx']))(x)
20          # 使用 LeakyReLU，而非 ReLU
21          if conv['leaky']: x = LeakyReLU(alpha=0.1, name='leaky_'
22                                  + str(conv['layer_idx']))(x)
23
24      return add([skip_connection, x]) if skip else x
```

3. 定義建立 YOLO v3 模型。

```
1   # 定義建立 YOLO v3 模型的函數
2   def make_yolov3_model():
3       input_image = Input(shape=(None, None, 3))
4       # Layer  0 => 4
5       x = _conv_block(input_image, [{'filter': 32, 'kernel': 3, 'stride': 1, 'bnorm': True, 'leaky': True, 'layer_idx': 0},
6                                     {'filter': 64, 'kernel': 3, 'stride': 2, 'bnorm': True, 'leaky': True, 'layer_idx': 1},
7                                     {'filter': 32, 'kernel': 1, 'stride': 1, 'bnorm': True, 'leaky': True, 'layer_idx': 2},
8                                     {'filter': 64, 'kernel': 3, 'stride': 1, 'bnorm': True, 'leaky': True, 'layer_idx': 3}])
9       # Layer  5 => 8
10      x = _conv_block(x, [{'filter': 128, 'kernel': 3, 'stride': 2, 'bnorm': True, 'leaky': True, 'layer_idx': 5},
11                          {'filter':  64, 'kernel': 1, 'stride': 1, 'bnorm': True, 'leaky': True, 'layer_idx': 6},
12                          {'filter': 128, 'kernel': 3, 'stride': 1, 'bnorm': True, 'leaky': True, 'layer_idx': 7}])
13      # Layer  9 => 11
14      x = _conv_block(x, [{'filter':  64, 'kernel': 1, 'stride': 1, 'bnorm': True, 'leaky': True, 'layer_idx': 9},
15                          {'filter': 128, 'kernel': 3, 'stride': 1, 'bnorm': True, 'leaky': True, 'layer_idx': 10}])
16      # Layer 12 => 15
17      x = _conv_block(x, [{'filter': 256, 'kernel': 3, 'stride': 2, 'bnorm': True, 'leaky': True, 'layer_idx': 12},
18                          {'filter': 128, 'kernel': 1, 'stride': 1, 'bnorm': True, 'leaky': True, 'layer_idx': 13},
19                          {'filter': 256, 'kernel': 3, 'stride': 1, 'bnorm': True, 'leaky': True, 'layer_idx': 14}])
20      # Layer 16 => 36
21      for i in range(7):
22          x = _conv_block(x, [{'filter': 128, 'kernel': 1, 'stride': 1, 'bnorm': True, 'leaky': True, 'layer_idx': 16+i*3},
23                              {'filter': 256, 'kernel': 3, 'stride': 1, 'bnorm': True, 'leaky': True, 'layer_idx': 17+i*3}])
24      skip_36 = x
25      # Layer 37 => 40
26      x = _conv_block(x, [{'filter': 512, 'kernel': 3, 'stride': 2, 'bnorm': True, 'leaky': True, 'layer_idx': 37},
27                          {'filter': 256, 'kernel': 1, 'stride': 1, 'bnorm': True, 'leaky': True, 'layer_idx': 38},
28                          {'filter': 512, 'kernel': 3, 'stride': 1, 'bnorm': True, 'leaky': True, 'layer_idx': 39}])
29      # Layer 41 => 61
30      for i in range(7):
31          x = _conv_block(x, [{'filter': 256, 'kernel': 1, 'stride': 1, 'bnorm': True, 'leaky': True, 'layer_idx': 41+i*3},
32                              {'filter': 512, 'kernel': 3, 'stride': 1, 'bnorm': True, 'leaky': True, 'layer_idx': 42+i*3}])
33      skip_61 = x
34      # Layer 62 => 65
35      x = _conv_block(x, [{'filter': 1024, 'kernel': 3, 'stride': 2, 'bnorm': True, 'leaky': True, 'layer_idx': 62},
36                          {'filter':  512, 'kernel': 1, 'stride': 1, 'bnorm': True, 'leaky': True, 'layer_idx': 63},
37                          {'filter': 1024, 'kernel': 3, 'stride': 1, 'bnorm': True, 'leaky': True, 'layer_idx': 64}])
```

```
38    # Layer 66 => 74
39    for i in range(3):
40        x = _conv_block(x, [{'filter':  512, 'kernel': 1, 'stride': 1, 'bnorm': True, 'leaky': True, 'layer_idx': 66+i*3},
41                            {'filter': 1024, 'kernel': 3, 'stride': 1, 'bnorm': True, 'leaky': True, 'layer_idx': 67+i*3}])
42    # Layer 75 => 79
43    x = _conv_block(x, [{'filter':  512, 'kernel': 1, 'stride': 1, 'bnorm': True, 'leaky': True, 'layer_idx': 75},
44                        {'filter': 1024, 'kernel': 3, 'stride': 1, 'bnorm': True, 'leaky': True, 'layer_idx': 76},
45                        {'filter':  512, 'kernel': 1, 'stride': 1, 'bnorm': True, 'leaky': True, 'layer_idx': 77},
46                        {'filter': 1024, 'kernel': 3, 'stride': 1, 'bnorm': True, 'leaky': True, 'layer_idx': 78},
47                        {'filter':  512, 'kernel': 1, 'stride': 1, 'bnorm': True, 'leaky': True, 'layer_idx': 79}], skip=Fal
48    # Layer 80 => 82
49    yolo_82 = _conv_block(x, [{'filter': 1024, 'kernel': 3, 'stride': 1, 'bnorm': True,  'leaky': True,  'layer_idx': 80},
50                              {'filter':  255, 'kernel': 1, 'stride': 1, 'bnorm': False, 'leaky': False, 'layer_idx': 81}],
```

```
51    # Layer 83 => 86
52    x = _conv_block(x, [{'filter': 256, 'kernel': 1, 'stride': 1, 'bnorm': True, 'leaky': True, 'layer_idx': 84}], skip=Fals
53    x = UpSampling2D(2)(x)
54    x = concatenate([x, skip_61])
55    # Layer 87 => 91
56    x = _conv_block(x, [{'filter': 256, 'kernel': 1, 'stride': 1, 'bnorm': True, 'leaky': True, 'layer_idx': 87},
57                        {'filter': 512, 'kernel': 3, 'stride': 1, 'bnorm': True, 'leaky': True, 'layer_idx': 88},
58                        {'filter': 256, 'kernel': 1, 'stride': 1, 'bnorm': True, 'leaky': True, 'layer_idx': 89},
59                        {'filter': 512, 'kernel': 3, 'stride': 1, 'bnorm': True, 'leaky': True, 'layer_idx': 90},
60                        {'filter': 256, 'kernel': 1, 'stride': 1, 'bnorm': True, 'leaky': True, 'layer_idx': 91}], skip=Fals
61    # Layer 92 => 94
62    yolo_94 = _conv_block(x, [{'filter': 512, 'kernel': 3, 'stride': 1, 'bnorm': True,  'leaky': True,  'layer_idx': 92},
63                              {'filter': 255, 'kernel': 1, 'stride': 1, 'bnorm': False, 'leaky': False, 'layer_idx': 93}], s
64    # Layer 95 => 98
65    x = _conv_block(x, [{'filter': 128, 'kernel': 1, 'stride': 1, 'bnorm': True, 'leaky': True,  'layer_idx': 96}], skip=Fa
66    x = UpSampling2D(2)(x)
67    x = concatenate([x, skip_36])
68    # Layer 99 => 106
69    yolo_106 = _conv_block(x, [{'filter': 128, 'kernel': 1, 'stride': 1, 'bnorm': True,  'leaky': True,  'layer_idx': 99},
70                               {'filter': 256, 'kernel': 3, 'stride': 1, 'bnorm': True,  'leaky': True,  'layer_idx': 100},
71                               {'filter': 128, 'kernel': 1, 'stride': 1, 'bnorm': True,  'leaky': True,  'layer_idx': 101},
72                               {'filter': 256, 'kernel': 3, 'stride': 1, 'bnorm': True,  'leaky': True,  'layer_idx': 102},
73                               {'filter': 128, 'kernel': 1, 'stride': 1, 'bnorm': True,  'leaky': True,  'layer_idx': 103},
74                               {'filter': 256, 'kernel': 3, 'stride': 1, 'bnorm': True,  'leaky': True,  'layer_idx': 104},
75                               {'filter': 255, 'kernel': 1, 'stride': 1, 'bnorm': False, 'leaky': False, 'layer_idx': 105}],
76    model = Model(input_image, [yolo_82, yolo_94, yolo_106])
77    return model
```

4. 讀取 YOLO 權重檔。

```
1     # 定義讀取 YOLO 權重檔的類別
2     class WeightReader:
3         def __init__(self, weight_file):
4             with open(weight_file, 'rb') as w_f:
5                 major,    = struct.unpack('i', w_f.read(4))
6                 minor,    = struct.unpack('i', w_f.read(4))
7                 revision, = struct.unpack('i', w_f.read(4))
8                 if (major*10 + minor) >= 2 and major < 1000 and minor < 1000:
9                     w_f.read(8)
10                else:
11                    w_f.read(4)
12                transpose = (major > 1000) or (minor > 1000)
13                binary = w_f.read()
14            self.offset = 0
15            self.all_weights = np.frombuffer(binary, dtype='float32')
16
17        def read_bytes(self, size):
18            self.offset = self.offset + size
19            return self.all_weights[self.offset-size:self.offset]
```

```
21        def load_weights(self, model):
22            for i in range(106):
23                try:
24                    conv_layer = model.get_layer('conv_' + str(i))
25                    print("loading weights of convolution #" + str(i))
26                    if i not in [81, 93, 105]:
27                        norm_layer = model.get_layer('bnorm_' + str(i))
28                        size = np.prod(norm_layer.get_weights()[0].shape)
29                        beta  = self.read_bytes(size) # bias
30                        gamma = self.read_bytes(size) # scale
31                        mean  = self.read_bytes(size) # mean
32                        var   = self.read_bytes(size) # variance
33                        weights = norm_layer.set_weights([gamma, beta, mean, var])
34                    if len(conv_layer.get_weights()) > 1:
35                        bias   = self.read_bytes(np.prod(conv_layer.get_weights()[1].shape))
36                        kernel = self.read_bytes(np.prod(conv_layer.get_weights()[0].shape))
37                        kernel = kernel.reshape(list(reversed(conv_layer.get_weights()[0].shape)))
38                        kernel = kernel.transpose([2,3,1,0])
39                        conv_layer.set_weights([kernel, bias])
40                    else:
41                        kernel = self.read_bytes(np.prod(conv_layer.get_weights()[0].shape))
42                        kernel = kernel.reshape(list(reversed(conv_layer.get_weights()[0].shape)))
43                        kernel = kernel.transpose([2,3,1,0])
44                        conv_layer.set_weights([kernel])
45                except ValueError:
46                    print("no convolution #" + str(i))
47
48        def reset(self):
49            self.offset = 0
```

5. 重建模型：結合模型結構和權重檔，並以 Keras 格式存檔。

```
1  # 建立模型
2  model = make_yolov3_model()
3  # 載入權重檔
4  weight_reader = WeightReader('./YOLO_weights/yolov3.weights')
5  weight_reader.load_weights(model)
6
7  # 轉為TensorFlow/Keras格式檔
8  model.save('yolov3.h5')
```

【範例 7】測試 Keras 格式檔(.h5)。

程式：08_06_YOLO_Keras_Test.ipynb、yolo_keras_utils.py。

1. 載入相關套件。

```
1  # 載入套件
2  from yolo_keras_utils import *
3  from tensorflow.keras.layers import Conv2D
4  from tensorflow.keras.layers import Input
5  from tensorflow.keras.layers import BatchNormalization
6  from tensorflow.keras.layers import LeakyReLU
7  from tensorflow.keras.layers import ZeroPadding2D
8  from tensorflow.keras.layers import UpSampling2D
9  from tensorflow.keras.layers import add, concatenate
10 from tensorflow.keras.models import Model
```

2. 測試：預測的輸出還須經過轉換。

```
1   # 測試
2   from tensorflow.keras.models import load_model
3
4   image_filename = './images_Object_Detection/zebra.jpg' # 測試圖像
5
6   model = load_model('./yolov3.h5')    # 載入模型
7   input_w, input_h = 416, 416          # YOLO v3 圖像尺寸
8   # 載入圖像，並縮放尺寸為 (416, 416)
9   image, image_w, image_h = load_image_pixels(image_filename, (input_w, input_h))
10  # 預測圖像
11  yhat = model.predict(image)
12  # 傳回偵測的物件資訊
13  print([a.shape for a in yhat])
```

執行結果：得到 3 個物件資訊。

```
[(1, 13, 13, 255), (1, 26, 26, 255), (1, 52, 52, 255)]
```

3. 輸出轉換：使用 NMS，移除重疊的 Bounding Box。

```
1   # 輸出轉換
2   # 每個陣列內前兩個值為grid寬/高，後四個為 anchors 的作標與尺寸
3   anchors = [[116,90, 156,198, 373,326], [30,61, 62,45, 59,119], [10,13, 16,30, 33,23]]
4
5   # 設定物件偵測的機率門檻
6   class_threshold = 0.6
7
8   # 依 anchors 的尺寸及機率門檻篩選 Bounding Box
9   boxes = list()
10  for i in range(len(yhat)):
11      boxes += decode_netout(yhat[i][0], anchors[i], class_threshold, input_h, input_w)
12
13  # 依原圖尺寸與縮放尺寸的比例，校正 Bounding Box 尺寸
14  correct_yolo_boxes(boxes, image_h, image_w, input_h, input_w)
15
16  #  使用 non-maximal suppress，移除重疊的 Bounding Box
17  do_nms(boxes, 0.5)
```

4. 取得 Bounding Box 資訊：座標、類別、機率，並進行繪圖。

```
1   # 取得 Bounding Box 資訊：座標、類別、機率
2   v_boxes, v_labels, v_scores = get_boxes(boxes, labels, class_threshold)
3
4   # 顯示執行結果
5   print(f'Bounding Box 個數：{len(v_boxes)}')
6   for i in range(len(v_boxes)):
7       print(f'類別：{v_labels[i]}, 機率：{v_scores[i]}')
8
9   # 繪圖
10  draw_boxes(image_filename, v_boxes, v_labels, v_scores)
```

執行結果：

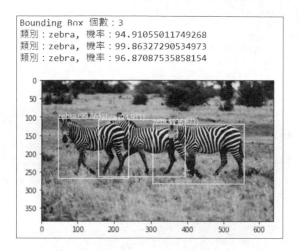

Bounding Box 個數：3
類別：zebra, 機率：94.91055011749268
類別：zebra, 機率：99.86327290534973
類別：zebra, 機率：96.87087535858154

【範例 8】直接使用 YOLO v4 權重檔。

程式：**08_07_ TensorFlow-Yolov4_Test.ipynb**、**YOLO4*.py**，此程式源自[22]。

1. 載入相關套件。

```
1  # 載入套件
2  import time
3  from absl import app, flags, logging
4  from absl.flags import FLAGS
5  import YOLO4.core.utils as utils
6  from YOLO4.core.yolov4 import YOLOv4, YOLOv3, YOLOv3_tiny, decode
7  from PIL import Image
8  import cv2
9  import numpy as np
10 import tensorflow as tf
```

2. 設定 YOLO 模型參數。

```
1  # 設定 YOLO 模型參數
2  STRIDES = np.array([8, 16, 32])
3  ANCHORS = utils.get_anchors("./YOLO4/data/anchors/yolov4_anchors.txt", False)
4  NUM_CLASS = len(utils.read_class_names("./YOLO4/data/classes/coco.names"))
5  XYSCALE = [1.2, 1.1, 1.05]  # 縮放比例
6  input_size = 608
7  yolov4_weights_path = "./YOLO_weights/yolov4.weights"
```

3. 讀取測試影像：拿 kite.jpg 進行測試。

```
1   # 讀取測試影像
2   import matplotlib.pyplot as plt
3   image_path = './YOLO4/data/kite.jpg'
4   original_image = cv2.imread(image_path)
5   original_image = cv2.cvtColor(original_image, cv2.COLOR_BGR2RGB)
6   original_image_size = original_image.shape[:2]
7   plt.imshow(original_image)
8
9   # 前置處理
10  image_data = utils.image_preporcess(np.copy(original_image),
11                                        [input_size, input_size])
12  image_data = image_data[np.newaxis, ...].astype(np.float32)
```

執行結果：

4. 結合模型結構與權重檔，建立模型。

```
1   # 結合模型結構與權重檔，建立模型
2   input_layer = tf.keras.layers.Input([input_size, input_size, 3])
3   feature_maps = YOLOv4(input_layer, NUM_CLASS)
4   bbox_tensors = []
5   for i, fm in enumerate(feature_maps):
6       bbox_tensor = decode(fm, NUM_CLASS, i)
7       bbox_tensors.append(bbox_tensor)
8   # 模型結構
9   model = tf.keras.Model(input_layer, bbox_tensors)
10
11  # 載入權重
12  utils.load_weights(model, yolov4_weights_path)
13
14  model.summary()
```

5. 預測並顯示結果：成功標示出人和飛行傘。

```
1  # 預測
2  pred_bbox = model.predict(image_data)
3
4  # 找出 Bounding Box
5  pred_bbox = utils.postprocess_bbbox(pred_bbox, ANCHORS, STRIDES, XYSCALE)
6  bboxes = utils.postprocess_boxes(pred_bbox, original_image_size, input_size, 0.25)
7
8  # 使用NMS，移除重疊的 Bounding Box
9  bboxes = utils.nms(bboxes, 0.213, method='nms')
10
11 # 原圖加框
12 image = utils.draw_bbox(original_image, bboxes)
13 image = Image.fromarray(image)
14
15 # 顯示結果
16 plt.imshow(image)
17 image.show()
```

執行結果：

8-8 YOLO 模型訓練

YOLO 預設模型是採用 COCO 資料集[23]，共有 80 個類別，如下表所示。

1	人	21	大象	41	紅酒杯	61	餐桌
2	自行車	22	熊	42	杯子	62	廁所
3	汽車	23	斑馬	43	叉子	63	電視
4	機車	24	長頸鹿	44	刀子	64	筆電
5	飛機	25	背包	45	湯匙	65	滑鼠
6	公車	26	傘	46	碗	66	遙控器
7	火車	27	手提包	47	香蕉	67	鍵盤
8	卡車	28	領帶	48	蘋果	68	手機
9	船	29	手提箱	49	三明治	69	微波爐
10	紅綠燈	30	飛盤	50	柳丁	70	烤箱
11	消防栓	31	雙板滑雪板	51	花椰菜	71	烤麵包機
12	停止標誌	32	單板滑雪板	52	紅蘿蔔	72	水槽
13	停車收費碼錶	33	運動類用球	53	熱狗	73	冰箱
14	長椅	34	風箏	54	比薩	74	書
15	鳥	35	棒球棒	55	甜甜圈	75	時鐘
16	貓	36	棒球手套	56	蛋糕	76	花瓶
17	狗	37	滑板	57	椅子	77	剪刀
18	馬	38	衝浪板	58	沙發	78	泰迪熊
19	羊	39	網球拍	59	植物盆栽	79	吹風機
20	牛	40	瓶子	60	床	80	牙刷

假若要偵測的物件不在這 80 類當中,則需自行訓練模型,大致步驟如下:

① 準備資料集 ➡ ② 標記(Labeling) ➡ ③ 模型訓練

1. 準備資料集:若只是要測試處理影像,不想製作資料集的話,可直接下載 COCO 資料集,內含影像與標註檔(Annotation),接著遵循 YOLO

步驟實作，但可能要訓練好多天，後續筆者使用「Open Images Dataset」，可以選擇部分類別，縮短測試時間。

2. 使用標記工具軟體，例如 LabelImg(https://github.com/tzutalin/labelImg)，產生 YOLO 格式的標註檔。LabelImg 安裝步驟如下：

    (1) conda install pyqt=5

    (2) conda install -c anaconda lxml

    (3) pyrcc5 -o libs/resources.py resources.qrc

    (4) 執行 LabelImg：

    python labelImg.py

圖 8.24 LabelImg 標記工具

3. 模型訓練：參閱官網的教學步驟 (https://github.com/AlexeyAB/darknet#how-to-train-to-detect-your-custom-objects)，訓練非常耗時，處理完成 300 張圖檔，大約需要 6 個小時。

【範例 9】使用自訂資料集訓練 YOLO 模型。內容參考「Create your own dataset for YOLOv4 object detection in 5 minutes」[24] 及 YOLO4 GitHub[25]這兩篇文章的做法。

1. 下載資料前置處理程式

```
git clone https://github.com/theAIGuysCode/OIDv4_ToolKit.git
```

2. 在 OIDv4_ToolKit 目錄開啟終端機(cmd)，並安裝相關套件：

```
pip install -r requirements.txt
```

3. 至「Open Images Dataset」網站(https://storage.googleapis.com/openimages/
   web/index.html)下載訓練資料，它包含 350 種類別可應用在實例分割
   (Instance Segmentation)上，我們只取三種類別來測試，避免訓練太
   久。在 OIDv4_ToolKit 目錄，執行下列指令，下載訓練資料：

```
python main.py downloader --classes Balloon Person Dog --type_csv
train --limit 200
```

注意，出現「missing files」錯誤訊息時，請輸入 y。

4. 執行下列指令，下載測試資料：

```
python main.py downloader --classes Balloon Person Dog --type_csv
test --limit 200
```

注意，出現「missing files」錯誤訊息時，請輸入 y。

5. 建立一個 classes.txt 檔案，內容如下：

```
Balloon
Person
Dog
```

6. 執行下列指令，產生 YOLO 標註檔(Annotation)，即每個影像檔都會有
   一個同名的標註檔(*.txt)：

```
python convert_annotations.py
```

標註檔的內容為：

<類別 ID> <標註框中心點 X 座標> <標註框中心點 Y 座標> <標註框寬度>
<標註框高度>。

7. 移除 OID\Dataset\train、OID\Dataset\test 子目錄下的 Label 目錄，包括：

```
OID\Dataset\train\Balloon\Label
OID\Dataset\train\Dog\Label
OID\Dataset\train\Person\Label
OID\Dataset\test\Balloon\Label
OID\Dataset\test\Dog\Label
OID\Dataset\test\Person\Label
```

8. 接著參考 YOLO4 GitHub 的說明，切換到之前建置的 darknet-master\build\darknet\x64 目錄。

9. 下載

```
https://github.com/AlexeyAB/darknet/releases/download/darknet_yolo_
v3_optimal/yolov4.conv.137
```

10. 複製 cfg/yolov4-custom.cfg 為 yolo-obj.cfg。並將 yolo-obj.cfg 進行下列更改：

- 修改第 6 行的 batch=64，改為 batch=16，筆者 GPU 記憶體只有 4GB，所以發生記憶體不足的錯誤，改為 16 即可順利執行，副作用是要花費更多的訓練時間。
- 修改第 7 行為 subdivisions=16。
- 修改第 20 行為「max_batches = 6000」，公式為類別數(3) x 2000=6000。
- 修改第 22 行為「steps=4800,5400」，為 6000 的 80%、90%。
- 修改第 8 行為 width=416，即輸入影像寬度。
- 修改第 9 行為 height=416，即輸入影像高度。
- 修改 [yolo]段落的 classes=80 改為 classes=3(第 970、1058、1146 行)。
- 修改 [yolo]段落前一個[convolutional]的 filters=255 改為 filters=24(第 963、1051、1139 行)，公式為(類別數+5) x 3=24。

11. 在 darknet-master\build\darknet\x64\data\目錄建立一個 obj.names 檔案，
    內容如下：

```
Balloon
Person
Dog
```

12. 在 darknet-master\build\darknet\x64\data\目錄建立一個 obj.data 檔案，
    內容如下：

```
classes = 3
train  = data/train.txt
valid  = data/test.txt
names = data/obj.names
backup = backup/
```

13. 複製 OID\Dataset\train、OID\Dataset\test 目錄至 darknet-master\build\
    darknet\x64\data\obj\目錄下。

14. 在 darknet-master\build\darknet\x64\data\目錄建立一個 train.txt 檔案，
    內容如下，並將每個訓練的影像檔案相對路徑放入：

```
data/obj/train/Balloon/0016f577f9811ad3.jpg
```

　　筆者寫了一支程式 gen_train.py，產生 train.txt 檔案：

```
python gen_train.py
```

15. 在 darknet-master\build\darknet\x64\data\目錄建立一個 test.txt 檔案，內
    容如下，將每個要訓練的影像檔案相對路徑放入：

```
data/obj/test/Balloon/00b585e025287555.jpg
```

　　執行程式 gen_train.py，產生 test.txt 檔案：

```
python gen_train.py test
```

16. 開啟終端機(cmd)，執行模型訓練：

```
darknet.exe detector train data/obj.data yolo-obj.cfg
yolov4.conv.137
```

- 在筆者的機器上大約執行了 8 個小時，真是苦啊，如果讀者要再自行標註影像的話，需有長期抗戰的準備。
- 若中途當機，可指定 backup 目錄下最大執行週期的檔案，繼續執行訓練：

```
darknet.exe detector train data/obj.data yolo-obj.cfg backup\yolo-obj_5000.weights
```

- 執行完成後，會產生 backup\yolo-obj_final.weights 權重檔。
- 訓練時會產生損失函數的變化，如下圖。

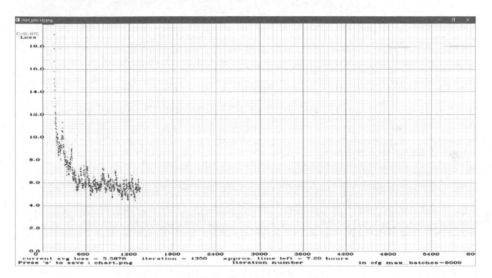

圖 8.25　YOLO 模型訓練時損失函數的變化

17. 自 test 目錄下或網路任取一檔案測試，執行下列指令：

```
darknet.exe detector test data/obj.data yolo-obj.cfg backup\yolo-obj_final.weights
```

輸入

```
D:\1\darknet-
master\build\darknet\x64\data\obj\test\Balloon\633dfe8635d30dad.jpg
```

效果不是很好，雖然有捕捉到人和氣球，不過氣球的機率(0.31)偏低，可能與第 10 步驟改成 batch=16 有關係，因原作者建議值為 64，又或者是 max_batches = 6000 應該加大吧。

圖 8.26 YOLO 模型測試結果

以上只是筆者簡單的實驗，相關的設定檔放在 code\YOLO_custom_datasets 目錄內，完整的目錄檔案過大，無法放入，請讀者見諒。上述訓練的步驟，在實際專案執行時，應該尚有一些改善空間，但最重要的還是要弄到一台高檔的 GPU 機器，用金錢換時間，畢竟人生苦短啊！

8-9 SSD 演算法

與 YOLO 齊名，Single Shot MultiBox Detector(SSD) 演算法也屬於一階段的演算法，在速度上比 R-CNN 系列演算法快，而在準確率(mAP)上比 YOLO v1 高，但卻好景不常，後來 YOLO 不斷的升級改良，SSD 網路聲量好像就變小了。

System	VOC2007 test *mAP*	FPS (Titan X)	Number of Boxes	Input resolution
Faster R-CNN (VGG16)	73.2	7	~6000	~1000 x 600
YOLO (customized)	63.4	45	98	448 x 448
SSD300* (VGG16)	77.2	46	8732	300 x 300
SSD512* (VGG16)	79.8	19	24564	512 x 512

圖 8.27 R-CNN、YOLO、SSD 比較表，資料來源：SSD 官網[26]

SSD 比較特別的地方是它採用 VGG 模型，並且在中間使用多個卷積層擷取特徵圖(Feature map)，同時進行預測。

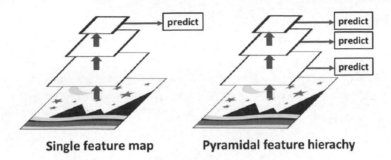

圖 8.28　左圖為 YOLO 模型，右圖為 SSD

詳細說明可參閱「一文看盡目標檢測演算法 SSD 的核心架構與設計思想」[27]。

它也是用 Caffe 架構開發的，SSD 官網[25]並未說明在 Windows 作業系統下要如何編譯，不過，TensorFlow 官方所提供的 TensorFlow Object Detection API，內含 SSD 模型，可直接使用。

8-10 TensorFlow Object Detection API

由於物件偵測應用廣泛，TensorFlow 與 PyTorch 套件都有特別提供 API，可直接呼叫相關模型，而 TensorFlow Object Detection API 就是 TensorFlow 所支援的版本。

API 包含各式的演算法，主要有 CenterNet、EfficientDet、SSD、Faster R-CNN：

Model name	Speed (ms)	COCO mAP	Outputs
CenterNet HourGlass104 512x512	70	41.9	Boxes
CenterNet HourGlass104 Keypoints 512x512	76	40.0/61.4	Boxes/Keypoints
CenterNet HourGlass104 1024x1024	197	44.5	Boxes
CenterNet HourGlass104 Keypoints 1024x1024	211	42.8/64.5	Boxes/Keypoints
CenterNet Resnet50 V1 FPN 512x512	27	31.2	Boxes
CenterNet Resnet50 V1 FPN Keypoints 512x512	30	29.3/50.7	Boxes/Keypoints
CenterNet Resnet101 V1 FPN 512x512	34	34.2	Boxes
CenterNet Resnet50 V2 512x512	27	29.5	Boxes
CenterNet Resnet50 V2 Keypoints 512x512	30	27.6/48.2	Boxes/Keypoints
CenterNet MobileNetV2 FPN 512x512	6	23.4	Boxes
CenterNet MobileNetV2 FPN Keypoints 512x512	6	41.7	Keypoints
EfficientDet D0 512x512	39	33.6	Boxes
EfficientDet D1 640x640	54	38.4	Boxes
EfficientDet D2 768x768	67	41.8	Boxes
EfficientDet D3 896x896	95	45.4	Boxes
EfficientDet D4 1024x1024	133	48.5	Boxes
EfficientDet D5 1280x1280	222	49.7	Boxes
EfficientDet D6 1280x1280	268	50.5	Boxes
EfficientDet D7 1536x1536	325	51.2	Boxes
SSD MobileNet v2 320x320	19	20.2	Boxes
SSD MobileNet V1 FPN 640x640	48	29.1	Boxes
SSD MobileNet V2 FPNLite 320x320	22	22.2	Boxes
SSD MobileNet V2 FPNLite 640x640	39	28.2	Boxes
SSD ResNet50 V1 FPN 640x640 (RetinaNet50)	46	34.3	Boxes
SSD ResNet50 V1 FPN 1024x1024 (RetinaNet50)	87	38.3	Boxes
SSD ResNet101 V1 FPN 640x640 (RetinaNet101)	57	35.6	Boxes
SSD ResNet101 V1 FPN 1024x1024 (RetinaNet101)	104	39.5	Boxes
SSD ResNet152 V1 FPN 640x640 (RetinaNet152)	80	35.4	Boxes
SSD ResNet152 V1 FPN 1024x1024 (RetinaNet152)	111	39.6	Boxes
Faster R-CNN ResNet50 V1 1024x1024	65	31.0	Boxes
Faster R-CNN ResNet50 V1 800x1333	65	31.6	Boxes
Faster R-CNN ResNet101 V1 640x640	55	31.8	Boxes
Faster R-CNN ResNet101 V1 1024x1024	72	37.1	Boxes

Faster R-CNN ResNet101 V1 800x1333	77	36.6	Boxes
Faster R-CNN ResNet152 V1 640x640	64	32.4	Boxes
Faster R-CNN ResNet152 V1 1024x1024	85	37.6	Boxes
Faster R-CNN ResNet152 V1 800x1333	101	37.4	Boxes
Faster R-CNN Inception ResNet V2 640x640	206	37.7	Boxes
Faster R-CNN Inception ResNet V2 1024x1024	236	38.7	Boxes
Mask R-CNN Inception ResNet V2 1024x1024	301	39.0/34.6	Boxes/Masks
ExtremeNet	--	--	Boxes

圖 8.29　TensorFlow Object Detection API 所支援的演算法

資料來源：TensorFlow Detection Model Zoo[28]

安裝環境需求請參考「TensorFlow Object Detection API 官網文件」[29]，
目前如下表所示：

Target Software versions	
OS	Windows, Linux
Python	3.8
TensorFlow	2.2.0
CUDA Toolkit	10.1
CuDNN	7.6.5
Anaconda	Python 3.7 (Optional)

圖 8.30　TensorFlow Object Detection API 的安裝環境需求

以下介紹在 Windows 作業環境下的安裝：

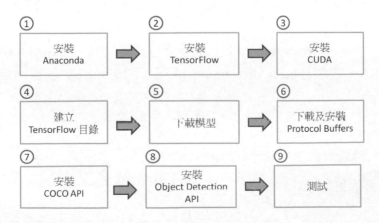

1. 安裝 Anaconda。

2. 安裝 TensorFlow：v2.2 以上版本。

3. 安裝 CUDA：GPU 不一定需要，如果要使用 GPU，需安裝 CUDA 10.1/cuDNN v7.6.5，詳細說明請參考「TensorFlow Object Detection API 教學網站」。

4. 建立 TensorFlow 目錄：在任意目錄下建立 TensorFlow 目錄。

5. 下載模型：自 TensorFlow Models GitHub (https://github.com/tensorflow/models)下載整個專案(repository)。並解壓縮至 TensorFlow 次目錄下，將 models-master 目錄改名為 models。

6. 下載 Protocol Buffers：自 https://github.com/protocolbuffers/protobuf/releases 下載最新版 protoc-3.xx.0-win64.zip，解壓縮至特定目錄，例如 "C:\Program Files\Google Protobuf"，將其下 bin 路徑加到環境變數 Path 中。

7. 安裝 Protobuf：在第 4 步驟的 TensorFlow\models\research 目錄開啟 cmd，執行：

```
protoc object_detection/protos/*.proto --python_out=.
```

8. 安裝 COCO API，執行下列指令：

```
pip install cython
pip install
git+https://github.com/philferriere/cocoapi.git#subdirectory=PythonAPI
```

9. 安裝 Object Detection API：更改目前目錄至 TensorFlow\models\research，複製 object_detection/packages/tf2/setup.py 至目前目錄，執行：

```
python -m pip install .
```

10. 安裝到此終於大功告成，執行測試指令：

```
python object_detection/builders/model_builder_tf2_test.py
```

```
...
[      OK ] ModelBuilderTF2Test.test_create_ssd_models_from_config
[ RUN      ] ModelBuilderTF2Test.test_invalid_faster_rcnn_batchnorm_update
[      OK ] ModelBuilderTF2Test.test_invalid_faster_rcnn_batchnorm_update
[ RUN      ] ModelBuilderTF2Test.test_invalid_first_stage_nms_iou_threshold
[      OK ] ModelBuilderTF2Test.test_invalid_first_stage_nms_iou_threshold
[ RUN      ] ModelBuilderTF2Test.test_invalid_model_config_proto
[      OK ] ModelBuilderTF2Test.test_invalid_model_config_proto
[ RUN      ] ModelBuilderTF2Test.test_invalid_second_stage_batch_size
[      OK ] ModelBuilderTF2Test.test_invalid_second_stage_batch_size
[ RUN      ] ModelBuilderTF2Test.test_session
[ SKIPPED ] ModelBuilderTF2Test.test_session
[ RUN      ] ModelBuilderTF2Test.test_unknown_faster_rcnn_feature_extractor
[      OK ] ModelBuilderTF2Test.test_unknown_faster_rcnn_feature_extractor
[ RUN      ] ModelBuilderTF2Test.test_unknown_meta_architecture
[      OK ] ModelBuilderTF2Test.test_unknown_meta_architecture
[ RUN      ] ModelBuilderTF2Test.test_unknown_ssd_feature_extractor
[      OK ] ModelBuilderTF2Test.test_unknown_ssd_feature_extractor
----------------------------------------------------------------------
Ran 20 tests in 68.510s

OK (skipped=1)
```

圖 8.31 TensorFlow Object Detection API 測試成功訊息

接下來，實作一個簡單的範例呼叫 TensorFlow Object Detection API，教學網站的範例寫得有點複雜，筆者把一些函數拿掉，盡量簡化程式。

【範例 10】使用 TensorFlow Object Detection API 進行物件偵測。

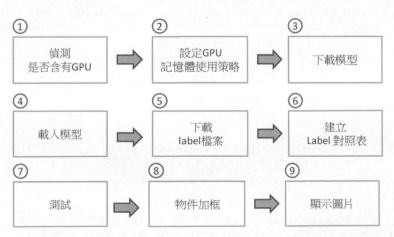

程式：08_08_TensorFlow_Object_Detection_API_Test.ipynb。

1. 載入相關套件。

```
1  # 載入套件
2  import os
3  import pathlib
4  import tensorflow as tf
5  import pathlib
```

2. 偵測機器是否含有 GPU：因速度比純靠 CPU 快上數倍的緣故，深度學習的模型訓練非常仰賴 GPU。一般而言，TensorFlow 對記憶體的回收並不是很理想，時常會發生記憶體不足(Out Of Memory, OOM)的狀況，所以我們通常會從這兩種策略中擇其一。

設定成記憶體動態調整(dynamic memory allocation)策略：避免記憶體爆掉。

```
1  # GPU 設定為 記憶體動態調整 (dynamic memory allocation)
2  gpus = tf.config.experimental.list_physical_devices('GPU')
3  for gpu in gpus:
4      tf.config.experimental.set_memory_growth(gpu, True)
```

設定成固定大小的記憶體，例如 2GB。

```
1   GPU 設定為固定為 2GB
2  gpus = tf.config.experimental.list_physical_devices('GPU')
3  if gpus:  # 1024*2 : 2048MB = 2GB
4      tf.config.experimental.set_virtual_device_configuration(gpus[0],
5          [tf.config.experimental.VirtualDeviceConfiguration(memory_limit=1024*2)])
```

3. 下載模型：內含多種模型，任選其中一種。

```
1  # 下載模型，並解壓縮
2  def download_model(model_name, model_date):
3      base_url = 'http://download.tensorflow.org/models/object_detection/tf2/'
4      model_file = model_name + '.tar.gz'
5      # 解壓縮
6      model_dir = tf.keras.utils.get_file(fname=model_name,
7                          origin=base_url + model_date + '/' + model_file,
8                          untar=True)
9      return str(model_dir)
10
11 MODEL_DATE = '20200711'
12 MODEL_NAME = 'centernet_hg104_1024x1024_coco17_tpu-32'
13 PATH_TO_MODEL_DIR = download_model(MODEL_NAME, MODEL_DATE)
14 PATH_TO_MODEL_DIR
```

4. 從下載的目錄載入模型。

```
1  # 從下載的目錄載入模型，耗時甚久
2  import time
3  from object_detection.utils import label_map_util
4  from object_detection.utils import visualization_utils as viz_utils
5
6  PATH_TO_SAVED_MODEL = PATH_TO_MODEL_DIR + "/saved_model"
7
8  print('載入模型...', end='')
9  start_time = time.time()
10
11 # 載入模型
12 detect_fn = tf.saved_model.load(PATH_TO_SAVED_MODEL)
13
14 end_time = time.time()
15 elapsed_time = end_time - start_time
16 print(f'共花費 {elapsed_time} 秒.')
```

執行結果：共花費 160.80 秒，相當緩慢，後面(步驟 10)會介紹另一種更快捷的方式。

5. 下載 label 檔案。

```
1  # 下載 labels file
2  def download_labels(filename):
3      base_url = 'https://raw.githubusercontent.com/tensorflow/models'
4      base_url += '/master/research/object_detection/data/'
5      label_dir = tf.keras.utils.get_file(fname=filename,
6                                          origin=base_url + filename,
7                                          untar=False)
8      label_dir = pathlib.Path(label_dir)
9      return str(label_dir)
10
11 LABEL_FILENAME = 'mscoco_label_map.pbtxt'
12 PATH_TO_LABELS = download_labels(LABEL_FILENAME)
13 PATH_TO_LABELS
```

執行結果：檔案會儲存在 C:\Users\<使用者登入帳號> \.keras\datasets
\mscoco_label_map.pbtxt。

6. 建立 Label 的對照表 (代碼與名稱)。

```
1  # 建立 Label 的對照表 (代碼與名稱)
2  category_index = label_map_util.create_category_index_from_labelmap(
3      PATH_TO_LABELS, use_display_name=True)
```

7. 任選一張圖片進行物件偵測。

```
1  # 任選一張圖片進行物件偵測
2  import numpy as np
3  from PIL import Image
4
5  # 開啟一張圖片
6  image_np = np.array(Image.open('./images_Object_Detection/zebra.jpg'))
7
8  # 轉為 TensorFlow tensor 資料型態
9  input_tensor = tf.convert_to_tensor(image_np)
10 # 加一維，變為 (筆數, 寬, 高, 顏色)
11 input_tensor = input_tensor[tf.newaxis, ...]
12
13 # detections : 物件資訊 內含 (候選框, 類別, 機率)
14 detections = detect_fn(input_tensor)
15 num_detections = int(detections.pop('num_detections'))
16 print(f'物件個數：{num_detections}')
17 detections = {key: value[0, :num_detections].numpy()
18                for key, value in detections.items()}
19
20 detections['num_detections'] = num_detections
21 # 轉為整數
22 detections['detection_classes'] = detections['detection_classes'].astype(np.int64)
23
24 print(f'物件資訊 (候選框, 類別, 機率)：')
25 for detection_boxes, detection_classes, detection_scores in \
26     zip(detections['detection_boxes'], detections['detection_classes'],
27         detections['detection_scores']):
28     print(np.around(detection_boxes,4), detection_classes,
29           round(detection_scores*100, 2))
```

部份執行結果：

```
物件個數：100
物件資訊 (候選框, 類別, 機率)：
[0.2647 0.5269 0.6977 0.8749] 24 98.77
[0.2243 0.2899 0.6752 0.6357] 24 98.19
[0.247  0.0723 0.6775 0.3546] 24 97.23
[0.2958 0.     0.4356 0.0021] 24 3.25
[0.2263 0.     0.4017 0.002 ] 24 3.06
[0.344  0.9967 0.478  1.    ] 24 2.85
[0.3139 0.9975 0.6764 1.    ] 24 2.7
[0.3315 0.9976 0.5939 0.9998] 24 2.68
[0.3326 0.9964 0.4505 1.    ] 16 2.3
[0.2882 0.9967 0.4172 1.    ] 24 2.3
```

8. 物件加框：掃描 Bounding Box，將圖片的物件加框。

```
1  import matplotlib.pyplot as plt
2
3  image_np_with_detections = image_np.copy()
4  # 加框
5  viz_utils.visualize_boxes_and_labels_on_image_array(
6        image_np_with_detections,
7        detections['detection_boxes'],
8        detections['detection_classes'],
9        detections['detection_scores'],
10       category_index,
11       use_normalized_coordinates=True,
12       max_boxes_to_draw=200,
13       min_score_thresh=.30,
14       agnostic_mode=False)
15
16 # 顯示，無效
17 plt.figure(figsize=(12,8))
18 plt.imshow(image_np_with_detections, cmap='viridis')
19 plt.show()
```

執行結果不如預期，無法顯示圖片，只好先存檔再顯示。

9. 顯示處理過後的圖片。

```
1  # 存檔
2  saved_file = './images_Object_Detection/zebra._detection1.png'
3  plt.savefig(saved_file)
4
5  # 顯示
6  from IPython.display import Image
7  Image(saved_file)
```

執行結果：

10. 另一種方法：從下載的目錄載入模型，執行速度大幅提升。

```python
1   # 快速從下載的目錄載入模型
2   import time
3   from object_detection.utils import label_map_util, config_util
4   from object_detection.utils import visualization_utils as viz_utils
5   from object_detection.builders import model_builder
6
7   # 組態檔及模型檔路徑
8   PATH_TO_CFG = PATH_TO_MODEL_DIR + "/pipeline.config"
9   PATH_TO_CKPT = PATH_TO_MODEL_DIR + "/checkpoint"
10
11  # 計時開始
12  print('Loading model... ', end='')
13  start_time = time.time()
14
15  # 載入組態檔，再建置模型
16  configs = config_util.get_configs_from_pipeline_file(PATH_TO_CFG)
17  model_config = configs['model']
18  detection_model = model_builder.build(model_config=model_config,
19                                         is_training=False)
20
21  # 還原模型
22  ckpt = tf.compat.v2.train.Checkpoint(model=detection_model)
23  ckpt.restore(os.path.join(PATH_TO_CKPT, 'ckpt-0')).expect_partial()
24
25  # 計時完成
26  end_time = time.time()
27  elapsed_time = end_time - start_time
28  print(f'共花費 {elapsed_time} 秒.')
```

執行結果：只花費 0.5 秒，與前一個方法相比真的超快。

11. 任選一張圖片進行物件偵測。

```python
1   # 任選一張圖片進行物件偵測
2   @tf.function
3   def detect_fn(image):
4       image, shapes = detection_model.preprocess(image)
5       prediction_dict = detection_model.predict(image, shapes)
6       detections = detection_model.postprocess(prediction_dict, shapes)
7
8       return detections
9
10  # 讀取圖檔
11  image_np = np.array(Image.open('./images_Object_Detection/zebra.jpg'))
12  input_tensor = tf.convert_to_tensor(image_np, dtype=tf.float32)
13  input_tensor = input_tensor[tf.newaxis, ...]
14  detections = detect_fn(input_tensor)
15  num_detections = int(detections.pop('num_detections'))
16
17  print(f'物件個數：{num_detections}')
```

```
18  detections = {key: value[0, :num_detections].numpy()
19                for key, value in detections.items()}
20  print(f'物件資訊 (候選框, 類別, 機率) : ')
21  for detection_boxes, detection_classes, detection_scores in \
22      zip(detections['detection_boxes'], detections['detection_classes'],
23          detections['detection_scores']):
24      print(np.around(detection_boxes,4), int(detection_classes)+1,
25          round(detection_scores*100, 2))
26
27  # 結果存入 detections 變數
28  detections['num_detections'] = num_detections
29  detections['detection_classes'] = detections['detection_classes'].astype(np.int64)
```

12. 重複步驟 8、9，執行結果相同。

【範例 11】使用 TensorFlow Object Detection API 進行影片測試。

程式：08_09_TensorFlow_Object_Detection_API_Video.ipynb。

由於前面 10 個步驟均相同，只說明差異的部份。

1. 影片物件偵測。

```
1   import numpy as np
2   import cv2
3
4   # 使用 webcam
5   #cap = cv2.VideoCapture(0)
6
7   # 讀取視訊檔案
8   cap = cv2.VideoCapture('./images_Object_Detection/pedestrians.mp4')
9   i=0
10  while True:
11      # 讀取一幀(frame) from camera or mp4
12      ret, image_np = cap.read()
13
14      # 加一維，變為 (筆數, 寬, 高, 顏色)
15      image_np_expanded = np.expand_dims(image_np, axis=0)
16
17      # 可測試水平翻轉
18      # image_np = np.fliplr(image_np).copy()
19
20      # 可測試灰階
21      # image_np = np.tile(
22      #     np.mean(image_np, 2, keepdims=True), (1, 1, 3)).astype(np.uint8)
23
24      # 轉為 TensorFlow tensor 資料型態
25      input_tensor = tf.convert_to_tensor(np.expand_dims(image_np, 0), dtype=tf.float32)
26
27      # detections : 物件資訊 內含 (候選框, 類別, 機率)
28      detections = detect_fn(input_tensor)
29      num_detections = int(detections.pop('num_detections'))
```

```
31        # 第一幀(Frame)才顯示物件個數
32        if i==0:
33            print(f'物件個數:{num_detections}')
34
35        # 結果存入 detections 變數
36        detections = {key: value[0, :num_detections].numpy()
37                      for key, value in detections.items()}
38        detections['detection_classes'] = detections['detection_classes'].astype(int)
39
40        # 將物件框起來
41        label_id_offset = 1
42        image_np_with_detections = image_np.copy()
43        viz_utils.visualize_boxes_and_labels_on_image_array(
44            image_np_with_detections,
45            detections['detection_boxes'],
46            detections['detection_classes'] + label_id_offset,
47            detections['detection_scores'],
48            category_index,
49            use_normalized_coordinates=True,
50            max_boxes_to_draw=200,
51            min_score_thresh=.30,
52            agnostic_mode=False)
53
54        # 顯示偵測結果
55        img = cv2.resize(image_np_with_detections, (800, 600))
56        cv2.imshow('object detection', img)
57
58        # 存檔
59        i+=1
60        if i==30:
61            cv2.imwrite('./images_Object_Detection/pedestrians.png', img)
```

```
63        # 按 q 可以結束
64        if cv2.waitKey(25) & 0xFF == ord('q'):
65            break
66
67 cap.release()
68 cv2.destroyAllWindows()
```

執行結果:影片中的車輛都可以偵測到,辨識度極高,也能夠使用 web cam,改為 cv2.VideoCapture(0),0 代表第一台攝影機。

使用另一段影片 night.mp4 試試看,該影片來自「Python Image Processing Cookbook GitHub」[30],內容為高速公路的夜景,辨識度也是相當好。

如果要像 YOLO 自訂資料集,偵測其他物件的話,TensorFlow Object Detection API 的官網文件[31]有非常詳盡的解說,讀者可依指示自行測試。

▌ 8-11 物件偵測的效能衡量指標

物件偵測的效能衡量指標是採「平均精確度均值」（mean Average Precision , mAP），YOLO 官網展示的圖表針對各種模型比較 mAP。

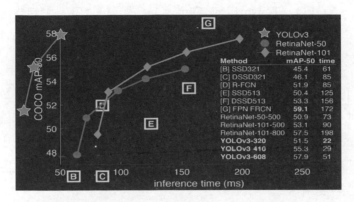

圖 8.29 YOLO 與其他模型比較，圖片來源：YOLO 官網[17]

第四章介紹過的 ROC/AUC 效能衡量指標，是以預測機率為基準，計算各種閾值(門檻值)下的真陽率與偽陽率，以偽陽率為 X 軸，真陽率為 Y 軸，繪製出 ROC 曲線。而 mAP 也類似 ROC/AUC，以 IoU 為基準，計算各種閾值(門檻值)下的精確率(Precision)與召回率(Recall)，以召回率為 X 軸，精確率為 Y 軸，繪製出 mAP 曲線。

不過，物件偵測模型通常是多分類，不是二分類，因此，採取計算各個種類的平均精確度，繪製後如下左方圖表，通常會調整成右方圖表的粗線，因為，在閾值低的精確率一定比閾值高的精確率更好，所以作此調整。

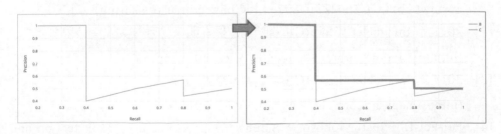

圖 8.30 mAP 曲線，左圖是實際計算的結果，右圖是調整後的結果

8-12 總結

這一章我們認識了許多物件偵測的演算法，包括 HOG、R-CNN、YOLO、SSD，同時也實作許多範例，像是傳統的影像金字塔、R-CNN、PyTorch Detectron2、YOLO、TensorFlow Object Detection API，還包含圖像和影片偵測，也可自訂資料集訓練模型，證明我們的確有能力，將物件偵測技術導入到專案中使用。

演算法各有優劣，Faster R-CNN 雖然較慢，但準確度高，儘管 YOLO 早期為了提升執行速度犧牲了準確度，但經過幾個版本升級後，準確度也已大幅提高。所以建議讀者在實際應用時，還是應該多方嘗試，找出最適合的模型，譬如在邊緣運算的場域使用輕量模型，不只要求辨識速度快，更要節省記憶體的使用。

現在許多學者開始研究動態物件偵測，例如姿態(Pose)偵測，可用來辨識體育運動姿勢是否標準，協助運動員提升成績，另外還有手勢偵測[32]、體感遊戲、製作皮影戲[33]等，也很好玩。

參考資料 (References)

[1] Joseph Redmon、Anelia Angelova,《Real-Time Grasp Detection Using Convolutional Neural Networks》, 2015
(https://docs.google.com/presentation/d/1Zc9-iR1eVz-zysinwb7bzLGC2no2ZiaD897_14dGbhw/edit?usp=sharing)

[2] 2011 年 ImageNet ILSVRC 挑戰賽比賽說明
(http://image-net.org/challenges/LSVRC/2011/index)

[3] 2017 年 ImageNet ILSVRC 挑戰賽比賽說明
(http://image-net.org/challenges/LSVRC/2017/)

[4] Fei-Fei Li、Justin Johnson、Serena Yeung,《Lecture 11: Detection and Segmentation》, 2017

(http://cs231n.stanford.edu/slides/2017/cs231n_2017_lecture11.pdf)

[5] Adrian Rosebrock,《Image Pyramids with Python and OpenCV》, 2015
(https://www.pyimagesearch.com/2015/03/16/image-pyramids-with-python-and-opencv/)

[6] IIPImage
(https://iipimage.sourceforge.io/documentation/images/)

[7] Adrian Rosebrock,《Sliding Windows for Object Detection with Python and OpenCV》, 2015
(https://www.pyimagesearch.com/2015/03/23/sliding-windows-for-object-detection-with-python-and-opencv/)

[8] 素娜 93,《方向梯度直方圖（HOG）》, 2017
(https://www.jianshu.com/p/6f69c751e9e7)

[9] Adrian Rosebrock,《Histogram of Oriented Gradients and Object Detection》, 2014
(https://www.pyimagesearch.com/2014/11/10/histogram-oriented-gradients-object-detection/)

[10] Adrian Rosebrock,《Non-Maximum Suppression for Object Detection in Python》, 2014
(https://www.pyimagesearch.com/2014/11/17/non-maximum-suppression-object-detection-python/)

[11] Tomasz Malisiewicz,《Ensemble of Exemplar-SVMs for Object Detection and Beyond》
(http://www.cs.cmu.edu/~tmalisie/projects/iccv11/index.html)

[12] Ross Girshick、Jeff Donahue、Trevor Darrell、Jitendra Malik,《Rich feature hierarchies for accurate object detection and semantic segmentation》, 2014
(https://arxiv.org/pdf/1311.2524.pdf)

[13] Lung-Ying Ling,《R-CNN 學習筆記, LaptrinhX》, 2019

(https://laptrinhx.com/r-cnn-xue-xi-bi-ji-1145354539/)

[14] Kaiming He、Xiangyu Zhang、Shaoqing Ren、Jian Sun,《Spatial Pyramid Pooling in Deep Convolutional Networks for Visual Recognition》, 2015

(https://arxiv.org/abs/1406.4729)

[15] v1_vivian,《SPP-Net 論文詳解》, 2017

(https://www.itread01.com/content/1542334444.html)

[16] Ross B. Girshick 於 GitHub 上放置的 Faster R-CNN 程式碼

(https://github.com/rbgirshick/py-faster-rcnn)

[17] YOLO 官網

(https://pjreddie.com/darknet/yolo/)

[18] Alexey Bochkovskiy、Chien-Yao Wang、Hong-Yuan Mark Liao,《YOLOv4: Optimal Speed and Accuracy of Object Detection》, 2020

(https://arxiv.org/abs/2004.10934)

[19] YOLO5 GitHub

(https://github.com/ultralytics/yolov5)

[20] Joseph Redmon、Santosh Divvala、Ross Girshick、Ali Farhadi,《You Only Look Once: Unified, Real-Time Object Detection》, 2016

(https://docs.google.com/presentation/d/1kAa7NOamBt4calBU9iHgT8a8 6RRHz9Yz2oh4-GTdX6M/edit?usp=sharing)

[21] Jason Brownlee,《How to Perform Object Detection With YOLOv3 in Keras》, 2019

(https://machinelearningmastery.com/how-to-perform-object-detection-with-yolov3-in-keras/)

[22] YOLO4 GitHub

(https://github.com/SoloSynth1/tensorflow-yolov4)

[23] COCO 資料集的 80 個類別

(https://github.com/amikelive/coco-labels/blob/master/coco-labels-2014_2017.txt)

[24] Aditya Chakraborty,《Create your own dataset for YOLOv4 object detection in 5 minutes》, 2020

(https://medium.com/analytics-vidhya/create-your-own-dataset-for-yolov4-object-detection-in-5-minutes-fdc988231088)

[25] YOLO4 GitHub

(https://github.com/AlexeyAB/darknet)

[26] SSD 官網

(https://github.com/weiliu89/caffe/tree/ssd)

[27] LoveMIss-Y,《一文看盡目標檢測演算法 SSD 的核心架構與設計思想》, 2019

(https://blog.csdn.net/qq_27825451/article/details/89137697)

[28] TensorFlow 2 Detection Model Zoo

(https://github.com/tensorflow/models/blob/master/research/object_detection/g3doc/tf2_detection_zoo.md)

[29] TensorFlow Object Detection API 的安裝環境需求

(https://tensorflow-object-detection-api-tutorial.readthedocs.io/en/latest/)

[30] Python Image Processing Cookbook GitHub

(https://github.com/PacktPublishing/Python-Image-Processing-Cookbook)

[31] TensorFlow Object Detection API 官網文件

(https://tensorflow-object-detection-api-tutorial.readthedocs.io/en/latest/training.html)

[32] Oz Ramos,《Introducing Handsfree.js - Integrate hand, face, and pose gestures to your frontend》

(https://dev.to/midiblocks/introducing-handsfree-js-integrate-hand-face-and-pose-gestures-to-your-frontend-4g3p)

[33] Jen Looper, 《Ombromanie: Creating Hand Shadow stories with Azure Speech and TensorFlow.js Handposes》
(https://dev.to/azure/ombromanie-creating-hand-shadow-stories-with-azure-speech-and-tensorflow-js-handposes-3cln)

進階的影像應用

除了物件偵測之外，CNN 還有許多影像方面的應用，譬如：

- 語義分割(Semantic segmentation)。
- 風格轉換(Style Transfer)。
- 影像標題(Image Captioning)。
- 姿態辨識(Pose Detection 或 Action Detection)。
- 生成對抗網路(GAN) 各式的應用。
- 深度偽造(Deep Fake)。

本章將繼續探討以上這些應用領域，其中生成對抗網路(GAN)的內容較多，會以專章來介紹。

9-1 語義分割(Semantic Segmentation)介紹

物件偵測是以整個物件作為標記(Label)，而語義分割(Semantic Segmentation)則以每個像素(Pixel)作為標記(Label)，區分物件涵蓋的區域，如下圖：

經語義分割後產生如下圖，各物件以不同顏色的像素表示。

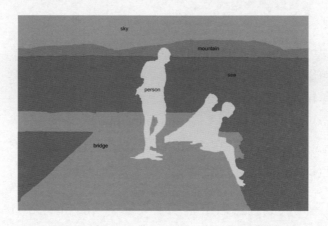

甚至更進一步，進行實例分割(Instance Segmentation)，相同類別的物件也以不同的顏色表示。

語義分割的應用非常廣泛，例如：

1. 自駕車的影像識別。
2. 醫療診斷：斷層掃描(CT)、核磁共振(MRI)的疾病區域標示。
3. 衛星照片。
4. 機器人的影像識別。

語義分割的原理是先利用 CNN 進行特徵萃取(Feature Extraction)，再運用萃取的特徵向量來重建影像，如下圖：

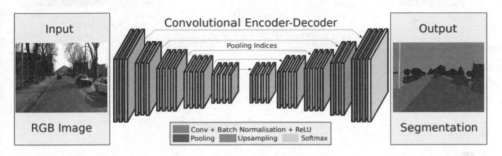

圖 9.1 語義分割的示意圖，圖片來源：SegNet: A Deep Convolutional Encoder-Decoder Architecture for Image Segmentation[1]

這種「原始影像 ➔ 特徵萃取 ➔ 重建影像」的作法，泛稱為「自動編碼器」(AutoEncoder, AE)架構，許多進階的演算法都以此架構為基礎，因此，我們先來探究 AutoEncoder 架構。

9-2 自動編碼器(AutoEncoder)

自動編碼器(AutoEncoder, AE)透過特徵萃取得到訓練資料的共同特徵，一些雜訊會被過濾掉，接著再依據特徵向量重建影像，這樣就可以達到「去雜訊」(Denosing)的目的，此作法也可以擴展到語義分割(Semantic segmentation)、風格轉換(Style Transfer)、U-net、生成對抗網路(GAN)等各式各樣的演算法。

AutoEncoder 由 Encoder 與 Decoder 組合而成：

- 編碼器(Encoder)：即為萃取特徵的過程，類似於 CNN 模型，但不含最後的分類層(Dense)。
- 解碼器(Decoder)：根據萃取的特徵來重建影像。

圖 9.2 自動編碼器(AutoEncoder)示意圖

接下來，我們實作 AutoEncoder，使用 MNIST 資料集，示範如何將雜訊去除。

【範例 1】實作 AutoEncoder，進行雜訊去除。

程式：09_01_MNIST_Autoencoder.ipynb。

1. 載入相關套件。

```python
1  # 載入相關套件
2  import numpy as np
3  import tensorflow as tf
4  import tensorflow.keras as K
5  import matplotlib.pyplot as plt
6  from tensorflow.keras.layers import Dense, Conv2D, MaxPooling2D, UpSampling2D
```

2. 超參數設定。

```
1  # 超參數設定
2  batch_size = 128        # 訓練批量
3  max_epochs = 50         # 訓練執行週期
4  filters = [32,32,16]  # 三層卷積層的輸出個數
```

3. 取得 MNIST 訓練資料，只取圖像(X)，不需要 Label(Y)，因為程式只要進行特徵萃取，不用辨識。

```
1  # 只取 X ，不需 Y
2  (x_train, _), (x_test, _) = K.datasets.mnist.load_data()
3
4  # 常態化
5  x_train = x_train / 255.
6  x_test = x_test / 255.
7
8  # 加一維：色彩
9  x_train = np.reshape(x_train, (len(x_train),28, 28, 1))
10 x_test = np.reshape(x_test, (len(x_test), 28, 28, 1))
```

4. 在圖像中加入雜訊：以利後續實驗，觀察 AutoEncoder 是否能去除雜訊。

```
1  # 在既有圖像加雜訊
2  noise = 0.5
3
4  # 固定隨機亂數
5  np.random.seed(11)
6  tf.random.set_seed(11)
7
8  # 隨機加雜訊
9  x_train_noisy = x_train + noise * np.random.normal(loc=0.0,
10                                   scale=1.0, size=x_train.shape)
11 x_test_noisy = x_test + noise * np.random.normal(loc=0.0,
12                                   scale=1.0, size=x_test.shape)
13
14 # 加完裁切數值，避免大於 1
15 x_train_noisy = np.clip(x_train_noisy, 0, 1)
16 x_test_noisy = np.clip(x_test_noisy, 0, 1)
17
18 # 轉換為浮點數
19 x_train_noisy = x_train_noisy.astype('float32')
20 x_test_noisy = x_test_noisy.astype('float32')
```

5. 建立編碼器(Encoder)模型：使用卷積層與池化層。

```
1   # 編碼器(Encoder)
2   class Encoder(K.layers.Layer):
3       def __init__(self, filters):
4           super(Encoder, self).__init__()
5           self.conv1 = Conv2D(filters=filters[0], kernel_size=3, strides=1,
6                               activation='relu', padding='same')
7           self.conv2 = Conv2D(filters=filters[1], kernel_size=3, strides=1,
8                               activation='relu', padding='same')
9           self.conv3 = Conv2D(filters=filters[2], kernel_size=3, strides=1,
10                              activation='relu', padding='same')
11          self.pool = MaxPooling2D((2, 2), padding='same')
12
13
14      def call(self, input_features):
15          x = self.conv1(input_features)
16          #print("Ex1", x.shape)
17          x = self.pool(x)
18          #print("Ex2", x.shape)
19          x = self.conv2(x)
20          x = self.pool(x)
21          x = self.conv3(x)
22          x = self.pool(x)
23          return x
```

6. 建立解碼器(Decoder)模型：使用卷積層和上採樣層(Up-sampling)，卷
積層的輸出個數與 Encoder 相反，代表把圖像還原，上採樣層與池化層
相反，則是將圖像放大。

```
1   # 解碼器(Decoder)
2   class Decoder(K.layers.Layer):
3       def __init__(self, filters):
4           super(Decoder, self).__init__()
5           self.conv1 = Conv2D(filters=filters[2], kernel_size=3, strides=1,
6                               activation='relu', padding='same')
7           self.conv2 = Conv2D(filters=filters[1], kernel_size=3, strides=1,
8                               activation='relu', padding='same')
9           self.conv3 = Conv2D(filters=filters[0], kernel_size=3, strides=1,
10                              activation='relu', padding='valid')
11          self.conv4 = Conv2D(1, 3, 1, activation='sigmoid', padding='same')
12          self.upsample = UpSampling2D((2, 2))
13
14      def call(self, encoded):
15          x = self.conv1(encoded)
16          # 上採樣
17          x = self.upsample(x)
18
19          x = self.conv2(x)
20          x = self.upsample(x)
21
22          x = self.conv3(x)
23          x = self.upsample(x)
24
25          return self.conv4(x)
```

7. 結合編碼器(Encoder)、解碼器(Decoder)，建立 AutoEncoder 模型。

```python
1  # 建立 Autoencoder 模型
2  class Autoencoder(K.Model):
3      def __init__(self, filters):
4          super(Autoencoder, self).__init__()
5          self.loss = []
6          self.encoder = Encoder(filters)
7          self.decoder = Decoder(filters)
8
9      def call(self, input_features):
10         #print(input_features.shape)
11         encoded = self.encoder(input_features)
12         #print(encoded.shape)
13         reconstructed = self.decoder(encoded)
14         #print(reconstructed.shape)
15         return reconstructed
```

8. 訓練模型。

```python
1  model = Autoencoder(filters)
2
3  model.compile(loss='binary_crossentropy', optimizer='adam')
4
5  loss = model.fit(x_train_noisy,
6                  x_train,
7                  validation_data=(x_test_noisy, x_test),
8                  epochs=max_epochs,
9                  batch_size=batch_size)
```

9. 繪製損失函數。

```python
1  # 繪製損失函數
2  plt.plot(range(max_epochs), loss.history['loss'])
3  plt.xlabel('Epochs')
4  plt.ylabel('Loss')
5  plt.show()
```

執行結果：損失隨著訓練次數越來越小，且趨於收斂。

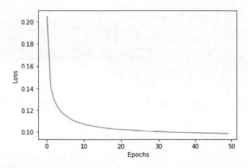

10. 比較含雜訊的圖像與去除雜訊後的圖像。

```
1   number = 10   # how many digits we will display
2   plt.figure(figsize=(20, 4))
3   for index in range(number):
4       # 加了雜訊的圖像
5       ax = plt.subplot(2, number, index + 1)
6       plt.imshow(x_test_noisy[index].reshape(28, 28), cmap='gray')
7       ax.get_xaxis().set_visible(False)
8       ax.get_yaxis().set_visible(False)
9
10      # 重建的圖像
11      ax = plt.subplot(2, number, index + 1 + number)
12      plt.imshow(tf.reshape(model(x_test_noisy)[index], (28, 28)), cmap='gray')
13      ax.get_xaxis().set_visible(False)
14      ax.get_yaxis().set_visible(False)
15  plt.show()
```

執行結果：效果相當好，雜訊被有效剔除。

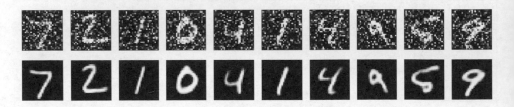

AutoEncoder 屬於非監督式學習算法，不需要標記(Labeling)。另外還有一個 AutoEncoder 的變形(Variants)，稱為 Variational AutoEncoders (VAE)，資料編碼不是一個輸出常數，而是一個常態機率分配，解碼時依據機率分配進行抽樣，取得輸出，利用此概念去除雜訊則會更穩健(Robust)，VAE常與生成對抗網路(GAN)相提並論，可以用來生成影像。

圖 9.3 Variational AutoEncoders (VAE)的架構

【範例 **2**】建立 VAE 模型，使用 MNIST 資料集，生成影像。

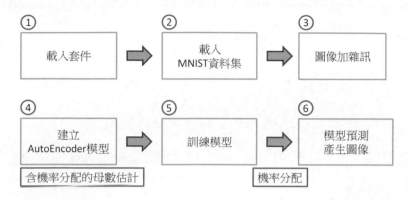

* VAE 的編碼器輸出不是特徵向量，而是機率分配的母數 μ 和 log(δ)。

程式：**09_02_MNIST_VAE.ipynb**，程式修改自 keras-mnist-VAE GitHub[2]。

1. 載入相關套件。

```
1  # 載入相關套件
2  import numpy as np
3  import matplotlib.pyplot as plt
4  import tensorflow as tf
5  from tensorflow.keras.datasets import mnist
6  from tensorflow.keras.layers import Input, Dense, Lambda
7  from tensorflow.keras.models import Model
8  from tensorflow.keras import backend as K
9  from scipy.stats import norm
```

2. 取得 MNIST 訓練資料。

```
1  # 取得 MNIST 訓練資料
2  (x_tr, y_tr), (x_te, y_te) = mnist.load_data()
3  x_tr, x_te = x_tr.astype('float32')/255., x_te.astype('float32')/255.
4  x_tr, x_te = x_tr.reshape(x_tr.shape[0], -1), x_te.reshape(x_te.shape[0], -1)
5  print(x_tr.shape, x_te.shape)
```

3. 超參數設定。

```
1  # 超參數設定
2  batch_size, n_epoch = 100, 100    # 訓練執行批量、週期
3  n_hidden, z_dim = 256, 2          # 編碼器隱藏層神經元個數、輸出層神經元個數
```

4. 定義編碼器模型。

```
1  # encoder
2  x = Input(shape=(x_tr.shape[1:]))
3  x_encoded = Dense(n_hidden, activation='relu')(x)
4  x_encoded = Dense(n_hidden//2, activation='relu')(x_encoded)
5
6  # encoder 後接 Dense，估算平均數 mu
7  mu = Dense(z_dim)(x_encoded)
8
9  # encoder 後接 Dense，估算 log 變異數 log_var
10 log_var = Dense(z_dim)(x_encoded)
```

5. 定義抽樣函數：根據平均數(mu)和 Log 變異數(log_var) 取隨機亂數。

```
1  # 定義抽樣函數
2  def sampling(args):
3      # 根據 mu, log_var 取隨機亂數
4      mu, log_var = args
5      eps = K.random_normal(shape=(batch_size, z_dim), mean=0., stddev=1.0)
6      return mu + K.exp(log_var) * eps
7
8  # 定義匿名函數，進行抽樣
9  z = Lambda(sampling, output_shape=(z_dim,))([mu, log_var])
```

6. 定義解碼器模型。

```
1  # decoder
2  z_decoder1 = Dense(n_hidden//2, activation='relu')
3  z_decoder2 = Dense(n_hidden, activation='relu')
4  y_decoder = Dense(x_tr.shape[1], activation='sigmoid')
5
6  # 解碼的輸入為匿名函數
7  z_decoded = z_decoder1(z)
8  z_decoded = z_decoder2(z_decoded)
9  y = y_decoder(z_decoded)
```

7. 以 KL 散度(Kullback-Leibler divergence, KL Loss)為損失函數(loss)：類似 MSE，主要用於衡量實際與理論機率分配之差。

```
1  # 定義特殊的損失函數(Loss)
2  reconstruction_loss = tf.keras.losses.binary_crossentropy(x, y) * x_tr.shape[1]
3  kl_loss = 0.5 * K.sum(K.square(mu) + K.exp(log_var) - log_var - 1, axis = -1)
4  vae_loss = reconstruction_loss + kl_loss
5
6  vae = Model(x, y)    # x:MNIST圖像， y:解碼器的輸出
7  vae.add_loss(vae_loss)
8  vae.compile(optimizer='rmsprop')
9
10 # 顯示模型彙總資訊
11 vae.summary()
```

編碼器模型的彙總資訊：

```
Model: "model"

Layer (type)            Output Shape        Param #      Connected to
==================================================================================
input_1 (InputLayer)    [(None, 784)]       0

dense (Dense)           (None, 256)         200960       input_1[0][0]

dense_1 (Dense)         (None, 128)         32896        dense[0][0]

dense_2 (Dense)         (None, 2)           258          dense_1[0][0]

dense_3 (Dense)         (None, 2)           258          dense_1[0][0]

lambda (Lambda)         (100, 2)            0            dense_2[0][0]
                                                         dense_3[0][0]
```

解碼器模型的彙總資訊：

```
dense_4 (Dense)                 (100, 128)      384       lambda[0][0]

dense_5 (Dense)                 (100, 256)      33024     dense_4[0][0]

dense_6 (Dense)                 (100, 784)      201488    dense_5[0][0]

tf.math.square (TFOpLambda)     (None, 2)       0         dense_2[0][0]

tf.math.exp (TFOpLambda)        (None, 2)       0         dense_3[0][0]

tf.__operators__.add (TFOpLambd (None, 2)       0         tf.math.square[0][0]
                                                          tf.math.exp[0][0]

tf.cast (TFOpLambda)            (None, 784)     0         input_1[0][0]

tf.convert_to_tensor (TFOpLambd (100, 784)      0         dense_6[0][0]

tf.math.subtract (TFOpLambda)   (None, 2)       0         tf.__operators__.add[0][0]
                                                          dense_3[0][0]

tf.keras.backend.binary_crossen (100, 784)      0         tf.cast[0][0]
                                                          tf.convert_to_tensor[0][0]

tf.math.subtract_1 (TFOpLambda) (None, 2)       0         tf.math.subtract[0][0]

tf.math.reduce_mean (TFOpLambda (100,)          0         tf.keras.backend.binary_crossentr
```

8. 訓練模型。

```
1  # 訓練模型
2  vae.fit(x_tr,
3          shuffle=True,
4          epochs=n_epoch,
5          batch_size=batch_size,
6          validation_data=(x_te, None), verbose=1)
```

9. 取得編碼器的輸出。

```
1  # 取得編碼器的輸出 mu
2  encoder = Model(x, mu)
3  encoder.summary()
```

10. 測試資料預測：將編碼器的輸出繪圖。

```
1  # 以測試資料預測，以編碼器的輸出繪圖
2  x_te_latent = encoder.predict(x_te, batch_size=batch_size)
3  plt.figure(figsize=(6, 6))
4  plt.scatter(x_te_latent[:, 0], x_te_latent[:, 1], c=y_te)
5  plt.colorbar()
6  plt.show()
```

執行結果：顯示 0~9 圖像的分佈，大致呈現分離，表示辨識度還不錯。

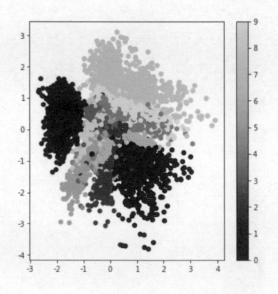

11. 取得解碼器的輸出。

```
1  # 取得解碼器的輸出
2  decoder_input = Input(shape=(z_dim,))
3  _z_decoded = z_decoder1(decoder_input)
4  _z_decoded = z_decoder2(_z_decoded)
5  _y = y_decoder(_z_decoded)
6  generator = Model(decoder_input, _y)
7  generator.summary()
```

12. 測試資料預測：以解碼器的輸出來生成圖像。

```
1   # 顯示 2D manifold
2   n = 15              # 顯示 15x15 視窗
3   digit_size = 28  # 圖像尺寸
4   figure = np.zeros((digit_size * n, digit_size * n))
5
6   #
7   grid_x = norm.ppf(np.linspace(0.05, 0.95, n))
8   grid_y = norm.ppf(np.linspace(0.05, 0.95, n))
9
10  # 取得各種機率下的生成的樣本
11  for i, yi in enumerate(grid_x):
12      for j, xi in enumerate(grid_y):
13          z_sample = np.array([[xi, yi]])
14          x_decoded = generator.predict(z_sample)
15          digit = x_decoded[0].reshape(digit_size, digit_size)
16          figure[i * digit_size: (i + 1) * digit_size,
17                 j * digit_size: (j + 1) * digit_size] = digit
18
19  plt.figure(figsize=(10, 10))
20  plt.imshow(figure, cmap='Greys_r')
21  plt.show()
```

執行結果：生成的樣本無雜訊且正確。

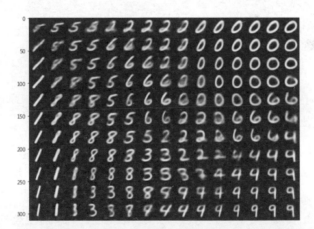

9-3 語義分割(Semantic segmentation)實作

語義分割(Semantic Segmentation)或稱影像分割(Image Segmentation)將每個像素(Pixel)作為標記(Label)，為避免抽樣造成像素資訊遺失，因此模型不使用池化層(Pooling)，學者提出許多演算法：

1. SegNet [2]，全名為影像分割的 Encoder-Decoder 架構(Deep Convolutional Encoder-Decoder Architecture for Image Segmentation)：使用反卷積放大特徵向量，還原圖像。

2. DeepLab [3]：以卷積作用在多種尺寸的圖像，得到 Score Map 後，再利用 Score Map 與 Conditional Random Field (CRF)演算法，以內插法 (interpolate)的方式還原圖像，詳細處理流程可參閱原文。

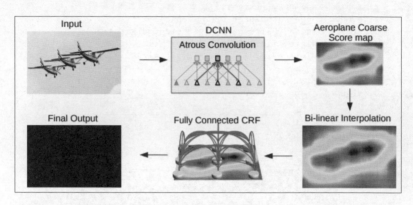

圖 9.4 DeepLab 的處理流程

3. RefiNet [4]：反卷積需佔大量記憶體，尤其是高解析度的圖像，所以 RefiNet 提出一種節省記憶體的方法。

4. PSPnet [5]：使用多種尺寸的池化層，稱為金字塔時尚(Pyramid Fashion)，金字塔掌握影像各個部份的圖像資料，利用此金字塔還原圖像。

5. U-Net [6]：廣泛應用於生物醫學的影像分割，這個模型很常被提到，且有許多的變形，所以，我們就來認識這個模型。

U-Net 是 AutoEncoder 的變形(Variant)，由於它的模型結構為 U 型而得名。

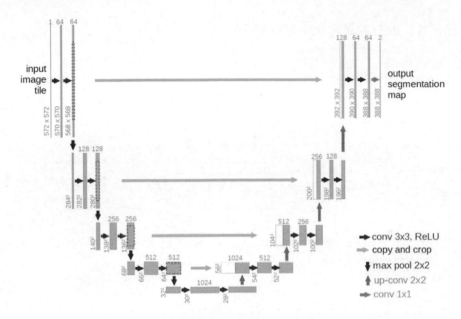

圖 9.5 U-Net 模型

圖片來源：U-Net: Convolutional Networks for Biomedical Image Segmentation [6]

傳統 AutoEncoder 的問題點發生在前半段的編碼器(Encoder)，由於它萃取特徵的過程，會使輸出的尺寸(Size)越變越小，接著解碼器(Decoder)再透過這些變小的特徵，重建出一個與原圖同樣大小的新圖像，因此原圖的很多資訊，像是前文所說的雜訊，就沒辦法傳遞到解碼器了。這個特點應用在去除雜訊上是十分恰當的，但假若目標是要偵測異常點(如檢測黃斑部病變)的話，那就糟糕了，經過模型過濾後，異常點通通都不見了。

所以，U-Net 在原有編碼器與解碼器的聯繫上，增加了一些連結，每一段編碼器的輸出都與其對面的解碼器相連接，使得編碼器每一層的資訊，都會額外輸入到一樣尺寸的解碼器，**如圖 9.5 橫跨 U 型兩側的中間長箭頭**，這樣在重建的過程中就比較不會遺失重要資訊。

【範例 3】以 U-Net 實作語義分割。

程式：09_03_Image_segmentation.ipynb，修改自 Keras 官網的範例[7]。

1. 載入相關套件。

```
1  # 載入相關套件
2  from tensorflow import keras
3  from tensorflow.keras.preprocessing.image import load_img
4  from tensorflow.keras.preprocessing.image import load_img
5  from tensorflow.keras import layers
6  import PIL
7  from PIL import ImageOps
8  import numpy as np
9  import os
10 from IPython.display import Image, display
```

2. 取得原圖與目標圖遮罩(Mask)的檔案路徑。自下列網址下載資料：

■ 原圖：http://www.robots.ox.ac.uk/~vgg/data/pets/data/images.tar.gz

■ 註解：http://www.robots.ox.ac.uk/~vgg/data/pets/data/annotations.tar.gz

```
1  # 訓練資料集路徑
2  root_path = "F:/0_DataMining/0_MY/Keras/ImageSegmentData/"
3  input_dir = root_path + "images/"                # 原圖目錄位置
4  target_dir = root_path + "annotations/trimaps/" # 遮罩圖(Mask)目錄位置
5
6  # 超參數設定
7  img_size = (160, 160) # 圖像寬高
8  num_classes = 4       # 類別個數
9  batch_size = 32       # 訓練批量
10
11 # 取得所有圖檔案路徑
12 input_img_paths = sorted(
13     [
14         os.path.join(input_dir, fname)
15         for fname in os.listdir(input_dir)
16         if fname.endswith(".jpg")
```

```
17        ]
18 )
19
20  # 取得所有遮罩圖檔案路徑
21  target_img_paths = sorted(
22      [
23          os.path.join(target_dir, fname)
24          for fname in os.listdir(target_dir)
25          if fname.endswith(".png") and not fname.startswith(".")
26      ]
27  )
28  print("樣本數:", len(input_img_paths))
```

執行結果：樣本數: 7390。

3. 檢查：顯示任一張圖。

```
1  # 顯示第10張圖
2  print(input_img_paths[9])
3  display(Image(filename=input_img_paths[9]))
4
5  # 調整對比，將最深的顏色當作黑色(0)，最淺的顏色當作白色(255)
6  print(target_img_paths[9])
7  img = PIL.ImageOps.autocontrast(load_img(target_img_paths[9]))
8  display(img)
```

執行結果：原圖和遮罩圖。

4.　建立 Iterator：Iterator 一次傳回一批圖像，不必一次全部載入至記憶
　　體。

```
1   # 建立圖像的 Iterator
2   class OxfordPets(keras.utils.Sequence):
3       """Helper to iterate over the data (as Numpy arrays)."""
4
5       def __init__(self, batch_size, img_size, input_img_paths, target_img_paths):
6           self.batch_size = batch_size
7           self.img_size = img_size
8           self.input_img_paths = input_img_paths
9           self.target_img_paths = target_img_paths
10
11      def __len__(self):
12          return len(self.target_img_paths) // self.batch_size
13
14      def __getitem__(self, idx):
15          """Returns tuple (input, target) correspond to batch #idx."""
16          i = idx * self.batch_size
17          batch_input_img_paths = self.input_img_paths[i : i + self.batch_size]
18          batch_target_img_paths = self.target_img_paths[i : i + self.batch_size]
19          x = np.zeros((batch_size,) + self.img_size + (3,), dtype="float32")
20          for j, path in enumerate(batch_input_img_paths):
21              img = load_img(path, target_size=self.img_size)
22              x[j] = img
23          y = np.zeros((batch_size,) + self.img_size + (1,), dtype="uint8")
24          for j, path in enumerate(batch_target_img_paths):
25              img = load_img(path, target_size=self.img_size, color_mode="grayscale")
26              y[j] = np.expand_dims(img, 2)
27          return x, y
```

5.　建立 U-Net 模型。

■　SeparableConv2D 神經層會針對色彩通道分別進行卷積。

■　模型 output 設為 4：以程式中的 num_classes 參數設定，一般 Filter 都
　　是設成 4 的倍數，作者利用後續的 display_mask 函數取最大值，判斷
　　遮罩的每一個像素是黑或是白。也有學者直接設為 1，詳情可參閱
　　「Understanding Semantic Segmentation with UNET」[8]。

```
1   def get_model(img_size, num_classes):
2       inputs = keras.Input(shape=img_size + (3,))
3
4       # 編碼器
5       x = layers.Conv2D(32, 3, strides=2, padding="same")(inputs)
6       x = layers.BatchNormalization()(x)
7       x = layers.Activation("relu")(x)
8       previous_block_activation = x  # Set aside residual
9
10      # 除了特徵圖大小，三個區塊均相同
11      for filters in [64, 128, 256]:
```

```
12          x = layers.Activation("relu")(x)
13          x = layers.SeparableConv2D(filters, 3, padding="same")(x)
14          x = layers.BatchNormalization()(x)
15
16          x = layers.Activation("relu")(x)
17          x = layers.SeparableConv2D(filters, 3, padding="same")(x)
18          x = layers.BatchNormalization()(x)
19
20          x = layers.MaxPooling2D(3, strides=2, padding="same")(x)
21
22          # 殘差層(residual)
23          residual = layers.Conv2D(filters, 1, strides=2, padding="same")(
24              previous_block_activation
25          )
26          x = layers.add([x, residual])  # Add back residual
27          previous_block_activation = x  # Set aside next residual
```

```
29      # 解碼器
30      for filters in [256, 128, 64, 32]:
31          x = layers.Activation("relu")(x)
32          x = layers.Conv2DTranspose(filters, 3, padding="same")(x)
33          x = layers.BatchNormalization()(x)
34
35          x = layers.Activation("relu")(x)
36          x = layers.Conv2DTranspose(filters, 3, padding="same")(x)
37          x = layers.BatchNormalization()(x)
38
39          x = layers.UpSampling2D(2)(x)
40
41          # 殘差層(residual)
42          residual = layers.UpSampling2D(2)(previous_block_activation)
43          residual = layers.Conv2D(filters, 1, padding="same")(residual)
44          x = layers.add([x, residual])  # Add back residual
45          previous_block_activation = x  # Set aside next residual
46
47      # per-pixel 卷積
48      outputs = layers.Conv2D(num_classes, 3, activation="softmax",
49                              padding="same")(x)
50
51      model = keras.Model(inputs, outputs)
52      return model
```

6. 建立模型。

```
1  # 釋放記憶體，以防執行多次造成記憶體的佔用
2  keras.backend.clear_session()
3
4  # 建立模型
5  model = get_model(img_size, num_classes)
6  model.summary()
```

7. 繪製模型結構：無法畫出 U 型結構，繪圖程式沒有那麼聰明，請參閱程式輸出。

```
1  import tensorflow as tf
2  tf.keras.utils.plot_model(model, to_file='Unet_model.png')
```

8. 資料切割為訓練資料和驗證資料。

```
1  import random
2
3  # Split our img paths into a training and a validation set
4  val_samples = 1000
5  random.Random(1337).shuffle(input_img_paths)
6  random.Random(1337).shuffle(target_img_paths)
7  train_input_img_paths = input_img_paths[:-val_samples]
8  train_target_img_paths = target_img_paths[:-val_samples]
9  val_input_img_paths = input_img_paths[-val_samples:]
10 val_target_img_paths = target_img_paths[-val_samples:]
11
12 # Instantiate data Sequences for each split
13 train_gen = OxfordPets(
14     batch_size, img_size, train_input_img_paths, train_target_img_paths
15 )
16 val_gen = OxfordPets(batch_size, img_size, val_input_img_paths,
17                      val_target_img_paths)
```

9. 訓練模型。

```
1  # 設定優化器(optimizer)、損失函數(loss)、效能衡量指標(metrics)的類別
2  model.compile(optimizer="rmsprop", loss="sparse_categorical_crossentropy")
3
4  # 設定檢查點 callbacks，模型存檔
5  callbacks = [
6      keras.callbacks.ModelCheckpoint("oxford_segmentation.h5", save_best_only=True)
7  ]
8
9  # 訓練 15 週期(epoch)
10 epochs = 15
11 model.fit(train_gen, epochs=epochs, validation_data=val_gen, callbacks=callbacks)
```

10. 預測並顯示圖像。

```
1  # 預測所有驗證資料
2  val_gen = OxfordPets(batch_size, img_size, val_input_img_paths,
3                       val_target_img_paths)
4  val_preds = model.predict(val_gen)
5
6  # 顯示遮罩(mask)
7  def display_mask(i):
8      """Quick utility to display a model's prediction."""
9      mask = np.argmax(val_preds[i], axis=-1)
```

```
10      mask = np.expand_dims(mask, axis=-1)
11      img = PIL.ImageOps.autocontrast(keras.preprocessing.image.array_to_img(mask))
12      display(img)
13
14  # 顯示驗證資料第11個圖檔
15  i = 10
16  # 顯示原圖
17  print('原圖：')
18  display(Image(filename=val_input_img_paths[i]))
19
20  # 顯示原圖遮罩(mask)
21  print('原遮罩圖：')
22  img = PIL.ImageOps.autocontrast(load_img(val_target_img_paths[i]))
23  display(img)
24
25  # 顯示預測結果
26  print('預測結果：')
27  display_mask(i)   # Note that the model only sees inputs at 150x150.
```

執行結果：圖(C)與圖(B)比較，效果還不錯。

原圖：

原遮罩圖：

預測結果：

(A)   (B)   (C)

9-4 實例分割(Instance Segmentation)

上一節的語義分割，同類別的物件只能夠以相同顏色呈現，如要做到同類別的物件以不同顏色呈現的話，就會輪到實例分割(Instance Segmentation)上場。

而實例分割所使用的 Mask R-CNN 演算法係由 Facebook AI Research 在 2018 年所發表[9]。Mask R-CNN 為 Faster R-CNN 的延伸，不只會框住物

件，更能產生遮罩(Mask)，如下圖所示。

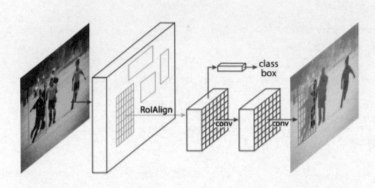

圖 9.6 Mask R-CNN 模型，圖片來源：Mask R-CNN [9]

除了辨識物件之外，實例分割還有以下延伸的應用：

1. 去背：偵測到物件後，將物件以外的背景全部去除。
2. 移除特殊的物件：將偵測到的物件移除後，根據周遭的顏色填補移除的區域。例如在觀光景點拍照時，最困擾的就是有陌生人一起入鏡，這時即可利用此技術將之移除，PhotoShop 就有提供類似的功能。

直接來看一個實例，使用 akTwelve Mask R-CNN 函數庫[10]，安裝程序如下：

1. 自 https://github.com/akTwelve/Mask_RCNN 下載整個專案，並解壓縮。
2. 切換至該專案的根目錄，執行下列指令：

```
pip install -r requirements.txt
python setup.py install
```

3. 執行下列指令，確認安裝成功：

```
pip show mask-rcnn
```

如下畫面即表示安裝成功：

```
Name: mask-rcnn
Version: 2.1
Summary: Mask R-CNN for object detection and instance segmentation
Home-page: https://github.com/matterport/Mask_RCNN
Author: Matterport
Author-email: waleed.abdulla@gmail.com
License: MIT
Location: c:\anaconda3\lib\site-packages\mask_rcnn-2.1-py3.8.egg
Requires:
Required-by:
```

4. 下載權重檔：https://github.com/matterport/Mask_RCNN/releases/download/ v2.0/mask_rcnn_coco.h5。

【範例 4】使用 Mask R-CNN 進行實例分割。

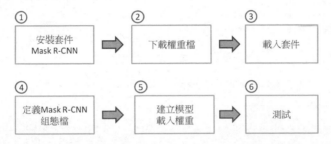

程式：09_04_Mask_R-CNN_Test.ipynb，修改自「How to Use Mask R-CNN in Keras for Object Detection in Photographs」所描述的例子[11]。

1. 載入相關套件。

```
1  # 載入相關套件
2  from tensorflow.keras.preprocessing.image import load_img
3  from tensorflow.keras.preprocessing.image import img_to_array
4  from mrcnn.config import Config
5  import matplotlib.pyplot as plt
6  from matplotlib.patches import Rectangle
7  from mrcnn.model import MaskRCNN
```

2. 定義在圖像上加框的函數。

```
1  # 定義函數，在圖像加框
2  def draw_image_with_boxes(filename, boxes_list):
3      # 讀取圖檔
4      data = plt.imread(filename)
5      # 顯示圖像
6      plt.imshow(data)
7
8      # 加框
```

```
9      ax = plt.gca()
10     for box in boxes_list:
11         # 上/下/左/右 座標
12         y1, x1, y2, x2 = box
13         # 計算框的寬高
14         width, height = x2 - x1, y2 - y1
15         # 畫框
16         rect = Rectangle((x1, y1), width, height, fill=False, color='red')
17         ax.add_patch(rect)
18
19     # 繪圖
20     plt.show()
```

3. 定義偵測的組態檔：Mask R-CNN 須指定各項參數。

```
1  # 定義偵測的組態檔
2  class TestConfig(Config):
3      NAME = "test1"          # 測試名稱，任意取名
4      GPU_COUNT = 1           # GPU 個數
5      IMAGES_PER_GPU = 1      # 每個 GPU 負責偵測的圖像數
6      NUM_CLASSES = 1 + 80 # 類別個數 + 1
```

4. 建立模型，載入權重，進行測試。

```
1  # 建立模型
2  rcnn = MaskRCNN(mode='inference', model_dir='./', config=TestConfig())
3
4  # 載入權重檔
5  rcnn.load_weights('./MaskRCNN_weights/mask_rcnn_coco.h5', by_name=True)
6
7  # 載入圖檔
8  img = load_img('./images_test/elephant.jpg')
9  img = img_to_array(img) # 影像轉陣列
10
11 # 預測
12 results = rcnn.detect([img], verbose=0)
13 # 加框、繪圖
14 draw_image_with_boxes('./images_test/elephant.jpg', results[0]['rois'])
```

執行結果：

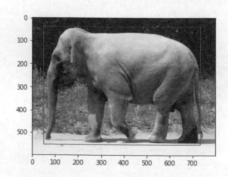

5. 定義類別名稱。

```
1  # 定義類別名稱
2  class_names = ['BG', 'person', 'bicycle', 'car', 'motorcycle', 'airplane',
3                 'bus', 'train', 'truck', 'boat', 'traffic light',
4                 'fire hydrant', 'stop sign', 'parking meter', 'bench', 'bird',
5                 'cat', 'dog', 'horse', 'sheep', 'cow', 'elephant', 'bear',
6                 'zebra', 'giraffe', 'backpack', 'umbrella', 'handbag', 'tie',
7                 'suitcase', 'frisbee', 'skis', 'snowboard', 'sports ball',
8                 'kite', 'baseball bat', 'baseball glove', 'skateboard',
9                 'surfboard', 'tennis racket', 'bottle', 'wine glass', 'cup',
10                'fork', 'knife', 'spoon', 'bowl', 'banana', 'apple',
11                'sandwich', 'orange', 'broccoli', 'carrot', 'hot dog', 'pizza',
12                'donut', 'cake', 'chair', 'couch', 'potted plant', 'bed',
13                'dining table', 'toilet', 'tv', 'laptop', 'mouse', 'remote',
14                'keyboard', 'cell phone', 'microwave', 'oven', 'toaster',
15                'sink', 'refrigerator', 'book', 'clock', 'vase', 'scissors',
16                'teddy bear', 'hair drier', 'toothbrush']
```

6. 顯示遮罩。

```
1  # 加載顯示遮罩的類別
2  from mrcnn.visualize import display_instances
3
4  # 取的第一個遮罩
5  r = results[0]
6  # 顯示框、遮罩、類別、機率
7  display_instances(img, r['rois'], r['masks'], r['class_ids'], class_names, r['scores'])
```

執行結果：觀察右圖發現整條尾巴都沒被遮罩到，所以要達到精準去背，應該需要更多的資料與訓練週期，甚至是改善演算法，才可能達到目的。

7. 測試含多個物件的圖檔。

```
1  # 載入另一圖檔
2  img = load_img('./images_Object_Detection/zebra.jpg')
3  img = img_to_array(img) # 影像轉陣列
4
5  # 預測
6  results = rcnn.detect([img], verbose=0)
7
8  # 取的第一個遮罩
9  r = results[0]
10 # 顯示框、遮罩、類別、機率
11 display_instances(img, r['rois'], r['masks'], r['class_ids'], class_names, r['scores'])
```

執行結果：三隻斑馬均可被遮罩到。

9-5 風格轉換(Style Transfer) --人人都可以是畢卡索

接著來認識另一個有趣的 AutoEncoder 變形，稱為「風格轉換」(Neural Style Transfer)，把一張照片轉換成某一幅畫的風格，如下圖。讀者可以在手機下載「Prisma」App 來玩玩，它能夠在拍照後，將照片風格即時轉換，內建近二十種的大師畫風可供選擇，只是轉換速度有點慢。

圖 9.7 風格轉換 (Style Transfer)，原圖+風格圖像=生成圖像，
圖片來源：fast-style-transfer GitHub [12]

之前有一則關於美圖影像實驗室(MTlab)的新聞，「催生全球首位 AI 繪師 Andy，美圖搶攻人工智慧卻面臨一大挑戰」[13]，該公司號稱投資了 1.99 億元人民幣，研發團隊超過 60 人，將風格轉換速度縮短到 3 秒鐘，開發成「美圖秀秀」App，大受歡迎，之後更趁勢推出專屬手機，狂銷 100 多萬台，算得上少數成功的 AI 商業模式。

風格轉換演算法由 Leon A. Gatys 等學者於 2015 年提出[14]，主要作法是重新定義損失函數，分為「內容損失」(Content Loss)與「風格損失」(Style Loss)，並利用 AutoEncoder 的解碼器合成圖像，隨著訓練週期，損失逐漸變小，亦即生成的圖像會越接近於原圖與風格圖的合成。

內容損失函數比較單純，即原圖與生成圖像的像素差異平方和，定義如下：

$$J_{content}(C, G) = \frac{1}{4 \times n_H \times n_W \times n_c} \sum \left(a^{(C)} - a^{(G)}\right)^2$$

n_H、n_W：原圖的寬、高。

n_C：色彩通道數。

$a^{(C)}$：原圖的像素。

$a^{(G)}$：生成圖像的像素。

風格損失函數為該演算法的重點，如何量化抽象的畫風是一大挑戰，Gatys 等學者想到的方法是，先定義 Gram 矩陣(Matrix)後，再利用 Gram 矩陣來定義風格損失。

Gram Matrix：兩個特徵向量進行點積，代表特徵的關聯性，顯現那些特徵是同時出現的，亦即風格。因此，風格損失就是要最小化風格圖像與生成圖像的 Gram 差異平方和，如下：

$$J_{style}(S, G) = \frac{1}{4 \times n_c^2 \times (n_H \times n_W)^2} \sum_{i=l}^{n_c} \sum_{j=1}^{n_c} \left(G^{(S)} - G^{(G)}\right)^2$$

$G^{(S)}$：風格圖像的 Gram。

$G^{(G)}$：生成圖像的 Gram。

上式只是單一神經層的風格損失，結合所有神經層的風格損失，定義如下：

$$J_{style}(S, G) = \sum_l \lambda^{(l)} J_{style}^{(l)}(S, G)$$

λ：每一層的權重。

總損失函數：

$$J(G) = \alpha J_{content}(C, G) + \beta J_{style}(S, G)$$

α、β：控制內容與風格的比重，可以控制生成圖像要偏重風格的比例。

接下來，我們就來進行實作。

【範例 5】使用風格轉換演算法進行圖檔的轉換。提醒一下，範例中的內容圖即是原圖的意思。

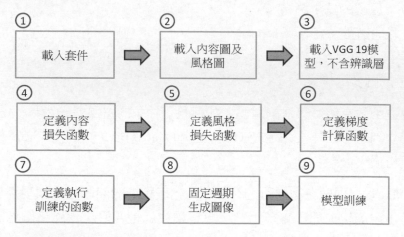

程式 **09_05_Neural_Style_Transfer.ipynb**，修改自 TensorFlow 官網提供的範例「Neural Style Transfer」[15]。

1. 載入相關套件。

```
1   # 載入相關套件
2   import os
3   import time
4   import sys
5   import matplotlib.pyplot as plt
6   from PIL import Image
7   import numpy as np
8   import tensorflow as tf
9   from tensorflow.keras.applications.vgg19 import VGG19
10  from tensorflow.keras.preprocessing.image import load_img
11  from tensorflow.keras.preprocessing.image import img_to_array
```

2. 載入內容圖檔。

```
1  # 載入內容圖檔
2  content_path = "./style_transfer/chicago.jpg"
3  content_image = load_img(content_path)
4  plt.imshow(content_image)
5  plt.axis('off')
6  plt.show()
```

執行結果：

3. 載入風格圖檔。

```
1  # 載入風格圖檔
2  style_path = "./style_transfer/wave.jpg"
3  style_image = load_img(style_path)
4  plt.imshow(style_image)
5  plt.axis('off')
6  plt.show()
```

執行結果：

4. 定義圖像前置處理的函數。

```
1   # 載入圖像並進行前置處理
2   def load_and_process_img(path_to_img):
3       img = load_img(path_to_img)
4       img = img_to_array(img)
5       img = np.expand_dims(img, axis=0)
6       img = tf.keras.applications.vgg19.preprocess_input(img)
7       # print(img.shape)
8
9       # 回傳影像陣列
10      return img
```

5. 定義由陣列還原成圖像的函數。

```
1   # 由陣列還原影像
2   def deprocess_img(processed_img):
3       x = processed_img.copy()
4       if len(x.shape) == 4:
5           x = np.squeeze(x, 0)
6
7       # 前置處理的還原
8       x[:, :, 0] += 103.939
9       x[:, :, 1] += 116.779
10      x[:, :, 2] += 123.68
11      x = x[:, :, ::-1]
12
13      # 裁切元素值在(0, 255)之間
14      x = np.clip(x, 0, 255).astype('uint8')
15      return x
```

6. 定義內容圖和風格圖輸出的神經層名稱。

```
1   # 定義內容圖輸出的神經層名稱
2   content_layers = ['block5_conv2']
3
4   # 定義風格圖輸出的神經層名稱
5   style_layers = ['block1_conv1',
6                   'block2_conv1',
7                   'block3_conv1',
8                   'block4_conv1',
9                   'block5_conv1'
10                  ]
11
12  num_content_layers = len(content_layers)
13  num_style_layers = len(style_layers)
```

7. 定義模型：載入 VGG 19 模型，不含辨識層，加上自訂的輸出。

```
1   from tensorflow.python.keras import models
2
3   def get_model():
4       # 載入 VGG19，不含辨識層
5       vgg = tf.keras.applications.vgg19.VGG19(include_top=False, weights='imagenet')
6       vgg.trainable = False  # 不重新訓練
7       for layer in vgg.layers:
8           layer.trainable = False
9
10      # 以之前定義的內容圖及風格圖神經層為輸出
11      style_outputs = [vgg.get_layer(name).output for name in style_layers]
12      content_outputs = [vgg.get_layer(name).output for name in content_layers]
13      model_outputs = style_outputs + content_outputs
14
15      # 建立模型
16      return models.Model(vgg.input, model_outputs)
```

8. 定義內容損失函數：為內容圖與生成圖特徵向量之差的平方和，也可
    以採用上述理論的公式。

```
1   # 內容損失函數
2   def get_content_loss(base_content, target):
3       # 下面可附加『/ (4. * (channels ** 2) * (width * height) ** 2)』
4       return tf.reduce_mean(tf.square(base_content - target))
```

9. 定義風格損失函數：先定義 Gram Matrix 計算函數，再定義風格損失函
    數。

```
1   # 計算 Gram Matrix 函數
2   def gram_matrix(input_tensor):
3       # We make the image channels first
4       channels = int(input_tensor.shape[-1])
5       a = tf.reshape(input_tensor, [-1, channels])
6       n = tf.shape(a)[0]
7       gram = tf.matmul(a, a, transpose_a=True)
8       return gram / tf.cast(n, tf.float32)
9
10  # 風格損失函數
11  def get_style_loss(base_style, gram_target):
12      # 取得風格圖的高、寬、色彩數
13      height, width, channels = base_style.get_shape().as_list()
14
15      # 計算 Gram Matrix
16      gram_style = gram_matrix(base_style)
17
18      # 計算風格損失
19      return tf.reduce_mean(tf.square(gram_style - gram_target))
```

10. 定義圖像的特徵向量計算函數。

```
1   # 計算內容圖及風格圖的特徵向量
2   def get_feature_representations(model, content_path, style_path):
3       # 載入圖檔
4       content_image = load_and_process_img(content_path)
5       style_image = load_and_process_img(style_path)
6
7       # 設定模型
8       style_outputs = model(style_image)
9       content_outputs = model(content_image)
10
11
12      # 取得特徵向量
13      style_features = [style_layer[0] for style_layer
14                        in style_outputs[:num_style_layers]]
15      content_features = [content_layer[0] for content_layer
16                          in content_outputs[num_style_layers:]]
17      return style_features, content_features
```

11. 定義梯度計算函數。

```
1   # 計算梯度
2   def compute_grads(cfg):
3       with tf.GradientTape() as tape:
4           # 累計損失
5           all_loss = compute_loss(**cfg)
6
7       # 取得梯度
8       total_loss = all_loss[0]
9       # cfg['init_image'] : 內容圖影像陣列
10      return tape.gradient(total_loss, cfg['init_image']), all_loss
```

12. 定義所有層的損失計算函數：按照內容圖與風格圖的權重比例，計算
    總損失。

```
1   # 計算所有層的損失
2   def compute_loss(model, loss_weights, init_image, gram_style_features, content_features):
3       # 內容圖及風格圖的權重比例
4       style_weight, content_weight = loss_weights
5
6       # 取得模型輸出
7       model_outputs = model(init_image)
8       style_output_features = model_outputs[:num_style_layers]
9       content_output_features = model_outputs[num_style_layers:]
10
11      # 累計風格分數
12      style_score = 0
13      weight_per_style_layer = 1.0 / float(num_style_layers)
14      for target_style, comb_style in zip(gram_style_features, style_output_features):
15          style_score += weight_per_style_layer * get_style_loss(comb_style[0], target_style)
16
17      # 累計內容分數
18      content_score = 0
19      weight_per_content_layer = 1.0 / float(num_content_layers)
20      for target_content, comb_content in zip(content_features, content_output_features):
21          content_score += weight_per_content_layer* get_content_loss(comb_content[0], target_content)
22
23      # 乘以權重比例
24      style_score *= style_weight
25      content_score *= content_weight
26
27      # 總損失
28      loss = style_score + content_score
29
30      return loss, style_score, content_score
```

13. 定義執行訓練的函數：這是程式的核心，在訓練的過程中，在固定週
    期生成圖像，可以看到圖像逐漸轉變的過程。

```
1   # 執行訓練的函數
2   import IPython.display
3
4   def run_style_transfer(content_path, style_path, num_iterations=1000,
5                          content_weight=1e3, style_weight=1e-2):
6       # 取得模型
7       model = get_model()
8
```

```
9      # 取得內容圖及風格圖的神經層輸出
10     style_features, content_features = get_feature_representations(model, content_path, style_path)
11     gram_style_features = [gram_matrix(style_feature) for style_feature in style_features]
12
13     # 載入內容圖
14     init_image = load_and_process_img(content_path)
15     init_image = tf.Variable(init_image, dtype=tf.float32)
16
17     # 指定優化器
18     opt = tf.optimizers.Adam(learning_rate=5, beta_1=0.99, epsilon=1e-1)
19
20     # 初始化變數
21     iter_count = 1
22     best_loss, best_img = float('inf'), None
23     loss_weights = (style_weight, content_weight)
24     cfg = {  # 組態
25             'model': model,
26             'loss_weights': loss_weights,
27             'init_image': init_image,
28             'gram_style_features': gram_style_features,
29             'content_features': content_features
```

```
30     }
31
32     # 參數設定
33     num_rows = 2  # 輸出小圖以 2 列顯示
34     num_cols = 5  # 輸出小圖以 5 行顯示
35     # 每N個週期數生成圖像，計算 N：display_interval
36     display_interval = num_iterations/(num_rows*num_cols)
37     start_time = time.time()    # 計時
38     global_start = time.time()
39
40     # RGB 三色中心值，輸入圖像以中心值為 0 作轉換
41     norm_means = np.array([103.939, 116.779, 123.68])
42     min_vals = -norm_means
43     max_vals = 255 - norm_means
44
45     # 開始訓練
46     imgs = []
47     for i in range(num_iterations):
48         grads, all_loss = compute_grads(cfg)
49         loss, style_score, content_score = all_loss
50         opt.apply_gradients([(grads, init_image)])
51         clipped = tf.clip_by_value(init_image, min_vals, max_vals)
52         init_image.assign(clipped)
53         end_time = time.time()
```

```
55         # 記錄最小損失時的圖像
56         if loss < best_loss:
57             best_loss = loss    # 記錄最小損失
58             best_img = deprocess_img(init_image.numpy()) # 生成圖像
59
60         # 每N個週期數生成圖像
61         if i % display_interval== 0:
62             start_time = time.time()
63
64             # 生成圖像
65             plot_img = init_image.numpy()
66             plot_img = deprocess_img(plot_img)
67             imgs.append(plot_img)
68
69             # IPython.display.clear_output(wait=True)  # 可清除之前的顯示
70             print(f'週期數: {i}')
71             elapsed_time = time.time() - start_time
72             print(f'總損失: {loss:.2e}, 風格損失: {style_score:.2e},' +
73                     f'內容損失: {content_score:.2e}, 耗時: {elapsed_time:.2f}s')
74             IPython.display.display_png(Image.fromarray(plot_img))
```

```
75
76      print(f'總耗時: {(time.time() - global_start):.2f}s')
77      # IPython.display.clear_output(wait=True)  # 可清除之前的顯示
78      # 顯示生成的圖像
79      plt.figure(figsize=(14,4))
80      for i,img in enumerate(imgs):
81              plt.subplot(num_rows,num_cols,i+1)
82              plt.imshow(img)
83              plt.axis('off')
84
85      return best_img, best_loss
```

14. 呼叫上述函數，執行模型訓練。

```
1   # 執行訓練
2   model = get_model()
3   best, best_loss = run_style_transfer(content_path,
4                               style_path, num_iterations=500)
```

執行結果：執行 500 個週期，剛開始畫面變化很大，之後轉換逐漸減少，
表示損失函數逐步收斂。

15. 顯示最佳的圖像。

```
1   # 顯示最佳的圖像
2   Image.fromarray(best)
```

執行結果：

16. 比較內容圖與生成圖。

```
1   # 定義顯示函數，比較原圖與生成圖像
2   def show_results(best_img, content_path, style_path, show_large_final=True):
3       plt.figure(figsize=(10, 5))
4       content = load_img(content_path)
5       style = load_img(style_path)
6
7       plt.subplot(1, 2, 1)
8       plt.axis('off')
9       plt.imshow(content)
10      plt.title('原圖')
11
12      plt.subplot(1, 2, 2)
13      plt.axis('off')
14      plt.imshow(style)
15      plt.title('風格圖')
16
```

```
17    if show_large_final:
18        plt.figure(figsize=(10, 10))
19        plt.axis('off')
20
21        plt.imshow(best_img)
22        plt.title('Output Image')
23        plt.show()
24
25  # 原圖與生成圖像的比較
26  show_results(best, content_path, style_path)
```

17. 拿另一張內容圖來測試。

```
1  # 以另一張圖測試
2  content_path = './style_transfer/Green_Sea_Turtle_grazing_seagrass.jpg'
3  style_path = './style_transfer/wave.jpg'
4
5  best_starry_night, best_loss = run_style_transfer(content_path, style_path)
6  show_results(best_starry_night, content_path, style_path)
```

執行結果:

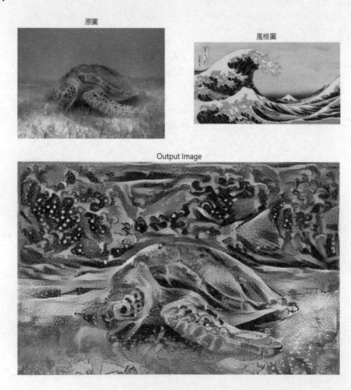

TensorFlow Hub 裡面有許多預先訓練好的模型可以直接套用,包括風格轉

換模型，以下範例使用 Fast style transfer 演算法預先訓練好的模型，可快速產生新圖像，但該模型每次產生的圖像均不相同。

【範例 6】使用 TensorFlow Hub 的 Fast style transfer 預先訓練模型完成風格轉換。

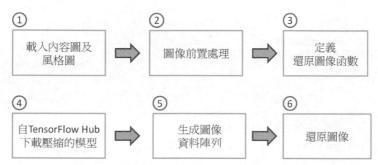

程式：請參閱 09_05_Neural_Style_Transfer.ipynb 的下半段。

1. 下載圖像：可從網路下載圖像。

```
1  # 下載圖像
2  storage_url = 'https://storage.googleapis.com/download.tensorflow.org/'
3  content_url = storage_url + 'example_images/YellowLabradorLooking_new.jpg'
4  style_url = storage_url + 'example_images/Vassily_Kandinsky%2C_1913_-_Composition_7.jpg'
5  content_path = tf.keras.utils.get_file('YellowLabradorLooking_new.jpg', content_url)
6  style_path = tf.keras.utils.get_file('kandinsky5.jpg', style_url)
```

2. 定義載入圖像並進行前置處理的函數：與前一個範例的處理方式相同。

```
1  # 定義載入圖像並進行前置處理的函數
2  def custom_load_img(path_to_img):
3      max_dim = 512
4      img = tf.io.read_file(path_to_img)
5      img = tf.image.decode_image(img, channels=3)
6      img = tf.image.convert_image_dtype(img, tf.float32)
7
8      shape = tf.cast(tf.shape(img)[:-1], tf.float32)
9      long_dim = max(shape)
10     scale = max_dim / long_dim
11
12     new_shape = tf.cast(shape * scale, tf.int32)
13
14     img = tf.image.resize(img, new_shape)
15     img = img[tf.newaxis, :]
16     return img
```

3. 定義顯示圖像的函數:與前一個範例的處理方式相同。

```
1  # 定義顯示圖像的函數
2  def custom_imshow(image, title=None):
3      if len(image.shape) > 3:
4          image = tf.squeeze(image, axis=0)
5
6      plt.axis('off')
7      plt.imshow(image)
8      if title:
9          plt.title(title)
```

4. 載入圖像並顯示。

```
1   # 載入圖像
2   content_image = custom_load_img(content_path)
3   style_image = custom_load_img(style_path)
4
5   # 繪圖
6   plt.subplot(1, 2, 1)
7   custom_imshow(content_image, '原圖')
8
9   plt.subplot(1, 2, 2)
10  custom_imshow(style_image, '風格圖')
```

執行結果:

原圖

風格圖

5. 定義還原圖像的函數。

```
1  # 定義還原圖像的函數
2  def tensor_to_image(tensor):
3      tensor = tensor*255
4      tensor = np.array(tensor, dtype=np.uint8)
5      if np.ndim(tensor)>3:
6          assert tensor.shape[0] == 1
7          tensor = tensor[0]
8      return Image.fromarray(tensor)
```

6. 自 TensorFlow Hub 下載壓縮的模型。

```
1  # 自 TensorFlow Hub 下載壓縮的模型
2  import tensorflow_hub as hub
3
4  os.environ['TFHUB_MODEL_LOAD_FORMAT'] = 'COMPRESSED'
5  hub_model = hub.load('https://tfhub.dev/google/magenta/arbitrary-image-stylization-v1-256/2')
```

7. 生成新的圖像。

```
1  # 生成圖像
2  stylized_image = hub_model(tf.constant(content_image), tf.constant(style_image))[0]
3  tensor_to_image(stylized_image)
```

執行結果：

TensorFlow 官網提供的範例「Neural Style Transfer」[15]，有提及一些改善的措施，譬如，基本的風格轉換演算法所生成的圖像經常會偏重高頻 (high frequency)，而高頻會造成邊緣特別明顯，可使用正則化 (Regularization)矯正損失函數，改善此一現象。

上述程式生成一張圖像需要花 290 秒，在如今的社群媒體時代，就算這酷炫的效果抓住了大眾的眼球，也難以造成流行，因此在網路上有許多的研究，討論如何加快演算法速度，有興趣的讀者可搜尋「Fast Style Transfer」。另外，也有同學問到，如果用同一張風格圖，對另一張新的內容圖進行風格轉換，也要重新訓練嗎？ 答案是不一定，這是一個值得研究的課題。開發美圖秀秀的公司砸了近兩億人民幣，才將速度縮短至 3 秒，可見技術難度頗高，所以，速度絕對是商業模式重要的考量因素。

風格轉換是一個非常有趣的應用，除了轉換成名畫風格之外，也可將照片卡通化，或是針對臉部美肌，凡此種種都值得一試。當然，不只有風格轉換演算法可以這樣玩，其他像 GAN 或 OpenCV 影像處理也都能做到類似的功能，大家一起天馬行空，胡思亂想吧！

▌ 9-6 臉部辨識(Facial Recognition)

臉部辨識(Facial Recognition)的應用面向非常廣泛，國內廠商不論是系統廠商、PC 廠商、NAS 廠商，甚至是電信業者，都已涉獵此一領域，推出各種五花八門的相關產品，已經有以下這些應用類型：

1. 智慧保全：結合門禁系統，運用在家庭、學校、員工宿舍、飯店、機場登機檢查、出入境比對、黑名單/罪犯/失蹤人口比對等方面。
2. 考勤系統：上下班臉部刷卡取代卡片。
3. 商店即時監控：即時辨識 VIP 和黑名單客戶的進出，進行客戶關懷、發送折扣碼、或記錄停留時間，作為商品陳列與改善經營效能的參考依據。
4. 快速結帳：以臉部辨識取代刷卡付帳。
5. 人流統計：針對有人數容量限制的公共場所，如百貨公司、遊樂園、體育場館，透過臉部辨識，進行人數控管。.
6. 情緒分析：辨識臉部情緒，發生意外時能迅速通報救援，或進行滿意度調查。

社群軟體上傳照片的辨識：標註朋友姓名等。

依據技術類別可細分為：

1. 臉部偵測(Face Detection)：與物件偵測類似，因此運用物件偵測技術即可做到此功能，偵測圖像中有那些臉部和其位置。
2. 臉部特徵點檢測(Facial Landmarks Detection)：偵測臉部的特徵點，用來比對兩張臉是否屬於同一人。

3. 臉部追蹤(Face Tracking)：在影片中追蹤移動中的臉部，可辨識人移動的軌跡。

4. 臉部辨識(Face Recognition)，分為兩種：
   * 臉部識別(Face Identification)：從 N 個人中找出最相似的人。
   * 臉部驗證(Face Verification)：驗證臉部是否相符，例如，出入境檢查旅客是否與其護照上的大頭照相符合。

各項臉部辨識技術及支援的套件，如下圖：

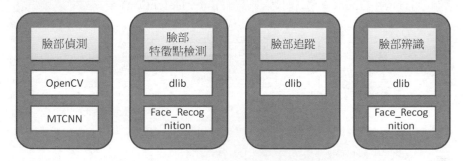

圖 9.8 臉部辨識的技術類別與相關套件支援

接下來，我們就逐一實作這些相關功能。

9-6-1 臉部偵測(Face Detection)

OpenCV 使用 Haar Cascades 演算法來進行各種物件的偵測，它會將各種物件的特徵記錄在 XML 檔案，稱為級聯分類器(Cascade File)，可在 OpenCV 或 OpenCV-Python 安裝目錄內找到(haarcascade_*.xml)，筆者已把相關檔案複製到範例程式目錄下。

Haar Cascades 技術發展較早，辨識速度快，能夠作到即時偵測，缺點則是準確度較差，容易造成偽陽性，即誤認臉部特徵。它的架構類似卷積，如下圖所示，以各種濾波器(Filters)掃描圖像，像是眼部比臉頰暗，鼻樑比臉頰亮等。

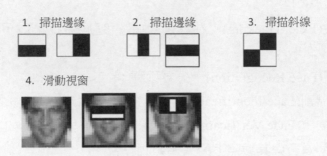

圖 9.9 Haar Cascades 以濾波器(Filters) 掃描圖像

【範例 1】使用 OpenCV 進行臉部偵測(Face Detection)。

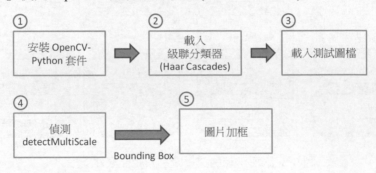

程式:09_06_臉部偵測_opencv.ipynb。

1. 載入相關套件,包含 OpenCV-Python。

```
1  # 載入相關套件
2  import cv2
3  from cv2 import CascadeClassifier
4  from cv2 import rectangle
5  import matplotlib.pyplot as plt
6  from cv2 import imread
```

2. 載入臉部的級聯分類器(face cascade file)。

```
1  # 載入臉部級聯分類器(face cascade file)
2  face_cascade = './cascade_files/haarcascade_frontalface_alt.xml'
3  classifier = cv2.CascadeClassifier(face_cascade)
```

3. 載入測試圖檔。

```
1  # 載入圖檔
2  image_file = "./images_face/teammates.jpg"
3  image = imread(image_file)
4
```

```
5   # OpenCV 預設為 BGR 色系，轉為 RGB 色系
6   im_rgb = cv2.cvtColor(image, cv2.COLOR_BGR2RGB)
7
8   # 顯示圖像
9   plt.imshow(im_rgb)
10  plt.axis('off')
11  plt.show()
```

執行結果：

4. 偵測臉部並顯示圖像。

```
1   # 偵測臉部
2   bboxes = classifier.detectMultiScale(image)
3   # 臉部加框
4   for box in bboxes:
5       # 取得框的座標及寬高
6       x, y, width, height = box
7       x2, y2 = x + width, y + height
8       # 加白色框
9       rectangle(im_rgb, (x, y), (x2, y2), (255,255,255), 2)
10
11  # 顯示圖像
12  plt.imshow(im_rgb)
13  plt.axis('off')
14  plt.show()
```

執行結果：全部人的臉都有被正確偵測到。

5. 載入另一圖檔。

```
1   # 載入圖檔
2   image_file = "./images_face/classmates.jpg"
3   image = imread(image_file)
4
5   # OpenCV 預設為 BGR 色系，轉為 RGB 色系
6   im_rgb = cv2.cvtColor(image, cv2.COLOR_BGR2RGB)
7
8   # 顯示圖像
9   plt.imshow(im_rgb)
10  plt.axis('off')
11  plt.show()
```

執行結果：臉部特寫。

6. 偵測臉部並顯示圖像。

```
1   # 偵測臉部
2   bboxes = classifier.detectMultiScale(image)
3   # 臉部加框
4   for box in bboxes:
5       # 取得框的座標及寬高
6       x, y, width, height = box
7       x2, y2 = x + width, y + height
8       # 加紅色框
9       rectangle(im_rgb, (x, y), (x2, y2), (255,0,0), 5)
10
11  # 顯示圖像
12  plt.imshow(im_rgb)
13  plt.axis('off')
14  plt.show()
```

執行結果： 就算圖像中的臉部占據畫面較大，亦可正確偵測。

7. 同時載入眼睛與微笑的級聯分類器。

```
1  # 載入眼睛級聯分類器(eye cascade file)
2  eye_cascade = './cascade_files/haarcascade_eye_tree_eyeglasses.xml'
3  classifier = cv2.CascadeClassifier(eye_cascade)
4
5  # 載入微笑級聯分類器(smile cascade file)
6  smile_cascade = './cascade_files/haarcascade_smile.xml'
7  smile_classifier = cv2.CascadeClassifier(smile_cascade)
```

8. 偵測臉部並顯示圖像。

```
1  # 偵測臉部
2  bboxes = classifier.detectMultiScale(image)
3  # 臉部加框
4  for box in bboxes:
5      # 取得框的座標及寬高
6      x, y, width, height = box
7      x2, y2 = x + width, y + height
8      # 加白色框
9      rectangle(im_rgb, (x, y), (x2, y2), (255,0,0), 5)
10
11  # 偵測微笑
12  # scaleFactor=2.5：掃描時每次縮減掃描視窗的尺寸比例。
13  # minNeighbors=20：每一個被選中的視窗至少要有鄰近且合格的視窗數
14  bboxes = smile_classifier.detectMultiScale(image, 2.5, 20)
15  #微笑加框
16  for box in bboxes:
17      # 取得框的座標及寬高
18      x, y, width, height = box
19      x2, y2 = x + width, y + height
20      # 加白色框
21      rectangle(im_rgb, (x, y), (x2, y2), (255,0,0), 5)
22  #     break
23
24  # 顯示圖像
25  plt.imshow(im_rgb)
26  plt.axis('off')
27  plt.show()
```

執行結果： 左邊人臉的眼睛少抓了一個，嘴巴誤抓好幾個。這是筆者調整 detectMultiScale 參數多次後，所能得到的較佳結果。

detectMultiScale 相關參數的介紹如下：

1. scaleFactor：設定每次掃描視窗縮小的尺寸比例，設定較小值，會偵測到較多合格的視窗。

2. minNeighbors：每一個被選中的視窗至少要有鄰近且合格的視窗數，設定較大值，會讓偽陽性降低，但會使偽陰性提高。

3. minSize：小於這個設定值，會被過濾掉，格式為(w, h)。

4. maxSize：大於這個設定值，會被過濾掉，格式為(w, h)。

9-6-2 MTCNN 演算法

Haar Cascades 技術發展較早，使用很簡單，但是要能準確偵測，必須因應圖像的色澤、光線、物件大小來調整參數，並不容易。因此，近幾年發展改用深度學習演算法進行臉部偵測，較知名的演算法 MTCNN 係由 Kaipeng Zhang 等學者於 2016 年「Joint Face Detection and Alignment using Multi-task Cascaded Convolutional Networks」發表[16]。

MTCNN 的架構是運用影像金字塔加上三個神經網路，如下圖所示，四個部分的功能各為：

1. 影像金字塔(Image Pyramid)：擷取不同尺寸的臉部。

2. 建議網路(Proposal Network or P-Net)：類似區域推薦，找出候選的區域。

3. 強化網路(Refine Network or R-Net)：找出合格框(bounding boxes)。

4. 輸出網路(Output Network or O-Net)：找出臉部特徵點(Landmarks)。

乍看下來，會不會覺得有些熟悉？其實 MTCNN 的作法與物件偵測演算法 Faster R-CNN 類似。

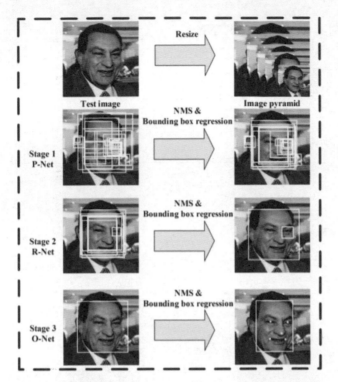

圖 9.10 MTCNN 使用影像金字塔加上三個神經網路

原作者使用 Caffe/C 開發[17]，許多人將以 Python 改寫，安裝指令如下：

```
pip install mtcnn
```

【範例 2】使用 MTCNN 進行臉部偵測。

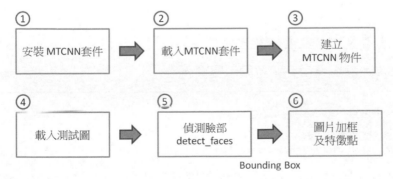

程式：09_07_臉部偵測_mtcnn.ipynb。

1. 載入相關套件,包含 MTCNN。

```
1  # 安裝套件 : pip install mtcnn
2  # 載入相關套件
3  import matplotlib.pyplot as plt
4  from matplotlib.patches import Rectangle, Circle
5  from mtcnn.mtcnn import MTCNN
```

2. 載入並顯示圖檔。

```
1  # 載入圖檔
2  image_file = "./images_face/classmates.jpg"
3  image = plt.imread(image_file)
4
5  # 顯示圖像
6  plt.imshow(image)
7  plt.axis('off')
8  plt.show()
```

執行結果:

3. 建立 MTCNN 物件,偵測臉部。

```
1  # 建立 MTCNN 物件
2  detector = MTCNN()
3
4  # 偵測臉部
5  faces = detector.detect_faces(image)
```

4. 臉部增加框與特徵點,並顯示圖像。

```
1  # 臉部加框
2  ax = plt.gca()
3  for result in faces:
4      # 取得框的座標及寬高
5      x, y, width, height = result['box']
6      # 加紅色框
7      rect = Rectangle((x, y), width, height, fill=False, color='red')
8      ax.add_patch(rect)
9
```

```
10      # 特徵點
11      for key, value in result['keypoints'].items():
12          # create and draw dot
13          dot = Circle(value, radius=5, color='green')
14          ax.add_patch(dot)
15
16  # 顯示圖像
17  plt.imshow(image)
18  plt.axis('off')
19  plt.show()
```

執行結果：特徵點包括眼睛、鼻子、嘴角，詳細內容請參閱程式。

5. 玩個小技巧，將每張臉個別顯示出來。

```
1  # 臉部加框
2  plt.figure(figsize=(8,6))
3  ax = plt.gca()
4
5  for i, result in enumerate(faces):
6      # 取得框的座標及寬高
7      x1, y1, width, height = result['box']
8      x2, y2 = x1 + width, y1 + height
9
10      # 顯示圖像
11      plt.subplot(1, len(faces), i+1)
12      plt.axis('off')
13      plt.imshow(image[y1:y2, x1:x2])
14  plt.show()
```

執行結果：

9-6-3 臉部追蹤(Face Tracking)

臉部追蹤(Face Tracking)可在影片中追蹤特定人的臉部，這裡使用的套件是 face-recognition，安裝指令如下：

```
pip install face-recognition
```

注意，face-recognition 是以 dlib 為基礎的套件，所以上述指令就是先安裝 dlib 套件，在 Windows 作業環境下，必須備妥下列工具：

1. Microsoft Visual Studio 2017/2019。
2. CMake for Windows：安裝後將 bin 路徑(例如 C:\Program Files\CMake\bin)加入環境變數 Path 中。

【範例 3】使用 Face-Recognition 套件進行臉部偵測。

程式：09_08_臉部偵測_Face_Recognition.ipynb。

1. 載入相關套件，包含 Face-Recognition。

```
1  # 安裝套件 : pip install face-recognition
2  # 載入相關套件
3  import matplotlib.pyplot as plt
4  from matplotlib.patches import Rectangle, Circle
5  import face_recognition
```

2. 載入並顯示圖檔。

```
1  # 載入圖檔
2  image_file = "./images_face/classmates.jpg"
3  image = plt.imread(image_file)
4
5  # 顯示圖像
6  plt.imshow(image)
7  plt.axis('off')
8  plt.show()
```

執行結果：

3. 呼叫 face_locations 函數偵測臉部。

```
1  # 偵測臉部
2  faces = face_recognition.face_locations(image)
```

4. 臉部加框，顯示圖像，注意，框的座標所代表的方向依序為上/左/下/右
   (逆時鐘)。

```
1   # 臉部加框
2   ax = plt.gca()
3   for result in faces:
4       # 取得框的座標
5       y1, x1, y2, x2 = result
6       width, height = x2 - x1, y2 - y1
7       # 加紅色框
8       rect = Rectangle((x1, y1), width, height, fill=False, color='red')
9       ax.add_patch(rect)
10
11  # 顯示圖像
12  plt.imshow(image)
13  plt.axis('off')
14  plt.show()
```

執行結果：

5. 偵測臉部特徵點並顯示。

```
1  # 偵測臉部特徵點並顯示
2  from PIL import Image, ImageDraw
3
4  # 載入圖檔
5  image = face_recognition.load_image_file(image_file)
6
7  # 轉為 Pillow 圖像格式
8  pil_image = Image.fromarray(image)
9
10 # 取得圖像繪圖物件
11 d = ImageDraw.Draw(pil_image)
12
13 # 偵測臉部特徵點
14 face_landmarks_list = face_recognition.face_landmarks(image)
15
16 for face_landmarks in face_landmarks_list:
17     # 顯示五官特徵點
18     for facial_feature in face_landmarks.keys():
19         print(f"{facial_feature} 特徵點: {face_landmarks[facial_feature]}\n")
20
21     # 繪製特徵點
22     for facial_feature in face_landmarks.keys():
23         d.line(face_landmarks[facial_feature], width=5, fill='green')
24
25 # 顯示圖像
26 plt.imshow(pil_image)
27 plt.axis('off')
28 plt.show()
```

執行結果如下：

■ 五官特徵點的座標。

```
chin 特徵點: [(958, 485), (968, 525), (982, 562), (999, 598), (1022, 630), (1054, 657), (1092, 677), (1135, 693), (1179, 689),
(1220, 670), (1249, 639), (1274, 606), (1291, 567), (1298, 524), (1296, 478), (1291, 433), (1283, 387)]

left_eyebrow 特徵點: [(969, 464), (978, 434), (1002, 417), (1032, 413), (1061, 415)]

right_eyebrow 特徵點: [(1119, 397), (1142, 373), (1172, 361), (1204, 364), (1228, 382)]

nose_bridge 特徵點: [(1098, 440), (1107, 477), (1115, 512), (1124, 548)]

nose_tip 特徵點: [(1092, 557), (1112, 562), (1133, 565), (1151, 552), (1167, 538)]

left_eye 特徵點: [(1006, 473), (1019, 458), (1038, 454), (1058, 461), (1042, 467), (1024, 472)]

right_eye 特徵點: [(1147, 436), (1160, 417), (1179, 409), (1201, 414), (1186, 423), (1167, 430)]

top_lip 特徵點: [(1079, 606), (1100, 595), (1121, 586), (1142, 585), (1160, 576), (1186, 570), (1215, 567), (1207, 571), (1164,
585), (1145, 593), (1125, 596), (1088, 605)]

bottom_lip 特徵點: [(1215, 567), (1197, 598), (1176, 619), (1155, 628), (1134, 631), (1109, 626), (1079, 606), (1088, 605), (112
8, 612), (1149, 610), (1168, 601), (1207, 571)]
```

■ 臉部輪廓畫線。

【範例 4】使用 Face-Recognition 套件進行影片臉部追蹤，程式修改自 face-recognition GitHub 的範例[18]。

程式 09_09_臉部追蹤_Face_Recognition.ipynb。

1. 載入相關套件。

```
1   # 安裝套件: pip install face-recognition
2   # 載入相關套件
3   import matplotlib.pyplot as plt
4   from matplotlib.patches import Rectangle, Circle
5   import face_recognition
6   import cv2
```

2. 載入影片檔。

```
1   # 載入影片檔
2   input_movie = cv2.VideoCapture("./images_face/short_hamilton_clip.mp4")
3   length = int(input_movie.get(cv2.CAP_PROP_FRAME_COUNT))
4   print(f'影片幀數：{length}')
```

執行結果：影片總幀數為 275。

3. 指定輸出檔名，注意，影片解析度設為(640, 360)，故輸入的影片不得低於此解析度，否則輸出檔將會無法播放。

```
1  # 指定輸出檔名
2  fourcc = cv2.VideoWriter_fourcc(*'XVID')
3  # 每秒幀數(fps):29.97，影片解析度(Frame Size):(640, 360)
4  output_movie = cv2.VideoWriter('./images_face/output.avi',
5                                 fourcc, 29.97, (640, 360))
```

4. 載入要辨識的圖像，範例設定這兩個人：Lin-Manuel Miranda(美國歌手)與 Barack Obama(美國總統)，需先編碼(Encode)為向量，以利臉部比對。

```
1  # 載入要辨識的圖像
2  image_file = 'lin-manuel-miranda.png' # 美國歌手
3  lmm_image = face_recognition.load_image_file("./images_face/"+image_file)
4  # 取得圖像編碼
5  lmm_face_encoding = face_recognition.face_encodings(lmm_image)[0]
6
7  # obama
8  image_file = 'obama.jpg' # 美國總統
9  obama_image = face_recognition.load_image_file("./images_face/"+image_file)
10 # 取得圖像編碼
11 obama_face_encoding = face_recognition.face_encodings(obama_image)[0]
12
13 # 設定陣列
14 known_faces = [
15     lmm_face_encoding,
16     obama_face_encoding
17 ]
18
19 # 目標名稱
20 face_names = ['lin-manuel-miranda', 'obama']
```

5. 變數初始化。

```
1  # 變數初始化
2  face_locations = [] # 臉部位置
3  face_encodings = [] # 臉部編碼
4  face_names = []     # 臉部名稱
5  frame_number = 0    # 幀數
```

6. 比對臉部並存檔。

```
1   # 偵測臉部並寫入輸出檔
2   while True:
3       # 讀取一幀影像
4       ret, frame = input_movie.read()
5       frame_number += 1
6
7       # 影片播放結束，即跳出迴圈
8       if not ret:
9           break
10
11      # 將 BGR 色系轉為 RGB 色系
12      rgb_frame = frame[:, :, ::-1]
13
14      # 找出臉部位置
15      face_locations = face_recognition.face_locations(rgb_frame)
16      # 編碼
17      face_encodings = face_recognition.face_encodings(rgb_frame, face_locations)
18
19      # 比對臉部
20      face_names = []
21      for face_encoding in face_encodings:
22          # 比對臉部編碼是否與圖檔符合
23          match = face_recognition.compare_faces(known_faces, face_encoding,
24                                      tolerance=0.50)
25
26          # 找出符合臉部的名稱
27          name = None
28          for i in range(len(match)):
```

```
29              if match[i] and 0 < i < len(face_names):
30                  name = face_names[i]
31                  break
32
33          face_names.append(name)
34
35      # 輸出影片標記臉部位置及名稱
36      for (top, right, bottom, left), name in zip(face_locations, face_names):
37          if not name:
38              continue
39
40          # 加框
41          cv2.rectangle(frame, (left, top), (right, bottom), (0, 0, 255), 2)
42
43          # 標記名稱
44          cv2.rectangle(frame, (left, bottom - 25), (right, bottom), (0, 0, 255)
45                      , cv2.FILLED)
46          font = cv2.FONT_HERSHEY_DUPLEX
47          cv2.putText(frame, name, (left + 6, bottom - 6), font, 0.5,
48                      (255, 255, 255), 1)
49
50      # 將每一幀影像存檔
51      print("Writing frame {} / {}".format(frame_number, length))
52      output_movie.write(frame)
53
54  # 關閉輸入檔
55  input_movie.release()
56  # 關閉所有視窗
57  cv2.destroyAllWindows()
```

執行結果：

```
Writing frame 4 / 275
Writing frame 5 / 275
Writing frame 6 / 275
Writing frame 7 / 275
Writing frame 8 / 275
Writing frame 9 / 275
Writing frame 10 / 275
Writing frame 11 / 275
Writing frame 12 / 275
Writing frame 13 / 275
Writing frame 14 / 275
Writing frame 15 / 275
Writing frame 16 / 275
Writing frame 17 / 275
Writing frame 18 / 275
Writing frame 19 / 275
Writing frame 20 / 275
Writing frame 21 / 275
```

輸出的影片為 images_face/output.avi 檔案：觀看影片後發現，偵測速度較慢，Obama 並未偵測到，因圖片檔是正面照，而影像檔則是側面的畫面。但瑕不掩瑜，大致上仍追蹤得到主要影像的動態。

【範例 5】改用 WebCam 進行臉部即時追蹤，程式修改自 face-recognition GitHub 的範例 [18]。

程式 09_10_臉部追蹤_webcam.ipynb。
由於步驟重疊的部分較多，所以只說明與範例 4 有差異的程式碼。

1. 以讀取 WebCam 取代載入影片檔。

```
1  # 指定第一台 webcam
2  video_capture = cv2.VideoCapture(0)
```

2. 讀取 WebCam 一幀影像：第 4 行。

```
1  # 偵測臉部並即時顯示
2  while True:
3      # 讀取一幀影像
4      ret, frame = video_capture.read()
```

3. 偵測臉部的處理均相同，但存檔改成即時顯示。

```
46      # 顯示每一幀影像
47      cv2.imshow('Video', frame)
```

4. 按 q 即可跳出迴圈。

```
49      # 按 q 即跳出迴圈
50      if cv2.waitKey(1) & 0xFF == ord('q'):
51          break
```

原作者也示範一個例子,可在 Raspberry pi 執行,即時進行臉部追蹤。

9-6-4 臉部特徵點偵測

偵測臉部特徵點可使用 Face-Recognition、dlib 或者 OpenCV 套件,以上這三種都可偵測到 68 個特徵點,如下圖:

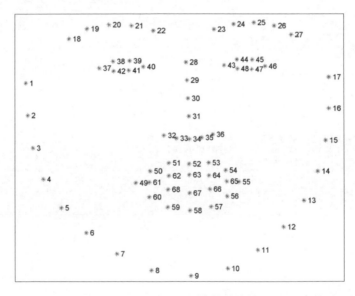

圖 9.11 臉部 68 個特徵點的位置

Face-Recognition 套件偵測臉部特徵點已在 09_08_臉部偵測_Face_Recognition.ipynb 實作過了,不再多作介紹。dlib 是另一套包含了機器學習、數值分析、計算機視覺、影像處理等功能的函數庫,它使用C++開發,只要安裝 Face-Recognition 套件就會自動安裝 dlib,如果要單獨安裝 dlib,可參考筆者撰寫的部落文「dlib 安裝心得 -- Windows 環境」[19]。

【範例 5】使用 dlib 實作臉部特徵點的偵測。

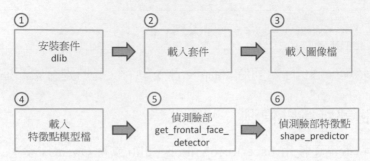

程式 09_11_臉部特徵點偵測.ipynb。

1. 載入相關套件，imutils 套件是一個簡易的影像處理函數庫，安裝指令如下：

```
pip install imutils
```

```
1  # 載入相關套件
2  import dlib
3  import cv2
4  import matplotlib.pyplot as plt
5  from matplotlib.patches import Rectangle, Circle
6  from imutils import face_utils
```

2. 載入並顯示圖檔。

```
1  # 載入圖檔
2  image_file = "./images_face/classmates.jpg"
3  image = plt.imread(image_file)
4
5  # 顯示圖像
6  plt.imshow(image)
7  plt.axis('off')
8  plt.show()
```

執行結果：

3. 偵測臉部特徵點並顯示：

- dlib 特徵點模型檔為 shape_predictor_68_face_landmarks.dat，可偵測 68 個點，如果只需要偵測 5 個點就好，可載入 shape_predictor_5_face_landmarks.dat。

- dlib.get_frontal_face_detector：偵測臉部。

- dlib.shape_predictor：偵測臉部特徵點。

```
1  # 載入 dlib 以 HOG 基礎的臉部偵測模型
2  model_file = "shape_predictor_68_face_landmarks.dat"
3  detector = dlib.get_frontal_face_detector()
4  predictor = dlib.shape_predictor(model_file)
5
6  # 偵測圖像的臉部
7  rects = detector(image)
8
9  print(f'偵測到{len(rects)}張臉部.')
10 # 偵測每張臉的特徵點
11 for (i, rect) in enumerate(rects):
12     # 偵測特徵點
13     shape = predictor(image, rect)
14
15     # 轉為 NumPy 陣列
16     shape = face_utils.shape_to_np(shape)
17
18     # 標示特徵點
19     for (x, y) in shape:
20         cv2.circle(image, (x, y), 10, (0, 255, 0), -1)
21
22 # 顯示圖像
23 plt.imshow(image)
24 plt.axis('off')
25 plt.show()
```

執行結果：

偵測到2張臉部.

4. 偵測影片檔也沒問題,按 Esc 鍵即可提前結束。

```
1   # 讀取視訊檔
2   cap = cv2.VideoCapture('./images_face/hamilton_clip.mp4')
3   while True:
4       # 讀取一幀影像
5       _, image = cap.read()
6
7       # 偵測圖像的臉部
8       rects = detector(image)
9       for (i, rect) in enumerate(rects):
10          # 偵測特徵點
11          shape = predictor(image, rect)
12          shape = face_utils.shape_to_np(shape)
13
14          # 標示特徵點
15          for (x, y) in shape:
16              cv2.circle(image, (x, y), 2, (0, 255, 0), -1)
17
18      # 顯示影像
19      cv2.imshow("Output", image)
20
21      k = cv2.waitKey(5) & 0xFF      # 按 Esc 跳離迴圈
22      if k == 27:
23          break
24
25  # 關閉輸入檔
26  cap.release()
27  # 關閉所有視窗
28  cv2.destroyAllWindows()
```

OpenCV 針對臉部特徵點的偵測,提供三種演算法:

1. FacemarkLBF:Shaoqing Ren 等學者於 2014 年發表「Face Alignment at 3000 FPS via Regressing Local Binary Features」所提出的[20]。

2. FacemarkAAM:Georgios Tzimiropoulos 等學者於 2013 年發表「Optimization problems for fast AAM fitting in-the-wild」所提出的[21]。

3. FacemarkKamezi:V.Kazemi 和 J. Sullivan 於 2014 年發表「One Millisecond Face Alignment with an Ensemble of Regression Trees」所提出的[22]。

我們分別實驗一下,看看有什麼差異。

【**範例 6**】使用 OpenCV 套件進行臉部特徵點偵測。

程式 09_12_臉部特徵點偵測_OpenCV.ipynb。

1. 載入相關套件：注意，只有 OpenCV 擴充版提供相關 API，所以，須執行下列指令，改安裝 OpenCV 擴充版：

   - 解除安裝： pip uninstall opencv-python opencv-contrib-python
   - 安裝套件： pip install opencv-contrib-python

```
1  # 解除安裝套件 : pip uninstall opencv-python opencv-contrib-python
2  # 安裝套件 :    pip install opencv-contrib-python
3  # 載入相關套件
4  import cv2
5  import numpy as np
6  from matplotlib import pyplot as plt
```

2. 載入並顯示圖檔：使用 Lena 圖像測試。

```
1  # 載入圖檔
2  image_file = "./images_Object_Detection/lena.jpg"
3  image = cv2.imread(image_file)
4
5  # 顯示圖像
6  image_RGB = cv2.cvtColor(image, cv2.COLOR_BGR2RGB)
7  plt.imshow(image_RGB)
8  plt.axis('off')
9  plt.show()
```

執行結果:

3. 使用 FacemarkLBF 偵測臉部特徵點。

```
1  # 偵測臉部
2  cascade = cv2.CascadeClassifier("./cascade_files/haarcascade_frontalface_alt2.xml")
3  faces = cascade.detectMultiScale(image , 1.5, 5)
4  print("faces", faces)
5
6  # 建立臉部特徵點偵測的物件
7  facemark = cv2.cv2.face.createFacemarkLBF()
8  # 訓練模型 lbfmodel.yaml 下載自:
9  # https://raw.githubusercontent.com/kurnianggoro/GSOC2017/master/data/lbfmodel.yaml
10 facemark .loadModel("lbfmodel.yaml")
11 # 偵測臉部特徵點
12 ok, landmarks1 = facemark.fit(image , faces)
13 print ("landmarks LBF", ok, landmarks1)
```

執行結果:顯示臉部和特徵點的座標。

```
faces [[225 205 152 152]]
landmarks LBF True [array([[[201.31314, 268.08807],
        [201.5153 , 293.1106 ],
        [204.91422, 317.07196],
        [210.71988, 340.4278 ],
        [222.97098, 360.37122],
        [240.34521, 375.51422],
        [260.10678, 386.35587],
        [280.64197, 392.04227],
        [298.6573 , 390.89835],
        [311.434  , 384.88406],
        [318.37827, 371.23538],
        [324.82266, 357.113  ],
        [331.87363, 342.1786 ],
        [339.7072 , 327.1501 ],
        [346.04462, 311.9719 ],
        [349.2847 , 296.59448],
        [348.95883, 280.12585],
        [236.43172, 252.06743],
```

4. 繪製特徵點並顯示圖像。

```
1  # 繪製特徵點
2  for p in landmarks1[0][0]:
3      cv2.circle(image, tuple(p), 5, (0, 255, 0), -1)
4
5  # 顯示圖像
6  image_RGB = cv2.cvtColor(image, cv2.COLOR_BGR2RGB)
7  plt.imshow(image_RGB)
8  plt.axis('off')
9  plt.show()
```

執行結果：很準確，可惜無法偵測到左上角被帽子遮蔽的部分。

5. 改用 FacemarkAAM 來偵測臉部特徵點。

```
1  # 建立臉部特徵點偵測的物件
2  facemark = cv2.face.createFacemarkAAM()
3  # 訓練模型 aam.xml 下載自：
4  # https://github.com/berak/tt/blob/master/aam.xml
5  facemark.loadModel("aam.xml")
6  # 偵測臉部特徵點
7  ok, landmarks2 = facemark.fit(image , faces)
8  print ("Landmarks AAM", ok, landmarks2)
```

6. 繪製特徵點並顯示圖像：過程與前面的程式碼相同。

```
1  # 繪製特徵點
2  for p in landmarks2[0][0]:
3      cv2.circle(image, tuple(p), 5, (0, 255, 0), -1)
4
5  # 顯示圖像
6  image_RGB = cv2.cvtColor(image, cv2.COLOR_BGR2RGB)
7  plt.imshow(image_RGB)
8  plt.axis('off')
9  plt.show()
```

執行結果：左上角反而多出一些錯誤的特徵點。

7. 換用 FacemarkKamezi 偵測臉部特徵點。

```
1  # 建立臉部特徵點偵測的物件
2  facemark = cv2.face.createFacemarkKazemi()
3  # 訓練模型 face_landmark_model.dat 下載自：
4  # https://github.com/opencv/opencv_3rdparty/tree/contrib_face_alignment_20170818
5  facemark.loadModel("face_landmark_model.dat")
6  # 偵測臉部特徵點
7  ok, landmarks2 = facemark.fit(image , faces)
8  print ("Landmarks Kazemi", ok, landmarks2)
```

8. 繪製特徵點並顯示圖像：過程與前面的程式碼相同。

```
1  # 繪製特徵點
2  for p in landmarks2[0][0]:
3      cv2.circle(image, tuple(p), 5, (0, 255, 0), -1)
4
5  # 顯示圖像
6  image_RGB = cv2.cvtColor(image, cv2.COLOR_BGR2RGB)
7  plt.imshow(image_RGB)
8  plt.axis('off')
9  plt.show()
```

執行結果：左上角也是多出一些錯誤的特徵點。

9-6-5 臉部驗證(Face Verification)

偵測完臉部特徵點後，利用線性代數的法向量比較多張臉的特徵點，就能
找出哪一張臉最相似，使用 Face-Recognition 或 dlib 套件都可以。

【範例 7】使用 Face-Recognition 或 dlib 套件，比對哪一張臉最相似。

程式 09_13_臉部驗證.ipynb。

1. 載入相關套件：使用 Face-Recognition 套件。

```
1  # 載入相關套件
2  import face_recognition
3  import numpy as np
4  from matplotlib import pyplot as plt
```

2. 載入所有要比對的圖檔。

```
1  # 載入圖檔
2  known_image_1 = face_recognition.load_image_file("./images_face/jared_1.jpg")
3  known_image_2 = face_recognition.load_image_file("./images_face/jared_2.jpg")
4  known_image_3 = face_recognition.load_image_file("./images_face/jared_3.jpg")
5  known_image_4 = face_recognition.load_image_file("./images_face/obama.jpg")
6
7  # 標記圖檔名稱
8  names = ["jared_1.jpg", "jared_2.jpg", "jared_3.jpg", "obama.jpg"]
9
10 # 顯示圖像
11 unknown_image = face_recognition.load_image_file("./images_face/jared_4.jpg")
12 plt.imshow(unknown_image)
13 plt.axis('off')
14 plt.show()
```

執行結果：

3. 圖像編碼：使用 face_recognition.face_encodings 函數編碼。

```
1  # 圖像編碼
2  known_image_1_encoding = face_recognition.face_encodings(known_image_1)[0]
3  known_image_2_encoding = face_recognition.face_encodings(known_image_2)[0]
4  known_image_3_encoding = face_recognition.face_encodings(known_image_3)[0]
5  known_image_4_encoding = face_recognition.face_encodings(known_image_4)[0]
6  known_encodings = [known_image_1_encoding, known_image_2_encoding,
7                     known_image_3_encoding, known_image_4_encoding]
8  unknown_encoding = face_recognition.face_encodings(unknown_image)[0]
```

4. 使用 face_recognition.compare_faces 進行比對。

```
1  # 比對
2  results = face_recognition.compare_faces(known_encodings, unknown_encoding)
3  print(results)
```

執行結果：[True, True, True, False]，前三筆符合，完全正確。

5. 載入相關套件：改用 dlib 套件。

```
1  # 載入相關套件
2  import dlib
3  import cv2
4  import numpy as np
5  from matplotlib import pyplot as plt
```

6. 載入模型：包括特徵點偵測、編碼、臉部偵測。

```
1  # 載入模型
2  pose_predictor_5_point = dlib.shape_predictor("shape_predictor_5_face_landmarks.dat")
3  face_encoder = dlib.face_recognition_model_v1("dlib_face_recognition_resnet_model_v1.dat")
4  detector = dlib.get_frontal_face_detector()
```

7. 定義臉部編碼與比對的函數：由於 dlib 無相關現成的函數，必須自行撰寫。

```
1  # 找出哪一張臉最相似
2  def compare_faces_ordered(encodings, face_names, encoding_to_check):
3      distances = list(np.linalg.norm(encodings - encoding_to_check, axis=1))
4      return zip(*sorted(zip(distances, face_names)))
5
6
7  # 利用線性代數的法向量比較兩張臉的特徵點
8  def compare_faces(encodings, encoding_to_check):
9      return list(np.linalg.norm(encodings - encoding_to_check, axis=1))
10
11 # 圖像編碼
12 def face_encodings(face_image, number_of_times_to_upsample=1, num_jitters=1):
13     # 偵測臉部
```

```
14      face_locations = detector(face_image, number_of_times_to_upsample)
15      # 偵測臉部特徵點
16      raw_landmarks = [pose_predictor_5_point(face_image, face_location)
17                          for face_location in face_locations]
18      # 編碼
19      return [np.array(face_encoder.compute_face_descriptor(face_image,
20                              raw_landmark_set, num_jitters)) for
21                              raw_landmark_set in raw_landmarks]
```

8. 載入圖檔並顯示。

```
1   # 載入圖檔
2   known_image_1 = cv2.imread("./images_face/jared_1.jpg")
3   known_image_2 = cv2.imread("./images_face/jared_2.jpg")
4   known_image_3 = cv2.imread("./images_face/jared_3.jpg")
5   known_image_4 = cv2.imread("./images_face/obama.jpg")
6   unknown_image = cv2.imread("./images_face/jared_4.jpg")
7   names = ["jared_1.jpg", "jared_2.jpg", "jared_3.jpg", "obama.jpg"]
8
9   # 轉換 BGR 為 RGB
10  known_image_1 = known_image_1[:, :, ::-1]
11  known_image_2 = known_image_2[:, :, ::-1]
12  known_image_3 = known_image_3[:, :, ::-1]
13  known_image_4 = known_image_4[:, :, ::-1]
14  unknown_image = unknown_image[:, :, ::-1]
```

9. 圖像編碼。

```
1   # 圖像編碼
2   known_image_1_encoding = face_encodings(known_image_1)[0]
3   known_image_2_encoding = face_encodings(known_image_2)[0]
4   known_image_3_encoding = face_encodings(known_image_3)[0]
5   known_image_4_encoding = face_encodings(known_image_4)[0]
6   known_encodings = [known_image_1_encoding, known_image_2_encoding,
7                       known_image_3_encoding, known_image_4_encoding]
8   unknown_encoding = face_encodings(unknown_image)[0]
```

10. 比對。

```
1   # 比對
2   computed_distances = compare_faces(known_encodings, unknown_encoding)
3   computed_distances_ordered, ordered_names = compare_faces_ordered(known_encodings,
4                                                   names, unknown_encoding)
5   print('比較兩張臉的法向量距離：', computed_distances)
6   print('排序：', computed_distances_ordered)
7   print('依相似度排序：', ordered_names)
```

執行結果：顯示兩張臉的法向量距離，數字愈小表示愈相似。

比較兩張臉的法向量距離： [0.3998327850880958, 0.4104153798439364, 0.3913189516694114, 0.9053701677487068]
排序： (0.3913189516694114, 0.3998327850880958, 0.4104153798439364, 0.9053701677487068)
依相似度排序： ('jared_3.jpg', 'jared_1.jpg', 'jared_2.jpg', 'obama.jpg')

9-7 光學文字辨識(OCR)

除了前面的介紹,另外還有很多其他類型的影像應用,例如:

1. 光學文字辨識(Optical Character Recognition, OCR)。
2. 影像修復(Image Inpainting):用周圍的影像將部分影像作修復,可用於抹除照片中不喜歡的物件。
3. 3D 影像的建構與辨識。

利用深度學習開發影像相關的應用系統,也是種類繁多,舉例來說:

1. 防疫:是否有戴口罩的偵測、社交距離的計算。
2. 交通:道路壅塞狀況的偵測、車速計算、車輛的違規(越線、闖紅燈)等。
3. 智慧製造:機器人與機器手臂的視覺輔助。
4. 企業運用:考勤、安全監控。

光學文字辨識,是把圖像中的印刷字辨識為文字,以節省大量的輸入時間或抄寫錯誤,可應用於支票號碼/金額辨識、車牌辨識(Automatic Number Plate Recognition, ANPR)等,但也有人拿來破解登入用的圖形碼驗證(Captcha),我們就來看看如何實作 OCR 辨識。

Tesseract OCR 是目前很盛行的 OCR 軟體,HP 公司於 2005 年開放原始程式碼(Open Source),以 C++開發而成的,可由原始程式碼建置,或直接安裝已建置好的程式,在這裡我們採取後者,自 https://github.com/UB-Mannheim/tesseract/wiki 下載最新版 exe 檔,直接執行即可。安裝完成後,將安裝路徑下的 bin 子目錄放入環境變數 path 內。若要以 Python 呼叫 Tesseract OCR,需額外安裝 pytesseract 套件,指令如下:

```
pip install pytesseract
```

最簡單的測試指令如下:

```
tesseract <圖檔> <辨識結果檔>
```

例如，辨識一張發票檔(./images_ocr/receipt.png)的指令為：

```
tesseract ./images_ocr/receipt.png ./images_ocr/result.txt -l eng --psm 6
```

其中 -l eng：為辨識英文，--psm 6：指單一區塊(a single uniform block of text)。相關的參數請參考「Tesseract OCR 官網」(https://github.com/tesseract-ocr/tesseract/blob/master/doc/tesseract.1.asc)。

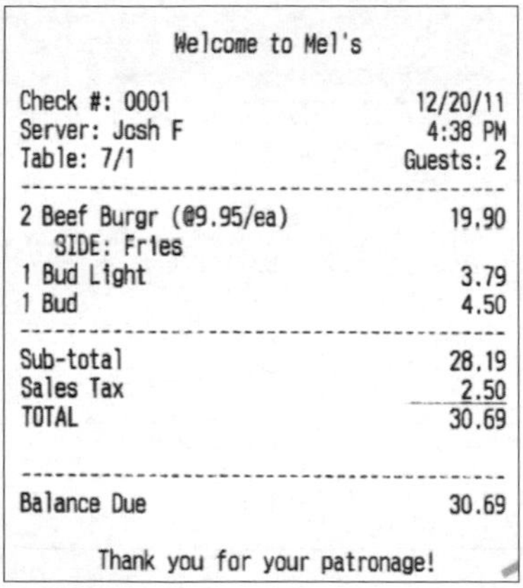

圖 9.12 發票，圖片來源：A comprehensive guide to OCR with Tesseract, OpenCV and Python [23]

執行結果：幾乎全對，只有特殊符號誤判。

```
 1 Welcome to Mel's
 2 Check #: 0001 12/20/11
 3 Server: Jesh F 4:38 PM
 4 Table: 7/1 Guests: 2
 5 2 Beef Burgr (€9.95/ea) 19,90
 6 SIDE: Fries
 7
 8 1 Bud Light 3.79
 9 1 Bud 4.50
10 Sub-total 28.19
11 Sales Tax 2.50
12 TOTAL 30.69
13 Balance Due 30.69
14
15 Thank you for your patronage! :
16
```

【**範例 8**】以 Python 呼叫 Tesseract OCR API，辨識中、英文。

程式 09_14_OCR.ipynb。

1. 載入相關套件。

```
1  # 載入相關套件
2  import cv2
3  import pytesseract
4  import matplotlib.pyplot as plt
```

2. 載入並顯示圖檔。

```
1  # 載入圖檔
2  image = cv2.imread('./images_ocr/receipt.png')
3
4  # 顯示圖檔
5  image_RGB = cv2.cvtColor(image, cv2.COLOR_BGR2RGB)
6  plt.figure(figsize=(10,6))
7  plt.imshow(image_RGB)
8  plt.axis('off')
9  plt.show()
```

3. OCR 辨識：呼叫 image_to_string 函數。

```
1  # 參數設定
2  custom_config = r'--psm 6'
3  # OCR 辨識
4  print(pytesseract.image_to_string(image, config=custom_config))
```

執行結果：與直接下指令的辨識結果大致相同。

4. 只辨識數字：參數設定加「outputbase digits」。

```
1  # 參數設定，只辨識數字
2  custom_config = r'--psm 6 outputbase digits'
3  # OCR 辨識
4  print(pytesseract.image_to_string(image, config=custom_config))
```

執行結果：

```
0001 122011
4338-
71 2
29.95 19.90
1 3.79
1 4.50
- 28.19
2.50
0 30.69
30.69
```

5. 只辨識有限字元。

```
1  # 參數設定白名單，只辨識有限字元
2  custom_config = r'-c tessedit_char_whitelist=abcdefghijklmnopqrstuvwxyz --psm 6'
3  # OCR 辨識
4  print(pytesseract.image_to_string(image, config=custom_config))
```

執行結果：

```
elcometoels
heck
erverdeshf
able uests
eefurgrea
efries

udight
ud
ubtotal
alesfax
a
alanceuc

hankyouforyourpatronage a
```

6. 設定黑名單：只辨識有限字元。

```
1  # 參數設定黑名單，只辨識有限字元
2  custom_config = r'-c tessedit_char_blacklist=abcdefghijklmnopqrstuvwxyz --psm 6'
3  # OCR 辨識
4  print(pytesseract.image_to_string(image, config=custom_config))
```

執行結果：

```
W  M]'
C #: 0001 12/20/11
S: J F 4:38 PM
T: 7/1 G: 2
2 B B (€9.95/) 19,90
SIDE: F

1 B L 3.79
1 B 4.50
S-] 28.19
S T 2.50
TOTAL 30.69
BI D 30.69

T  é! '
```

7. 辨識多國文字：先載入並顯示圖檔。

```
1  # 載入圖檔
2  image = cv2.imread('./images_ocr/chinese.png')
3
4  # 顯示圖檔
5  image_RGB = cv2.cvtColor(image, cv2.COLOR_BGR2RGB)
6  plt.figure(figsize=(10,6))
7  plt.imshow(image_RGB)
8  plt.axis('off')
9  plt.show()
```

圖檔如下：

> Tesseract OCR 是目前很盛行的 OCR 軟體，HP 公司於 2005 年開放原始程式碼
> (Open Source)，以 C++開發而成的，可由原始程式碼建置安裝，或直接安裝已建
> 置好的程式，我們採取後者，先自 https://github.com/UB-Mannheim/tesseract/wiki
> 下載最新版 exe 檔，直接執行即可，安裝完成後，將安裝路徑下 bin 子目錄放入
> 環境變數 Path 內。若要以 Python 呼叫 Tesseract OCR，需額外安裝 pytesseract 套
> 件，指令如下：
>
> pip install pytesseract

8. 辨識多國文字：先自 https://github.com/tesseract-ocr/tessdata_best 下載
   各國字形檔，放入安裝目錄的 tessdata 子目錄內 (C:\Program Files\
   Tesseract-OCR\tessdata)。

```
1  # 辨識多國文字，中文繁體、日文及英文
2  custom_config = r'-l chi_tra+jpn+eng --psm 6'
3  # OCR 辨識
4  print(pytesseract.image_to_string(image, config=custom_config))
```

執行結果：中文辨識的效果不如預期，即使改採新細明體字型，效果亦不
佳。

```
Tesseract OCR 苑 目 前 禎 盛 行 的 OCR 軟 · HP 公 司 於 2005 年 開 放 原 始 程 式 碼
(Open Souree)，以 CH+ 開 發 而 戒 的 · 可 田 原 委 程 式 殉 建 奢 安 裕 · 戒 直 接 安 襄 己 建
置 好 的 程 式，我 們 採 取 後 者，先 自 https://github.com' UB-NMIannheimtesseractwiki
下 載 最 新 版 exe 檔，直 接 執 行 即 可，安 裝 完 成 後，將 安 裝 路 徑 下 bin 子 目 錄 放 入
環 境 變 數 Path 內 · 若 要 以 Python 呼 司 Tesseract OCR，需 額 外 安 裝 pytesseract 会
件 · 拾 令 如 下 :

pip install pytesseract.
```

afr (Afrikaans), amh (Amharic), ara (Arabic), asm (Assamese), aze (Azerbaijani), aze_cyrl (Azerbaijani - Cyrilic), bel (Belarusian), ben (Bengali), bod (Tibetan), bos (Bosnian), bre (Breton), bul (Bulgarian), cat (Catalan; Valencian), ceb (Cebuano), ces (Czech), chi_sim (Chinese simplified), chi_tra (Chinese traditional), chr (Cherokee), cos (Corsican), cym (Welsh), dan (Danish), deu (German), div (Dhivehi), dzo (Dzongkha), ell (Greek, Modern, 1453-), eng (English), enm (English, Middle, 1100-1500), epo (Esperanto), equ (Math / equation detection module), est (Estonian), eus (Basque), fas (Persian), fao (Faroese), fil (Filipino), fin (Finnish), fra (French), frk (Frankish), frm (French, Middle, ca.1400-1600), fry (West Frisian), gla (Scottish Gaelic), gle (Irish), glg (Galician), grc (Greek, Ancient, to 1453), guj (Gujarati), hat (Haitian; Haitian Creole), heb (Hebrew), hin (Hindi), hrv (Croatian), hun (Hungarian), hye (Armenian), iku (Inuktitut), ind (Indonesian), isl (Icelandic), ita (Italian), ita_old (Italian - Old), jav (Javanese), jpn (Japanese), kan (Kannada), kat (Georgian), kat_old (Georgian - Old), kaz (Kazakh), khm (Central Khmer), kir (Kirghiz; Kyrgyz), kmr (Kurdish Kurmanji), kor (Korean), kor_vert (Korean vertical), lao (Lao), lat (Latin), lav (Latvian), lit (Lithuanian), ltz (Luxembourgish), mal (Malayalam), mar (Marathi), mkd (Macedonian), mlt (Maltese), mon (Mongolian), mri (Maori), msa (Malay), mya (Burmese), nep (Nepali), nld (Dutch; Flemish), nor (Norwegian), oci (Occitan post 1500), ori (Oriya), osd (Orientation and script detection module), pan (Panjabi; Punjabi), pol (Polish), por (Portuguese), pus (Pushto; Pashto), que (Quechua), ron (Romanian; Moldavian; Moldovan), rus (Russian), san (Sanskrit), sin (Sinhala; Sinhalese), slk (Slovak), slv (Slovenian), snd (Sindhi), spa (Spanish; Castilian), spa_old (Spanish; Castilian - Old), sqi (Albanian), srp (Serbian), srp_latn (Serbian - Latin), sun (Sundanese), swa (Swahili), swe (Swedish), syr (Syriac), tam (Tamil), tat (Tatar), tel (Telugu), tgk (Tajik), tha (Thai), tir (Tigrinya), ton (Tonga), tur (Turkish), uig (Uighur; Uyghur), ukr (Ukrainian), urd (Urdu), uzb (Uzbek), uzb_cyrl (Uzbek - Cyrilic), vie (Vietnamese), yid (Yiddish), yor (Yoruba)

圖 9.13 Tesseract 4 支援的語言，圖片來源：Tesseract 官網的語言列表[24]

9-8 車牌辨識(ANPR)

車牌辨識(Automatic Number Plate Recognition, ANPR)系統已行之有年了，早期用像素逐點辨識，或將數字細線化後，再比對線條，但最近幾年改採深度學習進行辨識，它已被應用到許多場域：

1. 機車檢驗：檢驗單位的電腦會先進行車牌辨識。
2. 停車場：當車輛進場時，系統會先辨識車牌並記錄，要出場時會辨識車牌，自動扣款。

【範例 9】以 OpenCV 及 Tesseract OCR 進行車牌辨識。

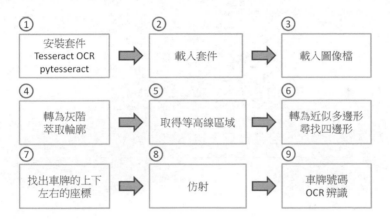

程式 **09_15_車牌辨識.ipynb**，程式修改自「Car License Plate Recognition using Raspberry Pi and OpenCV」[25]。

1. 載入相關套件。

```
1  # 載入相關套件
2  import cv2
3  import imutils
4  import numpy as np
5  import matplotlib.pyplot as plt
6  import pytesseract
7  from PIL import Image
```

2. 載入並顯示圖檔。

```
1  # 載入圖檔
2  image = cv2.imread('./images_ocr/2.jpg',cv2.IMREAD_COLOR)
3
4  # 顯示圖檔
5  image_RGB = cv2.cvtColor(image, cv2.COLOR_BGR2RGB)
6  plt.imshow(image_RGB)
7  plt.axis('off')
8  plt.show()
```

執行結果：此測試圖來自原程式。

3. 先轉為灰階，會比較容易辨識，再萃取輪廓。

```
1  # 萃取輪廓
2  gray = cv2.cvtColor(image, cv2.COLOR_BGR2GRAY) # 轉為灰階
3  gray = cv2.bilateralFilter(gray, 11, 17, 17)   # 模糊化，去除雜訊
4  edged = cv2.Canny(gray, 30, 200)               # 萃取輪廓
5
6  # 顯示圖檔
7  plt.imshow(edged, cmap='gray')
8  plt.axis('off')
9  plt.show()
```

執行結果：

4. 取得等高線區域，並排序，取前 10 個區域。

```
1  # 取得等高線區域，並排序，取前10個區域
2  cnts = cv2.findContours(edged.copy(), cv2.RETR_TREE, cv2.CHAIN_APPROX_SIMPLE)
3  cnts = imutils.grab_contours(cnts)
4  cnts = sorted(cnts, key = cv2.contourArea, reverse = True)[:10]
```

5. 找第一個含四個點的等高線區域：將等高線區域轉為近似多邊形，接著尋找四邊形的等高線區域。

```
1   # 找第一個含四個點的等高線區域
2   screenCnt = None
3   for i, c in enumerate(cnts):
4       # 計算等高線區域周長
5       peri = cv2.arcLength(c, True)
6       # 轉為近似多邊形
7       approx = cv2.approxPolyDP(c, 0.018 * peri, True)
8       # 等高線區域維度
9       print(c.shape)
10
11      # 找第一個含四個點的多邊形
12      if len(approx) == 4:
13          screenCnt = approx
14          print(i)
15          break
```

6. 在原圖上繪製多邊形，框住車牌。

```
1   # 在原圖上繪製多邊形，即車牌
2   if screenCnt is None:
3       detected = 0
4       print("No contour detected")
5   else:
6       detected = 1
7
8   if detected == 1:
9       cv2.drawContours(image, [screenCnt], -1, (0, 255, 0), 3)
10      print(f'車牌座標=\n{screenCnt}')
```

7. 去除車牌以外的圖像，找出車牌的上下左右的座標，計算車牌寬高。

```
1   # 去除車牌以外的圖像
2   mask = np.zeros(gray.shape,np.uint8)
3   new_image = cv2.drawContours(mask,[screenCnt],0,255,-1,)
4   new_image = cv2.bitwise_and(image, image, mask=mask)
5
6   # 轉為浮點數
7   src_pts = np.array(screenCnt, dtype=np.float32)
8
9   # 找出車牌的上下左右的座標
10  left = min([x[0][0] for x in src_pts])
```

```
11  right = max([x[0][0] for x in src_pts])
12  top = min([x[0][1] for x in src_pts])
13  bottom = max([x[0][1] for x in src_pts])
14
15  # 計算車牌寬高
16  width = right - left
17  height = bottom - top
18  print(f'寬度={width}, 高度={height}')
```

8. 仿射(affine transformation)，將車牌轉為矩形：仿射可將偏斜的梯形轉為矩形，筆者發現等高線區域的各點座標都是以**逆時針排列**，因此，當要找出第一點座標在哪個方向時，通常它會位在上方或左方，所以不需考慮右下角。

```
1   # 計算仿射(affine transformation)的目標區域座標，須與擷取的等高線區域座標順序相同
2   if src_pts[0][0][0] > src_pts[1][0][0] and src_pts[0][0][1] < src_pts[3][0][1]:
3       print('起始點為右上角')
4       dst_pts = np.array([[width, 0], [0, 0], [0, height], [width, height]], dtype=np.float32)
5   elif src_pts[0][0][0] < src_pts[1][0][0] and src_pts[0][0][1] > src_pts[3][0][1]:
6       print('起始點為左下角')
7       dst_pts = np.array([[0, height], [width, height], [width, 0], [0, 0]], dtype=np.float32)
8   else:
9       print('起始點為左上角')
10      dst_pts = np.array([[0, 0], [0, height], [width, height], [width, 0]], dtype=np.float32)
11
12  # 仿射
13  M = cv2.getPerspectiveTransform(src_pts, dst_pts)
14  Cropped = cv2.warpPerspective(gray, M, (int(width), int(height)))
```

9. 車牌號碼 OCR 辨識。

```
1   # 車牌號碼 OCR 辨識
2   text = pytesseract.image_to_string(Cropped, config='--psm 11')
3   print("車牌號碼：",text)
```

執行結果： /HR.26 BR 9044，有誤認一些非字母或數字的符號，可直接去除，這樣車牌辨識的結果就完全正確。

10. 顯示原圖和車牌。

```
1   # 顯示原圖及車牌
2   cv2.imshow('Orignal image',image)
3   cv2.imshow('Cropped image',Cropped)
4
5   # 車牌存檔
6   cv2.imwrite('Cropped.jpg', Cropped)
7
8   # 按 Enter 鍵結束
9   cv2.waitKey(0)
10
11  # 關閉所有視窗
12  cv2.destroyAllWindows()
```

執行結果：

車牌：

11. 再使用 images_ocr/1.jpg 測試，車牌為 NAX-6683，辨識為 NAY-6683，X 誤認為 Y，有可能是台灣車牌的字型不同，可使用台灣車牌字型供 Tesseract OCR 使用。

車牌：

另外，筆者實驗發現，若鏡頭拉遠或拉近，而造成車牌過大或過小的話，都有可能辨識錯誤，所以，實際進行時，鏡頭最好與車牌距離能固定，會

比較容易辨識。假如圖像的畫面太雜亂，取到的車牌區域也有可能是錯的，而這問題相對容易處理，當 OCR 辨識不到字或者字數不足時，就再找其他的等高線區域，即可解決。

從這個範例可以得知，通常一個實際的案例，並不會像內建的資料集一樣，可以直接套用，常常都需要進行前置處理，如灰階化、萃取輪廓、找等高線區域、仿射等等，資料清理(Data Clean)完才可餵入模型加以訓練，而且為了適應環境變化，這些工作還必須反覆進行。所以有人統計，光光收集資料、整理資料、特徵工程等工作就佔專案 85%的時間，只有把最boring、dirty 的工作處理好，才是專案成功的關鍵因素，這與參加 Kaggle競賽是截然不同的感受，魔鬼總是藏在細節裡。

9-9 卷積神經網路的缺點

CNN 的應用領域那麼多元，相當實用，但是它仍存在一些缺點：

1. 卷積不管特徵在圖像的所在位置，只針對局部視窗進行特徵辨識，因此，下列兩張圖，辨識結果是相同的，這種現象稱為「位置無差異性」(Position Invariant)。

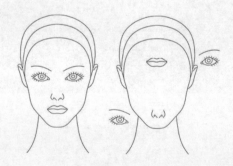

圖 9.14 左圖是正常的人臉，右圖五官移位，兩者對 CNN 來說是無差異的。

圖片來源：Disadvantages of CNN models[26]

2. 圖像中的物件如果經過旋轉或傾斜，CNN 就無法辨識了，如下圖：

圖 9.15 右圖為左圖側轉近 180 度的樣子，如此 CNN 辨識不了。
圖片來源：Disadvantages of CNN models [26]

3. 圖像座標轉換，人眼可以辨識不同的物件特徵，但對於 CNN 來說卻難以理解，如下圖：

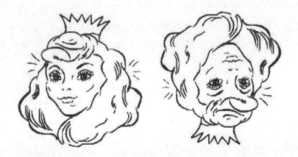

圖 9.16 右圖為上下顛倒的左圖，人眼可以看出年輕人與老年人，然而 CNN 就很難理解。圖片來源：Disadvantages of CNN models [26]

因此，Geoffrey Hinton 等學者就提出了「膠囊演算法」(Capsules)，用來改良 CNN 的問題，有興趣的讀者可以進一步研究 Capsules。

參考資料 (References)

[1] Vijay Badrinarayanan、Alex Kendall、Roberto Cipolla,《SegNet: A Deep Convolutional Encoder-Decoder Architecture for Image Segmentation》, 2015
(https://arxiv.org/abs/1511.00561)

[2] keras-mnist-VAE GitHub
(https://github.com/lyeoni/keras-mnist-VAE)

[3] Liang-Chieh Chen、George Papandreou 等人,《DeepLab: Semantic Image Segmentation with Deep Convolutional Nets, Atrous Convolution, and Fully Connected CRFs》, 2017
(https://arxiv.org/pdf/1606.00915.pdf)

[4] Guosheng Lin、Anton Milan、Chunhua Shen,《RefineNet: Multi-Path Refinement Networks for High-Resolution Semantic Segmentation》, 2016
(https://arxiv.org/pdf/1611.06612.pdf)

[5] Hengshuang Zhao、Jianping Shi、Xiaojuan Qi,《Pyramid Scene Parsing Network》, 2017
(https://arxiv.org/pdf/1612.01105.pdf)

[6] Olaf Ronneberger、Philipp Fischer、Thomas Brox,《U-Net: Convolutional Networks for Biomedical Image Segmentation》, 2015
(https://arxiv.org/pdf/1505.04597.pdf)

[7] Keras 官網提供的範例「Image segmentation with a U-Net-like architecture」
(https://keras.io/examples/vision/oxford_pets_image_segmentation/)

[8] Harshall Lamba,《Understanding Semantic Segmentation with UNET》, 2019
(https://towardsdatascience.com/understanding-semantic-segmentation-with-unet-6be4f42d4b47)

[9] Kaiming He、Georgia Gkioxari、Piotr Dollár 等人,《Mask R-CNN》, 2018 (https://arxiv.org/pdf/1703.06870.pdf)

[10] akTwelve Mask R-CNN 函數庫 (https://github.com/akTwelve/Mask_RCNN)

[11] Jason Brownlee,《How to Use Mask R-CNN in Keras for Object Detection in Photographs》, 2020 (https://machinelearningmastery.com/how-to-perform-object-detection-in-photographs-with-mask-r-cnn-in-keras/)

[12] fast-style-transfer GitHub (https://github.com/lengstrom/fast-style-transfer)

[13] 翁書婷,《催生全球首位 AI 繪師 Andy,美圖搶攻人工智慧卻面臨一大挑戰》, 2017 (https://www.bnext.com.tw/article/47330/ai-andy-meitu)

[14] Leon A. Gatys、Alexander S. Ecker、Matthias Bethge,《A Neural Algorithm of Artistic Style》, 2015 (https://arxiv.org/abs/1508.06576)

[15] TensorFlow 官網提供的範例「Neural Style Transfer」 (https://www.tensorflow.org/tutorials/generative/style_transfer)

[16] Kaipeng Zhang、Zhanpeng Zhang、Zhifeng Li、Yu Qiao,《Joint Face Detection and Alignment using Multi-task Cascaded Convolutional Networks》, 2016 (https://arxiv.org/abs/1604.02878)

[17] MTCNN_face_detection_alignment GitHub (https://github.com/kpzhang93/MTCNN_face_detection_alignment)

[18] face-recognition GitHub 的範例 (https://github.com/ageitgey/face_recognition)

[19] 陳昭明,《dlib 安裝心得 -- Windows 環境》, 2020 (https://ithelp.ithome.com.tw/articles/10231535)

[20] Shaoqing Ren、Xudong Cao、Yichen Wei 等人,《Face Alignment at 3000 FPS via Regressing Local Binary Features》, 2014 (http://www.jiansun.org/papers/CVPR14_FaceAlignment.pdf)

[21] Georgios Tzimiropoulos、Maja Pantic,《Optimization problems for fast AAM fitting in-the-wild》, 2013 (https://ibug.doc.ic.ac.uk/media/uploads/documents/tzimiro_pantic_iccv2013.pdf)

[22] V.Kazemi、J. Sullivan,《One Millisecond Face Alignment with an Ensemble of Regression Trees》, 2014 (http://www.csc.kth.se/~vahidk/face_ert.html)

[23]Filip Zelic、Anuj Sable,《A comprehensive guide to OCR with Tesseract, OpenCV and Python》, 2021 (https://nanonets.com/blog/ocr-with-tesseract/)

[24] Tesseract 官網的語言列表 (https://github.com/tesseract-ocr/tesseract/blob/master/doc/tesseract.1.asc#LANGUAGES)

[25] Aswinth Raj,《Car License Plate Recognition using Raspberry Pi and OpenCV》, 2019 (https://circuitdigest.com/microcontroller-projects/license-plate-recognition-using-raspberry-pi-and-opencv)

[26] Disadvantages of CNN models (https://iq.opengenus.org/disadvantages-of-cnn/)

生成對抗網路 (GAN)

話說水能載舟，亦能覆舟，AI 雖然給人類帶來了許多便利，但也造成不小的危害。近幾年氾濫的深度偽造(Deepfake)就是一例，它利用 AI 技術偽造政治人物與明星的影片，能夠做到真假難辨，一旦在網路上散播開來，就形成莫大的災難，根據統計，名人色情片 8 成都是偽造的。深度偽造的基礎演算法就是「生成對抗網路」(Generative Adversarial Network, GAN)，本章就來介紹此一課題。

Facebook 人工智慧研究院 Yann LeCun 在接受 Quora 專訪時說到：「GAN 與其變形是近十年最有趣的想法」(This, and the variations that are now being proposed is the most interesting idea in the last 10 years in ML, in my opinion.)，一句話造成 GAN 一炮而紅，其作者 Ian Goodfellow(真的是好傢伙!)也成為各界競相邀請演講的對象。

另外，2018 年 10 月紐約佳士得藝術拍賣會，也賣出第一幅以 GAN 演算法所繪製的肖像畫，最後得標價為$432,500 美金。有趣的是，畫作右下角還列出 GAN 的損失函數，相關報導可參見「全球首次！AI 創作肖像畫 10 月佳士得拍賣」[1] 及「Is artificial intelligence set to become art's next medium?」[2]。

圖 10.1 Edmond de Belamy 肖像畫，圖片來源：佳士得網站[3]

此後有人統計每 28 分鐘就有一篇與 GAN 相關的論文發表。

10-1 生成對抗網路介紹

關於生成對抗網路有一個很生動的比喻：它是由兩個神經網路所組成，一個網路扮演偽鈔製造者(Counterfeiter)，一直製造假鈔，另一個網路則扮演警察，不斷從偽造者那邊拿到假鈔，並判斷真假，然後，偽造者就根據警察判斷結果的回饋，不停改良，直到最後假鈔變成真假難辨(天啊！這是什麼電影情節！)，這就是 GAN 的概念。

偽鈔製造者稱為「生成模型」（Generative model），警察則是「判別模型」（Discriminative model），簡單的架構如下圖：

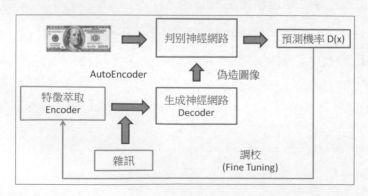

圖 10.2 生成對抗網路(Generative Adversarial Network, GAN)的架構

處理流程如下:

1. 先訓練判別神經網路:從訓練資料中抽取樣本,餵入判別神經網路,期望預測機率 D(x)≒1,相反的,判斷來自生成網路的偽造圖片,期望預測機率 D(G(z))≒0。

2. 訓練生成網路:剛開始以常態分配或均勻分配產生雜訊(z),餵入生成神經網路,生成偽造圖片。

3. 透過判別網路的反向傳導(Backpropagation),更新生成網路的權重,亦即改良偽造圖片的準確度(技術),反覆訓練,直到產生精準的圖片為止。

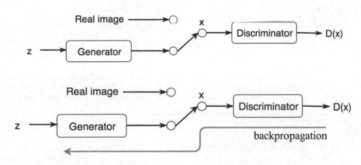

圖 10.3 判別神經網路的反向傳導(Backpropagation)

GAN 根據以上流程重新定義損失函數。

1. 判別神經網路的損失函數:前半段為真實資料的判別,後半段為偽造資料的判別。

$$\max_{D} V(D) = \mathbb{E}_{\boldsymbol{x} \sim p_{\text{data}}(\boldsymbol{x})}[\log D(\boldsymbol{x})] + \mathbb{E}_{\boldsymbol{z} \sim p_{\boldsymbol{z}}(\boldsymbol{z})}[\log(1 - D(G(\boldsymbol{z})))]$$

recognize real images better   recognize generated images better

- x:訓練資料,故預測機率 D(x)愈大愈好。
- E:為期望值,因為訓練資料並不完全相同,故預測機率有高有低。
- z:偽造資料,預測機率 D(G(z))愈小愈好,調整為 1- D(G(z)),變成愈大愈好。
- 兩者相加當然是愈大愈好。

■ 取 Log：並不會影響最大化求解，通常機率相乘會造成多次方，不容易
求解，故取 Log，變成一次方函數。

2. 生成神經網路的損失函數：即判別神經網路損失函數的右邊多項式，
生成神經網路期望偽造資料被分類為真的機率愈大愈好，故差距愈小
愈好。

$$\min_G V(G) = \mathbb{E}_{z \sim p_z(z)}[\log(1 - D(G(z)))]$$

3. 兩個網路損失函數合而為一的表示法如下：

$$\min_G \max_D V(D, G) = \mathbb{E}_{x \sim p_{data}}[\log D(x)] + \mathbb{E}_{z \sim p_z}[\log(1 - D(G(z)))]$$

因為函數左邊的多項式與生成神經網路的損失函數無關，故加上亦無礙。
整個演算法的虛擬碼如下，使用小批量梯度下降法，最小化損失函數。

Algorithm 1 Minibatch stochastic gradient descent training of generative adversarial nets. The number of steps to apply to the discriminator, k, is a hyperparameter. We used $k = 1$, the least expensive option, in our experiments.

for number of training iterations **do**

  for k steps **do**

  - Sample minibatch of m noise samples $\{z^{(1)}, \ldots, z^{(m)}\}$ from noise prior $p_g(z)$.
  - Sample minibatch of m examples $\{x^{(1)}, \ldots, x^{(m)}\}$ from data generating distribution $p_{\text{data}}(x)$.
  - Update the discriminator by ascending its stochastic gradient:

$$\nabla_{\theta_d} \frac{1}{m} \sum_{i=1}^{m} \left[\log D\left(x^{(i)}\right) + \log\left(1 - D\left(G\left(z^{(i)}\right)\right)\right) \right].$$

  end for

  - Sample minibatch of m noise samples $\{z^{(1)}, \ldots, z^{(m)}\}$ from noise prior $p_g(z)$.
  - Update the generator by descending its stochastic gradient:

$$\nabla_{\theta_g} \frac{1}{m} \sum_{i=1}^{m} \log\left(1 - D\left(G\left(z^{(i)}\right)\right)\right).$$

end for

The gradient-based updates can use any standard gradient-based learning rule. We used momentum in our experiments.

圖 10.4 GAN 演算法的虛擬碼

生成網路希望生成出來圖片越來越逼真，能通過判別網路的檢驗，而判別
網路則希望將生成網路所製造的圖片都判定為假資料，兩者目標相反，互
相對抗，故稱為「生成對抗網路」。

10-2　生成對抗網路種類

GAN 不是只有一種模型，其變形非常多，可以參閱「The GAN Zoo」
[4]，有上百種模型，其功能各有不同，譬如：

1. CGAN：參閱「Pose Guided Person Image Generation」[5]，可生成不同
   的姿勢。

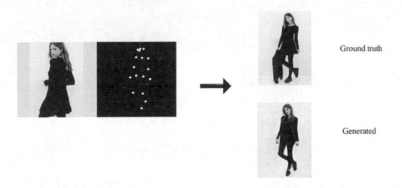

圖 10.5　CGAN 演算法的姿勢生成

2. ACGAN：參閱「Towards the Automatic Anime Characters Creation with
   Generative Adversarial Networks」[6]，可生成不同的動漫人物。

圖 10.6　ACGAN 演算法，從左邊的動漫角色，生成為右邊的新角色

作者有附一個展示的網站(https://make.girls.moe/#/)，可利用不同的模型與
參數，生成各種動漫人物。

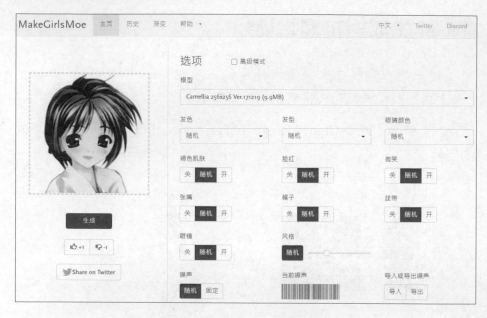

圖 10.7 ACGAN 展示網站

3. CycleGAN：風格轉換，作者也有附一個展示的網站(https://mil-tokyo.github.io/webdnn/)，可選擇不同的風格圖，生成各式風格的照片或影片。

圖 10.8 CycleGAN 的展示網站

4. StarGAN：參閱「StarGAN: Unified Generative Adversarial Networks for Multi-Domain Image-to-Image Translation」[7]，生成不同的臉部表情，轉換膚色、髮色或是性別，程式碼在 https://github.com/yunjey/stargan。

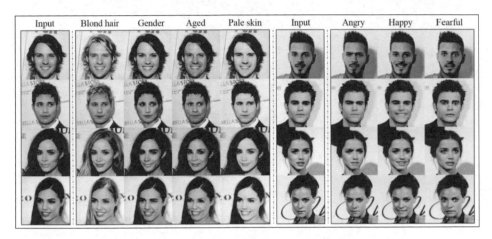

圖 10.9　StarGAN 展示，將左邊的臉轉換膚色、髮色、性別或表情

5. SRGAN：可以生成高解析度的圖像，參閱「Photo-Realistic Single Image Super-Resolution Using a Generative Adversarial Network」[8]。

圖 10.10　SRGAN 展示，由左而右從低解析度的圖像生成為高解析度的圖像

6. StyleGAN2：功能與語義分割(Image Segmentation)相反，它是從語義分割圖渲染成實景圖，參閱「Analyzing and Improving the Image Quality of StyleGAN」[9]，程式在 https://github.com/NVlabs/stylegan2。

圖 10.11 StyleGAN2 展示，從右邊的圖像生成為左邊的圖像

限於篇幅，僅介紹一小部分的演算法，讀者如有興趣可參閱「GAN — Some cool applications of GAN」[10]一文有更多種演算法的介紹。

只要修改原創者 GAN 的損失函數，即可產生不同的神奇效果，所以，根據「The GAN Zoo」[4]的統計，GAN 相關的論文數量呈現爆炸性的成長。

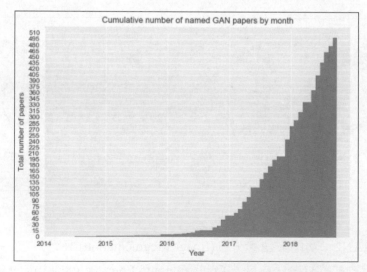

圖 10.12 與 GAN 有關的論文數量呈現爆炸性的成長

10-3 DCGAN

先來實作 DCGAN(Deep Convolutional Generative Adversarial Network)演算法。

【範例 1】以 MNIST 資料實作 DCGAN 演算法，產生手寫阿拉伯數字。

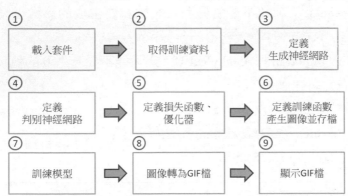

程式 10_01_DCGAN_MNIST.ipynb。

訓練資料集：MNIST。

1. 載入相關套件：需先安裝 imageio 套件，以利產生 GIF 動畫。

```
1   # 載入相關套件
2   import glob
3   import imageio
4   import matplotlib.pyplot as plt
5   import numpy as np
6   import os
7   import PIL
8   import tensorflow as tf
9   from tensorflow.keras import layers
10  import time
11  from IPython import display
```

2. 取得訓練資料，轉為 TensorFlow Dataset。

```
1   # 取得 MNIST 訓練資料
2   (train_images, train_labels), (_, _) = tf.keras.datasets.mnist.load_data()
3   # 像素標準化
4   train_images = train_images.reshape(train_images.shape[0], 28, 28, 1).astype('float32')
5   train_images = (train_images - 127.5) / 127.5  # 使像素值介於 [-1, 1]
6
```

```
 7  # 參數設定
 8  BUFFER_SIZE = 60000 # 緩衝區大小
 9  BATCH_SIZE = 256      # 訓練批量
10
11  # 轉為 Dataset
12  train_dataset = tf.data.Dataset.from_tensor_slices(train_images) \
13          .shuffle(BUFFER_SIZE).batch(BATCH_SIZE).prefetch(BUFFER_SIZE).cache()
```

3. 定義生成神經網路：

- use_bias=False：訓練不產生偏差項，因為要生成的影像盡量是像素所構成的。

- Activation Function 採 LeakyReLU：盡可能不要產生 0 的值，以免生成的影像有太多空白。

- Conv2DTranspose：反卷積層，進行上採樣(Up Sampling)，由小圖插補為大圖，strides=(2, 2)表示寬高各增大 2 倍。

- 最後產生寬高為(28,28)的單色向量。

```
 1  # 生成神經網路
 2  def make_generator_model():
 3      model = tf.keras.Sequential()
 4      model.add(layers.Dense(7*7*256, use_bias=False, input_shape=(100,)))
 5      model.add(layers.BatchNormalization())
 6      model.add(layers.LeakyReLU())
 7
 8      model.add(layers.Reshape((7, 7, 256)))
 9      assert model.output_shape == (None, 7, 7, 256)  # None 代表批量不檢查
10
11      model.add(layers.Conv2DTranspose(128, (5, 5), strides=(1, 1),
12                                       padding='same', use_bias=False))
13      assert model.output_shape == (None, 7, 7, 128)
14      model.add(layers.BatchNormalization())
15      model.add(layers.LeakyReLU())
16
17      model.add(layers.Conv2DTranspose(64, (5, 5), strides=(2, 2),
18                                       padding='same', use_bias=False))
19      assert model.output_shape == (None, 14, 14, 64)
20      model.add(layers.BatchNormalization())
21      model.add(layers.LeakyReLU())
22
23      model.add(layers.Conv2DTranspose(1, (5, 5), strides=(2, 2),
24                      padding='same', use_bias=False, activation='tanh'))
25      assert model.output_shape == (None, 28, 28, 1)
26
27      return model
```

4. 測試生成神經網路。

```
1   # 產生生成神經網路
2   generator = make_generator_model()
3
4   # 測試產生的雜訊
5   noise = tf.random.normal([1, 100])
6   generated_image = generator(noise, training=False)
7
8   # 顯示雜訊生成的圖像
9   plt.imshow(generated_image[0, :, :, 0], cmap='gray')
```

執行結果：

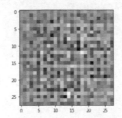

5. 定義判別神經網路：類似一般的 CNN 判別模型，但要去除池化層，避免資訊損失。

```
1   # 判別神經網路
2   def make_discriminator_model():
3       model = tf.keras.Sequential()
4       model.add(layers.Conv2D(64, (5, 5), strides=(2, 2), padding='same',
5                                       input_shape=[28, 28, 1]))
6       model.add(layers.LeakyReLU())
7       model.add(layers.Dropout(0.3))
8
9       model.add(layers.Conv2D(128, (5, 5), strides=(2, 2), padding='same'))
10      model.add(layers.LeakyReLU())
11      model.add(layers.Dropout(0.3))
12
13      model.add(layers.Flatten())
14      model.add(layers.Dense(1))
15
16      return model
```

6. 測試判別神經網路：真實影像的預測值會是較大的值，生成影像的預測值則會是較小的值，因為無明顯的線條特徵。

```
1   # 測試判別神經網路
2   discriminator = make_discriminator_model()
3
4   # 真實的影像預測值會是較大的值，生成的影像預測值會是較小的值
5   decision = discriminator(generated_image)
6   print (f'預測值={decision}')
```

執行結果：-0.00208456，為很小的值。

7. 測試真實的影像。

```
1  # 測試真實的影像 3 筆
2  decision = discriminator(train_images[0:3])
3  print (f'預測值={decision}')
```

執行結果：為絕對值較大的值。

```
[[ 0.8379558]
 [-0.5694627]
 [ 1.4122181]]
```

8. 定義損失函數為二分類交叉熵(BinaryCrossentropy)、優化器為 Adam。
   判別神經網路的損失函數為「真實影像」加上「生成影像」的損失函數和，因為判別神經網路會同時接收真實影像和生成影像。

```
1   # 定義損失函數為二分類交叉熵
2   cross_entropy = tf.keras.losses.BinaryCrossentropy(from_logits=True)
3
4   # 定義判別神經網路損失函數為 真實影像 + 生成影像 的損失函數
5   def discriminator_loss(real_output, fake_output):
6       real_loss = cross_entropy(tf.ones_like(real_output), real_output)
7       fake_loss = cross_entropy(tf.zeros_like(fake_output), fake_output)
8       total_loss = real_loss + fake_loss
9       return total_loss
10
11  # 定義生成神經網路損失函數為 生成影像 的損失函數
12  def generator_loss(fake_output):
13      return cross_entropy(tf.ones_like(fake_output), fake_output)
14
15  # 優化器均為 Adam
16  generator_optimizer = tf.keras.optimizers.Adam(1e-4)
17  discriminator_optimizer = tf.keras.optimizers.Adam(1e-4)
```

9. 設定檢查點：在檢查點將模型存檔，因為要花很長的時間訓練，萬一中途當掉，還可以從上次的中斷點繼續訓練。
   還原(restore)指令如下：

```
checkpoint.restore(tf.train.latest_checkpoint(checkpoint_dir))
```

```
1  # 在檢查點模型存檔
2  checkpoint_dir = './dcgan_training_checkpoints'
3  checkpoint_prefix = os.path.join(checkpoint_dir, "ckpt")
4  checkpoint = tf.train.Checkpoint(generator_optimizer=generator_optimizer,
5                                   discriminator_optimizer=discriminator_optimizer,
6                                   generator=generator,
7                                   discriminator=discriminator)
```

10. 參數設定。

```
1  # 參數設定
2  EPOCHS = 50                              # 訓練執行週期
3  noise_dim = 100                          # 雜訊向量大小
4  num_examples_to_generate = 16   # 生成筆數
5
6  # 產生亂數(雜訊)
7  seed = tf.random.normal([num_examples_to_generate, noise_dim])
```

11. 定義梯度下降函數：同時對生成神經網路和判別神經網路進行訓練，雜訊與真實圖像也一起餵入判別神經網路，計算損失及梯度，並更新權重。

@tf.function：會產生運算圖，使函數運算速度加快。

```
1  # 定義梯度下降，分別對判別神經網路、生成神經網路進行訓練
2  @tf.function  # 產生運算圖
3  def train_step(images):
4      noise = tf.random.normal([BATCH_SIZE, noise_dim])
5
6      with tf.GradientTape() as gen_tape, tf.GradientTape() as disc_tape:
7          # 生成神經網路進行訓練
8          generated_images = generator(noise, training=True)
9
10         # 判別神經網路進行訓練
11         real_output = discriminator(images, training=True)              # 真實影像
12         fake_output = discriminator(generated_images, training=True) # 生成影像
13
14         # 計算損失
15         gen_loss = generator_loss(fake_output)
16         disc_loss = discriminator_loss(real_output, fake_output)
17
18     # 梯度下降
19     gradients_of_generator = gen_tape.gradient(gen_loss, generator.trainable_variables)
20     gradients_of_discriminator = disc_tape.gradient(disc_loss,
21                                                 discriminator.trainable_variables)
22
23     # 更新權重
24     generator_optimizer.apply_gradients(zip(gradients_of_generator,
25                                             generator.trainable_variables))
26     discriminator_optimizer.apply_gradients(zip(gradients_of_discriminator,
27                                                 discriminator.trainable_variables))
```

12. 定義訓練函數：同時間會產生圖像並存檔。

```
1  # 定義訓練函數
2  def train(dataset, epochs):
3      for epoch in range(epochs):
4          start = time.time()
5
6          for image_batch in dataset:
7              train_step(image_batch)
```

```
8
9          # 產生圖像
10         display.clear_output(wait=True)
11         generate_and_save_images(generator, epoch + 1, seed)
12
13         # 每 10 個執行週期存檔一次
14         if (epoch + 1) % 10 == 0:
15             checkpoint.save(file_prefix = checkpoint_prefix)
16
17         print ('epoch {} 花費 {} 秒'.format(epoch + 1, time.time()-start))
18
19     # 顯示最後結果
20     display.clear_output(wait=True)
21     generate_and_save_images(generator, epochs, seed)
```

13. 定義產生圖像的函數

```
1   # 產生圖像並存檔
2   def generate_and_save_images(model, epoch, test_input):
3       # 預測
4       predictions = model(test_input, training=False)
5
6       # 顯示 4x4 的格子
7       fig = plt.figure(figsize=(4, 4))
8       for i in range(predictions.shape[0]):
9               plt.subplot(4, 4, i+1)
10              plt.imshow(predictions[i, :, :, 0] * 127.5 + 127.5, cmap='gray')
11              plt.axis('off')
12
13      # 存檔
14      plt.savefig('./GAN_result/image_at_epoch_{:04d}.png'.format(epoch))
15      plt.show()
```

14. 訓練模型：訓練過程會產生動畫效果。

```
1   train(train_dataset, EPOCHS)
```

執行結果：訓練了 50 個週期，數字已隱約成形。

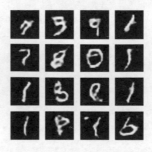

15. 將訓練過程中的存檔圖像轉為 GIF 檔，並顯示 GIF 檔。

```
1   # 產生 GIF 檔
2   anim_file = './GAN_result/dcgan.gif'
3   with imageio.get_writer(anim_file, mode='I') as writer:
4       filenames = glob.glob('./GAN_result/image*.png')
5       filenames = sorted(filenames)
6       for filename in filenames:
7           # print(filename)
8           image = imageio.imread(filename)
9           writer.append_data(image)
10
11  # 顯示 GIF 檔
12  import tensorflow_docs.vis.embed as embed
13  embed.embed_file(anim_file)
```

執行結果：注意，GIF 檔會不斷循環播放，所以也可以打開檔案總管點選
大圖顯示來檢視。

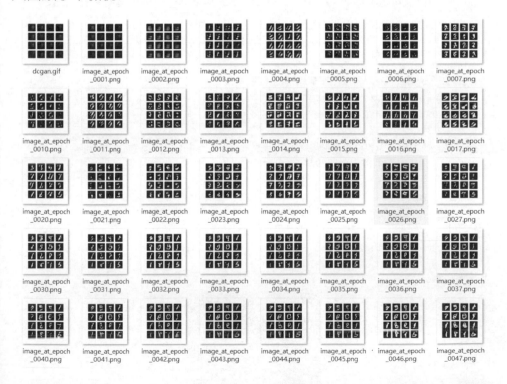

接著以名人臉部資料集，生成近似真實的圖像，程式修改自 Keras 官網
「DCGAN to generate face images」[11]。

【範例 2】以名人臉部資料集實作 DCGAN 演算法。

程式 10_02_DCGAN_Face.ipynb。

訓練資料集：名人臉部，約 1.3GB。

1. 載入相關套件。

```
1  # 載入相關套件
2  import tensorflow as tf
3  from tensorflow import keras
4  from tensorflow.keras import layers
5  import numpy as np
6  import matplotlib.pyplot as plt
7  import os
```

2. 自 https://drive.google.com/uc?id=1O7m1O1OEJjLE5QxLZiM9Fpjs7Oj6e684 下載 img_align_celeba.zip，解壓縮至 celeba_gan 目錄，產生資料集 (dataset)，圖像縮放為(64, 64)。

```
1  # 從celeba_gan目錄產生資料集
2  dataset = keras.preprocessing.image_dataset_from_directory(
3      "celeba_gan", label_mode=None, image_size=(64, 64), batch_size=32
4  )
5  # 像素標準化
6  dataset = dataset.map(lambda x: x / 255.0)
```

執行結果：共有 202599 個圖檔，同屬一個類別(以子目錄為類別名稱)。

3. 載入並顯示第一個圖檔。

```
1  # 載入並顯示第一個圖檔
2  image = next(iter(dataset))
3  plt.axis("off")
4  plt.imshow((image.numpy() * 255).astype("int32")[0])
```

執行結果：

4. 定義判別神經網路：使用一般的 CNN 結構，包含卷積層與完全連接層，用來辨識輸入的影像，但為了避免影像資訊減損，故不包含池化層。

```
1  # 判別神經網路
2  discriminator = keras.Sequential(
3      [
4          keras.Input(shape=(64, 64, 3)),
5          layers.Conv2D(64, kernel_size=4, strides=2, padding="same"),
6          layers.LeakyReLU(alpha=0.2),
7          layers.Conv2D(128, kernel_size=4, strides=2, padding="same"),
8          layers.LeakyReLU(alpha=0.2),
9          layers.Conv2D(128, kernel_size=4, strides=2, padding="same"),
10         layers.LeakyReLU(alpha=0.2),
11         layers.Flatten(),
12         layers.Dropout(0.2),
13         layers.Dense(1, activation="sigmoid"),
14     ],
15     name="discriminator",
16 )
17 discriminator.summary()
```

執行結果： 輸入圖像維度為(64, 64, 3)，寬高各為 64，RGB 3 顏色。

5. 定義生成神經網路：使用一般的解碼器結構，含完全連接層、反卷積層(Conv2DTranspose)與卷積層，可輸出特徵向量。

```
1  # 生成神經網路
2  latent_dim = 128
3
4  generator = keras.Sequential(
5      [
6          keras.Input(shape=(latent_dim,)),
7          layers.Dense(8 * 8 * 128),
8          layers.Reshape((8, 8, 128)),
9          layers.Conv2DTranspose(128, kernel_size=4, strides=2, padding="same"),
10         layers.LeakyReLU(alpha=0.2),
11         layers.Conv2DTranspose(256, kernel_size=4, strides=2, padding="same"),
12         layers.LeakyReLU(alpha=0.2),
13         layers.Conv2DTranspose(512, kernel_size=4, strides=2, padding="same"),
14         layers.LeakyReLU(alpha=0.2),
15         layers.Conv2D(3, kernel_size=5, padding="same", activation="sigmoid"),
16     ],
17     name="generator",
18 )
19 generator.summary()
```

執行結果： 輸入隨機向量大小為(8 * 8 * 128)，最後網路輸出為(64, 64, 3)，與判別神經網路輸入的維度一致。

```
Model: "generator"

Layer (type)                    Output Shape             Param #
=================================================================
dense_4 (Dense)                 (None, 8192)             1056768

reshape_1 (Reshape)             (None, 8, 8, 128)        0

conv2d_transpose_3 (Conv2DTr    (None, 16, 16, 128)      262272

leaky_re_lu_12 (LeakyReLU)      (None, 16, 16, 128)      0

conv2d_transpose_4 (Conv2DTr    (None, 32, 32, 256)      524544

leaky_re_lu_13 (LeakyReLU)      (None, 32, 32, 256)      0

conv2d_transpose_5 (Conv2DTr    (None, 64, 64, 512)      2097664

leaky_re_lu_14 (LeakyReLU)      (None, 64, 64, 512)      0

conv2d_10 (Conv2D)              (None, 64, 64, 3)        38403
=================================================================
Total params: 3,979,651
Trainable params: 3,979,651
Non-trainable params: 0
```

6. 定義 GAN：組合兩個網路。

```python
# 定義GAN，組合兩個網路
class GAN(keras.Model):
    def __init__(self, discriminator, generator, latent_dim):
        super(GAN, self).__init__()
        self.discriminator = discriminator
        self.generator = generator
        self.latent_dim = latent_dim

    # 編譯：定義損失函數
    def compile(self, d_optimizer, g_optimizer, loss_fn):
        super(GAN, self).compile()
        self.d_optimizer = d_optimizer
        self.g_optimizer = g_optimizer
        self.loss_fn = loss_fn
        self.d_loss_metric = keras.metrics.Mean(name="d_loss")
        self.g_loss_metric = keras.metrics.Mean(name="g_loss")

    # 效能指標：判別神經網路、生成神經網路
    @property
    def metrics(self):
        return [self.d_loss_metric, self.g_loss_metric]

    # 訓練
    def train_step(self, real_images):
        # 隨機抽樣 batch_size 筆，維度大小：latent_dim(128)
        batch_size = tf.shape(real_images)[0]
        random_latent_vectors = tf.random.normal(shape=(batch_size, self.latent_dim))
```

```
29          # 生成圖像
30          generated_images = self.generator(random_latent_vectors)
31
32          # 與訓練資料結合
33          combined_images = tf.concat([generated_images, real_images], axis=0)
34
35          # 訓練資料的標籤設為 1，生成圖像的標籤設為 0
36          labels = tf.concat(
37              [tf.ones((batch_size, 1)), tf.zeros((batch_size, 1))], axis=0
38          )
39
40          # 將標籤加入雜訊，此步驟非常重要
41          labels += 0.05 * tf.random.uniform(tf.shape(labels))
42
43          # 訓練判別神經網路
44          with tf.GradientTape() as tape:
45              predictions = self.discriminator(combined_images)
46              d_loss = self.loss_fn(labels, predictions)
47          grads = tape.gradient(d_loss, self.discriminator.trainable_weights)
48          self.d_optimizer.apply_gradients(
49              zip(grads, self.discriminator.trainable_weights)
50          )
51
52          # 隨機抽樣 batch_size 筆，維度大小：latent_dim(128)
53          random_latent_vectors = tf.random.normal(shape=(batch_size, self.latent_dim))
54
55          # 生成圖像的標籤設為 0
56          misleading_labels = tf.zeros((batch_size, 1))
```

```
57
58          # 訓練生成神經網路，注意，不可更新判別神經網路的權重，只更新生成神經網路的權重
59          with tf.GradientTape() as tape:
60              predictions = self.discriminator(self.generator(random_latent_vectors))
61              g_loss = self.loss_fn(misleading_labels, predictions)
62          grads = tape.gradient(g_loss, self.generator.trainable_weights)
63          self.g_optimizer.apply_gradients(zip(grads, self.generator.trainable_weights))
64
65          # 計算效能指標
66          self.d_loss_metric.update_state(d_loss)
67          self.g_loss_metric.update_state(g_loss)
68          return {
69              "d_loss": self.d_loss_metric.result(),
70              "g_loss": self.g_loss_metric.result(),
71          }
```

7. 自訂 Callback：在訓練過程中儲存圖像。

```
1   # 建立自訂的 Callback，在訓練過程中儲存圖像
2   class GANMonitor(keras.callbacks.Callback):
3       def __init__(self, num_img=3, latent_dim=128):
4           self.num_img = num_img
5           self.latent_dim = latent_dim
6
7       # 在每一執行週期結束時產生圖像
8       def on_epoch_end(self, epoch, logs=None):
9           random_latent_vectors = tf.random.normal(shape=(self.num_img, self.latent_dim))
10          generated_images = self.model.generator(random_latent_vectors)
11          generated_images *= 255
```

```
12          generated_images.numpy()
13          for i in range(self.num_img):
14              # 儲存圖像
15              img = keras.preprocessing.image.array_to_img(generated_images[i])
16              img.save("./GAN_generated/img_%03d_%d.png" % (epoch, i))
```

8. 訓練模型：在每個執行週期結束時產生 10 張圖像，依據原程式的註解，訓練週期正常需要 100 次，才能產生令人驚豔的圖片。

```
1  # 訓練模型
2  epochs = 1   # 訓練週期正常需要100次
3
4  gan = GAN(discriminator=discriminator, generator=generator, latent_dim=latent_dim)
5  gan.compile(
6      d_optimizer=keras.optimizers.Adam(learning_rate=0.0001),
7      g_optimizer=keras.optimizers.Adam(learning_rate=0.0001),
8      loss_fn=keras.losses.BinaryCrossentropy(),
9  )
10
11  # 產生10張圖像
12  gan.fit(
13      dataset, epochs=epochs, callbacks=[GANMonitor(num_img=10, latent_dim=latent_dim)]
14  )
```

執行結果：筆者的 PC 執行 1 週期就需花 2~3 個小時，如果單純使用 CPU，那可以先去睡個覺，隔天再來看結果了，但畢竟實際專案會有完成時間的壓力，這時候 GPU 卡就顯得格外重要。

執行 1 週期結果：還非常模糊。

執行 30 個週期的結果：如下圖，雖然有改善，但仍然不符預期。

10-4 Progressive GAN

Progressive GAN 也稱為 Progressive Growing GAN 或 PGGAN，它是
NVIDIA 2017 年發表的一篇文章「Progressive Growing of GANs for
Improved Quality, Stability, and Variation」[12]中提到的演算法，它可以生
成高畫質且穩定的圖像，小圖像透過層層的神經層不斷擴大，直到所要求
的尺寸為止，大部份是針對人臉的生成。

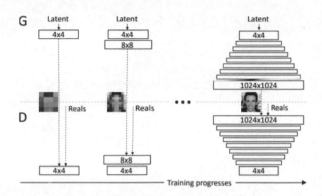

圖 10.13 Progressive GAN 的示意圖，要生成的圖像尺寸愈大，神經層就增加愈
多，圖片來源：「Progressive Growing of GANs for Improved Quality, Stability,
and Variation」[12]

它厲害的地方是演算法生成的尺寸可以大於訓練資料集的任何圖像，這稱
為「超解析度」(Super Resolution)。網路架構如下：

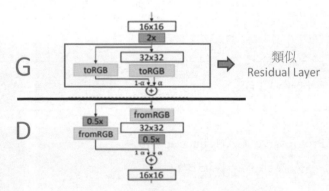

圖 10.14 Progressive GAN 的網路架構

生成網路(G)使用類似像 Residual 神經層，一邊輸入原圖像，另一邊輸入為反卷積層，使用權重 α，訂定兩個輸入層的比例。判別網路(D)與生成網路(G)做反向操作，進行辨識。使用名人臉部資料集進行模型訓練，根據論文估計，使用 8 顆 Tesla V100 GPU，大約要訓練 4 天左右，才可以得到不錯的效果。讀到這裡，對我們平民百姓來説，簡直是晴天霹靂，根本不用玩了。還好，TensorFlow Hub 提供預先訓練好的模型，我們馬上來測試一下吧。

【範例3】再拿名人臉部資料集來實作 Progressive GAN 演算法。
使用隨機向量，流程如下：

若使用自訂的圖像，流程如下：

程式 10_03_ Progressive_GAN_Face.ipynb。
訓練資料集：名人臉部，約 1.3GB。

1. 載入相關套件：需先安裝 scikit-image、imageio 套件。

```
1  # 載入相關套件
2  from absl import logging
3  import imageio
4  import PIL.Image
5  import matplotlib.pyplot as plt
6  import numpy as np
7  import tensorflow as tf
8  import tensorflow_hub as hub
9  import time
10 from IPython import display
11 from skimage import transform
```

2. 定義顯示圖像的函數、轉為動畫的函數。

```
1  # 生成網路的輸入向量尺寸
2  latent_dim = 512
3
4  # 定義顯示圖像的函數
5  def display_image(image):
6      image = tf.constant(image)
7      image = tf.image.convert_image_dtype(image, tf.uint8)
8      return PIL.Image.fromarray(image.numpy())
9
10 # 定義顯示一序列圖像轉為動畫的函數
11 from tensorflow_docs.vis import embed
12 def animate(images):
13     images = np.array(images)
14     converted_images = np.clip(images * 255, 0, 255).astype(np.uint8)
15     imageio.mimsave('./animation.gif', converted_images)
16     return embed.embed_file('./animation.gif')
17
18 # 工作記錄檔只記錄錯誤等級以上的訊息
19 logging.set_verbosity(logging.ERROR)
```

3. 插補兩個向量，產生動畫：採非線性插補，在超球面上插補兩個向量，這樣才能有較平滑的轉換，隨機抽取兩張圖像，由第一張圖像開始，漸變成第二張圖像。

```
1  # 在超球面上插補兩向量
2  def interpolate_hypersphere(v1, v2, num_steps):
3      v1_norm = tf.norm(v1)
4      v2_norm = tf.norm(v2)
5      v2_normalized = v2 * (v1_norm / v2_norm)
6
7      vectors = []
8      for step in range(num_steps):
9          interpolated = v1 + (v2_normalized - v1) * step / (num_steps - 1)
10         interpolated_norm = tf.norm(interpolated)
11         interpolated_normalized = interpolated * (v1_norm / interpolated_norm)
```

```
12        vectors.append(interpolated_normalized)
13    return tf.stack(vectors)
14
15 # 插補兩向量
16 def interpolate_between_vectors():
17    # 產生兩個雜訊向量
18    v1 = tf.random.normal([[latent_dim]])
19    v2 = tf.random.normal([[latent_dim]])
20
21    # 產生 25 個步驟的插補
22    vectors = interpolate_hypersphere(v1, v2, 50)
23
24    # 產生一序列的圖像
25    interpolated_images = progan(vectors)['default']
26    return interpolated_images
27
28 # 插補兩向量，產生動畫
29 interpolated_images = interpolate_between_vectors()
30 animate(interpolated_images)
```

執行結果：由左圖漸變為右圖。

4. 使用隨機亂數或自訂的圖像：由第 5 行的 image_from_module_space 變數來調控，設為 True 表示使用隨機亂數；False 表示使用自訂的圖像。

```
1 # 使用自訂的圖像
2 image_path='./images_face/face_test.png'
3
4 # True：使用隨機亂數，False：使用自訂的圖像
5 image_from_module_space = True
6
7 tf.random.set_seed(0)
8
9 def get_module_space_image():
10    vector = tf.random.normal([1, latent_dim])
11    images = progan(vector)['default'][0]
12    return images
13
14 # 使用隨機亂數
15 if image_from_module_space:
16    target_image = get_module_space_image()
17 else: # 使用自訂的圖像
```

```
18      image = imageio.imread(image_path)
19      target_image = transform.resize(image, [128, 128])
20
21  # 顯示圖像
22  display_image(target_image)
```

執行結果：目標圖像。

5. 以 PGGAN 處理的圖像作為起始圖像。

```
1  # 以PGGAN 處理的圖像作為起始圖像
2  initial_vector = tf.random.normal([1, latent_dim])
3
4  # 顯示圖像
5  display_image(progan(initial_vector)['default'][0])
```

執行結果：起始圖像。

6. 找到最接近的特徵向量，漸變成第二張圖像：模型訓練的過程中，每 5
   步驟就會產生一個圖像。

```
1  # 找到最接近的特徵向量
2  def find_closest_latent_vector(initial_vector, num_optimization_steps,
3      images = []
4      losses = []
5
6      vector = tf.Variable(initial_vector)
7      optimizer = tf.optimizers.Adam(learning_rate=0.01)
8      # 以 MAE 為損失函數
9      loss_fn = tf.losses.MeanAbsoluteError(reduction="sum")
10
11      # 訓練
```

```
12      for step in range(num_optimization_steps):
13          if (step % 100)==0:
14              print()
15          print('.', end='')
16
17          # 梯度下降
18          with tf.GradientTape() as tape:
19              image = progan(vector.read_value())['default'][0]
20
21              # 每 5 步驟產生一個圖像
22              if (step % steps_per_image) == 0:
23                  img = tf.keras.preprocessing.image.array_to_img(image.numpy())
24                  img.save(f"./PGGAN_generated/img_{step:03d}.png")
25                  images.append(image.numpy())
26
27              # 計算損失
28              target_image_difference = loss_fn(image, target_image[:,:,:3])
29              # 正則化
30              regularizer = tf.abs(tf.norm(vector) - np.sqrt(latent_dim))
31              loss = tf.cast(target_image_difference, tf.double) + tf.cast(regularizer, tf.double)
32              losses.append(loss.numpy())
```

```
33
34          # 根據梯度，更新權重
35          grads = tape.gradient(loss, [vector])
36          optimizer.apply_gradients(zip(grads, [vector]))
37
38      return images, losses
39
40
41  num_optimization_steps=200
42  steps_per_image=5
43  images, loss = find_closest_latent_vector(initial_vector, num_optimization_steps, steps_per_image)
```

7. 顯示訓練過程的損失函數。

```
1  # 顯示訓練過程的損失函數
2  plt.plot(loss)
3  plt.ylim([0,max(plt.ylim())])
```

8. 顯示動畫。

```
1  animate(np.stack(images))
```

執行結果：由左圖漸變為右圖。

修改步驟 4 第 5 行的 image_from_module_space 變數控制，更改為 False，使用自訂的圖像，臉部特寫必須一致，效果才會比較好。

起始圖像如下：

自訂的圖像如下，為目標圖像。

執行結果：

10-5 Conditional GAN

DCGAN 生成的圖片是隨機的，以 MNIST 資料集而言，生成圖像的確會是數字，但無法控制要生成哪一個數字。Conditional GAN 增加了一個條件 (Condition)，即目標變數 Y，用來控制生成的數字。

Conditional GAN 也稱為 cGAN，它是 Mehdi Mirz 等學者於 2014 年發表的一篇文章「Conditional Generative Adversarial Nets」[13]中所提出的演算法，它修改 GAN 損失函數如下：

$$\min_G \max_D V(D,G) = \mathbb{E}_{\boldsymbol{x} \sim p_{\text{data}}(\boldsymbol{x})}[\log D(\boldsymbol{x}|\boldsymbol{y})] + \mathbb{E}_{\boldsymbol{z} \sim p_z(\boldsymbol{z})}[\log(1 - D(G(\boldsymbol{z}|\boldsymbol{y})))]$$

將單純的 D(x)改為條件機率 D(x|y)。

【範例 4】以 Fashion MNIST 資料集實作 Conditional GAN 演算法。

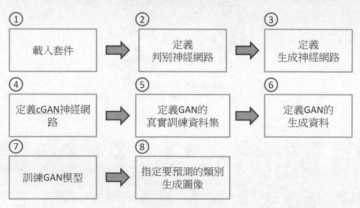

程式 10_04_CGAN_FashionMNIST.ipynb。

1. 載入相關套件。

```python
1  # 載入相關套件
2  from numpy import zeros, ones, expand_dims
3  from numpy.random import randn
4  from numpy.random import randint
5  from tensorflow.keras.datasets.fashion_mnist import load_data
6  from tensorflow.keras.optimizers import Adam
7  from tensorflow.keras.models import Model
8  from tensorflow.keras.layers import Input, Dense, Reshape, Flatten
9  from tensorflow.keras.layers import Dropout, Embedding, Concatenate, LeakyReLU
10 from tensorflow.keras.layers import Conv2D, Conv2DTranspose
```

2. 定義判別神經網路：把 Fashion MNIST 資料集的標記(Label)也視為特徵變數(X)，因為我們要指定生成圖像的類別，透過嵌入層(Embedding)轉成 50 個向量。

```
1   # 定義判別神經網路
2   def define_discriminator(in_shape=(28,28,1), n_classes=10):
3       # 輸入 Y
4       in_label = Input(shape=(1,))
5       li = Embedding(n_classes, 50)(in_label)
6       li = Dense(in_shape[0] * in_shape[1])(li)
7       li = Reshape((in_shape[0], in_shape[1], 1))(li)
8
9       # 輸入圖像
10      in_image = Input(shape=in_shape)
11
12      # 結合 Y 及圖像
13      merge = Concatenate()([in_image, li])
14
15      # 抽樣(downsampling)
16      fe = Conv2D(128, (3,3), strides=(2,2), padding='same')(merge)
17      fe = LeakyReLU(alpha=0.2)(fe)
18      fe = Conv2D(128, (3,3), strides=(2,2), padding='same')(fe)
19      fe = LeakyReLU(alpha=0.2)(fe)
20      fe = Flatten()(fe)
21      fe = Dropout(0.4)(fe)
22      out_layer = Dense(1, activation='sigmoid')(fe)
23
24      # 定義模型的輸入及輸出
25      model = Model([in_image, in_label], out_layer)
26
27      # 編譯
28      opt = Adam(lr=0.0002, beta_1=0.5)
29      model.compile(loss='binary_crossentropy', optimizer=opt, metrics=['accuracy'])
30      return model
```

3. 定義生成神經網路：一樣需要標記。

```
1   # 定義生成神經網路
2   def define_generator(latent_dim, n_classes=10):
3       # 輸入 Y
4       in_label = Input(shape=(1,))
5       # embedding for categorical input
6       li = Embedding(n_classes, 50)(in_label)
7       li = Dense(7 * 7)(li)
8       li = Reshape((7, 7, 1))(li)
9
10      # 輸入圖像
11      in_lat = Input(shape=(latent_dim,))
12      gen = Dense(128 * 7 * 7)(in_lat)
13      gen = LeakyReLU(alpha=0.2)(gen)
14      gen = Reshape((7, 7, 128))(gen)
15
16      # 結合 Y 及圖像
```

```
17    merge = Concatenate()([gen, li])
18
19    # 上採樣(upsampling)
20    gen = Conv2DTranspose(128, (4,4), strides=(2,2), padding='same')(merge)
21    gen = LeakyReLU(alpha=0.2)(gen)
22    gen = Conv2DTranspose(128, (4,4), strides=(2,2), padding='same')(gen)
23    gen = LeakyReLU(alpha=0.2)(gen)
24    out_layer = Conv2D(1, (7,7), activation='tanh', padding='same')(gen)
25
26    # 定義模型的輸入及輸出
27    model = Model([in_lat, in_label], out_layer)
28    return model
```

4. 定義 cGAN 神經網路：結合判別神經網路和生成神經網路。

```
1    # 定義cGAN神經網路
2    def define_gan(g_model, d_model):
3        d_model.trainable = False              # 判別神經網路不重新訓練
4        gen_noise, gen_label = g_model.input   # 取得生成神經網路的輸入
5        gen_output = g_model.output            # 取得生成神經網路的輸出
6
7        # 取得判別神經網路的輸出
8        gan_output = d_model([gen_output, gen_label])
9
10       # 定義模型的輸入及輸出
11       model = Model([gen_noise, gen_label], gan_output)
12
13       # 編譯
14       opt = Adam(lr=0.0002, beta_1=0.5)
15       model.compile(loss='binary_crossentropy', optimizer=opt)
16       return model
```

5. 定義 GAN 的真實訓練資料集：真實資料的標記均為 1。

```
1    # 載入 fashion mnist 資料集
2    def load_real_samples():
3        (trainX, trainy), (_, _) = load_data()
4
5        # 增加一維作為色彩通道
6        X = expand_dims(trainX, axis=-1)
7        X = X.astype('float32')
8        # 標準化，使像素值介於 [-1,1]
9        X = (X - 127.5) / 127.5
10       return [X, trainy]
11
12   # 定義模型的輸入 n 筆
13   def generate_real_samples(dataset, n_samples):
14       images, labels = dataset
15       # 隨機抽樣 n 筆
16       ix = randint(0, images.shape[0], n_samples)
17       # fashion mnist 資料集的 X、Y 均作為 GAN 的輸入
18       X, labels = images[ix], labels[ix]
19       # GAN 的 Y 均為 1
20       y = ones((n_samples, 1))
21       return [X, labels], y
```

6. 定義 GAN 的生成資料。

```python
1   # 生成隨機向量
2   def generate_latent_points(latent_dim, n_samples, n_classes=10):
3       # 隨機向量
4       x_input = randn(latent_dim * n_samples)
5       z_input = x_input.reshape(n_samples, latent_dim)
6       # 產生 n 個類別的 Y
7       labels = randint(0, n_classes, n_samples)
8       return [z_input, labels]
9
10  # 定義模型的生成資料 n 筆
11  def generate_fake_samples(generator, latent_dim, n_samples):
12      # 生成隨機向量
13      z_input, labels_input = generate_latent_points(latent_dim, n_samples)
14      # 生成圖像
15      images = generator.predict([z_input, labels_input])
16      # 產生均為 0 的 Y
17      y = zeros((n_samples, 1))
18      return [images, labels_input], y
```

7. 訓練模型：訓練 GAN 模型，內含判別神經網路、生成神經網路。

```python
1   # 訓練模型
2   def train(g_model, d_model, gan_model, dataset, latent_dim, n_epochs=100, n_batch=128):
3       bat_per_epo = int(dataset[0].shape[0] / n_batch)
4       half_batch = int(n_batch / 2)
5
6       # 訓練
7       for i in range(n_epochs):
8           for j in range(bat_per_epo):
9               # 隨機抽樣 n 筆
10              [X_real, labels_real], y_real = generate_real_samples(dataset, half_batch)
11              # 更新判別神經網路權重
12              d_loss1, _ = d_model.train_on_batch([X_real, labels_real], y_real)
13
14              # 生成隨機向量
15              [X_fake, labels], y_fake = generate_fake_samples(g_model, latent_dim, half_batch)
16              # 更新判別神經網路權重
17              d_loss2, _ = d_model.train_on_batch([X_fake, labels], y_fake)
18
19              # 生成一批的隨機向量
20              [z_input, labels_input] = generate_latent_points(latent_dim, n_batch)
21              # 產生均為 1 的 Y
22              y_gan = ones((n_batch, 1))
23
24              # 訓練模型
25              g_loss = gan_model.train_on_batch([z_input, labels_input], y_gan)
26
27              # 顯示損失
28              print('>%d, %d/%d, d1=%.3f, d2=%.3f g=%.3f' %
29                  (i+1, j+1, bat_per_epo, d_loss1, d_loss2, g_loss))
30
```

```python
31      # 模型存檔
32      g_model.save('cgan_generator.h5')
33
34  # 參數設定
35  latent_dim = 100                              # 隨機向量尺寸
```

```
36
37  d_model = define_discriminator()          # 建立判別神經網路
38  g_model = define_generator(latent_dim)    # 建立生成神經網路
39  gan_model = define_gan(g_model, d_model)   # 建立 GAN 神經網路
40
41  dataset = load_real_samples()              # 讀取訓練資料
42  # 訓練
43  train(g_model, d_model, gan_model, dataset, latent_dim)
```

執行結果：模型訓練需要幾個小時的時間，先去做其他事，再回來看結果。

8. 定義雜訊生成、顯示圖像的函數。

```
1   # 載入相關套件
2   from numpy import asarray
3   from tensorflow.keras.models import load_model
4   from matplotlib import pyplot
5
6   # 生成隨機資料
7   def generate_latent_points(latent_dim, n_samples, n_classes=10):
8       # generate points in the latent space
9       x_input = randn(latent_dim * n_samples)
10      # reshape into a batch of inputs for the network
11      z_input = x_input.reshape(n_samples, latent_dim)
12      # generate labels
13      labels = randint(0, n_classes, n_samples)
14      return [z_input, labels]
15
16  # 顯示圖像
17  def save_plot(examples, n):
18      # 繪製 n x n 個圖像
19      for i in range(n * n):
20          pyplot.subplot(n, n, 1 + i)
21          pyplot.axis('off')
22          # cmap='gray_r'：反轉黑白，因 MNIST 像素與 RGB 相反
23          pyplot.imshow(examples[i, :, :, 0], cmap='gray_r')
24      pyplot.show()
```

9. 預測並顯示結果。

```
1   # 載入模型
2   model = load_model('cgan_generator.h5')
3   # 生成 100 筆資料
4   latent_points, labels = generate_latent_points(100, 100)
5   # 標記 0~9
6   labels = asarray([x for _ in range(10) for x in range(10)])
7
8   # 預測並顯示結果
9   X = model.predict([latent_points, labels])
10  # 將像素範圍由 [-1,1] 轉換為 [0,1]
11  X = (X + 1) / 2.0
12  # 繪圖
13  save_plot(X, 10)
```

執行結果：結果非常理想，我們指定標記(Label)就可以生成該類別的圖像。

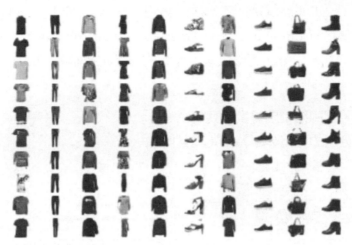

上例是運用 Conditional GAN 很簡單的例子，只是把標記(Label)一併當作 X，輸入模型中訓練，作為條件(Condition)或限制條件(Constraint)。另外還有很多延伸的作法，例如 ColorGAN，它把前置處理的輪廓圖作為條件，與雜訊一併當作 X，就可以生成與原圖相似的圖像，並且可以為灰階圖上色，相關細節可參閱「Colorization Using ConvNet and GAN」[14]或「 End-to-End Conditional GAN-based Architectures for Image Colourisation」[15]。

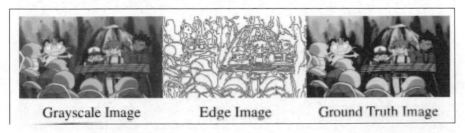

圖 10.15 ColorGAN，圖片來源：「Colorization Using ConvNet and GAN」[14]

10-6 Pix2Pix

Pix2Pix 為 Conditional GAN 演算法的應用，出自於 Phillip Isola 等學者在 2016 年發表的「Image-to-Image Translation with Conditional Adversarial Networks」[16]，它能夠將影像進行像素的轉換，故稱為 Pix2Pix，可應用於：

1. 將語義分割的街景圖轉換為真實圖像
2. 將語義分割的建築外觀轉換為真實圖像
3. 將衛星照轉換為地圖
4. 將白天圖像轉換為夜晚圖像
5. 將輪廓圖轉為實物圖像

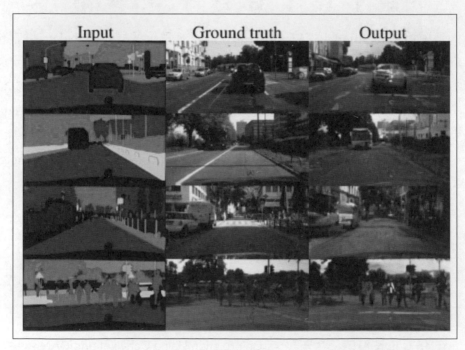

圖 10.16 將語義分割的街景圖轉換為真實圖像，以下圖片來源均來自「Image-to-Image Translation with Conditional Adversarial Networks」[16]

圖 10.17 將衛星照轉換為地圖,反之亦可

圖 10.18 將白天圖像轉換為夜晚圖像

圖 10.19 將輪廓圖轉為實物圖像

生成網路採用的 U-net 結構,引進了「Skip-connect」的技巧,即每一層反卷積層的輸入都是「前一層的輸出」加「與該層對稱的卷積層的輸出」,解碼時可從對稱的編碼器得到對應的資訊,使得生成的圖像保有原圖像的特徵。

判別網路額外考慮輸入圖像的判別,將真實圖像、生成圖像與輸入圖像合而為一,作為判別網路的輸入,進行辨識。原生的 GAN 在預測像素時,是以真實資料對應的「單一像素」進行辨識,然而 Pix2Pix 則引用

PatchGAN 的思維，利用卷積將圖像切成多個較小的區域，每個像素與對應的「區域」進行辨識(Softmax)，計算最大可能的輸出。PatchGAN 可參見「Image-to-Image Translation with Conditional Adversarial Networks」[16]一文。

【範例 5】以 CMP Facade Database 資料集實作 Pix2Pix GAN 演算法。CMP Facade Database[17]共有 12 類的建築物局部外型，如外觀(façade)、造型(molding)、屋簷(cornice)、柱子(pillar)、窗戶(window)、門(door)等。資料集自 https://people.eecs.berkeley.edu/~tinghuiz/projects/pix2pix/datasets/facades.tar.gz 下載。

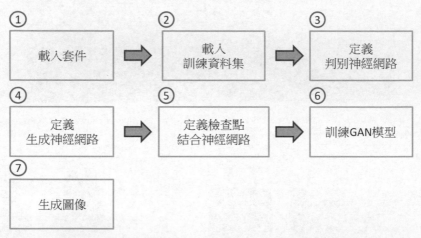

程式 10_05_Pix2Pix.ipynb，注意，執行此範例，請去睡個覺，再來看結果，原文估計使用單片 V100 GPU，訓練一個週期約需 15 秒。

1. 載入相關套件。

```
1  # 載入相關套件
2  import tensorflow as tf
3  import os
4  import time
5  from matplotlib import pyplot as plt
6  from IPython import display
```

2. 參數設定並定義圖像處理的函數。

```
1   # 參數設定
2   PATH = './CMP Facade Database/facades/'
3   BUFFER_SIZE = 400    # 緩衝區大小
4   BATCH_SIZE = 1       # 批量
5   IMG_WIDTH = 256      # 圖像寬度
6   IMG_HEIGHT = 256     # 圖像高度
7
8   # 載入圖像
9   def load(image_file):
10      # 讀取圖檔
11      image = tf.io.read_file(image_file)
12      image = tf.image.decode_jpeg(image)
13
14      w = tf.shape(image)[1]
15
16      # 高為寬的一半
17      w = w // 2
18      real_image = image[:, :w, :]
19      input_image = image[:, w:, :]
20
21      # 轉為浮點數
22      input_image = tf.cast(input_image, tf.float32)
23      real_image = tf.cast(real_image, tf.float32)
24
25      return input_image, real_image
```

```
26
27  # 縮放圖像
28  def resize(input_image, real_image, height, width):
29      input_image = tf.image.resize(input_image, [height, width],
30                          method=tf.image.ResizeMethod.NEAREST_NEIGHBOR)
31      real_image = tf.image.resize(real_image, [height, width],
32                          method=tf.image.ResizeMethod.NEAREST_NEIGHBOR)
33
34      return input_image, real_image
35
36  # 隨機裁切圖像
37  def random_crop(input_image, real_image):
38      stacked_image = tf.stack([input_image, real_image], axis=0)
39      cropped_image = tf.image.random_crop(
40        stacked_image, size=[2, IMG_HEIGHT, IMG_WIDTH, 3])
41
42      return cropped_image[0], cropped_image[1]
43
44  # 標準化，使像素值介於 [ 1, 1]
45  def normalize(input_image, real_image):
46      input_image = (input_image / 127.5) - 1
47      real_image = (real_image / 127.5) - 1
48
49      return input_image, real_image
50
```

```
51  # 隨機轉換
52  @tf.function()
53  def random_jitter(input_image, real_image):
54      # 縮放圖像為 286 x 286 x 3
55      input_image, real_image = resize(input_image, real_image, 286, 286)
56
57      # 隨機裁切至 256 x 256 x 3
58      input_image, real_image = random_crop(input_image, real_image)
59
60      if tf.random.uniform(()) > 0.5:
61          # 水平翻轉
62          input_image = tf.image.flip_left_right(input_image)
63          real_image = tf.image.flip_left_right(real_image)
64
65      return input_image, real_image
```

3. 顯示任一張訓練圖片。

```
1  # 隨意顯示一張訓練圖片
2  inp, re = load(PATH+'train/100.jpg')
3  # 顯示圖像
4  plt.figure()
5  plt.imshow(inp/255.0)
6  plt.figure()
7  plt.imshow(re/255.0)
```

執行結果：左為語義分割圖，右為實景圖。

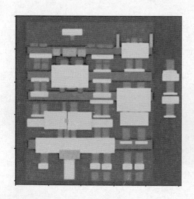 

4. 隨機轉換測試：也可作為資料增補。

```
1  # 隨機轉換測試
2  plt.figure(figsize=(6, 6))
3  for i in range(4):
4      rj_inp, rj_re = random_jitter(inp, re)
5      plt.subplot(2, 2, i+1)
6      plt.imshow(rj_inp/255.0)
7      plt.axis('off')
8  plt.show()
```

執行結果：

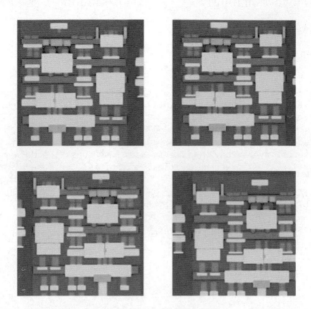

5. 載入訓練資料，轉換為 TensorFlow Dataset。

```
 1  def load_image_train(image_file):
 2      input_image, real_image = load(image_file)
 3      input_image, real_image = random_jitter(input_image, real_image)
 4      input_image, real_image = normalize(input_image, real_image)
 5
 6      return input_image, real_image
 7
 8  def load_image_test(image_file):
 9      input_image, real_image = load(image_file)
10      input_image, real_image = resize(input_image, real_image,
11                              IMG_HEIGHT, IMG_WIDTH)
12      input_image, real_image = normalize(input_image, real_image)
13
14      return input_image, real_image
15
16  train_dataset = tf.data.Dataset.list_files(PATH+'train/*.jpg')
17  train_dataset = train_dataset.map(load_image_train,
18                              num_parallel_calls=tf.data.AUTOTUNE)
19  train_dataset = train_dataset.shuffle(BUFFER_SIZE)
20  train_dataset = train_dataset.batch(BATCH_SIZE)
21
22  test_dataset = tf.data.Dataset.list_files(PATH+'test/*.jpg')
23  test_dataset = test_dataset.map(load_image_test)
24  test_dataset = test_dataset.batch(BATCH_SIZE)
```

6. 定義採樣(down sampling)、上採樣 (up sampling) 函數，並做簡單測試。

```
1  OUTPUT_CHANNELS = 3
2
3  # 定義採樣函數
4  def downsample(filters, size, apply_batchnorm=True):
5      initializer = tf.random_normal_initializer(0., 0.02)
6
7      result = tf.keras.Sequential()
8      result.add(
9          tf.keras.layers.Conv2D(filters, size, strides=2, padding='same',
10                         kernel_initializer=initializer, use_bias=False))
11
12     if apply_batchnorm:
13         result.add(tf.keras.layers.BatchNormalization())
14
15     result.add(tf.keras.layers.LeakyReLU())
16
17     return result
18
19 down_model = downsample(3, 4)
20 down_result = down_model(tf.expand_dims(inp, 0))
21 print(down_result.shape)
```

執行結果：(1, 128, 128, 3)。

```
1  # 定義上採樣函數
2  def upsample(filters, size, apply_dropout=False):
3      initializer = tf.random_normal_initializer(0., 0.02)
4
5      result = tf.keras.Sequential()
6      result.add(
7          tf.keras.layers.Conv2DTranspose(filters, size, strides=2,
8                                 padding='same',
9                                 kernel_initializer=initializer,
10                                use_bias=False))
11
12     result.add(tf.keras.layers.BatchNormalization())
13
14     if apply_dropout:
15         result.add(tf.keras.layers.Dropout(0.5))
16
17     result.add(tf.keras.layers.ReLU())
18
19     return result
20
21 up_model = upsample(3, 4)
22 up_result = up_model(down_result)
23 print(up_result.shape)
```

執行結果：(1, 256, 256, 3)。

7. 定義生成神經網路：U-Net 結構。

```
1   def Generator():
2       inputs = tf.keras.layers.Input(shape=[256, 256, 3])
3
4       down_stack = [
5           downsample(64, 4, apply_batchnorm=False),    # (bs, 128, 128, 64)
6           downsample(128, 4),      # (bs, 64, 64, 128)
7           downsample(256, 4),      # (bs, 32, 32, 256)
8           downsample(512, 4),      # (bs, 16, 16, 512)
9           downsample(512, 4),      # (bs, 8, 8, 512)
10          downsample(512, 4),      # (bs, 4, 4, 512)
11          downsample(512, 4),      # (bs, 2, 2, 512)
12          downsample(512, 4),      # (bs, 1, 1, 512)
13      ]
14
15      up_stack = [
16          upsample(512, 4, apply_dropout=True),    # (bs, 2, 2, 1024)
17          upsample(512, 4, apply_dropout=True),    # (bs, 4, 4, 1024)
18          upsample(512, 4, apply_dropout=True),    # (bs, 8, 8, 1024)
19          upsample(512, 4),    # (bs, 16, 16, 1024)
20          upsample(256, 4),    # (bs, 32, 32, 512)
21          upsample(128, 4),    # (bs, 64, 64, 256)
22          upsample(64, 4),     # (bs, 128, 128, 128)
23      ]
24
25      initializer = tf.random_normal_initializer(0., 0.02)
26      # (bs, 256, 256, 3)
27      last = tf.keras.layers.Conv2DTranspose(OUTPUT_CHANNELS, 4, strides=2,
28              padding='same', kernel_initializer=initializer, activation='tanh')
```

```
29
30      x = inputs
31
32      # Downsampling through the model
33      skips = []
34      for down in down_stack:
35          x = down(x)
36          skips.append(x)
37
38      skips = reversed(skips[:-1])
39
40      # Upsampling and establishing the skip connections
41      for up, skip in zip(up_stack, skips):
42          x = up(x)
43          x = tf.keras.layers.Concatenate()([x, skip])
44
45      x = last(x)
46
47      return tf.keras.Model(inputs=inputs, outputs=x)
```

8. 繪製生成神經網路模型。

```
1   # 建立生成神經網路
2   generator = Generator()
3
4   # 繪製生成神經網路模型
5   tf.keras.utils.plot_model(generator, show_shapes=True, dpi=64)
```

執行結果： 為 U 型結構。

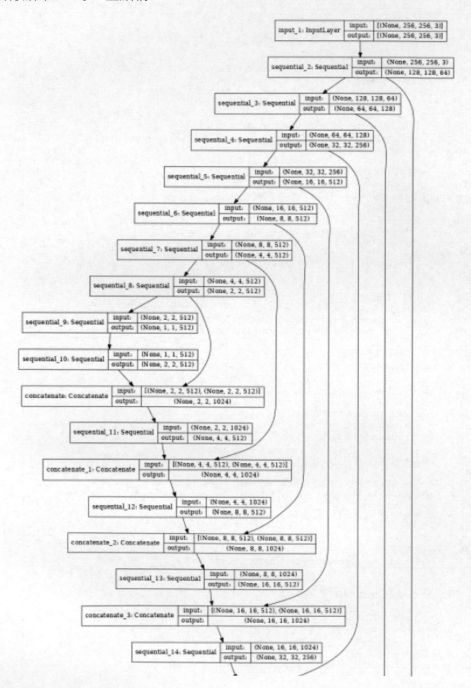

9. 測試生成神經網路：取第一筆資料。

```
1  # 測試生成神經網路
2  gen_output = generator(inp[tf.newaxis, ...], training=False)
3  plt.imshow(gen_output[0, ...])
```

執行結果：只有隱約的形狀。

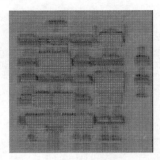

10. 定義判別神經網路。

```
1   # 定義判別神經網路
2   def Discriminator():
3       initializer = tf.random_normal_initializer(0., 0.02)
4
5       inp = tf.keras.layers.Input(shape=[256, 256, 3], name='input_image')
6       tar = tf.keras.layers.Input(shape=[256, 256, 3], name='target_image')
7
8       x = tf.keras.layers.concatenate([inp, tar])  # (bs, 256, 256, channels*2)
9
10      down1 = downsample(64, 4, False)(x)  # (bs, 128, 128, 64)
11      down2 = downsample(128, 4)(down1)  # (bs, 64, 64, 128)
12      down3 = downsample(256, 4)(down2)  # (bs, 32, 32, 256)
13
14      # (bs, 34, 34, 256)
15      zero_pad1 = tf.keras.layers.ZeroPadding2D()(down3)
16      # (bs, 31, 31, 512)
17      conv = tf.keras.layers.Conv2D(512, 4, strides=1,
18                          kernel_initializer=initializer,
19                          use_bias=False)(zero_pad1)
20
21      batchnorm1 = tf.keras.layers.BatchNormalization()(conv)
22
23      leaky_relu = tf.keras.layers.LeakyReLU()(batchnorm1)
24
25      # (bs, 33, 33, 512)
26      zero_pad2 = tf.keras.layers.ZeroPadding2D()(leaky_relu)
27
28      # (bs, 30, 30, 1)
29      last = tf.keras.layers.Conv2D(1, 4, strides=1,
30                          kernel_initializer=initializer)(zero_pad2)
31
32      return tf.keras.Model(inputs=[inp, tar], outputs=last)
```

11. 建立判別神經網路,並繪製模型。

```
1  # 建立判別神經網路
2  discriminator = Discriminator()
3
4  # 繪製生成神經網路模型
5  tf.keras.utils.plot_model(discriminator, show_shapes=True, dpi=64)
```

執行結果:主要為卷積層。

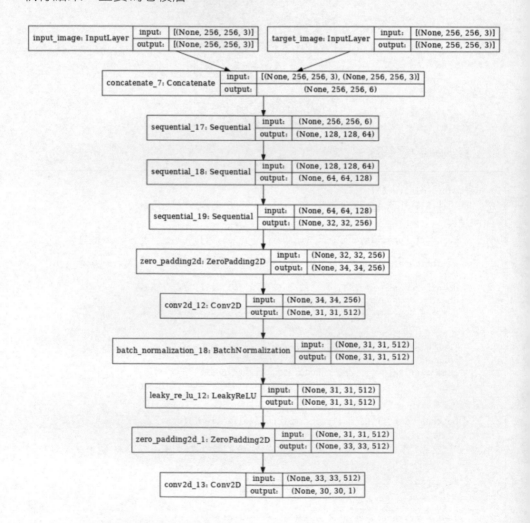

12. 測試判別神經網路：取第一筆資料。

```
1  # 測試判別神經網路
2  disc_out = discriminator([inp[tf.newaxis, ...], gen_output], training=False)
3  plt.imshow(disc_out[0, ..., -1], vmin=-20, vmax=20, cmap='RdBu_r')
4  plt.colorbar()
```

執行結果：

13. 定義損失函數。

```
1   # 定義損失函數為二分類交叉熵
2   LAMBDA = 100
3   loss_obj = tf.keras.losses.BinaryCrossentropy(from_logits=True)
4
5   # 定義判別網路損失函數
6   def discriminator_loss(disc_real_output, disc_generated_output):
7       real_loss = loss_object(tf.ones_like(disc_real_output), disc_real_output)
8
9       generated_loss = loss_object(tf.zeros_like(disc_generated_output), disc_generated_output)
10
11      total_disc_loss = real_loss + generated_loss
12
13      return total_disc_loss
14  # 定義生成網路損失函數
15  # 定義生成網路損失函數
16  def generator_loss(disc_generated_output, gen_output, target):
17      gan_loss = loss_object(tf.ones_like(disc_generated_output), disc_generated_output)
18
19      # mean absolute error
20      l1_loss = tf.reduce_mean(tf.abs(target - gen_output))
21
22      total_gen_loss = gan_loss + (LAMBDA * l1_loss)
23
24      return total_gen_loss, gan_loss, l1_loss
```

14. 定義優化器。

```
1  # 定義優化器
2  generator_optimizer = tf.keras.optimizers.Adam(2e-4, beta_1=0.5)
3  discriminator_optimizer = tf.keras.optimizers.Adam(2e-4, beta_1=0.5)
```

15. 定義檢查點：可設定以下函數，直接生成圖像。

```
1  # 定義檢查點
2  checkpoint_dir = './Pix2Pix_training_checkpoints'
3  checkpoint_prefix = os.path.join(checkpoint_dir, "ckpt")
4  checkpoint = tf.train.Checkpoint(generator_optimizer=generator_optimizer,
5                                   discriminator_optimizer=discriminator_optimizer,
6                                   generator=generator,
7                                   discriminator=discriminator)
```

16. 生成圖像，並顯示。

```
1   # 生成圖像，並顯示
2   def generate_images(model, test_input, tar):
3       prediction = model(test_input, training=True)
4       plt.figure(figsize=(15, 15))
5
6       display_list = [test_input[0], tar[0], prediction[0]]
7       title = ['Input Image', 'Ground Truth', 'Predicted Image']
8
9       # 顯示 輸入圖像、真實圖像、與生成圖像
10      for i in range(3):
11          plt.subplot(1, 3, i+1)
12          plt.title(title[i])
13          # 轉換像素質介於 [0, 1]
14          plt.imshow(display_list[i] * 0.5 + 0.5)
15          plt.axis('off')
16      plt.show()
17
18  # 取一筆資料測試
19  for example_input, example_target in test_dataset.take(1):
20      generate_images(generator, example_input, example_target)
```

執行結果： 還未訓練的結果。

17.定義訓練模型的函數。

```
1  import datetime
2  # 參數設定
3  EPOCHS = 150
4  log_dir="logs/"
5
6  summary_writer = tf.summary.create_file_writer(
7      log_dir + "fit/" + datetime.datetime.now().strftime("%Y%m%d-%H%M%S"))
8
9  # 定義訓練模型的函數
10 @tf.function
11 def train_step(input_image, target, epoch):
12     with tf.GradientTape() as gen_tape, tf.GradientTape() as disc_tape:
13         gen_output = generator(input_image, training=True)
14
15         disc_real_output = discriminator([input_image, target], training=True)
16         disc_generated_output = discriminator([input_image, gen_output], training=True)
17
18         gen_total_loss, gen_gan_loss, gen_l1_loss = generator_loss(disc_generated_output, gen_output, target)
19         disc_loss = discriminator_loss(disc_real_output, disc_generated_output)
20
21     generator_gradients = gen_tape.gradient(gen_total_loss, generator.trainable_variables)
22     discriminator_gradients = disc_tape.gradient(disc_loss, discriminator.trainable_variables)
23
24     generator_optimizer.apply_gradients(zip(generator_gradients, generator.trainable_variables))
25     discriminator_optimizer.apply_gradients(zip(discriminator_gradients, discriminator.trainable_variables))
26
27     with summary_writer.as_default():
28         tf.summary.scalar('gen_total_loss', gen_total_loss, step=epoch)
29         tf.summary.scalar('gen_gan_loss', gen_gan_loss, step=epoch)
30         tf.summary.scalar('gen_l1_loss', gen_l1_loss, step=epoch)
31         tf.summary.scalar('disc_loss', disc_loss, step=epoch)
```

```
1  def fit(train_ds, epochs, test_ds):
2      for epoch in range(epochs):
3          start = time.time()
4
5          display.clear_output(wait=True)
6
7          for example_input, example_target in test_ds.take(1):
8              generate_images(generator, example_input, example_target)
9          print("Epoch: ", epoch)
10
11         # Train
12         for n, (input_image, target) in train_ds.enumerate():
13             print('.', end='')
14             if (n+1) % 100 == 0:
15                 print()
16             train_step(input_image, target, epoch)
17         print()
18
19         # saving (checkpoint) the model every 20 epochs
20         if (epoch + 1) % 20 == 0:
21             checkpoint.save(file_prefix=checkpoint_prefix)
22
23         print ('Time taken for epoch {} is {} sec\n'.format(epoch + 1, time.time()-start))
24
25     checkpoint.save(file_prefix=checkpoint_prefix)
```

18. 訓練模型：可先啟動 Tensorboard 監看工作記錄，在 Notebook 中啟動如下。

```
1  # 啟動 Tensorboard
2  %load_ext tensorboard
3  %tensorboard --logdir {log_dir}
```

```
1  # 訓練模型
2  fit(train_dataset, EPOCHS, test_dataset)
```

19. 取 5 筆資料測試。

```
1  # 取 5 筆資料測試
2  for inp, tar in test_dataset.take(5):
3      generate_images(generator, inp, tar)
```

執行結果：以下僅只截圖 2 筆資料，經過訓練後，預測的圖像已沒有樹木或欄杆了。

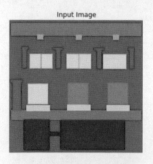

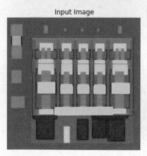

10-7 CycleGAN

前面 GAN 演算法處理的都是成對轉換資料，CycleGAN 則是針對非成對的
資料生成圖像。成對的意思是一張原始圖像對應一張目標圖像，下圖右方
表示多對多的資料，也就是給予不同的場域(Domain)，原始圖像就可以合
成指定場景的圖像。

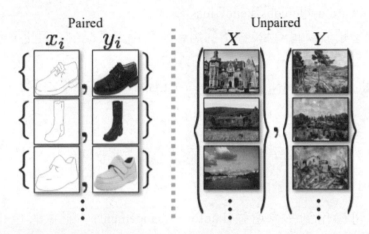

圖 10.20 成對的資料(左方) vs. 非成對的資料(右方)，以下圖片來源均來自
「Unpaired Image-to-Image Translation using Cycle-Consistent Adversarial
Networks」[18]

CycleGAN 或稱 Cycle-Consistent GAN，是 Jun-Yan Zhu 等學者於 2017 年
發表的一篇文章「Unpaired Image-to-Image Translation using Cycle-
Consistent Adversarial Networks」[18]中提出的演算法，概念如下圖：

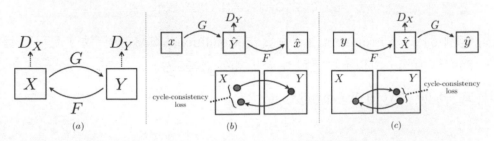

圖 10.21 CycleGAN 網路結構

- 上圖(a)：有兩個生成網路，G 將圖像由 X 場域(Domain)生成 Y 場域的圖像，F 網路則是相反功能，由 Y 場域(Domain)生成 X 場域的圖像。
- 上圖(b)：引進 cycle consistency losses 概念，可以做到 x → G(x) → F(G(x)) ≈ x，即 x 經過 G、F 轉換，可得到近似 x 的圖像，稱為 Forward cycle-consistency loss。
- 上圖(c)：從另一場域 y 開始，也可以做到 y → F(y) → G(F(y)) ≈ y，稱為 Backward cycle-consistency loss。
- 整個模型類似兩個 GAN 網路的組合，具備循環機制，因此，損失函數如下：

L(G, F, DX, DY) = LGAN(G, D$_Y$, X, Y) + LGAN(F, D$_X$, Y , X) +λLcyc(G, F)

其中：

$$\mathcal{L}_{cyc}(G, F) = \mathbb{E}_{x \sim p_{data}(x)}[\|F(G(x)) - x\|_1] + \mathbb{E}_{y \sim p_{data}(y)}[\|G(F(y)) - y\|_1]$$

λ 控制 G、F 損失函數的相對重要性。

這種機制可應用到影像增強(Photo Enhancement)、影像彩色化(Image Colorization)、風格轉換(Style Transfer)等功能，如下圖，幫一般的馬匹塗上斑馬紋。

圖 10.22 CycleGAN 的功能展示

【範例 6】以 horse2zebra 資料集實作 Conditional GAN 演算法。流程與上一節大致相同，所以不再複製/貼上，節省篇幅。

程式 10_06_ CycleGAN.ipynb。

1. 載入相關套件。

```
1   # 載入相關套件
2   import tensorflow as tf
3   import tensorflow_datasets as tfds
4   from tensorflow_examples.models.pix2pix import pix2pix
5
6   import os
7   import time
8   import matplotlib.pyplot as plt
9   from IPython.display import clear_output
10
11  # 類似 prefetch()，可以提升資料集存取效能
12  AUTOTUNE = tf.data.AUTOTUNE
```

2. 載入訓練資料：tfds.load 會載入內建資料集，相關資料可參考 TensorFlow 官網有關 CycleGAN 的描述[19]。

```
1   # 載入訓練資料
2   dataset, metadata = tfds.load('cycle_gan/horse2zebra',
3                                  with_info=True, as_supervised=True)
4
5   train_horses, train_zebras = dataset['trainA'], dataset['trainB']
6   test_horses, test_zebras = dataset['testA'], dataset['testB']
```

3. 定義圖像處理的函數：與上一節相同。

```
1   # 參數設定
2   BUFFER_SIZE = 1000 # 緩衝區大小
3   BATCH_SIZE = 1       # 批量
4   IMG_WIDTH = 256      # 圖像寬度
5   IMG_HEIGHT = 256     # 圖像高度
6
7   # 隨機裁切圖像
8   def random_crop(image):
9       cropped_image = tf.image.random_crop(
10              image, size=[IMG_HEIGHT, IMG_WIDTH, 3])
11
12      return cropped_image
13
14  # 標準化，使像素值介於 [-1, 1]
15  def normalize(image):
16      image = tf.cast(image, tf.float32)
17      image = (image / 127.5) - 1
18      return image
19
20  # 隨機轉換
21  def random_jitter(image):
22      # 縮放圖像為 286 x 286 x 3
23      image = tf.image.resize(image, [286, 286],
24              method=tf.image.ResizeMethod.NEAREST_NEIGHBOR)
25
```

```
26      # 隨機裁切至 256 x 256 x 3
27      image = random_crop(image)
28
29      # 水平翻轉
30      image = tf.image.random_flip_left_right(image)
31      return image
```

4. 定義資料前置處理的函數，設定資料集屬性。

```
1  # 訓練資料前置處理
2  def preprocess_image_train(image, label):
3      image = random_jitter(image)
4      image = normalize(image)
5      return image
6
7  # 測試資料前置處理
8  def preprocess_image_test(image, label):
9      image = normalize(image)
10     return image
11
12 # 設定資料集屬性
13 train_horses = train_horses.map(
14         preprocess_image_train, num_parallel_calls=AUTOTUNE).cache().shuffle(
15         BUFFER_SIZE).batch(1)
16
17 train_zebras = train_zebras.map(
18         preprocess_image_train, num_parallel_calls=AUTOTUNE).cache().shuffle(
19         BUFFER_SIZE).batch(1)
20
21 test_horses = test_horses.map(
22         preprocess_image_test, num_parallel_calls=AUTOTUNE).cache().shuffle(
23         BUFFER_SIZE).batch(1)
24
25 test_zebras = test_zebras.map(
26         preprocess_image_test, num_parallel_calls=AUTOTUNE).cache().shuffle(
27         BUFFER_SIZE).batch(1)
```

5. 資料測試。

```
1  # 各取一筆資料測試
2  sample_horse = next(iter(train_horses))
3  sample_zebra = next(iter(train_zebras))
4
5  plt.subplot(121)
6  plt.title('Horse')
7  # 轉換為 [0, 1]
8  plt.imshow(sample_horse[0] * 0.5 + 0.5)
9
10 plt.subplot(122)
11 plt.title('Horse with random jitter')
12 # 將像素值由 [-1, 1] 轉換為 [0, 1]，才能顯示
13 plt.imshow(random_jitter(sample_horse[0]) * 0.5 + 0.5)
```

執行結果：左為原圖，右為隨機轉換的圖。

6. 定義 CycleGAN 神經網路：借用上一節的 Pix2Pix 網路結構。

```
1   # 定義生成神經網路
2   OUTPUT_CHANNELS = 3
3
4   # 以 Pix2Pix 的模型建立 CycleGAN
5   generator_g = pix2pix.unet_generator(OUTPUT_CHANNELS, norm_type='instancenorm')
6   generator_f = pix2pix.unet_generator(OUTPUT_CHANNELS, norm_type='instancenorm')
7
8   discriminator_x = pix2pix.discriminator(norm_type='instancenorm', target=False)
9   discriminator_y = pix2pix.discriminator(norm_type='instancenorm', target=False)
10
11  # 生成圖像
12  to_zebra = generator_g(sample_horse)
13  to_horse = generator_f(sample_zebra)
14  plt.figure(figsize=(8, 8))
15  contrast = 8
16
17  # 顯示圖片
18  imgs = [sample_horse, to_zebra, sample_zebra, to_horse]
19  title = ['Horse', 'To Zebra', 'Zebra', 'To Horse']
20
21  for i in range(len(imgs)):
22      plt.subplot(2, 2, i+1)
23      plt.title(title[i])
24      if i % 2 == 0:
25          plt.imshow(imgs[i][0] * 0.5 + 0.5)
26      else:
27          plt.imshow(imgs[i][0] * 0.5 * contrast + 0.5)
28  plt.show()
```

執行結果：左為原圖，右為未訓練前的結果。

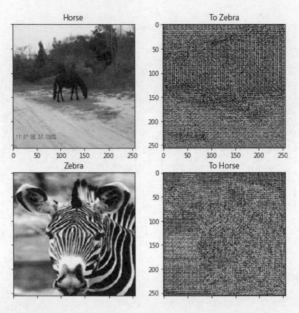

7. 判別網路圖像測試。

```
1  # 判別網路圖像測試
2  plt.figure(figsize=(8, 8))
3
4  plt.subplot(121)
5  plt.title('Is a real zebra?')
6  # 使用判別網路辨識圖像
7  plt.imshow(discriminator_y(sample_zebra)[0, ..., -1], cmap='RdBu_r')
8
9  plt.subplot(122)
10 plt.title('Is a real horse?')
11 # 使用判別網路辨識圖像
12 plt.imshow(discriminator_x(sample_horse)[0, ..., -1], cmap='RdBu_r')
13
14 plt.show()
```

執行結果：未訓練前的結果，左圖為 X 判別網路的結果，右為 Y 判別網路的結果。

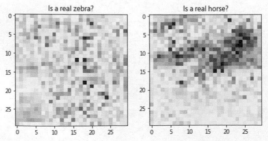

8. 定義損失函數為二分類交叉熵：包括判別網路、生成網路、循環損失
與 Identity 損失函數。Identity 損失函數為輸入 Y 與 Y 生成網路的差
異，X 亦同，正常的話差異應為 0。

```
1  # 定義損失函數為二分類交叉熵
2  LAMBDA = 10
3  loss_obj = tf.keras.losses.BinaryCrossentropy(from_logits=True)
4
5  # 定義判別網路損失函數
6  def discriminator_loss(real, generated):
7      # 真實資料的損失
8      real_loss = loss_obj(tf.ones_like(real), real)
9
10     # 生成資料的損失
11     generated_loss = loss_obj(tf.zeros_like(generated), generated)
12
13     # 總損失
14     total_disc_loss = real_loss + generated_loss
15
16     return total_disc_loss * 0.5
17
18 # 定義生成網路損失函數
19 def generator_loss(generated):
20     return loss_obj(tf.ones_like(generated), generated)
21
22 # 定義循環損失函數，參見圖 10.21 CycleGAN 網路結構
23 def calc_cycle_loss(real_image, cycled_image):
24     loss1 = tf.reduce_mean(tf.abs(real_image - cycled_image))
25     return LAMBDA * loss1
26
27 # 定義 Identity 損失函數， |G(Y)-Y| + |f(X)-X|
28 def identity_loss(real_image, same_image):
29     loss = tf.reduce_mean(tf.abs(real_image - same_image))
30     return LAMBDA * 0.5 * loss
```

9. 定義優化器：採用 Adam。

```
1  # 定義優化器
2  generator_g_optimizer = tf.keras.optimizers.Adam(2e-4, beta_1=0.5)
3  generator_f_optimizer = tf.keras.optimizers.Adam(2e-4, beta_1=0.5)
4
5  discriminator_x_optimizer = tf.keras.optimizers.Adam(2e-4, beta_1=0.5)
6  discriminator_y_optimizer = tf.keras.optimizers.Adam(2e-4, beta_1=0.5)
```

10. 定義檢查點：將以上網路函數放入。

```
1  # 定義檢查點
2  checkpoint_path = "./CycleGAN_checkpoints/train"
3
4  ckpt = tf.train.Checkpoint(generator_g=generator_g,
5                             generator_f=generator_f,
```

```
6                              discriminator_x=discriminator_x,
7                              discriminator_y=discriminator_y,
8                              generator_g_optimizer=generator_g_optimizer,
9                              generator_f_optimizer=generator_f_optimizer,
10                             discriminator_x_optimizer=discriminator_x_optimizer,
11                             discriminator_y_optimizer=discriminator_y_optimizer)
12
13  ckpt_manager = tf.train.CheckpointManager(ckpt, checkpoint_path, max_to_keep=5)
14
15  # 如果檢查點存在，回復至最後一個檢查點
16  if ckpt_manager.latest_checkpoint:
17      ckpt.restore(ckpt_manager.latest_checkpoint)
18      print ('Latest checkpoint restored!!')
```

11. 定義生成圖像與顯示的函數。

```
1   # 定義生成圖像及顯示的函數
2   def generate_images(model, test_input):
3       prediction = model(test_input)
4
5       plt.figure(figsize=(12, 12))
6
7       display_list = [test_input[0], prediction[0]]
8       title = ['Input Image', 'Predicted Image']
9
10      for i in range(2):
11          plt.subplot(1, 2, i+1)
12          plt.title(title[i])
13          # getting the pixel values between [0, 1] to plot it.
14          plt.imshow(display_list[i] * 0.5 + 0.5)
15          plt.axis('off')
16      plt.show()
```

12. 定義訓練模型的函數。

```
1   # 參數設定
2   EPOCHS = 40
3
4   # 定義訓練模型的函數
5   @tf.function
6   def train_step(real_x, real_y):
7       # persistent=True : 表示 tf.GradientTape 會重複使用
8       with tf.GradientTape(persistent=True) as tape:
9           # Generator G translates X -> Y
10          # Generator F translates Y -> X
11          fake_y = generator_g(real_x, training=True)
12          cycled_x = generator_f(fake_y, training=True)
13
14          fake_x = generator_f(real_y, training=True)
15          cycled_y = generator_g(fake_x, training=True)
16
17          # same_x and same_y are used for identity loss.
18          same_x = generator_f(real_x, training=True)
19          same_y = generator_g(real_y, training=True)
```

```
20
21          disc_real_x = discriminator_x(real_x, training=True)
22          disc_real_y = discriminator_y(real_y, training=True)
23
24          disc_fake_x = discriminator_x(fake_x, training=True)
25          disc_fake_y = discriminator_y(fake_y, training=True)
26
27          # 計算生成網路損失
28          gen_g_loss = generator_loss(disc_fake_y)
29          gen_f_loss = generator_loss(disc_fake_x)
30
```

```
31          # 計算循環損失
32          total_cycle_loss = calc_cycle_loss(real_x, cycled_x) + calc_cycle_loss(real_y, cycled_y)
33
34          # 計算總損失 Total generator loss = adversarial loss + cycle loss
35          total_gen_g_loss = gen_g_loss + total_cycle_loss + identity_loss(real_y, same_y)
36          total_gen_f_loss = gen_f_loss + total_cycle_loss + identity_loss(real_x, same_x)
37
38          disc_x_loss = discriminator_loss(disc_real_x, disc_fake_x)
39          disc_y_loss = discriminator_loss(disc_real_y, disc_fake_y)
40
41      # 計算生成網路梯度
42      generator_g_gradients = tape.gradient(total_gen_g_loss,
43                                              generator_g.trainable_variables)
44      generator_f_gradients = tape.gradient(total_gen_f_loss,
45                                              generator_f.trainable_variables)
46
47      # 計算判別網路梯度
48      discriminator_x_gradients = tape.gradient(disc_x_loss,
49                                                  discriminator_x.trainable_variables)
50      discriminator_y_gradients = tape.gradient(disc_y_loss,
51                                                  discriminator_y.trainable_variables)
52
53      # 更新權重
54      generator_g_optimizer.apply_gradients(zip(generator_g_gradients,
55                                              generator_g.trainable_variables))
56
57      generator_f_optimizer.apply_gradients(zip(generator_f_gradients,
58                                              generator_f.trainable_variables))
59
60      discriminator_x_optimizer.apply_gradients(zip(discriminator_x_gradients,
61                                                  discriminator_x.trainable_variables))
```

```
62
63      discriminator_y_optimizer.apply_gradients(zip(discriminator_y_gradients,
64                                                  discriminator_y.trainable_variables))
```

13. 訓練模型。

```
1   # 訓練模型
2   for epoch in range(EPOCHS):
3       start = time.time()
4
5       n = 0
6       for image_x, image_y in tf.data.Dataset.zip((train_horses, train_zebras)):
7           train_step(image_x, image_y)
8           if n % 10 == 0:
9               print ('.', end='')
10          n += 1
11
```

```
12      clear_output(wait=True)
13      # 產生圖像
14      generate_images(generator_g, sample_horse)
15
16      # 檢查點存檔
17      if (epoch + 1) % 5 == 0:
18          ckpt_save_path = ckpt_manager.save()
19          print ('Saving checkpoint for epoch {} at {}'.format(epoch+1, ckpt_save_path))
20
21      # 計時
22      print ('Time taken for epoch {} is {} sec\n'.format(epoch + 1, time.time()-start))
```

執行結果：訓練中的結果如下，有逐步的轉變，第一排右圖為第 9 週期的結果，第二排右圖為第 10 週期的結果，圖像已有明顯改善，第三排右圖為第 12 週期的結果，已集中在馬匹的處理上。

Time taken for epoch 9 is 1200.320835351944 sec

Saving checkpoint for epoch 10 at ./CycleGAN_checkpoints/train\ckpt-2
Time taken for epoch 10 is 1213.892056465149 sec

Time taken for epoch 12 is 1216.7267162799835 sec

訓練的最終結果：效果很好，馬匹已加上了斑馬紋。

每個訓練週期均執行約 1200 秒，即 20 分鐘，全部 40 個週期，筆者大概執行了 15 個小時。

14. 取 5 筆資料測試。

```
1   # 取 5 筆資料測試
2   for inp in test_horses.take(5):
3       generate_images(generator_g, inp)
```

執行結果：以下僅只截圖 2 筆資料，左側為原圖，右側為預測的圖像，效果比訓練樣本差，應該是因為訓練的執行週期不足，原文作者執行了 200 個週期，如果真的照做，這次就不用睡覺了，乾脆去日本玩個三天兩夜，再回來看結果。

10-8 GAN 挑戰

這一章我們認識了許多種不同的 GAN 演算法，由於大部分是由同一組學者發表的，因此可以看到演化的脈絡。原生 GAN 加上條件後，變成 Conditional GAN，再將生成網路改成對稱型的 U-Net 後，就變成 Pix2Pix GAN，接著再設定兩個 Pix2Pix 循環的網路，就衍生出 CycleGAN。除此之外，許多的演算法也會修改損失函數的定義，來產生各種意想不到的效果。本書介紹的演算法只是滄海一粟，更多的內容可參考李宏毅老師的 PPT「Introduction of Generative Adversarial Network (GAN)」[20]。

另一方面，GAN 不光是應用在圖像上，還可以結合自然語言處理(NLP)、強化學習(RL)等技術，擴大應用範圍，像是高解析圖像生成、虛擬人物的生成、資料壓縮、文字轉語音(Text To Speech, TTS)、醫療、天文、物理、遊戲等，可以參閱「Tutorial on Deep Generative Models」[21]一文。

縱使 GAN 應用廣泛，但仍然存在一些挑戰：

1. 生成的圖像模糊：因為神經網路是根據訓練資料求取迴歸，類似求取每個樣本在不同範圍的平均值，所以生成的圖像會是相似點的平均，導致圖像模糊。必須有非常大量的訓練資料，加上相當多的訓練週期，才能產生畫質較佳的圖像。另外，GAN 對超參數特別敏感，包括學習率、濾波器(Filter)尺寸，初始值設定得不好，造成生成的資料過差時，判別網路就會都判定為偽，到最後生成網路只能一直產生少數類別的資料了。

2. 梯度消失(Vanishing Gradient)：當生成的資料過差時，判別網路判定為真的機率接近 0，梯度會變得非常小，因此就無法提供良好的梯度來改善生成器，造成生成器梯度消失。發生這種情形時可以多使用 leaky ReLU activation function、簡化判別網路結構、或增加訓練週期加以改善。

3. 模式崩潰(Mode Collapse)：是指生成器生成的內容過於雷同，缺少變化。如果訓練資料的類別不只一種，生成網路則會為了讓判別網路辨識的準確率提高，而專注在比較擅長的類別，導致生成的類別缺乏多樣性。以製造偽鈔來舉例，假設鈔票分別有 100 元、500 元與 1000 元，若偽鈔製造者比較善於製作 500 元紙鈔，模型可能就會全部都製作 500 元的偽鈔。

4. 執行訓練時間過久：這是最大的問題了吧，反覆實驗的時候，假如沒有相當的硬體支援，每次調整個參數都要折磨好幾天，再多的耐心也會消磨殆盡。

10-9 深度偽造(Deepfake)

深度偽造(Deepfake)是目前很夯的技術，也是一個 AI 危害人類社會的明顯例子。BuzzFeed.com 在 2018 年放上一段影片，名叫「You Won't Believe What Obama Says In This Video!」(https://www.youtube.com/watch?v=

cQ54GDm1eL0)，影片中歐巴馬總統的演說全是偽造的，嘴型和聲音都十分逼真，震驚世人，自此以後，各界瘋狂製作各種深度偽造影片，使得網路上的影片真假難辨，造成非常嚴重的假新聞災難。

深度偽造大部分是在影片中換臉，由於人在説話時頭部會自然轉動，有各種角度的特寫，因此，必須要收集特定人 360 度的臉部圖像，才能讓演算法成功置換。從網路媒體中收集名人的各種影像是最容易的方式，所以，網路上流傳最多的大部分是偽造名人的影片，如政治人物、明星等。

深度偽造的技術基礎來自 GAN，類似於前面介紹的 CycleGAN，架構如下圖，也能結合臉部辨識的功能，在抓到臉部特徵點(Landmark)後，就可以進行原始臉部與要置換臉部的互換。

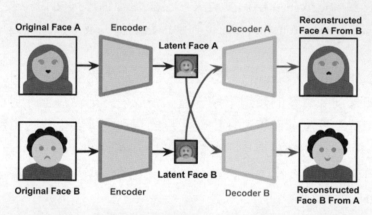

圖 10.23 深度偽造的架構示意圖

圖片來源：「Understanding the Technology Behind DeepFakes」[22]

Aayush Bansal 等學者在 2018 年發表「Recycle-GAN: Unsupervised Video Retargeting」[23]，RecycleGAN 是擴充 CycleGAN 的演算法，它的損失函數額外加上了時間同步的相關性(Temporal Coherence)，如下：

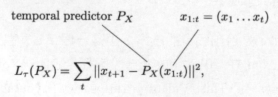

$$L_\tau(P_X) = \sum_t \|x_{t+1} - P_X(x_{1:t})\|^2,$$

類似時間序列(Time Series)，t+1 時間點的圖像應該是 1 至 t 時間點的圖像的延續。因此，生成網路的損失函數如下：

$$L_r(G_X, G_Y, P_Y) = \sum_t \|x_{t+1} - G_X(P_Y(G_Y(x_{1:t})))\|^2$$

其中：$G_y(x_i)$ 是將 x_i 轉成 y_i 的生成網路

圖 10.24 影片是連續的變化，因此 t+1 時間點的圖像應該是 1 至 t 時間點的圖像的延續，圖片來源：「Recycle-GAN: Unsupervised Video Retargeting」[21]

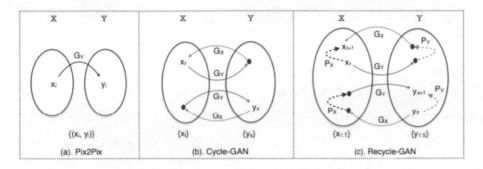

圖 10.25 RecycleGAN 演算法的演進

圖片來源：「Recycle-GAN: Unsupervised Video Retargeting」[21]

(a) Pix2Pix 是成對(Paired data)轉換。

(b) CycleGAN 是循環轉換，使用成對的網路架構。

(c) RecycleGAN 加上時間同步的相關性(Px、Py)。

因此整體的 RecycleGAN 的損失函數，包括以下部份：

$$
\min_{G,P} \max_{D} L_{rg}(G, P, D) = \overbrace{L_g(G_X, D_X) + L_g(G_Y, D_Y)}^{\text{GAN objective}} +
$$

$$
\underbrace{\lambda_{rx} L_r(G_X, G_Y, P_Y) + \lambda_{ry} L_r(G_Y, G_X, P_X)}_{\text{Recycle Loss}} + \underbrace{\lambda_{\tau x} L_\tau(P_X) + \lambda_{\tau y} L_\tau(P_Y)}_{\text{Recurrent loss}}
$$

另外，還有 Face2Face、嘴型同步技術(Lip-syncing technology)等演算法，有興趣的讀者可以參閱 Jonathan Hui 的「Detect AI-generated Images & Deepfakes」[24]系列文章，裡面有大量的圖片展示，十分有趣。

Deepfake 的實作可參閱「DeepFakes in 5 minutes」[25]一文，它介紹如何利用 DeepFaceLab 套件，在很短的時間內製作出深度偽造的影片，原始程式碼在「DeepFaceLab GitHub」(https://github.com/iperov/DeepFaceLab)，網頁附有一個影片「Mini tutorial」説明，只要按步驟執行腳本(Scripts)，就可以順利完成影片，不過，它比 GAN 需要更強的硬體設備，筆者就不敢測試了。

Deepfake 造成了嚴重的假新聞災難，許多學者及企業提出反制的方法來辨識真假，簡單的像是「Detect AI-generated Images & Deepfakes (Part 1)」[26]一文所述，可以從臉部邊緣是否模糊、是否有隨機的雜訊、臉部是否對稱等細節來辨別，當然也有大公司推出可辨識影片真假的工具，例如微軟的「Microsoft Video Authenticator」可參閱 ITHome 相關的報導[27]。

不管是 Deepfake 還是反制的演算法，未來發展都值得關注，這起事件也讓科學家留意到科學的發展必須兼顧倫理與道德，否則，好萊塢科幻片的劇情就不再只是幻想，人類有可能走向自我毀滅的道路。

參考資料 (References)

[1] 自由時報,《全球首次！AI 創作肖像畫 10 月佳士得拍賣》, 2018
(https://news.ltn.com.tw/news/world/breakingnews/2529174)

[2] 佳士得官網《Is artificial intelligence set to become art's next medium?》
(https://www.christies.com/features/A-collaboration-between-two-artists-one-human-one-a-machine-9332-1.aspx)

[3] 佳士得官網關於 Edmond de Belamy 肖像畫的介紹
(https://www.christies.com/lot/lot-edmond-de-belamy-from-la-famille-de-6166184)

[4] the-gan-zoo GitHub
(https://github.com/hindupuravinash/the-gan-zoo)

[5] Liqian Ma、Xu Jia、Qianru Sun 等人,《Pose Guided Person Image Generation》, 2018
(https://arxiv.org/pdf/1705.09368.pdf)

[6] Yanghua Jin、Jiakai Zhang 等人,《Towards the Automatic Anime Characters Creation with Generative Adversarial Networks》, 2017
(https://arxiv.org/pdf/1708.05509.pdf)

[7] Yunjey Choi、Minje Choi、Munyoung Kim 等人,《StarGAN: Unified Generative Adversarial Networks for Multi-Domain Image-to-Image Translation》, 2017
(https://arxiv.org/abs/1711.09020)

[8] Christian Ledig、Lucas Theis、Ferenc Huszár 等人,《Photo-Realistic Single Image Super-Resolution Using a Generative Adversarial Network》, 2017
(https://arxiv.org/pdf/1609.04802.pdf)

[9] Tero Karras、Samuli Laine、Miika Aittala 等人,《Analyzing and Improving the Image Quality of StyleGAN》, 2020

(https://arxiv.org/pdf/1912.04958.pdf)

[10] Jonathan Hui,《GAN — Some cool applications of GAN》, 2018

(https://jonathan-hui.medium.com/gan-some-cool-applications-of-gans-4c9ecca35900)

[11] Keras 官網範例「DCGAN to generate face images」

(https://keras.io/examples/generative/dcgan_overriding_train_step/)

[12] Tero Karras、Timo Aila、Samuli Laine 等人,《Progressive Growing of GANs for Improved Quality, Stability, and Variation》, 2017

(https://arxiv.org/abs/1710.10196)

[13] Mehdi Mirza、Simon Osindero,《Conditional Generative Adversarial Nets》, 2014

(https://arxiv.org/abs/1411.1784)

[14] Qiwen Fu、Wei-Ting Hsu、Mu-Heng Yang,《Colorization Using ConvNet and GAN》, 2017

(http://cs231n.stanford.edu/reports/2017/pdfs/302.pdf)

[15] ColorGAN GitHub

(https://github.com/bbc/ColorGAN#end-to-end-conditional-gan-based-architectures-for-image-colourisation)

[16] Phillip Isola、Jun-Yan Zhu、Tinghui Zhou,《 Image-to-Image Translation with Conditional Adversarial Networks》, 2016

(https://arxiv.org/abs/1611.07004)

[17] CMP Facade Database

(https://cmp.felk.cvut.cz/~tylecr1/facade/)

[18] Jun-Yan Zhu、Taesung Park、 Phillip Isola、Alexei A. Efros,《Unpaired Image-to-Image Translation using Cycle-Consistent Adversarial Networks》, 2017

(https://arxiv.org/abs/1703.10593)

[19] TensorFlow 官網有關 CycleGAN 的説明
(https://www.tensorflow.org/datasets/catalog/cycle_gan)

[20] 李宏毅老師的 PPT「Introduction of Generative Adversarial Network (GAN)」
(https://speech.ee.ntu.edu.tw/~tlkagk/slide/Tutorial_HYLee_GAN.pdf)

[21] Shakir Mohamed、Danilo Rezende,《Tutorial on Deep Generative Models》, 2017
(http://www.shakirm.com/slides/DeepGenModelsTutorial.pdf)

[22] Alan Zucconi,《Understanding the Technology Behind DeepFakes》, 2018
(https://www.alanzucconi.com/2018/03/14/understanding-the-technology-behind-deepfakes/)

[23] Aayush Bansal、Shugao Ma、Deva Ramanan、Yaser Sheikh, 《Recycle-GAN: Unsupervised Video Retargeting》, 2018
(https://arxiv.org/abs/1808.05174)

[24] Jonathan Hui,《Detect AI-generated Images & Deepfakes》, 2020
(https://jonathan-hui.medium.com/detect-ai-generated-images-deepfakes-part-1-b518ed5075f4)

[25] Louis (What's AI) Bouchard,《DeepFakes in 5 minutes》, 2020
(https://pub.towardsai.net/deepfakes-in-5-minutes-155c13d48fa3)

[26] Jonathan Hui,《Detect AI-generated Images & Deepfakes (Part 1)》, 2020 (https://jonathan-hui.medium.com/detect-ai-generated-images-deepfakes-part-1-b518ed5075f4)

[27] 林妍溱,《微軟開發能判別 Deepfake 影像及內容變造的技術》, 2020
(https://www.ithome.com.tw/news/139740)

│第四篇│自然語言處理

自然語言處理(Natural Language Processing, NLP)顧名思義，就是希望電腦能像人類一樣，看懂文字或聽懂人話，理解語意，並能給予適當的回答，以聊天機器人為例：

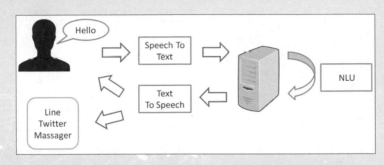

圖 11.1 聊天機器人概念示意圖

一個簡單的電腦與人類的對話，所涵蓋的技術就包括：

1. 當人對電腦說話，電腦會先把那句話轉成文字，稱之為「語音識別」(Speech recognition)或「語音轉文字」(Speech To Text, STT)。

2. 接著電腦對文字進行解析，瞭解意圖，稱為「自然語言理解」(Natural Language Understanding, NLU)。

3. 之後電腦依據對話回答，有兩種表達方式：

- 以文字回覆：從語料庫或常用問答(FAQ)中找出一段要回覆的文字，這部分稱為「文字生成」(Text Generation)。

- 以聲音回覆：將回覆文字轉為語音，稱為「語音合成」(Speech Synthesize)或「文字轉語音」(Text To Speech, TTS)。

整個過程看似容易，實則充滿了各種挑戰，接下來我們把相關技術仔細演練一遍吧！

自然語言處理的介紹

自然語言處理的發展非常早，大約從 1950 年就開始了，當年英國電腦科學家圖靈(Alan Mathison Turing) 已有先見之明，提出「圖靈測試」（Turing Test），目的是在測試電腦能否表現出像人類一樣的智慧，時至今日，許多聊天機器人如 Siri、小冰等產品的問世，才算得上真正啟動了 NLP 的熱潮，即便目前依然無法媲美人類的智慧，但相關技術仍然有許多方面的應用：

- 文本分類（Text classification）。
- 信息檢索（Information retrieval）。
- 文字校對（Text proofing）。
- 自然語言生成（Natural language generation）。
- 問答系統（Question answering）。
- 機器翻譯（Machine translation）。
- 自動摘要（Automatic summarization）。
- 情緒分析（Sentiment analysis）。
- 語音識別（Speech recognition）。
- 音樂方面的應用，比如曲風分類、自動編曲，聲音模仿等等。

11-1　詞袋(BOW)與 TF-IDF

人類的語言具高度曖昧性，一句話可能有多重的意思或隱喻，而電腦當前還無法真正理解語言或文字的意義，因此，現階段的作法與影像的處理方式類似，先將語音和文字轉換成向量，再對向量進行分析或使用深度學習建模，相關研究的進展非常快，這一節我們從最簡單的方法開始說起。

詞袋(Bag of Words, BOW)是把一篇文章進行詞彙的整理，然後統計每個詞彙出現次數，經由前幾名的詞彙猜測全文大意。

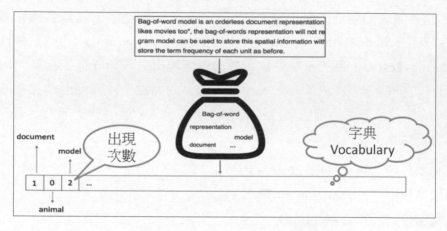

圖 11.2　詞袋(Bag of Words, BOW)

作法如下：

1. 分詞(Tokenization)：將整篇文章中的每個詞彙切開，整理成生字表或字典(Vocabulary)。英文較單純，以空白或句點隔開，中文較複雜，須以特殊方式處理。
2. 前置處理(Preprocessing)：將詞彙作詞形還原、轉換成小寫等。詞形還原是動詞轉為原形，複數轉為單數等，避免因為詞態不同，詞彙統計出現分歧。

3. 去除停用詞(Stop Word)： be 動詞、助動詞、代名詞、介系詞、冠詞等不具特殊意義的詞彙稱為停用詞(Stop Word)，將他們剔除，否則統計結果都是這些詞彙出現最多次。

4. 詞彙出現次數統計：計算每個詞彙在文章出現的次數，並由高至低排列。

【範例 1】以 BOW 實作自動摘要。

程式 11_01_BOW.ipynb。

1. 載入相關套件。

```
1  # 載入相關套件
2  import collections
```

2. 設定停用詞：這裡直接設定停用詞，許多套件有整理常用的停用詞，例如 NLTK、spaCy。

```
1  # 停用詞設定
2  stop_words=['\n', 'or', 'are', 'they', 'i', 'some', 'by', '-',
3             'even', 'the', 'to', 'a', 'and', 'of', 'in', 'on', 'for',
4             'that', 'with', 'is', 'as', 'could', 'its', 'this', 'other',
5             'an', 'have', 'more', 'at','don't', 'can', 'only', 'most']
```

3. 讀取文字檔 news.txt，統計詞彙出現的次數，資料來自「South Korea's Convenience Store Culture」[1]一文。

```
1  # 讀取文字檔 news.txt，統計字詞出現次數
2
3  # 參數設定
4  maxlen=1000          # 生字表最大個數
5
6  # 生字表的集合
7  word_freqs = collections.Counter()
8  with open('./NLP_data/news.txt','r+', encoding='UTF-8') as f:
9      for line in f:
10         # 轉小寫、分詞
11         words = line.lower().split(' ')
12         # 統計字詞出現次數
13         if len(words) > maxlen:
14             maxlen = len(words)
15         for word in words:
16             if not (word in stop_words):
17                 word_freqs[word] += 1
18
19 print(word_freqs.most_common(20))
```

執行結果：

```
[('stores', 15), ('convenience', 14), ('korean', 6), ('these', 6), ('one', 6), ('it's', 6), ('from', 5), ('my', 5), ('you', 5),
('their', 5), ('just', 5), ('has', 5), ('new', 4), ('do', 4), ('also', 4), ('which', 4), ('find', 4), ('would', 4), ('like',
4), ('up', 4)]
```

前 3 名分別為：

- stores：15 次。
- convenience：14 次。
- korean：6 次。

因此可以猜測這整篇文章應該是在討論「韓國便利商店」(Korea Convenience Store)，結果與標題契合。

BOW 方法十分簡單，效果也相當不錯，不過它有個缺點，有些詞彙不是停用詞，也經常出現，但對全文並不重要，譬如上文的 only、most，對猜測全文大意沒有幫助，所以，學者提出改良的演算法 TF-IDF(Term Frequency - Inverse Document Frequency)，它會針對跨文件常出現的詞彙給予較低的分數，例如 only 在每一個文件都出現的話，TF-IDF 對他的評分就相對較低，因此，TF-IDF 的公式定義如下：

$$tf\text{-}idf = tf \times idf$$

其中：

1. tf (詞頻, Term Frequency)：考慮詞彙出現在跨文件的次數，分母為在所有文件中出現的次數，分子為在目前文件中出現的次數。

$$tf_{i,j} = \frac{n_{i,j}}{\sum_k n_{k,j}}$$

2. idf (逆向檔案頻率, Inverse Document Frequency)：考慮詞彙出現的文件數，單一文件出現特定詞彙多次，也只視為 1，分子為總文件數，分母為詞彙出現的文件數，加 1 是避免分母為 0。

$$idf_{i,j} = \log \frac{|D|}{1 + |D_{r_i}|}$$

3. 除了以上的定義，TF-IDF 還有一些變形的公式，可參閱維基百科關於 tf-idf 的說明[2]。

除了猜測全文大意外，TF-IDF 也可以應用到文字分類(Text Classification) 或問答的配對。

【範例 2】以 TF-IDF 實作問答配對。
程式 11_02_TFIDF.ipynb 實作。

1. 載入相關套件。

```
1  # 載入相關套件
2  from sklearn.feature_extraction.text import CountVectorizer
3  from sklearn.feature_extraction.text import TfidfTransformer
4  import numpy as np
```

2. 設定輸入資料：最後一句為問題，其他的例句為回答。

```
1  # 語料：最後一句為問題，其他為回答
2  corpus = [
3      'This is the first document.',
4      'This is the second second document.',
5      'And the third one.',
6      'Is this the first document?',
7  ]
```

3. 將例句轉換為詞頻矩陣，計算各個詞彙出現的次數。

```
1  # 將語料轉換為詞頻矩陣，計算各個字詞出現的次數。
2  vectorizer = CountVectorizer()
3  X = vectorizer.fit_transform(corpus)
4
5  # 生字表
6  word = vectorizer.get_feature_names()
7  print ("Vocabulary：", word)
```

執行結果：

```
Vocabulary： ['and', 'document', 'first', 'is', 'one', 'second', 'the', 'third', 'this']
```

4. 查看四句話的 BOW。

```
1  # 查看四句話的 BOW
2  print ("BOW=\n", X.toarray())
```

執行結果：

```
BOW=
 [[0 1 1 1 0 0 1 0 1]
 [0 1 0 1 0 2 1 0 1]
 [1 0 0 0 1 0 1 1 0]
 [0 1 1 1 0 0 1 0 1]]
```

5. TF-IDF 轉換：將例句轉換為 TF-IDF。

```
1  # TF-IDF 轉換
2  transformer = TfidfTransformer()
3  tfidf = transformer.fit_transform(X)
4  print ("TF-IDF=\n", np.around(tfidf.toarray(), 4))
```

執行結果：每一個元素均介於[0, 1]，為了顯示整齊，取四捨五入，實際運算並不需要。

```
TF-IDF=
 [[0.     0.4388 0.542  0.4388 0.     0.     0.3587 0.     0.4388]
 [0.     0.2723 0.     0.2723 0.     0.8532 0.2226 0.     0.2723]
 [0.5528 0.     0.     0.     0.5528 0.     0.2885 0.5528 0.    ]
 [0.     0.4388 0.542  0.4388 0.     0.     0.3587 0.     0.4388]]
```

6. 比較最後一句與其他例句的相似度：以 cosine_similarity 比較向量的夾角，愈接近 1，表愈相似。

```
1  # 最後一句與其他句的相似度比較
2  from sklearn.metrics.pairwise import cosine_similarity
3  print (cosine_similarity(tfidf[-1], tfidf[:-1], dense_output=False))
```

執行結果：第一個例句與最後的問句最相似，結果與文意相符合。

```
  (0, 2)        0.1034849000930086
  (0, 1)        0.43830038447620107
  (0, 0)        1.0
```

11-2 詞彙前置處理

傳統上，我們會使用 NLTK(Natural Language Toolkit)套件來進行詞彙的前置處理，它具備非常多的功能，並內含超過 50 個語料庫(Corpora)可供測試，只可惜它沒有支援中文處理，比較新的 spaCy 套件有支援多國語系。

這裡先示範如何運用 NLTK 做一般詞彙的前置處理，之後再介紹可以處理中文資料的套件。

NLTK 分為程式和資料兩個部份。

1. 安裝 NLTK 程式：

```
pip install nltk
```

2. 安裝 NLTK 資料：
   先執行 python，再執行 import nltk; nltk.download()，出現畫面如下，包括套件與相關語料庫，可下載必要的項目。

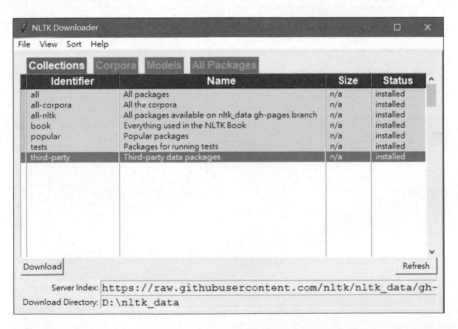

由於檔案眾多，下載時間很久，如需安裝至第二台 PC，可直接複製下載目錄至其他 PC 即可，NLTK 載入語料庫時，會自動檢查所有硬碟機的\nltk_data 目錄。

【範例 1】使用 NLTK 進行詞彙的前置處理。
程式 11_03_詞彙前置處理.ipynb。

1. 載入相關套件。

```
1  # 載入相關套件
2  import nltk
```

2. 輸入測試的文章段落如下：

```
1  # 測試文章段落
2  text="Today is a great day. It is even better than yesterday." + \
3       " And yesterday was the best day ever."
```

3. 將測試的文章段落分割成例句。

```
1  # 分割字句
2  nltk.sent_tokenize(text)
```

執行結果：分割成三句。

```
['Today is a great day.',
 'It is even better than yesterday.',
 'And yesterday was the best day ever.']
```

4. 分詞(Tokenize)。

```
1  # 分詞
2  nltk.word_tokenize(text)
```

執行結果：

```
['Today',
 'is',
 'a',
 'great',
 'day',
 '.',
 'It',
 'is',
 'even',
 'better',
 'than',
 'yesterday',
 '.',
 'And',
 'yesterday',
 'was',
 'the',
 'best',
 'day',
 'ever',
 '.']
```

5. 詞形還原有兩類方法：

■ 依字根作詞形還原(Stemming)：速度快，但不一定正確。

■ 依字典規則作詞形還原(Lemmatization)：速度慢，但準確率高。

6. 字根詞形還原(Stemming)：根據一般文法規則，不管字的涵義，直接進行字根詞形還原，比如 keeps 刪去 s，crashing 刪去 ing，這都正確，但 his 直接刪去 s，就會發生錯誤。

```
1  # 字根詞形還原(Stemming)
2  text = 'My system keeps crashing his crashed yesterday, ours crashes daily'
3  ps = nltk.porter.PorterStemmer()
4  ' '.join([ps.stem(word) for word in text.split()])
```

執行結果： his ➜ hi，daily ➜ daili。

```
'My system keep crash hi crash yesterday, our crash daili
```

7. 依字典規則的詞形還原(Lemmatization)：查詢字典，依單字的不同進行詞形還原，如此 his、daily 均不會改變。

```
1  # 依字典規則的詞形還原(Lemmatization)
2  text = 'My system keeps crashing his crashed yesterday, ours crashes daily'
3  lem = nltk.WordNetLemmatizer()
4  ' '.join([lem.lemmatize(word) for word in text.split()])
```

執行結果：完全正確。

```
'My system keep crashing his crashed yesterday, ours crash daily'
```

8. 分詞後剔除停用詞(Stopwords)：nltk.corpus.stopwords.words('english') 提供常用的停用詞，另外標點符號也可以列入停用詞。

```
1   # 標點符號(Punctuation)
2   import string
3   print('標點符號:', string.punctuation)
4
5   # 測試文章段落
6   text="Today is a great day. It is even better than yesterday." + \
7       " And yesterday was the best day ever."
8   # 讀取停用詞
9   stopword_list = set(nltk.corpus.stopwords.words('english')
10                      + list(string.punctuation))
11
12  # 移除停用詞(Removing Stopwords)
```

```
13  def remove_stopwords(text, is_lower_case=False):
14      if is_lower_case:
15          text = text.lower()
16      tokens = nltk.word_tokenize(text)
17      tokens = [token.strip() for token in tokens]
18      filtered_tokens = [token for token in tokens if token not in stopword_list]
19      filtered_text = ' '.join(filtered_tokens)
20      return filtered_text, filtered_tokens
21
22  filtered_text, filtered_tokens = remove_stopwords(text)
23  filtered_text
```

執行結果：

```
標點符號: !"#$%&'()*+,-./:;<=>?@[\]^_`{|}~

'Today great day It even better yesterday And yesterday best day ever'
```

9. 進行 BOW 統計。

```
1   # 測試文章段落
2   with open('./NLP_data/news.txt','r+', encoding='UTF-8') as f:
3       text = f.read()
4
5   filtered_text, filtered_tokens = remove_stopwords(text, True)
6
7   import collections
8   # 生字表的集合
9   word_freqs = collections.Counter()
10  for word in filtered_tokens:
11      word_freqs[word] += 1
12  print(word_freqs.most_common(20))
```

執行結果：同樣可以抓到文章大意是韓國便利超商。

```
[('`', 35), ('stores', 15), ('convenience', 14), ('one', 8), ('–', 8), ('even', 8), ('seoul', 8), ('city', 7), ('korea', 6),
('korean', 6), ('cities', 6), ('people', 5), ('summer', 4), ('new', 4), ('also', 4), ('find', 4), ('store', 4), ('would', 4),
('like', 4), ('average', 4)]
```

10. 改用正規表達式(Regular Expression)：上段程式還是有標點符號未剔除，正規表達式可完全剔除停用詞。

```
1   # 移除停用詞(Removing Stopwords)
2   lem = nltk.WordNetLemmatizer()
3   def remove_stopwords_regex(text, is_lower_case=False):
4       if is_lower_case:
5           text = text.lower()
6       tokenizer = nltk.tokenize.RegexpTokenizer(r'\w+') # 篩選文數字(Alphanumeric)
7       tokens = tokenizer.tokenize(text)
8       tokens = [lem.lemmatize(token.strip()) for token in tokens] # 詞形還原
9       filtered_tokens = [token for token in tokens if token not in stopword_list]
10      filtered_text = ' '.join(filtered_tokens)
```

```
11      return filtered_text, filtered_tokens
12
13  filtered_text, filtered_tokens = remove_stopwords_regex(text, True)
14  word_freqs = collections.Counter()
15  for word in filtered_tokens:
16      word_freqs[word] += 1
17  print(word_freqs.most_common(20))
```

11. 找出相似詞(Synonyms)：WordNet 語料庫內含相似詞、相反詞與簡短說明。

```
1  # 找出相似詞(Synonyms)
2  synonyms = nltk.corpus.wordnet.synsets('love')
3  synonyms
```

執行結果：列出前 10 名，是以例句顯示，故許多單字均相同。

```
[Synset('love.n.01'),
 Synset('love.n.02'),
 Synset('beloved.n.01'),
 Synset('love.n.04'),
 Synset('love.n.05'),
 Synset('sexual_love.n.02'),
 Synset('love.v.01'),
 Synset('love.v.02'),
 Synset('love.v.03'),
 Synset('sleep_together.v.01')]
```

12. 顯示相似詞說明。

```
1  # 單字說明
2  synonyms[0].definition()
```

執行結果：列出第一個相似詞的單字說明。

```
'a strong positive emotion of regard and affection'
```

13. 顯示相似詞的例句。

```
1  # 單字的例句
2  synonyms[0].examples()
```

執行結果：列出第一個相似詞的例句。

```
['his love for his work', 'children need a lot of love']
```

14. 找出相反詞(Antonyms)：須先呼叫 lemmas 進行詞形還原，再呼叫 antonyms。

```
1  # 找出相反詞(Antonyms)
2  antonyms=[]
3  for syn in nltk.corpus.wordnet.synsets('ugly'):
4      for l in syn.lemmas():
5          if l.antonyms():
6              antonyms.append(l.antonyms()[0].name())
7  antonyms
```

執行結果：ugly ➜ beautiful。

15. 分析詞性標籤(POS Tagging)：依照句子結構，顯示每個單字的詞性。

```
1  # 找出詞性標籤(POS Tagging)
2  text='I am a human being, capable of doing terrible things'
3  sentences=nltk.sent_tokenize(text)
4  for sent in sentences:
5      print(nltk.pos_tag(nltk.word_tokenize(sent)))
```

執行結果：

```
[('I', 'PRP'), ('am', 'VBP'), ('a', 'DT'), ('human', 'JJ'), ('being', 'VBG'), (',', ','), ('capable', 'JJ'), ('of', 'IN'), ('do
ing', 'VBG'), ('terrible', 'JJ'), ('things', 'NNS')]
```

詞性標籤(POS Tagging)列表如下：

- CC (Coordinating Conjunction)：並列連詞。
- CD (Cardinal Digit)：基數。
- DT (Determiner)：量詞。
- EX (Existential)：存在地，例如 There。
- FW (Foreign Word)：外來語。
- IN Preposition/Subordinating Conjunction.：介詞。
- JJ Adjective：形容詞。
- JJR Adjective, Comparative：比較級形容詞。
- JJS Adjective, Superlative：最高級形容詞。
- LS (List Marker) 1：清單標記。
- MD (Modal)：情態動詞。
- NN Noun, Singular：名詞單數。

- NNS Noun Plural：名詞複數。
- NNP Proper Noun, Singular：專有名詞單數。
- NNPS Proper Noun, Plural：專有名詞複數。
- PDT (Predeterminer)：放在量詞的前面，例如 both、a lot of。
- POS (Possessive Ending)：所有格，例如 parent's。
- PRP (Personal Pronoun)：代名詞，例如 I, he, she。
- PRP$ Possessive Pronoun：所有格代名詞，例如 my, his, her。
- RB Adverb：副詞，例如 very, silently。
- RBR Adverb, Comparative：比較級副詞，例如 better。
- RBS Adverb, Superlative：最高級副詞，例如 best。
- RP Particle：助詞，例如 give up。
- TO to：例如 go 'to' the store。
- UH Interjection：感嘆詞，例如 errrrrrrm。
- VB Verb, Base Form：動詞，例如 take。
- VBD Verb, Past Tense：動詞過去式，例如 took。
- VBG Verb, Gerund/Present Participle：動詞進行式，例如 taking。
- VBN Verb, Past Participle：動詞過去分詞，例如 taken。
- VBP Verb, Sing Present, non-3d：動詞現在式非第三人稱單數，例如 take。
- VBZ Verb, 3rd person sing. present：動詞現在式第三人稱單數，例如 takes。
- WDT wh-determiner：疑問代名詞，例如 which。
- WP wh-pronoun who, what：疑問代名詞。
- WP$ possessive wh-pronoun：疑問代名詞所有格，例如 whose。
- WRB wh-abverb：疑問副詞，例如 where, when。

spaCy 套件提供更強大的分析功能，但由於內容涉及詞向量(Word2Vec)，所以我們留待後續章節再討論。

11-3 詞向量(Word2Vec)

BOW 和 TF-IDF 都只著重於詞彙出現在文件中的次數，未考慮語言/文字有上下文的關聯，比如，「這間房屋有四扇？」，從上文大概可以推測出最後一個詞彙是「窗戶」，又譬如，我說喜歡吃辣，那我會點「麻婆豆腐」還是「家常豆腐」呢？相信聽到「吃辣」，應該都會猜是「麻婆豆腐」。另一方面，一個語系的單字數有限，中文大概就幾萬個字，我們是否也可以比照影像辨識，對所有的單字建構預先訓練的模型，之後是否就可以實現轉換學習(Transfer Learning)？

針對上下文的關聯，Google 研發團隊 Tomas Mikolov 等人於 2013 年提出「詞向量」(Word2Vec)，他們蒐集 1000 億個字(Word)加以訓練，將每個單字改以上下文表達，然後轉換為向量，而這就是「詞嵌入」(Word Embedding)的概念，與 TF-IDF 輸出是稀疏向量不同，詞嵌入的輸出是一個稠密的樣本空間。

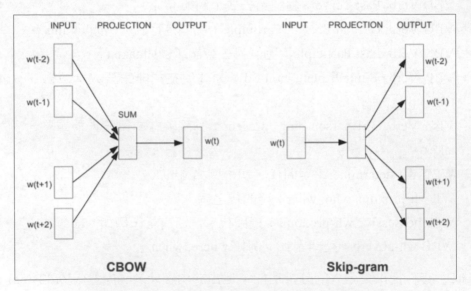

圖 11.3 CBOW 與 Continuous Skip-gram Model

圖片來源：Exploiting Similarities among Languages for Machine Translation [3]

詞向量有兩種作法：

1. 連續 BOW(Continuous Bag-of-Words, CBOW)：以單字的上下文預測單字。

2. Continuous Skip-gram Model：剛好相反，以單字預測上下文。

揭開 CBOW 演算法來看，它就是一個深度學習模型，如下圖：

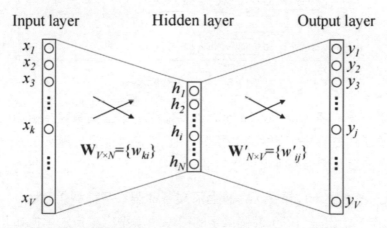

圖 11.3 CBOW 的網路結構，圖片來源：An Intuitive Understanding of Word Embeddings: From Count Vectors to Word2Vec [4]

以單字的上下文為輸入，以預測的單字為目標，如同下面 2-gram 的模型，例句為 " Hey, this is sample corpus using only one context word."，使用 One-hot encoding，輸出表格如下：

Input	Output		Hey	This	Is	sample	corpus	using	only	one	context	word
Hey	this	Datapoint 1	1	0	0	0	0	0	0	0	0	0
this	hey	Datapoint 2	0	1	0	0	0	0	0	0	0	0
is	this	Datapoint 3	0	0	1	0	0	0	0	0	0	0
is	sample	Datapoint 4	0	0	1	0	0	0	0	0	0	0
sample	is	Datapoint 5	0	0	0	1	0	0	0	0	0	0
sample	corpus	Datapoint 6	0	0	0	1	0	0	0	0	0	0
corpus	sample	Datapoint 7	0	0	0	0	1	0	0	0	0	0
corpus	using	Datapoint 8	0	0	0	0	1	0	0	0	0	0
using	corpus	Datapoint 9	0	0	0	0	0	1	0	0	0	0
using	only	Datapoint 10	0	0	0	0	0	1	0	0	0	0
only	using	Datapoint 11	0	0	0	0	0	0	1	0	0	0
only	one	Datapoint 12	0	0	0	0	0	0	1	0	0	0
one	only	Datapoint 13	0	0	0	0	0	0	0	1	0	0
one	context	Datapoint 14	0	0	0	0	0	0	0	1	0	0
context	one	Datapoint 15	0	0	0	0	0	0	0	0	1	0
context	word	Datapoint 16	0	0	0	0	0	0	0	0	1	0
word	context	Datapoint 17	0	0	0	0	0	0	0	0	0	1

圖 11.4 2-gram 與 One-hot encoding

2-gram 是每次取兩個單字,然後滑動視窗一個單字,輸出如下:

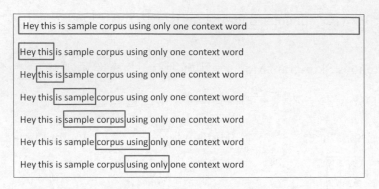

圖 11.5 2-gram 滑動視窗

接著以第一個單字 One-hot encoding 為輸入,第二個單字為預測目標,最後模型預測的是各單字的機率。這是一個簡略的說明,當然,實際的模型不會這麼簡單,還會額外考慮以下狀況:

1. 不會只考慮上一個單字,會將上下文各 n 個單字都納入考量。
2. 如此作,輸出是 1000 億個單字的機率,模型應該無法承擔如此多的類別,因此改用所謂的「負樣本抽樣」(Negative Sub-sampling),只推論輸出入是否為上下文,例如 orange, juice,從 P(juice|orange)改為預測 P(1|<orange, juice>),亦即從多類別(1000 億個)模型轉換成二分類(真/假)模型。

CBOW 的優點如下:

1. 簡單,而且比傳統確定性模型(Deterministic methods)的效能佳。
2. 對比相關矩陣,CBOW 對記憶體消耗節省很多。

CBOW 的缺點如下:

1. 像是 Apple 可能是指水果,但也可能是在指公司名稱,遇到這樣的情況,CBOW 會取平均值,造成失準,故 CBOW 無法處理一字多義。
2. CBOW 因為輸出高達 1000 億個單字機率,所以優化求解的收斂十分困難。

因此，後續發展出了 Skip-gram 模型，顛倒輸出與輸入，改由單字預測上下文，我們再以同樣的句子來舉例：

Input	Output(Context1)	Output(Context2)
Hey	this	\<padding\>
this	Hey	is
is	this	sample
sample	is	corpus
corpus	sample	corpus
using	corpus	only
only	using	one
one	only	context
context	one	word
word	context	\<padding\>

圖 11.6 Skip-gram 模型的輸出與輸入

Skip-gram 的優點如下：

1. 一個單字可以預測多個上下文，解決了一字多義的問題。
2. 結合「負樣本抽樣」(Negative Sub-sampling)技術，效能比其他模型佳。負樣本可以是任意單字的排列組合，如果要把所有負樣本放入訓練資料中，數量可能過於龐大，而且會造成不平衡資料(Imbalanced Data)，因此採用負樣本抽樣的方法。因我們不實作 Skip-gram 模型訓練，細節就不介紹，有興趣的讀者可參閱「NLP 102: Negative Sampling and GloVe」[5]。

我們可以利用預先訓練的模型來實驗一下，Gensim 和 spaCy 套件均提供 Word2Vec 模型。

先安裝 Gensim 套件：

pip install genism

【範例 1】運用 Gensim 進行相似性比較。
程式 11_04_gensim_相似性比較.ipynb。

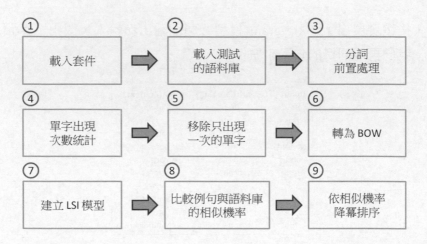

1. 載入相關套件。

```
1  # 載入相關套件
2  import pprint   # 較美觀的列印函數
3  import gensim
4  from collections import defaultdict
5  from gensim import corpora
```

2. 測試的文章段落如下。

```
1   # 語料庫
2   documents = [
3       "Human machine interface for lab abc computer applications",
4       "A survey of user opinion of computer system response time",
5       "The EPS user interface management system",
6       "System and human system engineering testing of EPS",
7       "Relation of user perceived response time to error measurement",
8       "The generation of random binary unordered trees",
9       "The intersection graph of paths in trees",
10      "Graph minors IV Widths of trees and well quasi ordering",
11      "Graph minors A survey",
12  ]
```

3. 分詞、前置處理。

```
1   # 任意設定一些停用詞
2   stoplist = set('for a of the and to in'.split())
3
4   # 分詞，轉小寫
5   texts = [
6       [word for word in document.lower().split() if word not in stoplist]
7       for document in documents
8   ]
9   texts
```

執行結果：

```
[['human', 'machine', 'interface', 'lab', 'abc', 'computer', 'applications'],
 ['survey', 'user', 'opinion', 'computer', 'system', 'response', 'time'],
 ['eps', 'user', 'interface', 'management', 'system'],
 ['system', 'human', 'system', 'engineering', 'testing', 'eps'],
 ['relation', 'user', 'perceived', 'response', 'time', 'error', 'measurement'],
 ['generation', 'random', 'binary', 'unordered', 'trees'],
 ['intersection', 'graph', 'paths', 'trees'],
 ['graph', 'minors', 'iv', 'widths', 'trees', 'well', 'quasi', 'ordering'],
 ['graph', 'minors', 'survey']]
```

4. 單字出現的次數統計。

```
1  # 單字出現次數統計
2  frequency = defaultdict(int)
3  for text in texts:
4      for token in text:
5          frequency[token] += 1
6  frequency
```

執行結果：顯示每個單字出現的次數。

```
defaultdict(int,
            {'human': 2,
             'machine': 1,
             'interface': 2,
             'lab': 1,
             'abc': 1,
             'computer': 2,
             'applications': 1,
             'survey': 2,
             'user': 3,
             'opinion': 1,
             'system': 4,
             'response': 2,
             'time': 2,
             'eps': 2,
             'management': 1,
             'engineering': 1,
             'testing': 1,
             'relation': 1,
             'perceived': 1,
             'error': 1,
             'measurement': 1,
             'generation': 1,
             'random': 1,
```

5. 移除只出現一次的單字：僅專注在較常出現的關鍵字。

```
1  # 移除只出現一次的單字
2  texts = [
3      [token for token in text if frequency[token] > 1]
4      for text in texts
5  ]
6  texts
```

執行結果：每句篩選的結果。

```
[['human', 'interface', 'computer'],
 ['survey', 'user', 'computer', 'system', 'response', 'time'],
 ['eps', 'user', 'interface', 'system'],
 ['system', 'human', 'system', 'eps'],
 ['user', 'response', 'time'],
 ['trees'],
 ['graph', 'trees'],
 ['graph', 'minors', 'trees'],
 ['graph', 'minors', 'survey']]
```

6. 轉為 BOW。

```
1  # 轉為字典
2  dictionary = corpora.Dictionary(texts)
3
4  # 轉為 BOW
5  corpus = [dictionary.doc2bow(text) for text in texts]
6  corpus
```

執行結果：

```
[[(0, 1), (1, 1), (2, 1)],
 [(0, 1), (3, 1), (4, 1), (5, 1), (6, 1), (7, 1)],
 [(2, 1), (5, 1), (7, 1), (8, 1)],
 [(1, 1), (5, 2), (8, 1)],
 [(3, 1), (6, 1), (7, 1)],
 [(9, 1)],
 [(9, 1), (10, 1)],
 [(9, 1), (10, 1), (11, 1)],
 [(4, 1), (10, 1), (11, 1)]]
```

7. 建立 LSI (Latent semantic indexing)模型：可指定議題的個數，每一項議題皆由所有單字加權組合而成。

```
1   # 建立 LSI (Latent semantic indexing) 模型
2   from gensim import models
3
4   # num_topics=2：取二維，即兩個議題
5   lsi = models.LsiModel(corpus, id2word=dictionary, num_topics=2)
6
7   # 兩個議題的 LSI 公式
8   lsi.print_topics(2)
```

執行結果：兩項議題的公式。

```
[(0,
  '0.644*"system" + 0.404*"user" + 0.301*"eps" + 0.265*"time" + 0.265*"response" + 0.240*"computer" + 0.221*"human" + 0.206*"su
rvey" + 0.198*"interface" + 0.036*"graph"'),
 (1,
  '0.623*"graph" + 0.490*"trees" + 0.451*"minors" + 0.274*"survey" + -0.167*"system" + -0.141*"eps" + -0.113*"human" + 0.107*"r
esponse" + 0.107*"time" + -0.072*"interface"')]
```

8. 測試 LSI (Latent semantic indexing)模型。

```
1   # 例句
2   doc = "Human computer interaction"
3
4   # 測試 LSI (Latent semantic indexing) 模型
5   vec_bow = dictionary.doc2bow(doc.lower().split())
6   vec_lsi = lsi[vec_bow]
7   print(vec_lsi)
```

執行結果：將例句帶入到兩項議題公式中，計算 LSI 值，結果比較接近第一項議題。

```
[(0, 0.4618210045327157), (1, -0.07002766527900067)]
```

9. 比較例句與文章段落內每一個句子的相似機率。

```
1    # 比較例句與語料庫每一句的相似機率
2    from gensim import similarities
3
4    # 比較例句與語料庫的相似性索引
5    index = similarities.MatrixSimilarity(lsi[corpus])
6
7    # 比較例句與語料庫的相似機率
8    sims = index[vec_lsi]
9
10   # 顯示語料庫的索引值及相似機率
11   print(list(enumerate(sims)))
```

執行結果：將例句帶入到兩項議題公式中，計算 LSI 值。

```
[(0, 0.998093), (1, 0.93748635), (2, 0.9984453), (3, 0.98658866), (4, 0.90755945), (5, -0.12416792), (6, -0.1063926), (7, -0.09
879464), (8, 0.05004177)]
```

10. 按照機率進行降冪排序。

```
1  # 依相似機率降冪排序
2  sims = sorted(enumerate(sims), key=lambda item: -item[1])
3  for doc_position, doc_score in sims:
4      print(doc_score, documents[doc_position])
```

執行結果：前兩句機率最大，依語意判斷結果正確無誤。

```
0.9984453 The EPS user interface management system
0.998093 Human machine interface for lab abc computer applications
0.98658866 System and human system engineering testing of EPS
0.93748635 A survey of user opinion of computer system response time
0.90755945 Relation of user perceived response time to error measurement
0.05004177 Graph minors A survey
-0.09879464 Graph minors IV Widths of trees and well quasi ordering
-0.1063926 The intersection graph of paths in trees
-0.12416792 The generation of random binary unordered trees
```

Gensim 不僅提供 Word2Vec 預先訓練的模型，也支援自訂資料訓練的功能，預先訓練的模型可提供一般內容的推論，但如果內容是屬於特殊領域，則應該自行訓練模型會比較恰當，Gensim Word2Vec 的用法請參考「Gensim 官網 Word2Vec 說明文件」[6]。

【範例 2】運用 Gensim 進行 Word2Vec 訓練與測試。

程式 11_05_gensim_Word2Vec.ipynb。

1. 載入相關套件。

```
1  # 載入相關套件
2  import gzip
3  import gensim
```

2. 以 Gensim 進行簡單測試：把 Gensim 內建的語料庫 common_texts 作為訓練資料，並且對"hello", "world", "michael"三個單字進行訓練，產生詞向量。

```
1  from gensim.test.utils import common_texts
2  # size：詞向量的大小，window：考慮上下文各自的長度
3  # min_count：單字至少出現的次數，workers：執行緒個數
4  model_simple = gensim.models.Word2Vec(sentences=common_texts, window=1,
5                                   min_count=1, workers=4)
6  # 傳回 有效的字數及總處理字數
7  model_simple.train([["hello", "world", "michael"]], total_examples=1, epochs=2)
```

執行結果：傳回兩個值(0, 6)，包括所有執行週期的有效字數與總處理字數，其中前者為內部處理的邏輯，不太理解，後者數字為 6=3 個單字 ＊ 2 個執行週期。

train()的參數有很多，可參閱上面所提的「Gensim 官網 Word2Vec 說明文件」，這裡僅摘錄此範例所用到的參數。

- sentences：訓練資料。
- size：產生的詞向量大小。
- window：考慮上下文各自的長度。
- min_count：單字至少出現的次數。
- workers：執行緒的個數。

3. 另一個例子。

```
1   sentences = [["cat", "say", "meow"], ["dog", "say", "woof"]]
2
3   model_simple = gensim.models.Word2Vec(min_count=1)
4   model_simple.build_vocab(sentences)   # 建立生字表(vocabulary)
5   model_simple.train(sentences, total_examples=model_simple.corpus_count
6                      , epochs=model_simple.epochs)
```

執行結果：傳回(1, 30)，其中 30=6 個單字 ＊ 5 個執行週期。

4. 實例測試：載入 OpinRank 語料庫，文章內容是關於車輛與旅館的評論。

```
1   # 載入 OpinRank 語料庫：關於車輛與旅館的評論
2   data_file="./Word2Vec/reviews_data.txt.gz"
3
4   with gzip.open (data_file, 'rb') as f:
5       for i,line in enumerate (f):
6           print(line)
7           break
```

執行結果：

```
b"Oct 12 2009 \tNice trendy hotel location not too bad.\tI stayed in this hotel for one night. As this is a fairly new place so
me of the taxi drivers did not know where it was and/or did not want to drive there. Once I have eventually arrived at the hote
l, I was very pleasantly surprised with the decor of the lobby/ground floor area. It was very stylish and modern. I found the r
eception's staff geeting me with 'Aloha' a bit out of place, but I guess they are briefed to say that to keep up the coroporate
image.As I have a Starwood Preferred Guest member, I was given a small gift upon-check in. It was only a couple of fridge magne
ts in a gift box, but nevertheless a nice gesture.My room was nice and roomy, there are tea and coffee facilities in each room
and you get two complimentary bottles of water plus some toiletries by 'bliss'.The location is not great. It is at the last met
ro stop and you then need to take a taxi, but if you are not planning on going to see the historic sites in Beijing, then you w
ill be ok.I chose to have some breakfast in the hotel, which was really tasty and there was a good selection of dishes. There a
re a couple of computers to use in the communal area, as well as a pool table. There is also a small swimming pool and a gym ar
ea.I would definitely stay in this hotel again, but only if I did not plan to travel to central Beijing, as it can take a long
time. The location is ok if you plan to do a lot of shopping, as there is a big shopping centre just few minutes away from the
hotel and there are plenty of eating options around, including restaurants that serve a dog meat!\t\r\n"
```

5. 讀取 OpinRank 語料庫，並進行前置處理，如分詞。

```
1  # 讀取 OpinRank 語料庫，並作前置處理
2  def read_input(input_file):
3      with gzip.open (input_file, 'rb') as f:
4          for i, line in enumerate (f):
5              # 前置處理
6              yield gensim.utils.simple_preprocess(line)
7
8  # 載入 OpinRank 語料庫，分詞
9  documents = list(read_input(data_file))
10 documents
```

執行結果： 為一個 List。

```
[['oct',
  'nice',
  'trendy',
  'hotel',
  'location',
  'not',
  'too',
  'bad',
  'stayed',
  'in',
  'this',
  'hotel',
  'for',
  'one',
  'night',
  'as',
  'this',
```

6. Word2Vec 模型訓練：約需 10 分鐘。

```
1  # Word2Vec 模型訓練，約10分鐘
2  model = gensim.models.Word2Vec(documents, size=150, window=10,
3                                 min_count=2, workers=10)
4  model.train(documents,total_examples=len(documents),epochs=10)
```

執行結果： (303,484,226, 415,193,580)，處理數億個單字。

接下來進行各種測試。

7. 測試「dirty」的相似詞。

```
1  # 測試『骯髒』相似詞
2  w1 = "dirty"
3  model.wv.most_similar(positive=w1) # positive：相似詞
```

執行結果： 顯示 10 個最相似的單字。

```
[('filthy', 0.8602699041366577),
 ('stained', 0.7798251509666443),
 ('dusty', 0.7683317065238953),
 ('unclean', 0.7638086676597595),
 ('grubby', 0.757234513759613),
 ('smelly', 0.7431163787841797),
 ('dingy', 0.7304496169090271),
 ('disgusting', 0.7111263275146484),
 ('soiled', 0.7099645733833313),
 ('mouldy', 0.706375241279602)]
```

8. 測試「france」的相似詞：topn 可指定列出前 n 名。

```
1  # 測試『法國』相似詞
2  w1 = ["france"]
3  model.wv.most_similar (positive=w1, topn=6) # topn：只列出前 n 名
```

執行結果： 顯示 6 個最相似的單字。

```
[('germany', 0.6627413034439087),
 ('canada', 0.6545147895812988),
 ('spain', 0.644172728061676),
 ('england', 0.6122641563415527),
 ('mexico', 0.6106705665588379),
 ('rome', 0.6044377684593201)]
```

9. 同時測試多個詞彙：「床、床單、枕頭」的相似詞與「長椅」的相反詞。

```
1  # 測試『床、床單、枕頭』相似詞及『長椅』相反詞
2  w1 = ["bed",'sheet','pillow']
3  w2 = ['couch']
4  model.wv.most_similar (positive=w1, negative=w2, topn=10) # negative：相反詞
```

執行結果： 顯示 10 個最適合的單字。

```
[('duvet', 0.7157680988311768),
 ('blanket', 0.7036269903182983),
 ('mattress', 0.7003698348999023),
 ('quilt', 0.7003640532493591),
 ('matress', 0.6967926621437073),
 ('pillowcase', 0.665346086025238),
 ('sheets', 0.6376352310180664),
 ('pillows', 0.6317484378814697),
 ('comforter', 0.6119856834411621),
 ('foam', 0.6095048785209656)]
```

10. 比較兩個詞彙的相似機率。

```
1  # 比較兩詞相似機率
2  model.wv.similarity(w1="dirty",w2="smelly")
```

執行結果: 相似機率為 0.7431163。

11. 挑選出較不相似的詞彙。

```
1  # 選出較不相似的字詞
2  model.wv.doesnt_match(["cat","dog","france"])
```

執行結果: france。

12. 接著測試載入預先訓練模型,有兩種方式:程式直接下載或者手動下載後再讀取檔案。

程式直接下載。

```
1  # 下載預先訓練的模型
2  import gensim.downloader as api
3  wv = api.load('word2vec-google-news-300')
```

手動下載後載入,預先訓練模型的下載網址為 https://drive.google.com /file/d/0B7XkCwpI5KDYNlNUTTlSS21pQmM/edit。

```
1  # 載入本機的預先訓練模型
2  from gensim.models import KeyedVectors
3
4  # 每個詞向量有 300 個元素
5  model = KeyedVectors.load_word2vec_format(
6      './Word2Vec/GoogleNews-vectors-negative300.bin', binary=True)
```

接下來進行各種測試。

13. 取得 dog 的詞向量。

```
1  # 取得 dog 的詞向量(300個元素)
2  model['dog']
```

執行結果：共有 300 個元素。

```
array([ 5.12695312e-02, -2.23388672e-02, -1.72851562e-01,  1.61132812e-01,
       -8.44726562e-02,  5.73730469e-02,  5.85937500e-02, -8.25195312e-02,
       -1.53808594e-02, -6.34765625e-02,  1.79687500e-01, -4.23828125e-01,
       -2.25830078e-02, -1.66015625e-01, -2.51464844e-02,  1.07421875e-01,
       -1.99218750e-01,  1.59179688e-01, -1.87500000e-01, -1.20117188e-01,
        1.55273438e-01, -9.91210938e-02,  1.42578125e-01, -1.64062500e-01,
       -8.93554688e-02,  2.00195312e-01, -1.49414062e-01,  3.20312500e-01,
        3.28125000e-01,  2.44140625e-02, -9.71679688e-02, -8.20312500e-02,
       -3.63769531e-02, -8.59375000e-02, -9.86328125e-02,  7.78198242e-03,
       -1.34277344e-02,  5.27343750e-02,  1.48437500e-01,  3.33984375e-01,
```

14. 測試「woman, king」的相似詞和「man」的相反詞。

```
1  # 測試『woman, king』相似詞及『man』相反詞
2  model.most_similar(positive=['woman', 'king'], negative=['man'])
```

執行結果：這就是有名的 king - man + woman = queen。

```
[('queen', 0.7118192911148071),
 ('monarch', 0.6189674139022827),
 ('princess', 0.5902431011199951),
 ('crown_prince', 0.5499460697174072),
 ('prince', 0.5377321243286133),
 ('kings', 0.5236844420433044),
 ('Queen_Consort', 0.5235945582389832),
 ('queens', 0.518113374710083),
 ('sultan', 0.5098593235015869),
 ('monarchy', 0.5087411999702454)]
```

15. 挑選出較不相似的詞彙。

```
1  # 選出較不相似的字詞
2  model.doesnt_match("breakfast cereal dinner lunch".split())
```

執行結果：cereal(麥片)與三餐較不相似。

16. 比較兩詞相似機率。

```
1  # 比較兩詞相似機率
2  model.similarity('woman', 'man')
```

執行結果：機率為 0.76640123，'woman', 'man'是相似的。

由上面測試可以知道,對於一般的文字判斷,使用預先訓練的模型都相當準確,但是,如果要判斷特殊領域的相關內容,效果可能就會打折,舉例來說,Kaggle 上有一個很有趣的資料集「辛普生對話」(Dialogue Lines of The Simpsons),是有關辛普生家庭的卡通劇情問答,像是詢問劇中人物 Bart 與 Nelson 的相似度,結果只有 0.5,因為在卡通裡面他們雖然是朋友,但不是很親近,假如使用預先訓練的模型,來推論問題的話,答案應該就不會如此精確,除此之外,還有很多其他的例子,讀者有空不妨測試看看此範例程式「Gensim Word2Vec Tutorial」(https://www.kaggle.com/pierremegret/gensim-word2vec-tutorial)。

另外 TensorFlow 官網也有提供一個範例,能夠直接使用 TensorFlow 實現 Skip-gram 模型,網址為 https://www.tensorflow.org/tutorials/text/word2vec。

之前都是比較單字的相似度,然而更常見的需求是對「語句」(Sentence)的比對,譬如常見問答集(FAQ)或是對話機器人,系統會先比對問題的相似度,再將答案回覆給使用者,Gensim 支援 Doc2Vec 演算法,可進行語句相似度比較,程式碼如下:

1. 筆者從 Starbucks 官網抓了一段 FAQ 的標題當作測試語料庫。

```
1  import numpy as np
2  import nltk
3  import gensim
4  from gensim.models import Word2Vec
5  from gensim.models.doc2vec import Doc2Vec, TaggedDocument
6  from sklearn.metrics.pairwise import cosine_similarity
7
8  # 測試語料
9  f = open('./FAQ/starbucks_faq.txt', 'r', encoding='utf8')
10 corpus = f.readlines()
11 # print(corpus)
12
13 # 參數設定
14 MAX_WORDS_A_LINE = 30   # 每行最多字數
15
16 # 標點符號(Punctuation)
17 import string
18 print('標點符號:', string.punctuation)
19
20 # 讀取停用詞
21 stopword_list = set(nltk.corpus.stopwords.words('english')
22                     + list(string.punctuation) + ['\n'])
```

2. 訓練 Doc2Vec 模型。

```
1   # 分詞函數
2   def tokenize(text, stopwords, max_len = MAX_WORDS_A_LINE):
3       return [token for token in gensim.utils.simple_preprocess(text
4                           , max_len=max_len) if token not in stopwords]
5
6   # 分詞
7   document_tokens=[] # 整理後的字詞
8   for line in corpus:
9       document_tokens.append(tokenize(line, stopword_list))
10
11  # 設定為 Gensim 標籤文件格式
12  tagged_corpus = [TaggedDocument(doc, [i]) for i, doc in
13                  enumerate(document_tokens)]
14
15  # 訓練 Doc2Vec 模型
16  model_d2v = Doc2Vec(tagged_corpus, vector_size=MAX_WORDS_A_LINE, epochs=200)
17  model_d2v.train(tagged_corpus, total_examples=model_d2v.corpus_count,
18                  epochs=model_d2v.epochs)
```

3. 比較語句的相似度。

```
1   # 測試
2   questions = []
3   for i in range(len(document_tokens)):
4       questions.append(model_d2v.infer_vector(document_tokens[i]))
5   questions = np.array(questions)
6   # print(questions.shape)
7
8   # 測試語句
9   # text = "find allergen information"
10  text = "mobile pay"
11  filtered_tokens = tokenize(text, stopword_list)
12  # print(filtered_tokens)
13
14  # 比較語句相似度
15  similarity = cosine_similarity(model_d2v.infer_vector(
16      filtered_tokens).reshape(1, -1), questions, dense_output=False)
17
18  # 選出前 10 名
19  top_n = np.argsort(np.array(similarity[0]))[::-1][:10]
20  print(f'前 10 名 index:{top_n}\n')
21  for i in top_n:
22      print(round(similarity[0][i], 4), corpus[i].rstrip('\n'))
```

執行結果：以 "mobile pay"(手機支付)尋找前 10 名相似的語句，結果還不錯。讀者可再試試其他語句，筆者測試其他的結果並不理想，後面改用 BERT 模型時，準確率會提升許多。

另外，TensorFlow 也提供一個詞嵌入的視覺化工具 Embedding Projector

(https://projector.tensorflow.org/)，支援 3D 的向量空間，讀者可以按下列步驟操作：

1. 在右方的搜尋欄位輸入單字後，系統就會顯示候選字。
2. 選擇其中一個候選字，接著系統會顯示相似字，且利用各種演算法 (PCA、T-SNE、UMAP)來降維，以 3D 介面顯示單字間的距離。
3. 點選「Isolate 101 points」：只顯示距離最近的 101 個單字。
4. 也可以修改詞嵌入的模型：Word2Vec All、Word2Vec 10K、GNMT(全球語言神經機器翻譯)等。

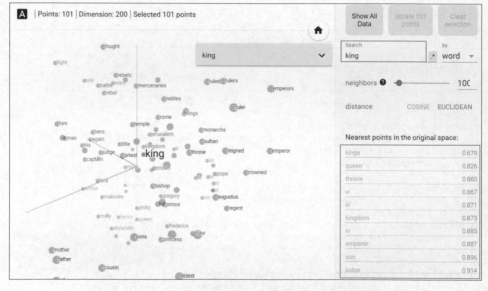

圖 11.7　TensorFlow Embedding Projector 視覺化工具

11-4　GloVe 模型

GloVe(Global Vectors)是由史丹佛大學 Jeffrey Pennington 等學者於 2014 所提出的另一套詞嵌入模型，與 Word2Vec 齊名，他們認為 Word2Vec 並未考慮全局的機率分配，只以移動視窗內的詞彙為樣本，沒有掌握全文的資訊，因此，提出了「詞彙共現矩陣」(word-word cooccurrence matrix)，考慮詞彙同時出現的機率，解決 Word2Vec 只看局部的缺陷以及 BOW 稀疏向量空間的問題，詳細內容可參閱「GloVe: Global Vectors for Word Representation」[7]。

GloVe 有 4 個預先訓練好的模型：

1. glove.42B.300d.zip(https://nlp.stanford.edu/data/wordvecs/glove.42B.300d.zip)：430 億詞彙，300 維向量，佔 1.75 GB 的檔案。

2. glove.840B.300d.zip(https://nlp.stanford.edu/data/wordvecs/glove.840B.300d.zip)：8400 億詞彙，300 維向量，佔 2.03 GB 的檔案。

3. glove.6B.300d.zip(https://nlp.stanford.edu/data/wordvecs/glove.6B.zip)：60 億詞彙，300 維向量，佔 822 MB 的檔案。

4. glove.twitter.27B.zip(https://nlp.stanford.edu/data/wordvecs/glove.twitter.27B.zip)：270 億詞彙，200 維向量，佔 1.42 GB 的檔案。

GloVe 詞向量模型檔的格式十分簡單，每列是一個單字，每個欄位以空格隔開，第一欄為單字，第二欄以後為該單字的詞向量。所以，通常把模型檔讀入後，轉為字典(dict)的資料型態，以利查詢。

【範例 3】GloVe 測試。
程式 11_06_GloVe.ipynb。

1. 載入 GloVe 詞向量檔 glove.6B.300d.txt。

```
1  # 載入相關套件
2  import numpy as np
3
4  # 載入GloVe詞向量檔 glove.6B.300d.txt
5  embeddings_dict = {}
6  with open("./glove/glove.6B.300d.txt", 'r', encoding="utf-8") as f:
7      for line in f:
8          values = line.split()
9          word = values[0]
10         vector = np.asarray(values[1:], "float32")
11         embeddings_dict[word] = vector
```

2. 取得 GloVe 的詞向量：任選一個單字(love)測試，取得 GloVe 的詞向量。

```
1  # 隨意測試一個單字(love)，取得 GloVe 的詞向量
2  embeddings_dict['love']
```

部份執行結果：

```
array([-4.5205e-01, -3.3122e-01, -6.3607e-02,  2.8325e-02, -2.1372e-01,
        1.6839e-01, -1.7186e-02,  4.7309e-02, -5.2355e-02, -9.8706e-01,
        5.3762e-01, -2.6893e-01, -5.4294e-01,  7.2487e-02,  6.6193e-02,
       -2.1814e-01, -1.2113e-01, -2.8832e-01,  4.8161e-01,  6.9185e-01,
       -2.0022e-01,  1.0082e+00, -1.1865e-01,  5.8710e-01,  1.8482e-01,
        4.5799e-02, -1.7836e-02, -3.3952e-01,  2.9314e-01, -1.9951e-01,
       -1.8930e-01,  4.3267e-01, -6.3181e-01, -2.9510e-01, -1.0547e+00,
        1.8231e-01, -4.5040e-01, -2.7800e-01, -1.4021e-01,  3.6785e-02,
        2.6487e-01, -6.6712e-01, -1.5204e-01, -3.5001e-01,  4.0864e-01,
       -7.3615e-02,  6.7630e-01,  1.8274e-01, -4.1660e-02,  1.5014e-02,
        2.5216e-01, -1.0109e-01,  3.1915e-02, -1.1298e-01, -4.0147e-01,
        1.7274e-01,  1.8497e-03,  2.4456e-01,  6.8777e-01, -2.7019e-01,
        8.0728e-01, -5.8296e-02,  4.0550e-01,  3.9893e-01, -9.1688e-02,
       -5.2080e-01,  2.4570e-01,  6.3001e-02,  2.1421e-01,  3.3197e-01,
       -3.4299e-01, -4.8735e-01,  2.2264e-02,  2.7862e-01,  2.3881e-01,
```

3. 指定以歐基里德(euclidean)距離計算相似性：找出最相似的 10 個單字。

```
1  # 以歐基里德(euclidean)距離計算相似性
2  from scipy.spatial.distance import euclidean
3
4  def find_closest_embeddings(embedding):
5      return sorted(embeddings_dict.keys(),
6                  key=lambda word: euclidean(embeddings_dict[word], embedding))
7
8  print(find_closest_embeddings(embeddings_dict["king"])[1:10])
```

執行結果：大部份與「king」的意義相似。

```
'queen', 'monarch', 'prince', 'kingdom', 'reign', 'ii', 'iii',
'brother', 'crown'
```

4. 任選 100 個單字，並以散佈圖觀察單字的相似度。

```
1   # 任意選 100 個單字
2   words =  list(embeddings_dict.keys())[100:200]
3   # print(words)
4
5   from sklearn.manifold import TSNE
6   import matplotlib.pyplot as plt
7
8   # 以 T-SNE 降維至二個特徵
9   tsne = TSNE(n_components=2)
10  vectors = [embeddings_dict[word] for word in words]
11  Y = tsne.fit_transform(vectors)
12
13  # 繪製散佈圖，觀察單字相似度
14  plt.figure(figsize=(12, 10))
15  plt.scatter(Y[:, 0], Y[:, 1])
16  for label, x, y in zip(words, Y[:, 0], Y[:, 1]):
17      plt.annotate(label, xy=(x, y), xytext=(0, 0), textcoords="offset points")
```

執行結果：每次的執行結果均不相同，可以看到相似詞都集中在局部區域。

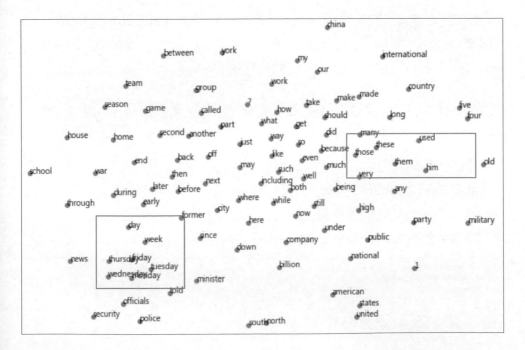

11-5 中文處理

前面介紹的都是英文語料，中文是否也可以比照辦理呢？答案是肯定的，NLP 所有作法都有考慮非英語系的支援。Jieba 套件提供中文分詞的功能，而 spaCy 套件則有支援中文語料的模型，現在我們就來介紹這兩個套件的用法。

Jieba 的主要功能包括：

1. 分詞(Tokenization)。
2. 關鍵字萃取(Keyword Extraction)。
3. 詞性標註(POS)。

Jieba 安裝指令如下：

```
pip install jieba
```

預設為簡體語詞字典，須自 https://github.com/APCLab/jieba-tw/tree/master /jieba 下載繁體字典，可覆蓋安裝的檔案，也可以於程式中使用 set_dictionary()設定繁體字典，我們使用後者。

【範例 4】以 Jieba 套件進行中文分詞。
程式 11_07_中文_NLP.ipynb。

1. 簡體字分詞：包含三種模式。
- 全模式(Full Mode)：顯示所有可能的片語。
- 精確模式：只顯示最有可能的片語，此為預設模式。
- 搜索引擎模式：使用隱馬可夫鏈(HMM)模型。

```
 1  # 載入相關套件
 2  import numpy as np
 3  import jieba
 4
 5  # 分詞
 6  text = "小明碩士毕业于中国科学院计算所，后在日本京都大学深造"
 7  # cut_all=True：全模式
 8  seg_list = jieba.cut(text, cut_all=True)
 9  print("全模式: " + "/ ".join(seg_list))
10
```

```
11  # cut_all=False：精確模式
12  seg_list = jieba.cut(text, cut_all=False)
13  print("精確模式: " + "/ ".join(seg_list))
14
15  # cut_for_search：搜索引擎模式
16  seg_list = jieba.cut_for_search(text)
17  print('搜索引擎模式: ', ', '.join(seg_list))
```

執行結果：

```
全模式: 小/ 明/ 碩士/ 畢業/ 于/ 中国/ 中国科学院/ 科学/ 科学院/ 学院/ 计算/ 计算所/ ，/ 后/ 在/ 日本/ 日本京都大学/ 京都/ 京都大学/ 大
学/ 深造
精確模式: 小明/ 碩士/ 畢業/ 于/ 中国科学院/ 计算所/ ，/ 后/ 在/ 日本京都大学/ 深造
搜索引擎模式:  小明, 碩士, 畢業, 于, 中国, 科学, 学院, 科学院, 中国科学院, 计算, 计算所, ，, 后, 在, 日本, 京都, 大学, 日本京都大学, 深
造
```

2. 繁體字分詞：先呼叫 set_dictionary()，設定繁體字典 dict.txt。

```
1   # 設定繁體字典
2   jieba.set_dictionary('./jieba/dict.txt')
3
4   # 分詞
5   text = "新竹的交通大學在新竹的大學路上"
6
7   # cut_all=True：全模式
8   seg_list = jieba.cut(text, cut_all=True)
9   print("全模式: " + "/ ".join(seg_list))
10
11  # cut_all=False：精確模式
12  seg_list = jieba.cut(text, cut_all=False)
13  print("精確模式: " + "/ ".join(seg_list))
14
15  # cut_for_search：搜索引擎模式
16  seg_list = jieba.cut_for_search(text)
17  print('搜索引擎模式: ', ', '.join(seg_list))
```

執行結果：

```
全模式：新竹/ 的/ 交通/ 交通大/ 大學/ 在/ 新竹/ 的/ 大學/ 大學路/ 學路/ 路上
精確模式：新竹/ 的/ 交通/ 大學/ 在/ 新竹/ 的/ 大學路/ 上
搜索引擎模式： 新竹, 的, 交通, 大學, 在, 新竹, 的, 大學, 學路, 大學路, 上
```

3. 分詞後，顯示詞彙的位置。

```
1   text = "新竹的交通大學在新竹的大學路上"
2   result = jieba.tokenize(text)
3   print("單字\t開始位置\t結束位置")
4   for tk in result:
5       print(f"{tk[0]}\t{tk[1]:-2d}\t{tk[2]:-2d}")
```

執行結果：

單字	開始位置	結束位置
新竹	0	2
的	2	3
交通	3	5
大學	5	7
在	7	8
新竹	8	10
的	10	11
大學路	11	14
上	14	15

4. 加詞：假如詞彙不在預設的字典中，可使用 add_word()將詞彙加入字典中，各行各業的專門術語都可以利用此方式加入，不必直接修改 dict.txt。

```
1  # 測試語句
2  text = "張惠妹在演唱會演唱三天三夜"
3
4  # 加詞前的分詞
5  seg_list = jieba.cut(text, cut_all=False)
6  print("加詞前的分詞: " + "/ ".join(seg_list))
7
8  # 加詞
9  jieba.add_word('三天三夜')
10
11 seg_list = jieba.cut(text, cut_all=False)
12 print("加詞後的分詞: " + "/ ".join(seg_list))
```

執行結果：原本「三天三夜」分為兩個詞「三天三」、「夜」，加詞後，分詞就正確了。

```
加詞前的分詞: 張惠妹/ 在/ 演唱會/ 演唱/ 三天三/ 夜
加詞後的分詞: 張惠妹/ 在/ 演唱會/ 演唱/ 三天三夜
```

5. 關鍵字萃取：呼叫 extract_tags()函數，萃取關鍵字，參數 topk 可指定顯示的筆數。測試語句來自近日缺水的新聞[8]。

```
1  # 測試語句來自新聞 https://news.ltn.com.tw/news/life/breakingnews/3497315
2  with open('./jieba/news.txt', encoding='utf8') as f:
3      text = f.read()
4
5  # 加詞前的分詞
6  import jieba.analyse
7
8  jieba.analyse.extract_tags(text, topK=10)
```

執行結果： 新聞標題為「中市明第二輪分區限水 百貨業買 20 個水塔桶」，以下萃取的關鍵字還算不錯。

'百貨公司', '水車', '中友', '用水', '限水', '封閉', '數間', '公廁', '因應', '20'

6. 設定停用詞改進：呼叫 stop_words(file_name)函數。

```
1   # 測試語句來自新聞 https://news.ltn.com.tw/news/life/breakingnews/3497315
2   with open('./jieba/news.txt', encoding='utf8') as f:
3       text = f.read()
4
5   import jieba.analyse
6
7   # 設定停用詞
8   jieba.analyse.set_stop_words('./jieba/stop_words.txt')
9
10  # 加詞前的分詞
11  jieba.analyse.extract_tags(text, topK=10)
```

執行結果： 設定停用詞為「20、因應、52」，萃取的關鍵字調整如下。

'百貨公司', '水車', '中友', '用水', '限水', '封閉', '數間', '公廁', '百貨', '週二'

7. 取得詞性(POS)標註：呼叫 posseg.cut 函數，可使用 POSTokenizer 自訂分詞器。

```
1   # 測試語句
2   text = "張惠妹在演唱會演唱三天三夜"
3
4   # 詞性(POS) 標註
5   words = jieba.posseg.cut(text)
6   for word, flag in words:
7       print(f'{word} {flag}')
```

執行結果：

```
張惠妹 N
在 P
演唱會 N
演唱 Vt
三天三夜 x
```

詞性代碼表可參閱「彙整中文與英文的詞性標註代號」[9]一文，內文有完整的說明與範例。

11-6　spaCy 套件

spaCy 套件支援超過 64 種語言，不只有 Wod2Vec 詞向量模型，也支援 BERT 預先訓練的模型，主要的功能包括：

項次	功能	説明
1.	分詞(Tokenization)	詞彙切割。
2.	詞性標籤(POS Tagging)	分析語句中每個單字的詞性。
3.	文法解析(Dependency Parsing)	依文法解析單字的相依性。
4.	詞性還原(Lemmatization)	還原成詞彙的原型。
5.	語句切割(Sentence Boundary Detection)	將文章段落切割成多個語句。
6.	命名實體識別(Named Entity Recognition)	識別語句中的命名實體，例如人名、地點、機構名稱等。
7.	實體連結(Entity Linking)	根據知識圖譜連結實體。
8.	相似性比較(Similarity)	單字或語句的相似性比較。
9.	文字分類(Text Classification)	對文章或語句進行分類。
10.	語意標註(Rule-based Matching)	類似 Regular expression，依據語意找出詞彙的順序。
11.	模型訓練(Training)	
12.	模型存檔(Serialization)	

可利用 spaCy 網頁(https://spacy.io/usage)的選單產生安裝指令，產生的指令如下：

- pip install spacy
- 支援 GPU，須配合 CUDA 版本：pip install -U spacy[cuda111]
- cuda111：為 cuda v11.1 版。

下載詞向量模型，spaCy 稱為 pipeline，指令如下：

- 英文：python -m spacy download en_core_web_sm
- 中文：python -m spacy download zh_core_web_sm

- 其他語系亦可參考「spaCy Quickstart 網頁」(https://spacy.io/usage/ models)。

詞向量模型分成大型(lg)、中型(md) 、小型(sm)。

中文分詞有三個選項，可在組態檔(config.cfg)選擇：

- char：預設選項。
- jieba：使用 Jieba 套件分詞。
- pkuseg：支援多領域分詞，可參閱「pkuseg GitHub」[10]，依照文件說明，pkuseg 的各項效能(Precision、Recall、F1)比 jieba 來得好。

spaCy 相關功能的展示，可參考 spaCy 官網「spaCy 101: Everything you need to know」[11]的說明，以下就依照該文測試相關的功能。

【範例 5】spaCy 相關功能測試。
程式 11_08_spaCy_test.ipynb。

1. 載入相關套件。

```
1  # 載入相關套件
2  import spacy
```

2. 載入小型詞向量模型。

```
1  # 載入詞向量模型
2  nlp = spacy.load("en_core_web_sm")
```

3. 分詞及取得詞性標籤(POS Tagging)：

- token 的屬性可參閱 https://spacy.io/api/token
- 詞性標籤表則請參考 https://github.com/explosion/spaCy/blob/master/ spacy/glossary.py。

```
1  # 分詞及取得詞性標籤(POS Tagging)
2  doc = nlp("Apple is looking at buying U.K. startup for $1 billion")
3  for token in doc:
4      print(token.text, token.pos_, token.dep_)
```

執行結果：

```
Apple PROPN nsubj
is AUX aux
looking VERB ROOT
at ADP prep
buying VERB pcomp
U.K. PROPN dobj
startup NOUN advcl
for ADP prep
$ SYM quantmod
1 NUM compound
billion NUM pobj
```

4. 取得詞性標籤詳細資訊。

```
1  # 取得詳細的詞性標籤(POS Tagging)
2  for token in doc:
3      print(token.text, token.lemma_, token.pos_, token.tag_, token.dep_,
4              token.shape_, token.is_alpha, token.is_stop)
```

執行結果：

```
Apple Apple PROPN NNP nsubj Xxxxx True False
is be AUX VBZ aux xx True True
looking look VERB VBG ROOT xxxx True False
at at ADP IN prep xx True True
buying buy VERB VBG pcomp xxxx True False
U.K. U.K. PROPN NNP dobj X.X. False False
startup startup NOUN NN advcl xxxx True False
for for ADP IN prep xxx True True
$ $ SYM $ quantmod $ False False
1 1 NUM CD compound d False False
billion billion NUM CD pobj xxxx True False
```

5. 以 displaCy Visualizer 顯示語意分析圖，displacy.serve 的參數請參閱
「displaCy visualizer 的說明文件」(https://spacy.io/api/top-level#display)。

```
1  # 顯示語意分析圖
2  from spacy import displacy
3
4  displacy.serve(doc, style="dep")
```

執行結果：可使用網頁瀏覽 http://127.0.0.1:5000。箭頭表示依存關係，例
如 looking 的主詞是 Apple，buying 的受詞是 UK。

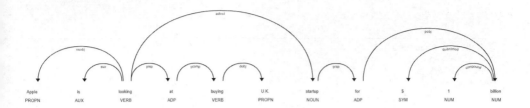

6. 以 displaCy visualizer 標示命名實體(Named Entity)。

```
1  # 標示實體
2  text = "When Sebastian Thrun started working on self-driving cars " + \
3         "at Google in 2007, few people outside of the company took him seriously."
4
5  doc = nlp(text)
6  # style="ent" : 實體
7  display.serve(doc, style="ent")
```

執行結果：可使用網頁瀏覽 http://127.0.0.1:5000。

When Sebastian Thrun **PERSON** started working on self-driving cars at Google in 2007 **DATE** , few people outside of the company took him seriously.

7. 繁體中文分詞。

```
1  # 繁體中文分詞
2  import spacy
3
4  nlp = spacy.load("zh_core_web_sm")
5  doc = nlp("清華大學位於新竹")
6  for token in doc:
7      print(token.text, token.pos_, token.dep_)
```

執行結果：大學被切割成兩個詞，結果不太正確，建議實際執行時可以先用簡體分詞後，再轉回繁體。

```
清華 NOUN compound:nn
大 ADJ amod
學位 NOUN nsubj
於 ADP case
新竹 PROPN ROOT
```

8. 簡體中文分詞。

```
1  # 簡體中文分詞
2  import spacy
3
4  nlp = spacy.load("zh_core_web_sm")
5  doc = nlp("清华大学位于北京")
6  for token in doc:
7      print(token.text, token.pos_, token.dep_)
```

執行結果：

```
清华 PROPN compound:nn
大学 NOUN nsubj
位于 VERB ROOT
北京 PROPN dobj
```

9. 顯示中文語意分析圖。

```
1  # 顯示中文語意分析圖
2  from spacy import displacy
3
4  displacy.serve(doc, style="dep")
```

執行結果：可使用網頁瀏覽 http://127.0.0.1:5000。

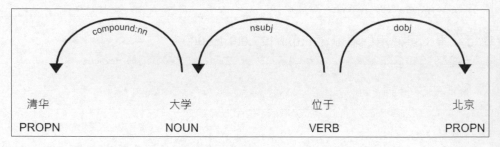

10. 分詞，並判斷是否不在字典中(Out of Vocabulary, OOV)。

```
1  # 分詞，並判斷是否不在字典中(Out of Vocabulary, OOV)
2  nlp = spacy.load("en_core_web_md")
3  tokens = nlp("dog cat banana afskfsd")
4
5  for token in tokens:
6      print(token.text, token.has_vector, token.vector_norm, token.is_oov)
```

執行結果：afskfsd 不在字典中，注意，必須使用中型(md)以上的模型，小型(sm)會出現錯誤。

```
dog True 7.0336733 False
cat True 6.6808186 False
banana True 6.700014 False
afskfsd False 0.0 True
```

11. 相似度比較。

```
 1  # 相似度比較
 2  nlp = spacy.load("en_core_web_md")
 3
 4  # 測試兩語句
 5  doc1 = nlp("I like salty fries and hamburgers.")
 6  doc2 = nlp("Fast food tastes very good.")
 7
 8  # 兩語句的相似度比較
 9  print(doc1, "<->", doc2, doc1.similarity(doc2))
10
11  # 關鍵字的相似度比較
12  french_fries = doc1[2:4]
13  burgers = doc1[5]
14  print(french_fries, "<->", burgers, french_fries.similarity(burgers))
```

執行結果：

```
I like salty fries and hamburgers. <-> Fast food tastes very good. 0.7799485853415737
salty fries <-> hamburgers 0.7304624
```

參考資料 (References)

[1] Sebastian Andrei,《South Korea's Convenience Store Culture》, 2018
     (https://medium.com/@sebastian_andrei/south-koreas-convenience-store-
     culture-187c33a649a6)

[2] 維基百科關於 tf-idf 的說明
     (https://en.wikipedia.org/wiki/Tf%E2%80%93idf)

[3] Tomas Mikolov、Quoc V. Le、Ilya Sutskever,《Exploiting Similarities
     among Languages for Machine Translation》, 2013
     (https://arxiv.org/pdf/1309.4168v1.pdf)

[4] NSS,《An Intuitive Understanding of Word Embeddings: From Count
     Vectors to Word2Vec》, 2017
     (https://www.analyticsvidhya.com/blog/2017/06/word-embeddings-
     count-word2veec/)

[5] Ria Kulshrestha,《NLP 102: Negative Sampling and GloVe》, 2019
     (https://towardsdatascience.com/nlp-101-negative-sampling-and-glove-
     936c88f3bc68)

[6] Gensim 官網關於 Word2Vec 的說明
     (https://radimrehurek.com/gensim/models/word2vec.html)

[7] Jeffrey Pennington、Richard Socher、Christopher D. Manning,《GloVe:
     Global Vectors for Word Representation》, 2014
     (https://www.aclweb.org/anthology/D14-1162.pdf)

[8] 自由時報 蘇金鳳,《中市明第二輪分區限水 百貨業買 20 個水塔桶》,
     2021
     (https://news.ltn.com.tw/news/life/breakingnews/3497315)

[9] 布丁布丁吃布丁,《彙整中文與英文的詞性標註代號》, 2017
     (http://blog.pulipuli.info/2017/11/fasttag-identify-part-of-speech-in.html)

[10] pkuseg GitHub
     (https://github.com/explosion/spacy-pkuseg)

[11] spaCy 101 官網「spaCy 101: Everything you need to know」
     (https://spacy.io/usage/spacy-101)

自然語言處理的演算法

上一章我們認識了自然語言處理的前置處理和詞向量應用，我們接著探討自然語言處理相關的深度學習演算法。

自然語言的推斷(Inference)不僅需要考慮語文上下文的關聯，還要考量人類特殊的能力 -- 記憶力，譬如，我們從小就學習歷史，講到治水，第一個可能想到治水的老祖宗「大禹」，講到台灣嘉南水庫，就會聯想到「八田與一」，這就是記憶力的影響，就算時間再久遠，都會深印在腦中。因此，NLP 相關的深度學習演算法要能夠提升預測準確率，模型就必須額外添加上下文關聯與記憶力的功能。

我們會依照循環神經網路發展的軌跡依序研究，從簡單的 RNN、LSTM、注意力機制(Attention)、到 Transformer 等演算法，包括目前最夯的 BERT 模型。

12-1 循環神經網路(RNN)

一般神經網路以迴歸為基礎，以特徵(x)預測目標(y)，但 NLP 的特徵並不互相獨立，他們有上下文的關聯，因此，循環神經網路(Recurrent Neural Network, 以下簡稱 RNN)就像自迴歸(Auto-regression)模型一樣，會考慮同一層前面的神經元影響。可以用數學式表示兩者的差異：

- 迴歸：y=Wx+b

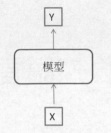

圖 12.1 迴歸的示意圖

- RNN：

$h_t = W * h_{t-1} + U * x_t + b$

$y = V * h_t$

其中 W、U、V 都是權重，h 為隱藏層的輸出。

可以看到時間點 t 的 h 會受到前一時間點的 h_{t-1} 影響，如下圖：

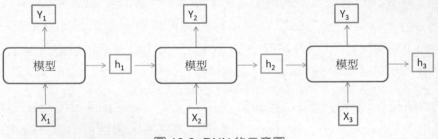

圖 12.2 RNN 的示意圖

由於每一個時間點的模型都類似，因此又可簡化為下圖的循環網路，這不僅有助於理解，在開發時也可簡化為遞迴結構：

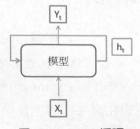

圖 12.3 RNN 循環

歸納上述説明，一般神經網路假設同一層的神經元是互相獨立的，而 RNN 則將同一層的前一個神經元也視為輸入。

深度學習框架如 TensorFlow、PyTorch 均直接支援 RNN 神經層，以下我們就以 TensorFlow/Keras 實作此一模型，看看效果如何。

【範例 1】簡單的 RNN 測試。

程式 12_01_RNN_test.ipynb。

1. 載入相關套件。

```
1  # 載入相關套件
2  import numpy as np
3  import tensorflow as tf
4  from tensorflow import keras
5  from tensorflow.keras import layers
```

2. 嵌入層測試：模型只含「嵌入層」 (Embedding layer)，注意，RNN 系列的神經網路模型的第一層必須為嵌入層，它會將輸入轉為稠密的向量空間(dense vector)。嵌入層的參數如下：

- input_dim：字典的尺寸。
- output_dim：輸出的向量尺寸。
- input_length：每筆輸入語句的詞彙長度，如果後面接 Flatten 和 Dense 層，則此參數是必填的。
- mask_zero：作長度不足時是否補 0，使用時須有許多配合措施，請參閱 Keras 中文官網(https://keras.io/zh/layers/embeddings/)。

```
1  # 建立模型
2  model = tf.keras.Sequential()
3
4  # 模型只含嵌入層(Embedding Layer)
5  # 字彙表最大為1000，輸出維度為 64，輸入的字數為 10
6  model.add(layers.Embedding(input_dim=1000, output_dim=64))
7
8  # 產生亂數資料，32筆資料，每筆 10 個數字
9  input_array = np.random.randint(1000, size=(32, 10))
10
11 # 指定優化器、損失函數
12 model.compile('rmsprop', 'mse')
13
14 # 預測
15 output_array = model.predict(input_array)
```

```
16  print(output_array.shape)
17  output_array[0]
```

執行結果：嵌入層(Embedding layer) 輸入尺寸(input_dim)為 1000，代表字典的尺寸，嵌入層輸入必須為 2 維，包含筆數(32)、語句字數(10)，輸出會加上一維，向量空間尺寸設定為 64，故輸出維度大小為(32, 10, 64)。

```
(32, 10, 64)

array([[-4.09067757e-02,  4.13169377e-02,  3.79419327e-03,
        -8.12249258e-03,  3.98785211e-02,  4.84695174e-02,
        -3.52774151e-02, -2.07844265e-02, -3.32484469e-02,
         1.15686059e-02, -2.05504298e-02,  4.01307456e-02,
        -3.33517343e-02,  4.53372933e-02, -1.14959478e-02,
        -3.42349410e-02, -2.31464747e-02,  4.93111499e-02,
         3.65070440e-02,  1.29793398e-02,  3.98182534e-02,
        -4.83712554e-02,  2.58997716e-02,  3.76032479e-02,
         4.48194407e-02, -3.18442471e-02,  1.50911510e-05,
         4.13540117e-02, -1.83008537e-02, -3.48059647e-02,
         4.89773043e-02, -2.05516815e-04,  6.68109581e-03,
         2.11245939e-03, -4.50933240e-02,  7.08359480e-03,
        -3.61134633e-02, -3.95359285e-02,  4.99451868e-02,
```

3. 使用真實的資料轉換，觀察嵌入層轉換的結果。

```
1   from tensorflow.keras.preprocessing.text import one_hot
2   from tensorflow.keras.preprocessing.sequence import pad_sequences
3
4   # 測試資料
5   docs = ['Well done!',
6           'Good work',
7           'Great effort',
8           'nice work',
9           'Excellent!',
10          'Weak',
11          'Poor effort!',
12          'not good',
13          'poor work',
14          'Could have done better.']
15
16  # 轉成 one-hot encoding
17  vocab_size = 50 # 字典最大字數
18  maxlen = 4      # 語句最大字數
19  encoded_docs = [one_hot(d, vocab_size) for d in docs]
20
21  # 轉成固定長度，長度不足則後面補空白
22  padded_docs = pad_sequences(encoded_docs, maxlen=maxlen, padding='post')
23
24  # 模型只有 Embedding
25  model = tf.keras.Sequential()
26  model.add(layers.Embedding(vocab_size, 64, input_length=maxlen))
27  model.compile('rmsprop', 'mse')
```

```
28
29  # 預測
30  output_array = model.predict(padded_docs)
31  output_array.shape
```

嵌入層輸入必須為固定尺寸，否則無法做向量運算，故長度不足時，則後面需補 0。

執行結果：(輸出筆數、語句字數、單字向量維度) = (10, 4, 64)

4. 觀察 one-hot encoding 轉換結果與補 0 後的輸入維度。

```
1  # one-hot encoding 轉換結果
2  print(encoded_docs[0])
3
4  # 補空白後的輸入維度
5  print(padded_docs.shape)
```

執行結果：[34, 33]為 one-hot encoding 的兩個單字編碼，補 0 後的輸入維度為(10, 4)。

5. 模型接上完全連接層(Dense)，進行分類預測。

```
1  # 定義 10 個語句的正面(1)或負面(0)的情緒
2  labels = np.array([1,1,1,1,1,0,0,0,0,0])
3
4  vocab_size = 50
5  maxlen = 4
6  encoded_docs = [one_hot(d, vocab_size) for d in docs]
7  padded_docs = pad_sequences(encoded_docs, maxlen=maxlen, padding='post')
8
9  model = tf.keras.Sequential()
10 model.add(layers.Embedding(vocab_size, 8, input_length=maxlen))
11 model.add(layers.Flatten())
12
13 # 加上完全連接層(Dense)
14 model.add(layers.Dense(1, activation='sigmoid'))
15
16 # 指定優化器、損失函數
17 model.compile(optimizer='adam', loss='binary_crossentropy',
18               metrics=['accuracy'])
19
20 print(model.summary())
21
22 # 模型訓練
23 model.fit(padded_docs, labels, epochs=50, verbose=0)
24
25 # 模型評估
26 loss, accuracy = model.evaluate(padded_docs, labels, verbose=0)
27 print('Accuracy: %f' % (accuracy*100))
```

執行結果：準確率為 90%，還不錯，畢竟還未引進詞向量模型。

```
Model: "sequential_2"

Layer (type)                  Output Shape              Param #
=================================================================
embedding_2 (Embedding)       (None, 4, 8)              400

flatten (Flatten)             (None, 32)                0

dense (Dense)                 (None, 1)                 33
=================================================================
Total params: 433
Trainable params: 433
Non-trainable params: 0

None
Accuracy: 80.000001
```

6. 測試資料預測。

```
1  model.predict(padded_docs)
```

執行結果：機率均在 50%上下，答案並不肯定。

```
array([[0.5838079 ],
       [0.5428731 ],
       [0.50959533],
       [0.52323276],
       [0.53539276],
       [0.50386965],
       [0.49095556],
       [0.49666357],
       [0.5119646 ],
       [0.41784462]], dtype=float32)
```

7. 加上 RNN 神經層：simple_rnn() 也稱為簡單(Vanilla) RNN，第一個參
   數為輸出的神經元個數，另一個參數 unroll=True 時，會使網路以圖
   12.2 的架構求解，好處是速度可加快，壞處是需要占用較多的記憶
   體，反之，則以遞迴的架構執行，如圖 12.3。

```
1  model = tf.keras.Sequential()
2  model.add(layers.Embedding(vocab_size, 8, input_length=maxlen))
3
4  # 加上 RNN 神經層，輸出 128 個神經元
5  model.add(layers.SimpleRNN(128))
6
7  # 加上完全連接層(Dense)
```

```
 8  model.add(layers.Dense(1, activation='sigmoid'))
 9
10  # 指定優化器、損失函數
11  model.compile(optimizer='adam', loss='binary_crossentropy',
12                metrics=['accuracy'])
13
14  print(model.summary())
15  # 模型訓練
16  model.fit(padded_docs, labels, epochs=50, verbose=0)
17
18  # 模型評估
19  loss, accuracy = model.evaluate(padded_docs, labels, verbose=0)
20  print('Accuracy: %f' % (accuracy*100))
```

執行結果：準確率為 100%，完美。

```
Model: "sequential_9"

Layer (type)                 Output Shape              Param #
=================================================================
embedding_9 (Embedding)      (None, 4, 8)              400

simple_rnn_1 (SimpleRNN)     (None, 128)               17536

dense_3 (Dense)              (None, 1)                 129
=================================================================
Total params: 18,065
Trainable params: 18,065
Non-trainable params: 0
_____

None
Accuracy: 100.000000
```

8. 觀察機率預測。

```
1  model.predict(padded_docs)
```

執行結果：正面預測機率接近 1，負面預測機率接近 0，答案很肯定。

```
array([[9.99999642e-01],
       [9.99999881e-01],
       [9.97938693e-01],
       [9.97859895e-01],
       [9.99974608e-01],
       [2.61648721e-03],
       [1.31143715e-05],
       [1.73492054e-03],
       [7.56993222e-06],
       [8.76635386e-06]], dtype=float32)
```

9. 改用詞向量(Word2Vec)：讀取 GloVe 300 維的詞向量，產生字典資料
   型變數，方便搜尋。

```
1  # load the whole embedding into memory
2  embeddings_index = dict()
3  f = open('./GloVe/glove.6B.300d.txt', encoding='utf8')
4  for line in f:
5      values = line.split()
6      word = values[0]
7      coefs = np.array(values[1:], dtype='float32')
8      embeddings_index[word] = coefs
9  f.close()
```

10. 分詞、轉為序列整數、補 0。

```
1  # 分詞
2  from tensorflow.keras.preprocessing.text import Tokenizer
3  t = Tokenizer()
4  t.fit_on_texts(docs)
5
6  # 轉為序列整數
7  encoded_docs = t.texts_to_sequences(docs)
8
9  # 補 0
10 padded_docs = pad_sequences(encoded_docs, maxlen=maxlen, padding='post')
11 padded_docs
```

執行結果：正面預測機率接近 1，負面預測機率接近 0，答案很肯定。

```
array([[ 6,  2,  0,  0],
       [ 3,  1,  0,  0],
       [ 7,  4,  0,  0],
       [ 8,  1,  0,  0],
       [ 9,  0,  0,  0],
       [10,  0,  0,  0],
       [ 5,  4,  0,  0],
       [11,  3,  0,  0],
       [ 5,  1,  0,  0],
       [12, 13,  2, 14]])
```

11. 轉換為 GloVe 300 維的詞向量。

```
1  # 轉換為 GloVe 300維的詞向量
2  # 初始化輸出
3  embedding_matrix = np.zeros((vocab_size, 300))
4
5  # 讀取詞向量值
6  for word, i in t.word_index.items():
7      embedding_vector = embeddings_index.get(word)
8      if embedding_vector is not None:
9          embedding_matrix[i] = embedding_vector
10
11 # 任取一筆觀察
12 embedding_matrix[2]
```

執行結果：整數轉為詞向量。

```
array([ 0.19205999,  0.16459   ,  0.060122  ,  0.17696001, -0.27405   ,
        0.079646  , -0.25292999, -0.11763  ,  0.17614   , -1.97870004,
        0.10707   , -0.028088  ,  0.093991  ,  0.48135   , -0.037581  ,
        0.0059231 , -0.11118   , -0.099847  , -0.22189   ,  0.0062044 ,
        0.17721   ,  0.25786   ,  0.42120999, -0.13085   , -0.32839   ,
        0.39208999, -0.050214  , -0.46766999, -0.063107  , -0.0023065 ,
        0.21005   ,  0.26982   , -0.22652   , -0.42958999, -0.89682001,
        0.21932   , -0.0020377 ,  0.1358    , -0.12661999, -0.058927  ,
        0.0049502 , -0.28457999, -0.29530999, -0.29295999, -0.24212   ,
        0.091915  ,  0.01977   ,  0.14503001,  0.26495999,  0.10817   ,
        0.029115  ,  0.075254  ,  0.16463999,  0.12097   , -0.37494001,
        0.52671999,  0.094318  , -0.054813  , -0.021008  ,  0.081353  ,
        0.18735   , -0.14458001, -0.031203  ,  0.31753999,  0.027703  ,
       -0.28657001,  0.34630999, -0.27772   ,  0.18669   , -0.11684   ,
        0.21551999, -0.21927001,  0.19778   ,  0.68763   , -0.076211  ,
       -0.06296   ,  0.13236   ,  0.55324   ,  0.15331   , -0.17332999,
       -0.35551   ,  0.16426   ,  0.34196001, -0.13568   ,  0.071228  ,
        0.49147001, -0.45590001,  0.28874999, -0.14091   , -0.025825  ,
       -0.55035001,  0.4946    , -0.2378    , -0.10571   ,  0.06842   ,
```

12. Embedding 層設為不需訓練(trainable=False)，直接使用詞向量作為權重。

```
1   model = tf.keras.Sequential()
2
3   # trainable=False：不需訓練，直接輸入轉換後的向量
4   model.add(layers.Embedding(vocab_size, 300, weights=[embedding_matrix],
5                              input_length=maxlen, trainable=False))
6   model.add(layers.SimpleRNN(128))
7   model.add(layers.Dense(1, activation='sigmoid'))
8
9   # 指定優化器、損失函數
10  model.compile(optimizer='adam', loss='binary_crossentropy',
11              metrics=['accuracy'])
12
13  print(model.summary())
14
15  # 模型訓練
16  model.fit(padded_docs, labels, epochs=50, verbose=0)
17
18  # 模型評估
19  loss, accuracy = model.evaluate(padded_docs, labels, verbose=0)
20  print('Accuracy: %f' % (accuracy*100))
```

執行結果：準確率為 100%，完美。

```
Model: "sequential_14"

Layer (type)                Output Shape           Param #
=================================================================
embedding_13 (Embedding)    (None, 4, 300)         4500

simple_rnn_5 (SimpleRNN)    (None, 128)            54912

dense_7 (Dense)             (None, 1)              129
=================================================================
Total params: 59,541
Trainable params: 55,041
Non-trainable params: 4,500
```

13. 觀察預測結果。

```
1  list(model.predict_classes(padded_docs).reshape(-1))
```

執行結果：完全正確。

```
[1, 1, 1, 1, 1, 0, 0, 0, 0, 0]
```

14. 觀察機率預測。

```
1  model.predict(padded_docs)
```

執行結果：正面預測機率接近 1，負面預測機率接近 0，答案很肯定。

```
array([[9.9977702e-01],
       [9.9972540e-01],
       [9.9990332e-01],
       [9.9989653e-01],
       [9.9989903e-01],
       [1.1761398e-04],
       [1.1944198e-04],
       [2.3762533e-04],
       [1.7523505e-04],
       [1.6099006e-04]], dtype=float32)
```

12-2　長短期記憶網路(LSTM)

簡單(Vanilla) RNN 只考慮上文(上一個神經元)，如果要同時考慮下文，可以直接將 simple_rnn 包在 Bidirectional() 函數內即可。

簡單(Vanilla) RNN 有一個重大的瑕疵，它跟 CNN 一樣是簡化模型，均假設「權值共享」(Shared Weights)，因為

$h_t = W * h_{t-1} + U * x_t + b$

$h_{t-1} = W * h_{t-2} + U * x_{t-1} + b$

➜ $h_t = W * (W * h_{t-2} + U * x_{t-1} + b) + U * x_t + b$

➜ $h_t = (W^2 * h_{t-2} + W * U * x_{t-1} + W * b) + U * x_t + b$

在優化求解時，進行反向傳導(Backpropagation)，偏微分後

- 若 W<1，則越前面的神經層 W^n 會愈來愈小，造成影響力越小，這種現象稱為「梯度消失」(Vanishing Gradient)。
- 反之，若 W>1，則越前面的神經層 W^n 會愈來愈大，造成「梯度爆炸」 (Exploding Gradient)，優化求解無法收斂。

梯度消失導致考慮的上文長度有限，因此，Hochreiter 和 Schmidhuber 於 1997 年提出「長短期記憶網路」 (Long Short Term Memory Network, LSTM)演算法，額外維護一條記憶網路，以下示意圖為比較 RNN 與 LSTM 的差別：

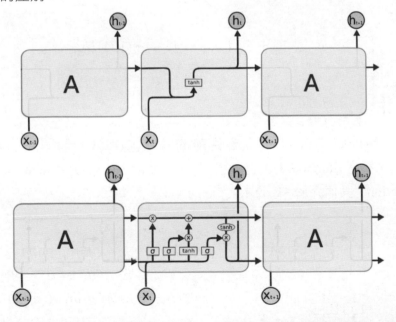

圖 12.4 RNN 與 LSTM 內部結構的比較，上圖為 RNN，下圖為 LSTM，圖片來源：Understanding LSTM Networks [1]

我們將圖 12.4 的 LSTM 進行拆解，就可以瞭解 LSTM 的運算機制：

1. 額外維護一條記憶線(Cell state)。

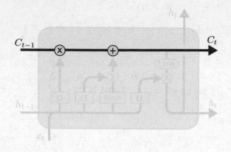

2. LSTM 多了四個閥(Gate)，用來維護記憶網路與預測網路：即圖 12.4 中的⊗、⊕(原圖粉紅色標誌)。

3. 遺忘閥(Forget Gate)：決定之前記憶是否刪除，σ 為 sigmoid 神經層，輸出為 0 時，乘以原記憶，表示刪除，反之則為保留記憶。

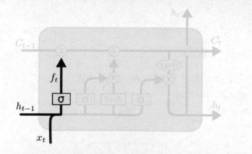

$$f_t = \sigma\left(W_f \cdot [h_{t-1}, x_t] + b_f\right)$$

4. 輸入閥(Input Gate)：輸入含目前的特徵(x_t)加 t-1 時間點的隱藏層 (h_{t-1})，透過 σ(sigmoid)，得到輸出(i_t) ，而記憶(C_t) 使用 tanh activation function，其值介於(-1, 1)。

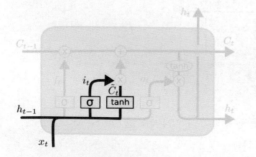

$$i_t = \sigma\left(W_i \cdot [h_{t-1}, x_t] + b_i\right)$$
$$\tilde{C}_t = \tanh(W_C \cdot [h_{t-1}, x_t] + b_C)$$

5. 更新閥(Update Gate)：更新記憶(Ct)，為之前的記憶加上目前增加的資訊。

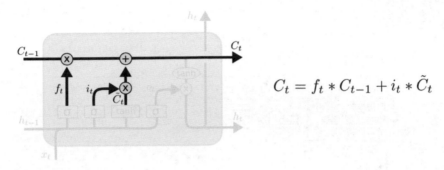

$$C_t = f_t * C_{t-1} + i_t * \tilde{C}_t$$

6. 輸出閥(Output Gate)：輸出包括目前的正常輸出，乘以更新的記憶。

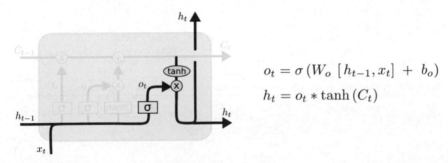

$$o_t = \sigma \left(W_o \left[h_{t-1}, x_t \right] + b_o \right)$$
$$h_t = o_t * \tanh \left(C_t \right)$$

依照前面的拆解，大概就能知道 LSTM 是怎麼保存記憶及使用記憶了，網路上也有人直接用 NumPy 開發 LSTM，不過，既然 TensorFlow/Keras 已經直接定義 LSTM 神經層了，我們可直接拿來使用。

【範例 2】以 LSTM 實作情緒分析(Sentiment Analysis)。情緒分析是預測一段評論為正面或負面的情緒。

資料集：影評資料集(IMDB movie review)，注意，TensorFlow 已將資料轉為索引值，而非文字，如果要取得文字資料集，可自 https://ai.stanford.edu/~amaas/data/sentiment/aclImdb_v1.tar.gz 下載。

程式 12_02_LSTM_IMDB.ipynb。

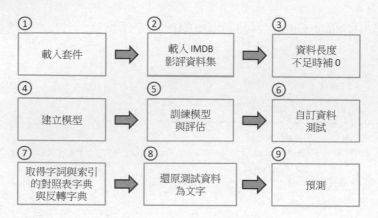

1. 載入相關套件。

```
1  # 載入相關套件
2  import tensorflow as tf
3  from tensorflow.keras.datasets import imdb
4  from tensorflow.keras.layers import Embedding, Dense, LSTM
5  from tensorflow.keras.losses import BinaryCrossentropy
6  from tensorflow.keras.models import Sequential
7  from tensorflow.keras.optimizers import Adam
8  from tensorflow.keras.preprocessing.sequence import pad_sequences
```

2. 參數設定。

```
1  # 參數設定
2  batch_size = 128                    # 批量
3  embedding_output_dims = 15          # 嵌入層輸出維度
4  max_sequence_length = 300           # 句子最大字數
5  num_distinct_words = 5000           # 字典
6  number_of_epochs = 5                # 訓練執行週期
7  validation_split = 0.20             # 驗證資料比例
8  verbosity_mode = 1                  # 訓練資料訊息顯示程度
```

3. 建立模型：使用 Embedding + LSTM + Dense 神經層。

```
1  # 載入 IMDB 影評資料集，TensorFlow 已將資料轉為索引值
2  (x_train, y_train), (x_test, y_test) = imdb.load_data(
3      num_words=num_distinct_words)
4  print(x_train.shape)
5  print(x_test.shape)
6
7  # 長度不足時補 0
8  padded_inputs = pad_sequences(x_train, maxlen=max_sequence_length
9                              , value = 0.0)
10 padded_inputs_test = pad_sequences(x_test, maxlen=max_sequence_length
11                              , value = 0.0)
12
13 # 建立模型
```

```
14  model = Sequential()
15  model.add(Embedding(num_distinct_words, embedding_output_dims,
16                      input_length=max_sequence_length))
17  model.add(LSTM(10))
18  model.add(Dense(1, activation='sigmoid'))
19
20  # 指定優化器、損失函數
21  model.compile(optimizer=Adam(), loss=BinaryCrossentropy, metrics=['accuracy'])
22
23  # 模型彙總資訊
24  model.summary()
```

4. 訓練模型與評估。

```
1  # 訓練模型
2  history = model.fit(padded_inputs, y_train, batch_size=batch_size,
3              epochs=number_of_epochs, verbose=verbosity_mode,
4              validation_split=validation_split)
5
6  # 模型評估
7  test_results = model.evaluate(padded_inputs_test, y_test, verbose=False)
8  print(f'Loss: {test_results[0]}, Accuracy: {100*test_results[1]}%')
```

執行結果：訓練執行 5 週期，損失為 0.3370，準確率為 86.63。

5. 雖然，模型準確率很高，但我們還是希望能以自訂的資料測試一下。由於 IMDB 的影評內容都非常長，假使以短句測試，如"I like the movie"或"I hate the movie"，並不能得到正確的預測值，因此，我們使用測試資料的前兩句來測試。首先我們要取得詞彙與索引的對照表字典。

```
1  # 取得字詞與索引的對照表字典
2  imdb_dict = imdb.get_word_index()
3  list(imdb_dict.keys())[:10]
```

6. 反轉字典，變成索引與詞彙的對照表：要將資料集的資料還原成文字。

```
1  # 反轉字典，變成索引與字詞的對照表
2  imdb_dict_reversed = {}
3  for k, v in imdb_dict.items():
4      imdb_dict_reversed[v] = k
```

7. 還原測試資料前兩筆為文字：由於資料有許多 0，而空白的索引值也是 0，為避免以空白切割詞彙時，將空白均刪除，故以「,」隔開單字。

```
1  # 還原測試資料前兩筆為文字
2  text = []
3  for i, line in enumerate(padded_inputs_test[:2]):
4      text.append('')
5      for j, word in enumerate(line):
6          if word != 0:
7              text[i] += imdb_dict_reversed[word]+','
8          else:
9              text[i] += ' ,'
10 text
```

執行結果：

```
[" , , , , , , , , , , , , , , , , , , , , , , , , , , , , , , , , , , , , , , , ,
 , , , , , , , , , , , , , , , , , , , , , , , , , , , , , ,the,wonder,own,as,by,is,sequence,i,i,and,an
d,to,of,hollywood,br,of,down,and,getting,boring,of,ever,it,sadly,sadly,sadly,i,i,was,then,does,don't,close,and,after,one,carry,
as,by,are,be,and,all,family,turn,in,does,as,three,part,in,another,some,to,be,probably,with,world,and,her,an,have,and,beginning,
own,as,is,sequence,",
```

8. 預測。

```
1  # 長度不足時補 0
2  padded_inputs = pad_sequences(X_index, maxlen=max_sequence_length,
3                      padding=pad_type, truncating=trunc_type, value = 0.0)
4
5  # 預測
6  model.predict_classes(padded_inputs)
```

9. 以原資料預測：確認答案相同。

```
1  # 以原資料預測，確認答案相同
2  model.predict_classes(padded_inputs_test[:2])
```

【範例 3】以文字資料集，實作 LSTM 情緒分析(Sentiment Analysis)。

資料集：文字型的影評資料集(IMDB movie review)，可自 https://ai.stanford.edu/~amaas/data/sentiment/aclImdb_v1.tar.gz 下載。

程式 12_03_LSTM_IMDB_Text.ipynb。

與範例 2 不同的要點如下：

1. 以文字資料集作為測試資料：以 TensorFlow Dataset 資料形態處理。

2. 也以卷積模型(Conv1D)預測，比較模型差異。

3. 以雙向 LSTM 模型(Bidirectional、LSTM)測試。

其他程序與 12_02_LSTM_IMDB.ipynb 大同小異，就不浪費篇幅介紹了，讀者可直接測試程式。

▌ 12-3 LSTM 重要參數與多層 LSTM

LSTM 可以像 CNN 一樣使用多層的卷積層嗎？答案是肯定的，但使用多層 LSTM，必須注意一些重要的參數，可參閱「keras 官網 LSTM 的說明」[2]：

1. return_sequences：預設為 False，只傳回最後一個神經元的輸出(y)，若為 True，則表示每一個神經元的輸出(y)都會傳回。
2. return_state：預設為 False，不會傳回隱藏層狀態(Hidden State)及記憶狀態(Cell State)，反之，則會傳回最後一個神經元的狀態，當作下一週期或批次的輸入。
3. stateful：預設為 False，前一批的隱藏層及記憶狀態不會傳給下一批訓練，延續記憶，反之，則會傳回前一批的隱藏層及記憶狀態，作為下一批訓練的輸入。
4. 若兩個 LSTM 層相串連，前面的 LSTM 必須設定參數 return_sequences=True，表示每一神經元的輸出(y)都會傳回，使用批次(Batch)時，須設定 stateful=True，表示訓練時每一批的最後輸出會接到下一批的輸入，同時每一個訓練週期(Epoch)後要呼叫 reset_states 重置狀態，避免接到上一個訓練週期的輸出。

以上的說明應該有些複雜，我們以實作幫助理解，這部份需要專心一點，大家喝杯咖啡提神一下吧。

【範例 2】LSTM 重要參數測試。
程式 12_04_LSTM_參數測試.ipynb。

1. 載入相關套件。

```
1  # 載入相關套件
2  import tensorflow as tf
3  from tensorflow.keras.layers import Dense, LSTM, Input
4  from tensorflow.keras.models import Sequential
5  import numpy as np
```

2. 定義模型，內含一個 LSTM，參數均為預設值。

```
1  # 定義模型，參數均為預設值
2  model = Sequential()
3  model.add(LSTM(1, input_shape=(3, 1)))
4
5  # 測試資料
6  data = np.array([0.1, 0.2, 0.3]).reshape((1,3,1))
7  # 預測：只傳回最後的輸出(y)
8  print(model.predict(data))
```

執行結果：LSTM 輸入 3 個數值，輸出 1 個數值為 [[0.10563352]]。

3. 加一個參數 return_sequences=True。

```
1  # 定義模型，參數 return_sequences=True
2  model = Sequential()
3  model.add(LSTM(1, input_shape=(3, 1), return_sequences=True))
4
5  # 測試資料
6  data = np.array([0.1, 0.2, 0.3]).reshape((1,3,1))
7  # 預測：傳回每一節點的輸出(y)
8  print(model.predict(data))
```

執行結果：LSTM 每一個神經元的輸出均傳回，3 個輸入就會有 3 個輸出：

```
[[[0.02329711], [0.06432742], [0.11739781]]]。
```

4. 加一個參數 return_state=True。

```
1  # 定義模型，參數 return_state=True
2  # 多個輸出必須使用 Function API
3  from keras.models import Model
4
5  inputs1 = Input(shape=(3, 1))
6  lstm1 = LSTM(1, return_state=True)(inputs1)
7  model = Model(inputs=inputs1, outputs=lstm1)
8
9  # 測試資料
10  data = np.array([0.1, 0.2, 0.3]).reshape((1,3,1))
11  # 預測：傳回 輸出(y), state_h, state_c
12  print(model.predict(data))
```

執行結果：除了一個輸出外，還會傳回隱藏層狀態(Hidden State)及記憶狀態(Cell State)，總共 3 個數值：

```
[array([[-0.04487724]], dtype=float32), array([[-0.04487724]],
dtype=float32), array([[-0.07851784]], dtype=float32)]。
```

注意，多個輸出必須使用 Function API，不能使用 Sequential 模型。

5. return_sequences=True、return_state=True

```
1   # 定義模型，參數 return_sequences=True、return_state=True
2   # 多個輸出必須使用 Function API
3   from keras.models import Model
4
5   inputs1 = Input(shape=(3, 1))
6   lstm1 = LSTM(1, return_sequences=True, return_state=True)(inputs1)
7   model = Model(inputs=1nputs1, outputs=lstm1)
8
9   # 測試資料
10  data = np.array([0.1, 0.2, 0.3]).reshape((1,3,1))
11  # 預測：傳回 輸出(y), state_h, state_c
12  print(model.predict(data))
```

執行結果：輸出有 3 個數值，還有隱藏層狀態(Hidden State)及記憶狀態(Cell State)，總共 3 個數值：

```
[array([[-0.04487724]], dtype=float32), array([[-0.04487724]],
dtype=float32), array([[-0.07851784]], dtype=float32)]。
```

6. 模型包含 2 個 LSTM 神經層。

```
1   # 定義模型，參數 return_sequences=True、return_state=True
2   # 多個輸出必須使用 Function API
3   from keras.models import Model
4
5   inputs1 = Input(shape=(3, 1))
6   lstm1 = LSTM(1, return_sequences=True)(inputs1)
7   lstm2 = LSTM(1, return_sequences=True, return_state=True)(lstm1)
8   model = Model(inputs=inputs1, outputs=lstm2)
9
10  # 測試資料
11  data = np.array([0.1, 0.2, 0.3]).reshape((1,3,1))
12  # 預測：傳回 輸出(y), state_h, state_c
13  print(model.predict(data))
```

第 1 個 LSTM 神經層的參數 return_sequences 須為 True，否則會出現錯誤如下：

```
incompatible with the layer: expected ndim=3, found ndim=2.
```

執行結果：輸出有 3 個數值，還有隱藏層狀態(Hidden State)及記憶狀態(Cell State)，總共 3 個數值：

```
[array([[[-0.00471755], [-0.01753834], [-0.04042896]]],
dtype=float32),
array([[-0.04042896]], dtype=float32)
array([[-0.07566848]], dtype=float32)]。
```

藉由以上的實驗，我們就比較能理解 LSTM 參數的影響。接下來看另一個實例，使用多層 LSTM，即所謂堆疊 LSTM (Stacked LSTM)，可以建立更多層的神經網路，與 CNN 的卷積層一樣，主要目的是萃取特徵，常被運用在語音辨識方面。

【範例 3】時間序列(Time Series)預測，以 LSTM 演算法預測航空公司的未來營收，包括以下各種模型測試：

- 前期資料為 X，當期資料為 Y。
- 取前 3 期的資料作為 X，即以 t-3、t-2、t-1 期預測 t 期。
- 以 t-3 期(單期)預測 t 期。
- 將每個週期再分多批訓練。
- Stacked LSTM：使用多層的 LSTM。

程式 **12_05_Stacked_LSTM.ipynb**，修改自「Time Series Prediction with LSTM Recurrent Neural Networks in Python with Keras」[3]。

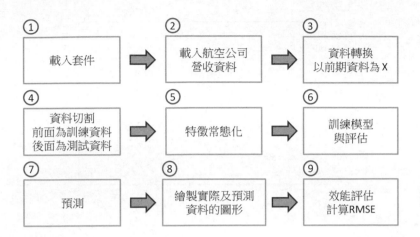

1. 載入相關套件。

```
1  # 載入相關套件
2  import tensorflow as tf
3  from tensorflow.keras.datasets import imdb
4  from tensorflow.keras.layers import Embedding, Dense, LSTM
5  from tensorflow.keras.losses import BinaryCrossentropy
6  from tensorflow.keras.models import Sequential
7  from tensorflow.keras.optimizers import Adam
8  from tensorflow.keras.preprocessing.sequence import pad_sequences
9  import pandas as pd
10 import numpy as np
11 import matplotlib.pyplot as plt
```

2. 載入航空公司的營收資料：這份資料年代久遠，每月一筆資料，自 1949 年 1 月至 1960 年 12 月。

```
1  # 載入測試資料
2  df = pd.read_csv('./RNN/monthly-airline-passengers.csv')
3  df.head()
```

3. 繪圖：透過圖表可以發現，航空公司的營收除了有淡旺季之分以外，還有逐步上升的成長趨勢。

```
1  # 繪圖
2  df2 = df.set_index('Month')
3  df2.plot(legend=None)
4  plt.xticks(rotation=30)
```

執行結果：

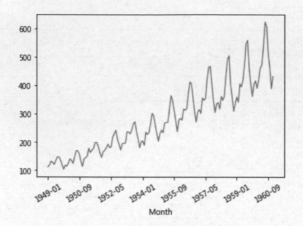

4. 轉換資料：以前期資料為 X，當期資料為 Y，以前期營收預測當期營收。因此訓練資料和測試資料不採取隨機切割，前面 2/3 為訓練資料，後面 1/3 為測試資料，並對特徵進行常態化。

```
1   # 轉換資料
2   from sklearn.preprocessing import MinMaxScaler
3
4   # 函數：以前期資料為 X，當前期資料為 Y
5   def create_dataset(dataset, look_back=1):
6       dataX, dataY = [], []
7       for i in range(len(dataset)-look_back-1):
8           a = dataset[i:(i+look_back), 0]
9           dataX.append(a)
10          dataY.append(dataset[i + look_back, 0])
11      return np.array(dataX), np.array(dataY)
12
13  dataset = df2.values
14  dataset = dataset.astype('float32')
15
16  # X 常態化
17  scaler = MinMaxScaler(feature_range=(0, 1))
18  dataset = scaler.fit_transform(dataset)
19
20  # 資料分割
21  train_size = int(len(dataset) * 0.67)
22  test_size = len(dataset) - train_size
23  train, test = dataset[0:train_size,:], dataset[train_size:len(dataset),:]
24
25  # 以前期資料為 X，當期資料為 Y
26  look_back = 1
27  trainX, trainY = create_dataset(train, look_back)
28  testX, testY = create_dataset(test, look_back)
29
30  # 轉換為三維 [筆數, 落後期數, X維度]
31  trainX = np.reshape(trainX, (trainX.shape[0], 1, trainX.shape[1]))
32  testX = np.reshape(testX, (testX.shape[0], 1, testX.shape[1]))
```

5. 模型訓練與評估：由於訓練資料較少，故訓練較多執行週期(100)。模型為 LSTM+Dense(1)，會輸出一個數值，即下期營收。LSTM 的 input_shape 為(筆數, 落後期數, 特徵個數)三維。

```
1   # 訓練模型
2   model = Sequential()
3   model.add(LSTM(4, input_shape=(1, look_back)))
4   model.add(Dense(1))
5   model.compile(loss='mean_squared_error', optimizer='adam')
6   model.fit(trainX, trainY, epochs=100, batch_size=1, verbose=1)
7
8   # 模型評估
9   trainPredict = model.predict(trainX)
10  testPredict = model.predict(testX)
```

6. 預測後還原常態化：繪製實際資料和預測資料的圖表。

```
1   from sklearn.metrics import mean_squared_error
2   import math
3
4   # 還原常態化的訓練及測試資料
5   trainPredict = scaler.inverse_transform(trainPredict)
6   trainY = scaler.inverse_transform([trainY])
7   testPredict = scaler.inverse_transform(testPredict)
8   testY = scaler.inverse_transform([testY])
9
10  # 計算 RMSE
11  trainScore = math.sqrt(mean_squared_error(trainY[0], trainPredict[:,0]))
12  print('Train Score: %.2f RMSE' % (trainScore))
13  testScore = math.sqrt(mean_squared_error(testY[0], testPredict[:,0]))
14  print('Test Score: %.2f RMSE' % (testScore))
15
16  # 訓練資料的 X/Y
17  trainPredictPlot = np.empty_like(dataset)
18  trainPredictPlot[:, :] = np.nan
19  trainPredictPlot[look_back:len(trainPredict)+look_back, :] = trainPredict
20
21  # 測試資料 X/Y
22  testPredictPlot = np.empty_like(dataset)
23  testPredictPlot[:, :] = np.nan
24  testPredictPlot[len(trainPredict)+(look_back*2)+1:len(dataset)-1, :] = testPredict
25
26  # 繪圖
27  plt.plot(scaler.inverse_transform(dataset))
28  plt.plot(trainPredictPlot)
29  plt.plot(testPredictPlot)
30  plt.show()
```

執行結果：訓練資料和測試資料的 RMSE 分別為 22.84、47.63。

圖表請參考程式，藍色線條為實際值，橘色為訓練資料的預測值，綠色為測試資料的預測值。

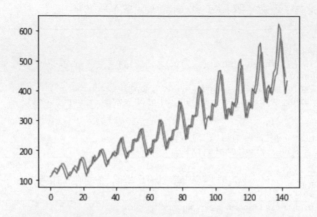

7. Loopback 改為 3：X 由前 1 期改為前 3 期，即以 t-3、t-2、t-1 期預測 t 期。

```
1   # 載入測試資料
2   df = pd.read_csv('./RNN/monthly-airline-passengers.csv', usecols=[1])
3
4   dataset = df.values
5   dataset = dataset.astype('float32')
6
7   # X 常態化
8   scaler = MinMaxScaler(feature_range=(0, 1))
9   dataset = scaler.fit_transform(dataset)
10
11  # 資料分割
12  train_size = int(len(dataset) * 0.67)
13  test_size = len(dataset) - train_size
14  train, test = dataset[0:train_size,:], dataset[train_size:len(dataset),:]
15
16  # 以前期資料為 X，當前期資料為 Y
17  look_back = 3
18  trainX, trainY = create_dataset(train, look_back)
19  testX, testY = create_dataset(test, look_back)
20
21  # 轉換為三維 [筆數, 落後期數, X維度]
22  trainX = np.reshape(trainX, (trainX.shape[0], 1, trainX.shape[1]))
23  testX = np.reshape(testX, (testX.shape[0], 1, testX.shape[1]))
```

執行結果：訓練資料的特徵(筆數, 落後期數, X 維度) = (92, 1, 3) 三維度。

8. 模型訓練與評估： LSTM 的參數 input_shape 設為(1, 3)，因為特徵有 3 個(期)。

```
1   # 訓練模型
2   model = Sequential()
3   model.add(LSTM(4, input_shape=(1, look_back)))
4   model.add(Dense(1))
5   model.compile(loss='mean_squared_error', optimizer='adam')
6   model.fit(trainX, trainY, epochs=100, batch_size=1, verbose=1)
7
8   # 模型評估
9   trainPredict = model.predict(trainX)
10  testPredict = model.predict(testX)
```

9. 預測後還原常態化：繪製實際資料和預測資料的圖表。

```
1   from sklearn.metrics import mean_squared_error
2   import math
3
4   # 還原常態化的訓練資料
5   trainPredict = scaler.inverse_transform(trainPredict)
6   trainY = scaler.inverse_transform([trainY])
7   testPredict = scaler.inverse_transform(testPredict)
8   testY = scaler.inverse_transform([testY])
9
10  # 計算 RMSE
11  trainScore = math.sqrt(mean_squared_error(trainY[0], trainPredict[:,0]))
12  print('Train Score: %.2f RMSE' % (trainScore))
13  testScore = math.sqrt(mean_squared_error(testY[0], testPredict[:,0]))
14  print('Test Score: %.2f RMSE' % (testScore))
15
16  # 訓練資料的 X/Y
17  trainPredictPlot = np.empty_like(dataset)
18  trainPredictPlot[:, :] = np.nan
19  trainPredictPlot[look_back:len(trainPredict)+look_back, :] = trainPredict
20
21  # 測試資料 X/Y
22  testPredictPlot = np.empty_like(dataset)
23  testPredictPlot[:, :] = np.nan
24  testPredictPlot[len(trainPredict)+(look_back*2)+1:len(dataset)-1, :] = testPredict
25
26  # 繪圖
27  plt.plot(scaler.inverse_transform(dataset))
28  plt.plot(trainPredictPlot)
29  plt.plot(testPredictPlot)
30  plt.show()
```

執行結果：訓練資料和測試資料的 RMSE 分別為 22.83、62.55，測試 RMSE 比只以 1 期為特徵預測差，因為使用多期，有移動平均的效果，預測曲線會比較平緩，比較不容易受到激烈變化的樣本點影響。

圖表請參考程式，藍色線條為實際值，橘色為訓練資料的預測值，綠色為測試資料的預測值。

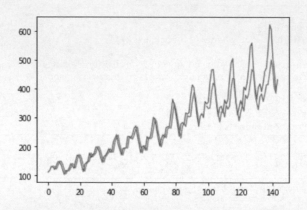

10. 改變落後期數(Time Steps)為 3：即以 t-3 期預測 t 期。

```
1  # 載入測試資料
2  df = pd.read_csv('./RNN/monthly-airline-passengers.csv', usecols=[1])
3
4  dataset = df.values
5  dataset = dataset.astype('float32')
6
7  # X 常態化
8  scaler = MinMaxScaler(feature_range=(0, 1))
9  dataset = scaler.fit_transform(dataset)
10
11 # 資料分割
12 train_size = int(len(dataset) * 0.67)
13 test_size = len(dataset) - train_size
14 train, test = dataset[0:train_size,:], dataset[train_size:len(dataset),:]
15
16 # 以前期資料為 X，當前期資料為 Y
17 look_back = 3
18 trainX, trainY = create_dataset(train, look_back)
19 testX, testY = create_dataset(test, look_back)
20
21 # 轉換為三維 [筆數, 落後期數, X維度]
22 # trainX = np.reshape(trainX, (trainX.shape[0], 1, trainX.shape[1]))
23 # testX = np.reshape(testX, (testX.shape[0], 1, testX.shape[1]))
24 trainX = np.reshape(trainX, (trainX.shape[0], trainX.shape[1], 1))
25 testX = np.reshape(testX, (testX.shape[0], testX.shape[1], 1))
```

執行結果：訓練資料的特徵三維度(筆數, 落後期數, X 維度) = (92, 3, 1)。

11. 模型訓練與評估： LSTM 的參數 input_shape 設為(3, 1)，因為特徵只有 1 個，落後 3 期(t-3)。

```
1   # 訓練模型
2   model = Sequential()
3   # (1, look_back) 改為 (look_back, 1)
4   model.add(LSTM(4, input_shape=(look_back, 1)))
5   model.add(Dense(1))
6   model.compile(loss='mean_squared_error', optimizer='adam')
7   model.fit(trainX, trainY, epochs=100, batch_size=1, verbose=1)
8
9   # 模型評估
10  trainPredict = model.predict(trainX)
11  testPredict = model.predict(testX)
```

12. 預測後還原常態化：繪製實際資料和預測資料的圖表。

```
1   from sklearn.metrics import mean_squared_error
2   import math
3
4   # 還原常態化的訓練資料
5   trainPredict = scaler.inverse_transform(trainPredict)
6   trainY = scaler.inverse_transform([trainY])
7   testPredict = scaler.inverse_transform(testPredict)
8   testY = scaler.inverse_transform([testY])
9
10  # 計算 RMSE
11  trainScore = math.sqrt(mean_squared_error(trainY[0], trainPredict[:,0]))
12  print('Train Score: %.2f RMSE' % (trainScore))
13  testScore = math.sqrt(mean_squared_error(testY[0], testPredict[:,0]))
14  print('Test Score: %.2f RMSE' % (testScore))
15
16  # 訓練資料的 X/Y
17  trainPredictPlot = np.empty_like(dataset)
18  trainPredictPlot[:, :] = np.nan
19  trainPredictPlot[look_back:len(trainPredict)+look_back, :] = trainPredict
20
21  # 測試資料 X/Y
22  testPredictPlot = np.empty_like(dataset)
23  testPredictPlot[:, :] = np.nan
24  testPredictPlot[len(trainPredict)+(look_back*2)+1:len(dataset)-1, :] = testPredict
25
26  # 繪圖
27  plt.plot(scaler.inverse_transform(dataset))
28  plt.plot(trainPredictPlot)
29  plt.plot(testPredictPlot)
30  plt.show()
```

執行結果：訓練資料和測試資料的 RMSE 分別為 27.19、66.19，測試 RMSE 也比「以前期為特徵預測」差。

圖表請參考程式，藍色線條為實際值，橘色為訓練資料的預測值，綠色為測試資料的預測值。

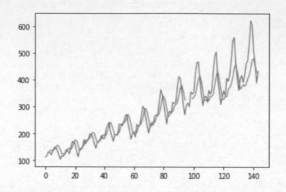

13. 將每個週期細分多批訓練：須設定 LSTM 參數 stateful=True，表示訓練時每一批的最後輸出會接到下一批的輸入，故每一個訓練週期(Epoch)要呼叫 reset_states 重置狀態(cell state)，避免接到上一個訓練週期的輸出。LSTM 的 input_shape 為(每批筆數, 落後期數, 特徵個數)三維，前置處理不變。

```
1   # 載入測試資料
2   df = pd.read_csv('./RNN/monthly-airline-passengers.csv', usecols=[1])
3
4   dataset = df.values
5   dataset = dataset.astype('float32')
6
7   # X 常態化
8   scaler = MinMaxScaler(feature_range=(0, 1))
9   dataset = scaler.fit_transform(dataset)
10
11  # 資料分割
12  train_size = int(len(dataset) * 0.67)
13  test_size = len(dataset) - train_size
14  train, test = dataset[0:train_size,:], dataset[train_size:len(dataset),:]
15
16  # 以前期資料為 X，當前期資料為 Y
17  look_back = 3
18  trainX, trainY = create_dataset(train, look_back)
19  testX, testY = create_dataset(test, look_back)
20
21  # 轉換為三維 [筆數, 落後期數, X維度]
22  # trainX = np.reshape(trainX, (trainX.shape[0], 1, trainX.shape[1]))
23  # testX = np.reshape(testX, (testX.shape[0], 1, testX.shape[1]))
24  trainX = np.reshape(trainX, (trainX.shape[0], trainX.shape[1], 1))
25  testX = np.reshape(testX, (testX.shape[0], testX.shape[1], 1))
```

14. 模型訓練與評估：以批量設為 1 測試，也可以加大批量。

```
1  # 訓練模型
2  model = Sequential()
3  # (1, look_back) 改為 (look_back, 1)
4  # model.add(LSTM(4, input_shape=(look_back, 1)))
5  batch_size = 1
6  model.add(LSTM(4, batch_input_shape=(batch_size, look_back, 1), stateful=True))
7  model.add(Dense(1))
8  model.compile(loss='mean_squared_error', optimizer='adam')
9  # model.fit(trainX, trainY, epochs=100, batch_size=1, verbose=1)
10 for i in range(100):
11     model.fit(trainX, trainY, epochs=1, batch_size=batch_size,
12             shuffle=False, verbose=2)
13     # 重置狀態(cell state)
14     model.reset_states()
15
16 # 模型評估
17 trainPredict = model.predict(trainX, batch_size=batch_size)
18 model.reset_states()
19 testPredict = model.predict(testX, batch_size=batch_size)
```

15. 預測後還原常態化：繪製實際資料和預測資料的圖表。

```
1  from sklearn.metrics import mean_squared_error
2  import math
3
4  # 還原常態化的訓練資料
5  trainPredict = scaler.inverse_transform(trainPredict)
6  trainY = scaler.inverse_transform([trainY])
7  testPredict = scaler.inverse_transform(testPredict)
8  testY = scaler.inverse_transform([testY])
9
10 # 計算 RMSE
11 trainScore = math.sqrt(mean_squared_error(trainY[0], trainPredict[:,0]))
12 print('Train Score: %.2f RMSE' % (trainScore))
13 testScore = math.sqrt(mean_squared_error(testY[0], testPredict[:,0]))
14 print('Test Score: %.2f RMSE' % (testScore))
15
16 # 訓練資料的 X/Y
17 trainPredictPlot = np.empty_like(dataset)
18 trainPredictPlot[:, :] = np.nan
19 trainPredictPlot[look_back:len(trainPredict)+look_back, :] = trainPredict
20
21 # 測試資料 X/Y
22 testPredictPlot = np.empty_like(dataset)
23 testPredictPlot[:, :] = np.nan
24 testPredictPlot[len(trainPredict)+(look_back*2)+1:len(dataset)-1, :] = testPredict
25
26 # 繪圖
27 plt.plot(scaler.inverse_transform(dataset))
28 plt.plot(trainPredictPlot)
29 plt.plot(testPredictPlot)
30 plt.show()
```

執行結果：訓練資料和測試資料的 RMSE 分別為 25.91、51.74，測試 RMSE 比「不分批次」有好一點，但不明顯，主要是示範 stateful 參數的用法。

16. Stacked LSTM：使用多層的 LSTM，以下前置處理不變。

```
1   # 載入測試資料
2   df = pd.read_csv('./RNN/monthly-airline-passengers.csv', usecols=[1])
3
4   dataset = df.values
5   dataset = dataset.astype('float32')
6
7   # X 常態化
8   scaler = MinMaxScaler(feature_range=(0, 1))
9   dataset = scaler.fit_transform(dataset)
10
11  # 資料分割
12  train_size = int(len(dataset) * 0.67)
13  test_size = len(dataset) - train_size
14  train, test = dataset[0:train_size,:], dataset[train_size:len(dataset),:]
15
16  # 以前期資料為 X，當前期資料為 Y
17  look_back = 3
18  trainX, trainY = create_dataset(train, look_back)
19  testX, testY = create_dataset(test, look_back)
20
21  # 轉換為三維 [筆數, 落後期數, X維度]
22  # trainX = np.reshape(trainX, (trainX.shape[0], 1, trainX.shape[1]))
23  # testX = np.reshape(testX, (testX.shape[0], 1, testX.shape[1]))
24  trainX = np.reshape(trainX, (trainX.shape[0], trainX.shape[1], 1))
25  testX = np.reshape(testX, (testX.shape[0], testX.shape[1], 1))
```

17. 模型訓練與評估：因為 LSTM 需接收上一個神經元的隱藏層輸出 (h_{t-1})，因此，若兩個 LSTM 層串連，前面的 LSTM 必須設定參數 return_sequences=True，表示每一個神經元的輸出(y)都會傳給下一個 LSTM。訓練資料與測試資料預測之間也要重置狀態(cell state)，避免訓練資料預測的狀態傳給測試資料預測。

```
1   # 訓練模型
2   model = Sequential()
3   # (1, look_back) 改為 (look_back, 1)
4   # model.add(LSTM(4, input_shape=(look_back, 1)))
5   batch_size = 1
6
7   # Stacked LSTM
8   model.add(LSTM(4, batch_input_shape=(batch_size, look_back, 1),
9               stateful=True, return_sequences=True))
```

```
10  model.add(LSTM(4, batch_input_shape=(batch_size, look_back, 1),
11              stateful=True))
12
13  model.add(Dense(1))
14  model.compile(loss='mean_squared_error', optimizer='adam')
15  # model.fit(trainX, trainY, epochs=100, batch_size=1, verbose=1)
16  for i in range(100):
17      model.fit(trainX, trainY, epochs=1, batch_size=batch_size,
18              shuffle=False, verbose=2)
19      # 重置狀態(cell state)
20      model.reset_states()
21
22  # 模型評估
23  trainPredict = model.predict(trainX, batch_size=batch_size)
24  # 重置狀態(cell state)
25  model.reset_states()
26  testPredict = model.predict(testX, batch_size=batch_size)
```

18. 預測後還原常態化：繪製實際資料和預測資料的圖表。

```
 1  from sklearn.metrics import mean_squared_error
 2  import math
 3
 4  # 還原常態化的訓練資料
 5  trainPredict = scaler.inverse_transform(trainPredict)
 6  trainY = scaler.inverse_transform([trainY])
 7  testPredict = scaler.inverse_transform(testPredict)
 8  testY = scaler.inverse_transform([testY])
 9
10  # 計算 RMSE
11  trainScore = math.sqrt(mean_squared_error(trainY[0], trainPredict[:,0]))
12  print('Train Score: %.2f RMSE' % (trainScore))
13  testScore = math.sqrt(mean_squared_error(testY[0], testPredict[:,0]))
14  print('Test Score: %.2f RMSE' % (testScore))
15
16  # 訓練資料的 X/Y
17  trainPredictPlot = np.empty_like(dataset)
18  trainPredictPlot[:, :] = np.nan
19  trainPredictPlot[look_back:len(trainPredict)+look_back, :] = trainPredict
20
21  # 測試資料 X/Y
22  testPredictPlot = np.empty_like(dataset)
23  testPredictPlot[:, :] = np.nan
24  testPredictPlot[len(trainPredict)+(look_back*2)+1:len(dataset)-1, :] = testPredict
25
26  # 繪圖
27  plt.plot(scaler.inverse_transform(dataset))
28  plt.plot(trainPredictPlot)
29  plt.plot(testPredictPlot)
30  plt.show()
```

執行結果：訓練資料和測試資料的 RMSE 分別為 24.28、86.85，測試 RMSE
比較差，應該是因為資料很單純，使用太複雜的網路結構反而沒有助益。

圖表請參考程式，藍色線條為實際值，橘色為訓練資料的預測值，綠色為測試資料的預測值。

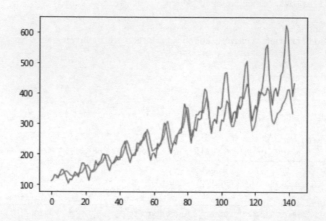

▍ 12-4 Gate Recurrent Unit (GRU)

Gate Recurrent Unit (GRU)也是 RNN 變形的演算法，由 Kyunghyun Cho 在 2014 年提出的，可參閱「Empirical Evaluation of Gated Recurrent Neural Networks on Sequence Modeling」[4]，主要就是要改良 LSTM 缺陷：

1. LSTM 計算過慢，GRU 可改善訓練速度。
2. 簡化 LSTM 模型，節省記憶體的空間。

LSTM 是由遺忘閥(Forget gate)與輸入閥(Input gate)來維護記憶狀態(Cell State)，然而因為這部分太過耗時，所以 GRU 廢除記憶狀態，直接使用隱藏層輸出(h_t)，並且將前述兩個閥改由更新閥(Update gate)替代，兩個模型的架構比較圖如下。

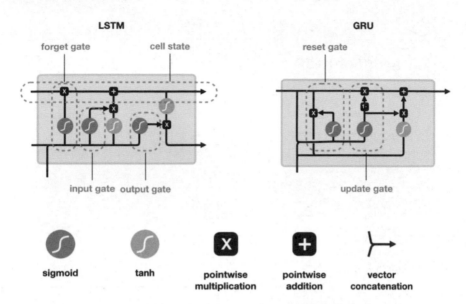

圖 12.5 LSTM 與 GRU 內部結構的比較，左圖為 LSTM，右圖為 GRU，圖片來源：
「Illustrated Guide to LSTM's and GRU's: A step by step explanation」[5]

雖然原作者提出效能測試圖表，說明 GRU 的效能比 LSTM 好，不過，稍
微吐槽一下，筆者實際測試的結果，差異並不明顯，而且網路上也比較少
提到 GRU，大多仍以 LSTM 為主流，因此，就不詳細研究了。

12-5 股價預測

由於 LSTM 與時間序列模型很類似，所以網路上有許多文章探討以 LSTM
預測股票價格，以下我們就來實作看看。

【範例 4】以 LSTM/GRU 演算法預測股價。
程式 **12_06_Stock_Forecast.ipynb**，修改自「Predicting stock prices with
LSTM」[6]。

資料集：本範例使用亞馬遜企業股票(https://www.kaggle.com/szrlee/stock-
time-series-20050101-to-20171231)，也可以使用台股。

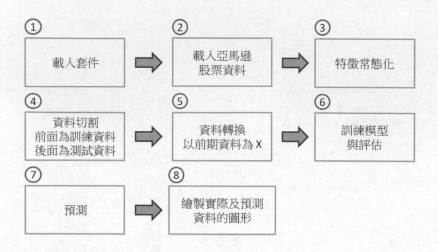

1. 載入相關套件。

```
1  # 載入相關套件
2  import tensorflow as tf
3  from tensorflow.keras.datasets import imdb
4  from tensorflow.keras.layers import Embedding, Dense, LSTM, Dropout
5  from tensorflow.keras.losses import BinaryCrossentropy
6  from tensorflow.keras.models import Sequential
7  from tensorflow.keras.optimizers import Adam
8  import pandas as pd
9  import numpy as np
10 import matplotlib.pyplot as plt
```

2. 載入測試資料。

```
1  # 載入測試資料 -- 亞馬遜
2  df = pd.read_csv('./RNN/AMZN_2006-01-01_to_2018-01-01.csv',
3                   index_col='Date', parse_dates=['Date'])
4  df.head()
```

執行結果：

Date	Open	High	Low	Close	Volume	Name
2006-01-03	47.47	47.85	46.25	47.58	7582127	AMZN
2006-01-04	47.48	47.73	46.69	47.25	7440914	AMZN
2006-01-05	47.16	48.20	47.11	47.65	5417258	AMZN
2006-01-06	47.97	48.58	47.32	47.87	6154285	AMZN
2006-01-09	46.55	47.10	46.40	47.08	8945056	AMZN

3. 繪圖。

```
1  # 只使用收盤價
2  df = df['Close']
3
4  # 繪圖
5  plt.figure(figsize = (12, 6))
6  plt.plot(df, label='Stock Price')
7  plt.legend(loc='best')
8  plt.show()
```

執行結果：

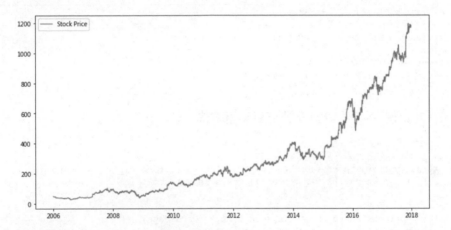

4. 參數設定：以過去 40 期為特徵(X)，一次預測 10 天(y)，測試資料量設定 20 期。

```
1  # 參數設定
2  look_back = 40     # 以過去 40 期為特徵(X)
3  forward_days = 10  # 一次預測 10 天 (y)
4  num_periods = 20   # 測試資料量設定 20 期
```

5. 特徵常態化。

```
1  # 特徵常態化
2  from sklearn.preprocessing import MinMaxScaler
3  scl = MinMaxScaler()
4  array = df.values.reshape(df.shape[0],1)
5  array = scl.fit_transform(array)
```

6. 前置處理函數：取得模型輸入的格式。

```
1  # 前置處理函數，取得模型輸入的格式
2  # look_back：特徵(X)個數，forward_days：目標(y)個數，jump：移動視窗
3  def processData(data, look_back, forward_days,jump=1):
4      X,Y = [],[]
5      for i in range(0,len(data) -look_back -forward_days +1, jump):
6          X.append(data[i:(i+look_back)])
7          Y.append(data[(i+look_back):(i+look_back+forward_days)])
8      return np.array(X),np.array(Y)
```

7. 資料切割成訓練資料和測試資料。

```
1  # 資料切割訓練資料及測試資料
2  # 一次預測 10 天，共 20 期
3  division = len(array) - num_periods*forward_days
4
5  # 再往前推 40 天當第一筆的 X
6  array_test = array[division-look_back:]
7  array_train = array[:division]
```

8. 前置處理、資料切割成訓練資料和驗證資料。

```
1   # 前置處理、資料切割
2   # 測試資料前置處理，注意最後一個參數，一次預測 10天，不重疊
3   X_test,y_test = processData(array_test,look_back,forward_days,forward_days)
4   y_test = np.array([list(a.ravel()) for a in y_test])
5
6   # 訓練資料前置處理
7   X,y = processData(array_train,look_back,forward_days)
8   y = np.array([list(a.ravel()) for a in y])
9
10  # 資料切割成訓練資料及驗證資料
11  from sklearn.model_selection import train_test_split
12  X_train, X_validate, y_train, y_validate = train_test_split(X, y, test_size=0.20)
```

9. 訓練模型。

```
1   # 訓練模型
2   NUM_NEURONS_FirstLayer = 50
3   NUM_NEURONS_SecondLayer = 30
4   EPOCHS = 10
5
6   # 模型
7   model = Sequential()
8   model.add(LSTM(NUM_NEURONS_FirstLayer,input_shape=(look_back,1), return_sequences=True))
9   model.add(LSTM(NUM_NEURONS_SecondLayer,input_shape=(NUM_NEURONS_FirstLayer,1)))
10  model.add(Dense(forward_days))
11  model.compile(loss='mean_squared_error', optimizer='adam')
12
13  # 訓練
14  history = model.fit(X_train,y_train,epochs=EPOCHS,validation_data=(X_validate,y_validate)
15                      ,shuffle=True,batch_size=2, verbose=2)
```

10. 繪製損失函數。

```
1  # 繪製損失函數
2  plt.figure(figsize = (12, 6))
3  plt.plot(history.history['loss'], label='loss')
4  plt.plot(history.history['val_loss'], label='val_loss')
5  plt.legend(loc='best')
6  plt.show()
```

執行結果：

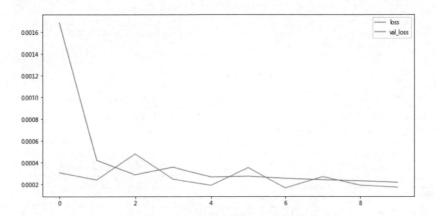

11. 若一次預測 1 天：jump=1。

```
1   # 前置處理，資料切割
2   # 測試資料前置處理，注意最後一個參數，一次預測 1 天
3   X_test,y_test = processData(array_test,look_back,1,1)
4   y_test = np.array([list(a.ravel()) for a in y_test])
5
6   # 測試資料預測
7   Xt = model.predict(X_test)
8   print(Xt.shape)
9
10  Xt = Xt[:, 0]
11
12  # 繪製測試資料預測值
13  plt.figure(figsize = (12, 6))
14  # 繪製 1 條預測值，scl.inverse_transform：還原常態化
15  plt.plot(scl.inverse_transform(Xt.reshape(-1,1)), color='r', label='Prediction')
16
17  # 繪製實際值
18  plt.plot(scl.inverse_transform(y_test.reshape(-1,1)), label='Target')
19  plt.legend(loc='best')
20  plt.show()
```

執行結果：若一次只預測 1 天，結果看起來還不錯。

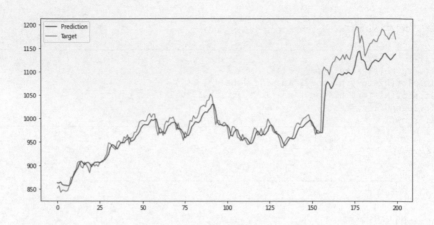

12. 一次預測 10 天，移動視窗不重疊。

```
1  # 測試資料前置處理，注意最後一個參數，一次預測 10天，移動視窗不重疊
2  X_test,y_test = processData(array_test,look_back,forward_days,forward_days)
3  y_test = np.array([list(a.ravel()) for a in y_test])
4
5  # 測試資料預測
6  Xt = model.predict(X_test)
7  Xt.shape
```

執行結果：(20, 10)表示 20 期，每期預測 10 天。

13. 繪製測試資料的預測值與實際值的比較。

```
1  # 繪製測試資料預測值
2  plt.figure(figsize = (12, 6))
3  # 繪製 20 條預測值，scl.inverse_transform：還原常態化
4  for i in range(0,len(Xt)):
5      plt.plot([x + i*forward_days for x in range(len(Xt[i]))],
6              scl.inverse_transform(Xt[i].reshape(-1,1)), color='r')
7
8  # 指定預測值 label
9  plt.plot(0, scl.inverse_transform(Xt[i].reshape(-1,1))[0], color='r'
10          , label='Prediction')
11
12  # 繪製實際值
13  plt.plot(scl.inverse_transform(y_test.reshape(-1,1)), label='Target')
14  plt.legend(loc='best')
15  plt.show()
```

執行結果：測試資料的預測值與實際值的比較，如下圖，預測並不理想。

第 9 行程式只是要指定標籤(Label)，並不是要畫線，缺這一行的話，預測值圖表會有 20 個。

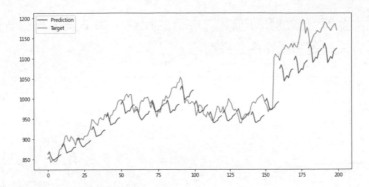

14. 全部資料預測：將訓練資料一併預測。

```
1   # 全部資料預測
2   division = len(array) - num_periods*forward_days
3   array_test = array[division-look_back:]
4
5   # 去掉不能整除的資料，取完整的訓練資料
6   leftover = division%forward_days+1
7   array_train = array[leftover:division]
8
9   Xtrain,ytrain = processData(array_train,look_back,forward_days,forward_days)
10  Xtest,ytest = processData(array_test,look_back,forward_days,forward_days)
11
12  # 預測
13  Xtrain = model.predict(Xtrain)
14  Xtrain = Xtrain.ravel() # 轉成一維
15
16  Xtest = model.predict(Xtest)
17  Xtest = Xtest.ravel() # 轉成一維
18
19  # 合併訓練資料與測試資料
20  y = np.concatenate((ytrain, ytest), axis=0)
```

15. 繪製測試資料的預測值與實際值的比較。

```
1   # 繪製訓練資料預測值
2   plt.figure(figsize = (12, 6))
3   plt.plot([x for x in range(look_back+leftover, len(Xtrain)+look_back+leftover)],
4            scl.inverse_transform(Xtrain.reshape(-1,1)), color='r', label='Train')
5   # 繪製測試資料預測值
6   plt.plot([x for x in range(look_back +leftover+ len(Xtrain),
7            len(Xtrain)+len(Xtest)+look_back+leftover)],
8            scl.inverse_transform(Xtest.reshape(-1,1)), color='y', label='Test')
9
10  # 繪製實際值
11  plt.plot([x for x in range(look_back+leftover,
12                           look_back+leftover+len(Xtrain)+len(Xtest))],
13           scl.inverse_transform(y.reshape(-1,1)), color='b', label='Target')
14
15  plt.legend(loc='best')
16  plt.show()
```

執行結果：全部資料的預測值與實際值的比較，如下圖，預測看似相當理想，但其實只是鏡頭拉遠的效果而已。

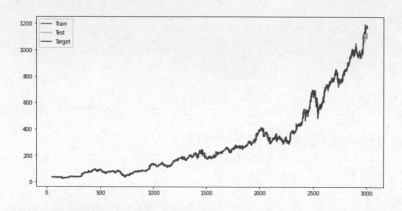

16. 拉近看，只觀察測試資料的預測值。

```
1  # 繪製測試資料預測值
2  plt.figure(figsize = (12, 6))
3  # 全部連成一線
4  plt.plot(scl.inverse_transform(Xtest.reshape(-1,1)))
5  # 畫20條線
6  for i in range(0,len(Xt)):
7      plt.plot([x + i*forward_days for x in range(len(Xt[i]))],
8               scl.inverse_transform(Xt[i].reshape(-1,1)), color='r')
```

執行結果：測試資料的預測值與實際值的比較，如下圖，在移動視窗之間畫線就會造成錯覺。

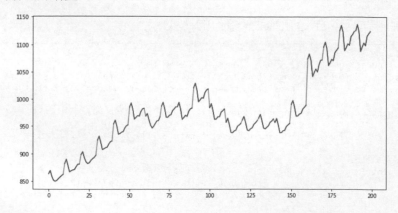

17. 改用 GRU 模型：GRU 的用法/參數與 LSTM 的完全相同。

```
1  from tensorflow.keras.layers import GRU
2
3  model_GRU = Sequential()
4  model_GRU.add(GRU(NUM_NEURONS_FirstLayer,input_shape=(look_back,1)
5                   , return_sequences=True))
6  model_GRU.add(GRU(NUM_NEURONS_SecondLayer
7                   ,input_shape=(NUM_NEURONS_FirstLayer,1)))
8  model_GRU.add(Dense(forward_days))
9  model_GRU.compile(loss='mean_squared_error', optimizer='adam')
10
11 history = model_GRU.fit(X_train,y_train,epochs=EPOCHS
12                   ,validation_data=(X_validate,y_validate)
13                   ,shuffle=True,batch_size=2, verbose=2)
```

使用 LSTM 預測股價並不準確，有下列的原因：

1. 股價非穩態(Non-stationary)：每個時間點的股價平均數與標準差都不一致，從圖表觀察，股價資料有一股趨勢影響力，並非隨機跳動。因此，時間序列預測通常會將股價轉換為收益率(Return Rate)，使資料呈現穩態。

2. 股價變化非常大，以長期歷史資料預測未來股價並不合理。

3. 只預測 1 天還算準確，但是意義不大，因為 1 天股價漲跌幅度有限，預測微幅波動並無意義，假若要預測多天來掌握長期趨勢，依上述測試，LSTM 的預測結果並不準確。

以 LSTM 預測股價，網路上有很多的討論，有興趣的讀者可以自行搜尋相關資料來閱讀，並以回測(Back testing)的方式測試個股，觀察策略是否奏效，回測的方式可參考筆者撰寫的「演算法交易(Algorithmic Trading) 實作」[7]一文。

12-6 注意力機制(Attention Mechanism)

RNN 從之前的隱藏層狀態取得上文的資訊，LSTM 則額外維護一條記憶線(Cell State)，目的都是希望能藉由記憶的方式來提高預測的準確性，但是，兩個演算法都侷限於上文的序列順序，導致越靠近預測目標的資訊，權重越大。實際上當我們在閱讀一篇文章時，往往會對文中的標題、人事時地物或強烈的形容詞特別注意,這就是所謂的注意力機制(Attention Mechanism),對於圖像也是如此,例如下圖。

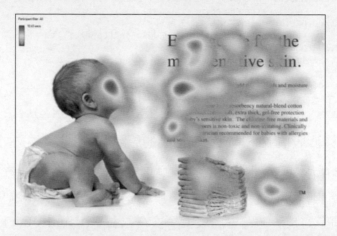

圖 12.6 人類的視覺注意力分佈,比如圖片左方,嬰兒的臉部就是注意力熱區,圖右下方的紙尿褲也是注意力熱區,圖片來源:「深度學習中的注意力機制(2017 版)」[8]

只要透過注意力機制,在預測時可額外把重點單字或部位納入考量,而不只是上下文。注意力機制常被應用到神經機器翻譯(Neural Machine Translation, NMT),下面我們就來瞧瞧 NMT 的作法。

機器翻譯是一個序列生成的模型,它是一種 Encoder-Decoder 的變形,稱為序列到序列(Sequence to Sequence, Seq2Seq)模型,結構如下圖,其中的 Context Vector 是 Encoder 輸出的上下文向量,類似 CNN AutoEncoder 萃取的特徵向量。

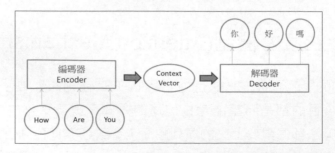

圖 12.7 Seq2Seq 模型,「How are you」翻譯為「你好嗎」

也可以應用於對話問答。

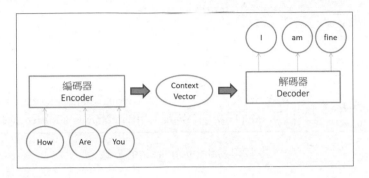

圖 12.8 對話問答，問「How are you」，回答「I am fine」

而注意力機制就是把要輸入到解碼器(decoder)的詞彙都乘上一個權重(weight)，與 Context Vector 混合計算成 Attention Vector，以預測下一個詞彙，這個機制會應用到解碼器的每一層，如下圖所示。

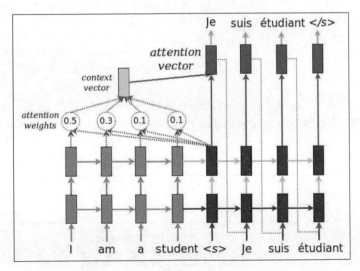

圖 12.8 注意力(Attention)機制

權重(weight)的計算的方式有兩種，分別由 Luong 和 Bahdanau 提出的乘法與加法的公式，它利用完全連接層(Dense)及 Softmax activation function，優化求得整個語句內每個詞彙可能的機率。

$$\alpha_{ts} = \frac{\exp\left(\text{score}(\boldsymbol{h}_t, \bar{\boldsymbol{h}}_s)\right)}{\sum_{s'=1}^{S} \exp\left(\text{score}(\boldsymbol{h}_t, \bar{\boldsymbol{h}}_{s'})\right)} \qquad \text{[Attention weights]}$$

$$\boldsymbol{c}_t = \sum_s \alpha_{ts} \bar{\boldsymbol{h}}_s \qquad \text{[Context vector]}$$

$$\boldsymbol{a}_t = f(\boldsymbol{c}_t, \boldsymbol{h}_t) = \tanh(\boldsymbol{W}_c[\boldsymbol{c}_t; \boldsymbol{h}_t]) \qquad \text{[Attention vector]}$$

$$\text{score}(\boldsymbol{h}_t, \bar{\boldsymbol{h}}_s) = \begin{cases} \boldsymbol{h}_t^\top \boldsymbol{W} \bar{\boldsymbol{h}}_s & \text{[Luong's multiplicative style]} \\ \boldsymbol{v}_a^\top \tanh\left(\boldsymbol{W_1} \boldsymbol{h}_t + \boldsymbol{W_2} \bar{\boldsymbol{h}}_s\right) & \text{[Bahdanau's additive style]} \end{cases}$$

圖 12.9 Attention weight 公式

虛擬程式碼如下：

```
score = FC(tanh(FC(EO) + FC(H)))，其中
FC：完全連接層
EO：Encoder 輸出(output)
H：所有隱藏層輸出
tanh：tanh activation function
Attention weights = softmax(score, axis = 1)
Context vector = sum(Attention weights * EO, axis = 1)
Embedding output = 解碼器的 input 經嵌入層(Embedding layer)處理後的輸出。
Attention Vector = concat(embedding output, context vector)。
```

推薦各位「淺談神經機器翻譯 & 用 Transformer 與 TensorFlow 2 英翻中」[9]一文的流程圖動畫，比較兩個模型的差異，從另一角度觀察，會有更多的收穫。

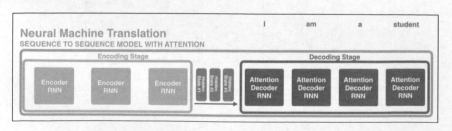

圖 12.10 Seq2Seq 模型加上注意力機制，圖片來源：同上[9]

Seq2Seq 模型加上注意力機制，動畫網址為 https://leemeng.tw/images/transformer/seq2seq-unrolled-with-attention.mp4

單純的 Seq2Seq 模型，未加上注意力機制，動畫網址為 https://leemeng.tw/images/transformer/seq2seq-unrolled-no-attention.mp4

接下來，我們就來實作神經機器翻譯(NMT)。

【**範例 5**】使用 Seq2Seq 架構，加上注意力(Attention)機制，實作神經機器翻譯(NMT)。

程式 **12_07_機器翻譯_attention.ipynb**，修改自 TensorFlow 官網所提供的範例「Neural machine translation with attention」[10]。

資料集：自 http://www.manythings.org/anki/spa-eng.zip 下載西班牙文/英文對照檔 spa-eng.zip。

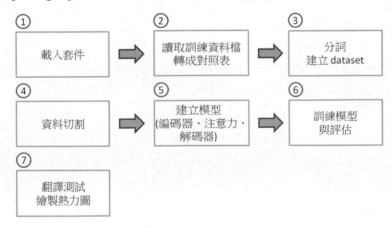

1. 載入相關套件。

```
1   # 載入相關套件
2   import tensorflow as tf
3   import matplotlib.pyplot as plt
4   import matplotlib.ticker as ticker
5   from sklearn.model_selection import train_test_split
6   import unicodedata
7   import re
8   import numpy as np
9   import os
10  import io
11  import time
```

2. 資料前置處理函數。

```
1   # 將 unicode 轉為 ascii 內碼的函數
2   def unicode_to_ascii(s):
3       return ''.join(c for c in unicodedata.normalize('NFD', s)
4                       if unicodedata.category(c) != 'Mn')
5
6   # 去除特殊符號的函數
7   def preprocess_sentence(w):
8       w = unicode_to_ascii(w.lower().strip())
9
10      w = re.sub(r"([?.!,¿])", r" \1 ", w)
11      w = re.sub(r'[" "]+', " ", w)
12      w = re.sub(r"[^a-zA-Z?.!,¿]+", " ", w)
13      w = w.strip()
14
15      # 前後加特殊字串，讓模型知道語句的開頭與結尾
16      w = '<start> ' + w + ' <end>'
17      return w
```

3. 測試資料前置處理函數。

```
1   # 測試
2   en_sentence = u"May I borrow this book?"
3   sp_sentence = u"¿Puedo tomar prestado este libro?"
4   print(preprocess_sentence(en_sentence))
5   print(preprocess_sentence(sp_sentence).encode('utf-8'))
```

執行結果：

```
<start> may i borrow this book ? <end>
b'<start> \xc2\xbf puedo tomar prestado este libro ? <end>'
```

4. 讀取訓練資料檔，轉成對照表。

```
1   # 讀取訓練資料檔，轉成對照表
2   def create_dataset(path, num_examples):
3       lines = io.open(path, encoding='UTF-8').read().strip().split('\n')
4
5       word_pairs = [[preprocess_sentence(w) for w in line.split('\t')[0:2]]
6                   for line in lines[:num_examples]]
7
8       return zip(*word_pairs)
9
10  # 讀取訓練資料檔
11  path_to_file='./RNN/spa.txt'
12  en, sp = create_dataset(path_to_file, None)
13
14  # 顯示最後一句對照
15  print(en[-1])
16  print(sp[-1])
```

執行結果：

<start> if you want to sound like a native speaker , you must be willing to practice saying the same sentence over and over in the same way that banjo players practice the same phrase over and over until they can play it correctly and at the desired tempo . <end>
<start> si quieres sonar como un hablante nativo , debes estar dispuesto a practicar diciendo la misma frase una y otra vez de la misma manera en que un musico de banjo practica el mismo fraseo una y otra vez hasta que lo puedan tocar correctamente y en el tiempo esperado . <end>

5. 分詞、建立 dataset。

```
1  # 分詞
2  def tokenize(lang):
3      lang_tokenizer = tf.keras.preprocessing.text.Tokenizer(filters='')
4      lang_tokenizer.fit_on_texts(lang)
5      # 文字轉索引值
6      tensor = lang_tokenizer.texts_to_sequences(lang)
7      # 長度不足補 0
8      tensor = tf.keras.preprocessing.sequence.pad_sequences(tensor,
9                                                      padding='post')
10
11     return tensor, lang_tokenizer
12
13 # 建立 dataset
14 def load_dataset(path, num_examples=None):
15     targ_lang, inp_lang = create_dataset(path, num_examples)
16
17     input_tensor, inp_lang_tokenizer = tokenize(inp_lang)
18     target_tensor, targ_lang_tokenizer = tokenize(targ_lang)
19
20     return input_tensor, target_tensor, inp_lang_tokenizer, targ_lang_tokenizer
21
22 # 限制 30000 筆訓練資料
23 num_examples = 30000
24 input_tensor, target_tensor, inp_lang, targ_lang = load_dataset(path_to_file,
25                                                      num_examples)
26
27 # 計算語句最大長度
28 max_length_targ, max_length_inp = target_tensor.shape[1], input_tensor.shape[1]
```

6. 資料切割。

```
1  # 資料切割
2  input_tensor_train, input_tensor_val, target_tensor_train, target_tensor_val = \
3      train_test_split(input_tensor, target_tensor, test_size=0.2)
4
5  len(input_tensor_train), len(input_tensor_val)
```

7. 參數設定。

```
1  # 參數設定
2  BUFFER_SIZE = len(input_tensor_train) # Dataset的緩衝區大小
3  BATCH_SIZE = 64                        # 批量
4  steps_per_epoch = len(input_tensor_train)//BATCH_SIZE # 每週期包含的步驟數
5  embedding_dim = 256                    # 嵌入層的輸出維度
6  units = 1024                           # GRU 輸出維度
7  vocab_inp_size = len(inp_lang.word_index)+1  # 原始語言的字彙表大小
8  vocab_tar_size = len(targ_lang.word_index)+1 # 目標語言的字彙表大小
```

8. 建立 dataset。

```
1  # 建立 dataset
2  dataset = tf.data.Dataset.from_tensor_slices(
3      (input_tensor_train, target_tensor_train)).shuffle(BUFFER_SIZE)
4  dataset = dataset.batch(BATCH_SIZE, drop_remainder=True)
```

9. 建立模型編碼器。

```
1  # 建立模型編碼器
2  class Encoder(tf.keras.Model):
3      def __init__(self, vocab_size, embedding_dim, enc_units, batch_sz):
4          super(Encoder, self).__init__()
5          self.batch_sz = batch_sz
6          self.enc_units = enc_units
7          self.embedding = tf.keras.layers.Embedding(vocab_size, embedding_dim)
8          self.gru = tf.keras.layers.GRU(self.enc_units,
9                          return_sequences=True, return_state=True,
10                         recurrent_initializer='glorot_uniform')
11
12     def call(self, x, hidden):
13         x = self.embedding(x)
14         output, state = self.gru(x, initial_state=hidden)
15         return output, state
16
17     def initialize_hidden_state(self):
18         return tf.zeros((self.batch_sz, self.enc_units))
19
20 encoder = Encoder(vocab_inp_size, embedding_dim, units, BATCH_SIZE)
21
22 # 測試
23 sample_hidden = encoder.initialize_hidden_state()
24 sample_output, sample_hidden = encoder(example_input_batch, sample_hidden)
25 print('編碼器輸出維度: (批量, 語句長度, 輸出)=', sample_output.shape)
26 print('編碼器隱藏層輸出維度: (批量, 輸出)=', sample_hidden.shape)
```

執行結果:

編碼器輸出維度: (批量, 語句長度, 輸出) = (64, 19, 1024)
編碼器隱藏層輸出維度: (批量, 輸出) = (64, 1024)

10. 建立注意力(Attention)機制：Bahdanau 作法。

```
1   # 建立模型注意力(Attention)機制：Bahdanau 作法
2   class BahdanauAttention(tf.keras.layers.Layer):
3       def __init__(self, units):
4           super(BahdanauAttention, self).__init__()
5           self.W1 = tf.keras.layers.Dense(units)
6           self.W2 = tf.keras.layers.Dense(units)
7           self.V = tf.keras.layers.Dense(1)
8
9       def call(self, query, values):
10          # 依虛擬程式碼的公式計算
11          query_with_time_axis = tf.expand_dims(query, 1)
12          score = self.V(tf.nn.tanh(
13                  self.W1(query_with_time_axis) + self.W2(values)))
14          attention_weights = tf.nn.softmax(score, axis=1)
15
16          context_vector = attention_weights * values
17          context_vector = tf.reduce_sum(context_vector, axis=1)
18
19          return context_vector, attention_weights
20
21  attention_layer = BahdanauAttention(10)
22  attention_result, attention_weights = attention_layer(sample_hidden,
23                                                        sample_output)
24
25  print("Attention維度: (批量, 輸出)=", attention_result.shape)
26  print("Attention權重: (批量, 語句長度, 1)=", attention_weights.shape)
```

執行結果：

```
Attention 維度: (批量, 輸出) = (64, 1024)
Attention 權重: (批量, 語句長度, 1) = (64, 19, 1)
```

11. 建立模型解碼器。

```
1   # 建立模型解碼器
2   class Decoder(tf.keras.Model):
3       def __init__(self, vocab_size, embedding_dim, dec_units, batch_sz):
4           super(Decoder, self).__init__()
5           self.batch_sz = batch_sz
6           self.dec_units = dec_units
7           self.embedding = tf.keras.layers.Embedding(vocab_size, embedding_dim)
8           self.gru = tf.keras.layers.GRU(self.dec_units,
9                       return_sequences=True, return_state=True,
10                      recurrent_initializer='glorot_uniform')
11          self.fc = tf.keras.layers.Dense(vocab_size)
12          self.attention = BahdanauAttention(self.dec_units)
13
14      def call(self, x, hidden, enc_output):
15          context_vector, attention_weights = self.attention(hidden, enc_output)
16          x = self.embedding(x)
```

```
17          x = tf.concat([tf.expand_dims(context_vector, 1), x], axis=-1)
18          output, state = self.gru(x)
19          output = tf.reshape(output, (-1, output.shape[2]))
20          x = self.fc(output)
21
22          return x, state, attention_weights
23
24  decoder = Decoder(vocab_tar_size, embedding_dim, units, BATCH_SIZE)
25  sample_decoder_output, _, _ = decoder(tf.random.uniform((BATCH_SIZE, 1)),
26                                          sample_hidden, sample_output)
27
28  print('解碼器維度: (批量, 字彙表尺寸)=', sample_decoder_output.shape)
```

執行結果：解碼器維度為(批量, 字彙表尺寸) = (64, 4807)

12. 指定優化器、損失函數：語句長度不足補 0 的部分，損失不予計算。

```
1   # 指定優化器、損失函數
2   optimizer = tf.keras.optimizers.Adam()
3   loss_object = tf.keras.losses.SparseCategoricalCrossentropy(
4       from_logits=True, reduction='none')
5
6   # 損失函數
7   def loss_function(real, pred):
8       mask = tf.math.logical_not(tf.math.equal(real, 0))
9       loss_ = loss_object(real, pred)
10
11      mask = tf.cast(mask, dtype=loss_.dtype)
12      loss_ *= mask
13
14      return tf.reduce_mean(loss_)
```

13. 檢查點存檔：由於訓練需要 10 多分鐘，避免中途當機要重來，故設檢查點存檔，萬一發生當機的狀況，即可自最後一個檢查點還原。

```
1   # 檢查點存檔
2   checkpoint_dir = './training_checkpoints'
3   checkpoint_prefix = os.path.join(checkpoint_dir, "ckpt")
4   checkpoint = tf.train.Checkpoint(optimizer=optimizer,
5                           encoder=encoder, decoder=decoder)
```

14. 定義梯度下降函數。

```
1  # 訓練函數
2  @tf.function
3  def train_step(inp, targ, enc_hidden):
4      loss = 0
5
6      # 梯度下降
7      with tf.GradientTape() as tape:
8          enc_output, enc_hidden = encoder(inp, enc_hidden)
9          dec_hidden = enc_hidden
10         dec_input = tf.expand_dims([targ_lang.word_index['<start>']] * BATCH_SIZE, 1)
11
12         # Teacher forcing：以解碼器目標(y)為下一節點的輸入
13         for t in range(1, targ.shape[1]):
14             # 解碼器有三個輸入：目標語言的輸入、解碼器隱藏層的輸出、編碼器的輸出
15             predictions, dec_hidden, _ = decoder(dec_input, dec_hidden, enc_output)
16             loss += loss_function(targ[:, t], predictions)
17             dec_input = tf.expand_dims(targ[:, t], 1)
18
19     # 梯度計算
20     batch_loss = (loss / int(targ.shape[1]))
21     variables = encoder.trainable_variables + decoder.trainable_variables
22     gradients = tape.gradient(loss, variables)
23     optimizer.apply_gradients(zip(gradients, variables))
24
25     return batch_loss
```

15. 訓練模型：在筆者的 PC 上執行，每一個執行週期約需 80 秒的時間。

```
1  # 模型訓練
2  EPOCHS = 10
3  for epoch in range(EPOCHS):
4      start = time.time()
5
6      enc_hidden = encoder.initialize_hidden_state()
7      total_loss = 0
8
9      for (batch, (inp, targ)) in enumerate(dataset.take(steps_per_epoch)):
10         batch_loss = train_step(inp, targ, enc_hidden)
11         total_loss += batch_loss
12
13         if batch % 100 == 0:
14             print(f'Epoch {epoch+1} Batch {batch} Loss {batch_loss.numpy():.4f}')
15
16     # 每2個訓練週期存檔一次
17     if (epoch + 1) % 2 == 0:
18         checkpoint.save(file_prefix=checkpoint_prefix)
19
20     print(f'Epoch {epoch+1} Loss {total_loss/steps_per_epoch:.4f}')
21     print(f'Time taken for 1 epoch {time.time()-start:.2f} sec\n')
```

執行結果：

```
Epoch 8 Batch 0 Loss 0.1553
Epoch 8 Batch 100 Loss 0.1314
Epoch 8 Batch 200 Loss 0.1630
Epoch 8 Batch 300 Loss 0.1755
Epoch 8 Loss 0.1607
Time taken for 1 epoch 79.50 sec

Epoch 9 Batch 0 Loss 0.0995
Epoch 9 Batch 100 Loss 0.1169
Epoch 9 Batch 200 Loss 0.1294
Epoch 9 Batch 300 Loss 0.1857
Epoch 9 Loss 0.1259
Time taken for 1 epoch 75.70 sec

Epoch 10 Batch 0 Loss 0.0781
Epoch 10 Batch 100 Loss 0.0725
Epoch 10 Batch 200 Loss 0.0826
Epoch 10 Batch 300 Loss 0.0960
Epoch 10 Loss 0.1016
Time taken for 1 epoch 79.90 sec
```

16. 定義預測函數。

```
1   # 定義預測函數
2   def evaluate(sentence):
3       attention_plot = np.zeros((max_length_targ, max_length_inp))
4       sentence = preprocess_sentence(sentence)
5
6       # 分詞，長度不足補空白
7       inputs = [inp_lang.word_index[i] for i in sentence.split(' ')]
8       inputs = tf.keras.preprocessing.sequence.pad_sequences([inputs],
9                              maxlen=max_length_inp, padding='post')
10      inputs = tf.convert_to_tensor(inputs)
11
12      # 依模型訓練的程序計算
13      result = ''
14      hidden = [tf.zeros((1, units))]
15      enc_out, enc_hidden = encoder(inputs, hidden)
16      dec_hidden = enc_hidden
17      dec_input = tf.expand_dims([targ_lang.word_index['<start>']], 0)
18      for t in range(max_length_targ):
19          predictions, dec_hidden, attention_weights = decoder(dec_input,
20                                          dec_hidden, enc_out)
21          attention_weights = tf.reshape(attention_weights, (-1, ))
22          attention_plot[t] = attention_weights.numpy()
23          predicted_id = tf.argmax(predictions[0]).numpy()
24          result += targ_lang.index_word[predicted_id] + ' '
25
26          if targ_lang.index_word[predicted_id] == '<end>':
27              return result, sentence, attention_plot
28          dec_input = tf.expand_dims([predicted_id], 0)
29
30      return result, sentence, attention_plot
```

17. 定義 attention weights 繪圖函數。

```
1  # 定義 attention weights 繪圖函數
2  def plot_attention(attention, sentence, predicted_sentence):
3      fig = plt.figure(figsize=(10, 10))
4      ax = fig.add_subplot(1, 1, 1)
5      ax.matshow(attention, cmap='viridis')
6
7      fontdict = {'fontsize': 14}
8
9      ax.set_xticklabels([''] + sentence, fontdict=fontdict, rotation=90)
10     ax.set_yticklabels([''] + predicted_sentence, fontdict=fontdict)
11
12     ax.xaxis.set_major_locator(ticker.MultipleLocator(1))
13     ax.yaxis.set_major_locator(ticker.MultipleLocator(1))
14
15     plt.show()
```

18. 定義翻譯函數。

```
1  # 定義翻譯函數
2  def translate(sentence):
3      result, sentence, attention_plot = evaluate(sentence)
4
5      print('Input:', sentence)
6      print('Predicted translation:', result)
7
8      attention_plot = attention_plot[:len(result.split(' ')),
9                                      :len(sentence.split(' '))]
10     plot_attention(attention_plot,
11                    sentence.split(' '), result.split(' '))
```

19. 翻譯測試，並繪製熱力圖。

```
1  # 翻譯測試，並繪製熱力圖
2  translate(u'hace mucho frio aqui.')
```

執行結果：可自 Google 翻譯查證測試結果。

```
Input: <start> hace mucho frio aqui . <end>
Predicted translation: it s very cold here . <end>
```

觀察下圖，如果原始語言與目標語言的單字是一一對照的，則熱力圖的對
角線應該會是關聯最大。

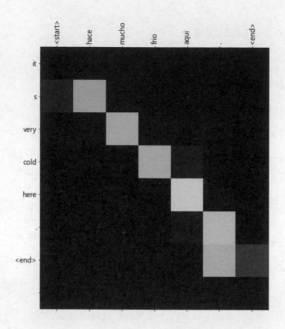

20. 可下載其他語言試試看,包括中英文對照檔。

除了翻譯功能之外,Seq2Seq 模型還有其他各種型態和應用,如下圖所示。

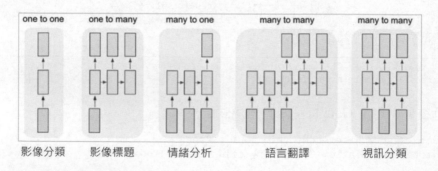

圖 12.11 Seq2Seq 模型的各種型態和應用,圖片來源:「The Unreasonable Effectiveness of Recurrent Neural Networks」[11]。

1. 一對一(one to one):固定長度的輸入(input)與輸出(output),即一般的神經網路模型。例如影像分類,輸入一張影像後,預測這張影像所屬的類別。

2. 一對多(one to many)：單一輸入，多個輸出。例如影像標題（Image Captioning），輸入一個影像後，接著偵測影像內的多個物件，並一一給予標題，這稱之為「Sequence output」。

3. 多對一(many to one)：多個輸入，單一輸出。例如情緒分析(Sentiment Analysis)，輸入一大段話後，判斷這段話是正面或負面的情緒表達，這稱之為「Sequence input」。

4. 多對多(many to many)：多個輸入，多個輸出。例如語言翻譯（Machine Translation），輸入一段英文句子後，翻譯成中文，這稱之為「Sequence input and sequence output」。

5. 另一種多對多(many to many)：多個輸入，多個輸出「同步」（Synchronize）。例如影片分類（Video Classification），輸入一段影片後，每一幀（Frame）都各產生一個標題，這稱之為「Synced sequence input and output」。

這裡大家先有個基本概念就好，在下一章介紹應用時，我們會再多看一些範例。

12-7 Transformer 架構

Google 的學者 Ashish Vaswani 等人於 2017 年依照 Seq2Seq 模型加上注意力機制，提出了 Transformer 架構，如下圖所示。架構一推出後，馬上躍身為 NLP 近年來最夯的演算法，而「Attention Is All You Need」一文也被公認是必讀的文章，各種改良的演算法也紛紛出籠，例如 BERT、GPT-2、GPT-3、XLNet、ELMo、T5 等，幾乎搶佔了 NLP 大部份的版面，可說是當紅炸子雞，接下來我們就來好好認識 Transformer 與其相關的演算法。

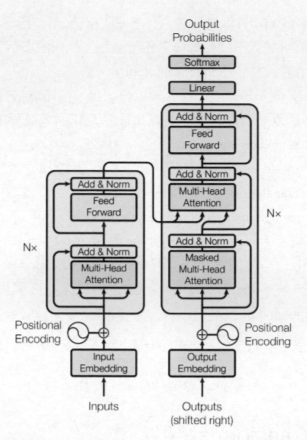

圖 12.12 Transformer 架構，圖片來源：「Attention Is All You Need」[12]

12-7-1 Transformer 原理

RNN/LSTM/GRU 有個最大的缺點，因為要以上文預測目前的目標，必須以序列的方式，「依序」執行每一個節點的訓練，進而造成執行效能過慢。而 Transformer 克服了此一問題，提出「自注意力機制」（Self-Attention mechanism），能夠平行計算出所有的輸出，計算步驟如下，請同時參考上圖：

1. 首先輸入向量(Input Vector)被表徵為 Q、K、V 向量(Vector)。

■ K：Key Vector，為 Encoder 隱藏層狀態的鍵值，即上下文的詞向量。

- V：Value Vector，為 Encoder 隱藏層狀態的輸出值。
- Q：Query Vector，為 Decoder 的前一期輸出。
- 故自注意力機制對應 Q、K、V，共有三種權重，而單純的注意力機制則只有一種權重—Attention Weight。
- 利用神經網路優化可以找到三種權重的最佳值，然後，以輸入向量個別乘以三種權重，即可求得 Q、K、V 向量。

2. Q、K、V 再經過下圖的運算即可得到自注意力矩陣。

- 點積(Dot product)運算：Q x K，計算輸入向量與上下文詞彙的相似度。
- 特徵縮放：Q、K 維度開根號，通常 Q、K 維度是 64，故 $\sqrt{64} = 8$。
- Softmax 運算：將上述結果轉為機率。
- 找出要重視的上下文詞彙：以 Value Vector 乘以上述機率，較大值為要重視的上下文詞彙。

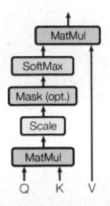

圖 12.13 「自注意力機制」運算，圖片來源：「Attention Is All You Need」[12]

上圖運算過程以數學式表達如下，即自注意力矩陣公式：

$$Attention\,(Q, K, V) = softmax \left(\frac{QK^T}{\sqrt{d_k}} \right) V$$

其中 d_k 為 Key Vector 維度，通常是 64。

3. 自注意力機制是多頭(Multi-Head)的，通常是 8 個頭，如圖 12.13 的機制，經過內積(Dot product)運算，串聯這 8 個頭，如下圖：

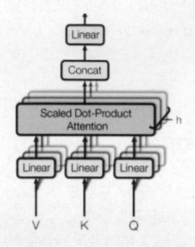

圖 12.14 「自注意力機制」多頭運算，圖片來源：「Attention Is All You Need」[12]

多頭自注意力矩陣公式如下：

$$MultiHead\,(Q,K,V) = concat\,(head_1\,head_2...head_n)\,W_O$$
$$where, head_i = Attention\left(QW_i^Q, KW_i^K, VW_i^V\right)$$

4. 最後加上其他的神經層，如圖 12.12，就構成了 Transformer 網路架構。

要瞭解詳細的計算過程請參考「Illustrated: Self-Attention」[13]一文，它還有附精美的動畫，另外，「The Illustrated Transformer」[14]也值得一讀。

總而言之，自注意力機制就是要找出應該關注的上下文詞彙，舉例來說：

The animal didn't cross the street because it was too tired.

其中的 it 是代表 animal 還是 street？

透過自注意力機制，可以幫我們找出 it 與上下文詞彙的關聯度，進而判斷出 it 所代表的是 animal。

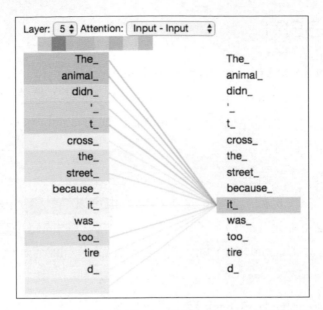

圖 12.15 「自注意力機制」示意圖，it 與上下文詞彙的關聯度

圖片來源：「The Illustrated Transformer」[14]

12-7-2 Transformer 效能

依「Self-attention in NLP」[15]一文中的實驗，上述的 Transformer 網路在一台 8 顆 NVIDIA P100 GPU 的伺服器上運作，大約要 3.5 天才能完成訓練，英/德文翻譯的準確率(BLEU)約 28.4 分，英/法文翻譯約 41.8 分。BLEU(Bilingual Evaluation Understudy)是專為雙語言翻譯所設定的效能衡量指標，是根據 n-gram 的相符數目(不考慮順序)，乘以對應的權重而得到的分數，詳細的計算可參考「A Gentle Introduction to Calculating the BLEU Score for Text in Python」[16]一文。

由於，筆者沒有這麼好的設備可以訓練模型，再往下鑽研細部原理也沒什麼意義，因此，我們只簡單闡述其精神，集中火力在應用面。

目前以 BERT、GPT 最為普遍，GPT 須在 AWS 執行，我們只介紹 BERT 與其應用。

12-8 BERT

BERT (Bidirectional Encoder Representations from Transformers)顧名思義，就是雙向的 Transformer，由 Google Jacob Devlin 等學者於 2018 年發表，參閱「BERT: Pre-training of Deep Bidirectional Transformers for Language Understanding」[17]一文。

Word2Vec/GloVe 一個單字只以一個詞向量表示，但是，一詞多義是所有語系共有的現象，例如，Apple 是「水果」也可以是「蘋果公司」，Bank 是「銀行」也可能是「岸邊」，BERT 就能解決這個問題，它是「上下文相關」(Context Dependent)，輸入的是一個句子，而不是一個單字，例如：

We go to the river bank. ➜ bank 是「岸邊」。

I need to go to bank to make a deposit. ➜ bank 是「銀行」。

BERT 演算法比 Transformer 更複雜，要花更多的時間訓練，為什麼還要介紹呢？有以下兩點原因：

1. 它跟 CNN 一樣有預先訓練好的模型(Pre-trained model)，可以進行轉移學習。BERT 分為兩階段，首先是一般的模型訓練，之後依不同應用領域進行模型的效能微調(Fine tuning)，類似 CNN 預先訓練模型接上自訂的完全連接層，就可以符合各應用領域的需求。所以，沒有昂貴設備的我們，只要知道模型的調校就夠了。
2. BERT 支援中文模型。

雖然沒辦法訓練模型，但為了在實務上能靈活運用，我們還是要稍微理解一下 BERT 的運作原理，免得誤用，還不知道為什麼錯，那就糗大了。

BERT 使用兩個訓練策略：

1. Masked LM (MLM)。
2. Next Sentence Prediction (NSP)

12-8-1　Masked LM

RNN/LSTM/GRU 都是以序列的方式，逐一產生輸出，導致訓練速度過慢，而 Masked LM (MLM)則可以克服這個問題，訓練資料在餵進模型前，有 15%的詞彙先以[MASK]符號取代，即所謂的「遮罩」(Mask)，之後演算法就試圖用未遮罩的詞彙來預測被遮罩的詞彙。Masked LM 的架構如下：

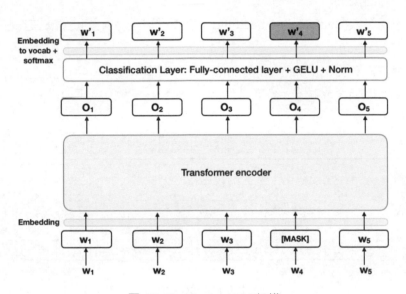

圖 12.16　Masked LM 架構

圖片來源：「BERT Explained: State of the art language model for NLP」[18]

計算過程如下：

1. 完全連接層(Full-connected layer)：對 Encoder 的輸出進行分類。
2. 上一步驟的輸出乘以詞嵌入矩陣，得到字彙表的維度(Vocabulary dimension.)。
3. 以 Softmax 換算字彙表內每個詞彙的機率。

12-8-2 Next Sentence Prediction

Next Sentence Prediction(NSP)訓練時會收到兩個字句，NSP 預測第 2 句是否是第 1 句的接續下文。訓練時會取樣正負樣本各佔 50%，進行下列的前置處理：

1. 符號詞嵌入(Token embedding)：[CLS]插在第 1 句的前面，[SEP] 插在每一句的後面。
2. 字句詞嵌入(Sentence embedding)：在每個符號(詞彙)上加註它是屬於第 1 句或第 2 句。
3. 位置詞嵌入(Positional embedding)：在每個符號(詞彙)上加註它是在合併字句中的第幾個位置。

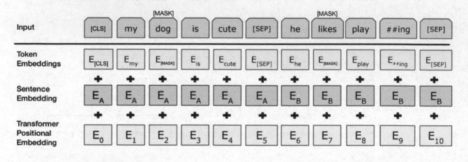

圖 12.17 Next Sentence Prediction(NSP)

圖片來源：「BERT Explained: State of the art language model for NLP」[18]

這三種詞嵌入就類似於前面「自注意力機制」的 Q、K、V。

預測第 2 句是否為第 1 句的接續下文，處理步驟如下：

1. 所有輸入序列餵入 Transformer 模型。
2. 將字句進行簡單的分類。
3. 使用 Softmax，判斷是否為接續下文(IsNextSequence)。

BERT 訓練時會結合兩個演算法，目標是最小化兩個策略的合併損失函數。

根據 BERT GitHub [19]說明，模型訓練(Pre-trained model)在 4~16 個 TPU 的伺服器上要訓練 4 天的時間，這又是身為平凡小民的筆者，不可承受之重，因此，我們還是乖乖的下載預先訓練好的模型，然後，集中火力在效能微調(Fine-tuning)上，比較實際。

12-8-3 BERT 效能微調(Fine-tuning)

效能微調就是根據不同應用領域，加入各行業別的知識，使 BERT 能更聰明，有下列應用類型：

1. 分類(Classification)：加一個分類層(Dense)，進行情緒分析(Sentiment analysis)的判別，可參考 BERT GitHub 的程式 run_classifier.py。
2. 問答(Question Answering)：比如 SQuAD 資料集，輸入一個問題後，能夠在全文中標示出答案的開頭與結束的位置，可參考 BERT GitHub 的程式 run_squad.py。
3. 命名實體識別(Named Entity Recognition, NER)：輸入一段文字後，可以標註其中的實體，如人名、組織、日期等。

以分類為例，測試處理程序如下：

1. 自 GLUE(https://gluebenchmark.com/tasks)下載資料集，較有名的是 Quora Question Pairs，它是科技問答網站的問題配對，標籤是相似與否。
2. 下載預先訓練模型 BERT-Base(https://storage.googleapis.com/bert_models /2018_10_18/uncased_L-12_H-768_A-12.zip)，解壓縮至一目錄，例如 BERT_BASE_DIR 所指向的目錄。
3. 效能微調：下列是 Linux 指令，Windows 作業環境下可直接把變數帶入。

```
export BERT_BASE_DIR=/path/to/bert/uncased_L-12_H-768_A-12
export GLUE_DIR=/path/to/glue

python run_classifier.py \
  --task_name=MRPC \
  --do_train=true \
  --do_eval=true \
  --data_dir=$GLUE_DIR/MRPC \
  --vocab_file=$BERT_BASE_DIR/vocab.txt \
  --bert_config_file=$BERT_BASE_DIR/bert_config.json \
  --init_checkpoint=$BERT_BASE_DIR/bert_model.ckpt \
  --max_seq_length=128 \
  --train_batch_size=32 \
  --learning_rate=2e-5 \
  --num_train_epochs=3.0 \
  --output_dir=/tmp/mrpc_output/
```

4. 得到結果如下：

```
***** Eval results *****
  eval_accuracy = 0.845588
  eval_loss = 0.505248
  global_step = 343
  loss = 0.505248
```

5. 預測：參數 do_predict=true，輸入放在 input/test.tsv，執行結果則在 output/test_results.tsv。

```
export BERT_BASE_DIR=/path/to/bert/uncased_L-12_H-768_A-12
export GLUE_DIR=/path/to/glue
export TRAINED_CLASSIFIER=/path/to/fine/tuned/classifier

python run_classifier.py \
  --task_name=MRPC \
  --do_predict=true \
  --data_dir=$GLUE_DIR/MRPC \
  --vocab_file=$BERT_BASE_DIR/vocab.txt \
  --bert_config_file=$BERT_BASE_DIR/bert_config.json \
  --init_checkpoint=$TRAINED_CLASSIFIER \
  --max_seq_length=128 \
  --output_dir=/tmp/mrpc_output/
```

問答(Question Answering)的話，以 SQuAD 資料集為例，程序與上述程序類似。

注意，效能微調使用 GPU 時，BERT GitHub 建議為 Titan X 或 GTX 1080，否則容易發生記憶體不足的情形。看到這裡，可能一堆讀者(包括筆者)臉上又三條線了，還好，有一些套件可以讓我們直接實驗，不須使用上述程序。

12-9 Transformers 套件

Transformers 套件的功能十分強大，它支援數十種的模型，包括 BERT、GPT、T5、XLNet、XLM 等架構，詳情請參閱 Transformers GitHub[20]。

接下來我們就拿 Transformers 這個套件當例子，做一些實驗，它包含下列功能：

- 情緒分析(Sentiment analysis)。
- 文字生成(Text generation)：限英文。
- 命名實體識別(Named Entity Recognition, NER)。
- 問題回答(Question Answering)。
- 克漏字填空(Filling masked text)。
- 文字摘要(Text Summarization)：將文章節錄出大意。
- 翻譯(Translation)。
- 特徵萃取(Feature extraction)：類似詞向量，將文字轉換為向量。

12-9-1 Transformers 套件範例

先安裝套件，指令如下：

```
pip install transformers
```

【範例 1】情緒分析(Sentiment analysis)。
程式：**12_08_BERT_情緒分析.ipynb**，修改自 Transformers 官網的「Quick tour」[21]。

1. 載入相關套件。

```
1  # 載入相關套件
2  from transformers import pipeline
```

2. 載入模型：BERT 有許多變型，下列指令預設下載 distilbert-base-uncased-finetuned-sst-2-english 模型，使用「The Stanford Sentiment Treebank」(SST-2)資料集進行效能微調。

```
1  # 載入模型
2  classifier = pipeline('sentiment-analysis')
```

3. 情緒分析測試。

```
1  # 正面
2  print(classifier('We are very happy to show you the 🤗 Transformers library.'))
3
4  # 負面
5  print(classifier('I hate this movie.'))
6
7  # 否定句也可以正確分類
8  print(classifier('the movie is not bad.'))
```

執行結果：非常準確，否定句也可以正確分類，不像之前的 RNN/LSTM/GRU 碰到否定句都無法正確分類。

```
[{'label': 'POSITIVE', 'score': 0.9997795224189758}]
[{'label': 'NEGATIVE', 'score': 0.9996869564056396}]
[{'label': 'POSITIVE', 'score': 0.999536395072937}]
```

4. 一次測試多筆。

```
1  # 一次測試多筆
2  results = classifier(["We are very happy.",
3                        "We hope you don't hate it."])
4  for result in results:
5      print(f"label: {result['label']}, with score: {round(result['score'], 4)}")
```

執行結果：非常準確，就連否定句有可能是中性的這點也能夠分辨，像是 don't hate 不討厭，但不意味是喜歡，所以分數只有 0.5。

```
label: POSITIVE, with score: 0.9999
label: NEGATIVE, with score: 0.5309
```

5. 多語系支援：BERT 支援 100 多種語系，提供 24 種模型，以 BERT-base 的檔名 uncased_L-12_H-768_A-12.zip 為例來說明，L-12：12 層神經層，H-768：768 個隱藏層神經元，A-12：12 個頭。

6. 西班牙文(Spanish)測試。

```
1  # 西班牙文(Spanish)
2  # 負面, I hate this movie
3  print(classifier('Odio esta pelicula.'))
4
5  # the movie is not bad.
6  print(classifier('la pelicula no esta mal.'))
```

執行結果：不是很準確，分數都非極端值。

```
[{'label': '1 star', 'score': 0.4615824222564697}]
[{'label': '3 stars', 'score': 0.6274545788764954}]
```

7. 法文(French)測試。

```
1  # 法文(French)
2  # 負面, I hate this movie
3  print(classifier('Je déteste ce film.'))
4
5  # the movie is not bad.
6  print(classifier('le film n\'est pas mal.'))
```

執行結果：不是很準確，分數都非極端值。

```
[{'label': '1 star', 'score': 0.631117582321167}]
[{'label': '3 stars', 'score': 0.5710769295692444}]
```

【範例 2】問題回答(Question Answering)。

程式：**12_09_BERT_問題回答.ipynb**，修改自 Transformers 官網「Summary of the tasks」的 Extractive Question Answering [22]。

1. 載入相關套件。

```
1  # 載入相關套件
2  from transformers import pipeline
```

2. 載入模型：參數須設為 question-answering。

```
1  # 載入模型
2  nlp = pipeline("question-answering")
```

3. 設定訓練資料。

```
1  # 訓練資料
2  context = r"Extractive Question Answering is the task of extracting an answer " + \
3  "from a text given a question. An example of a question answering " + \
4  "dataset is the SQuAD dataset, which is entirely based on that task. " + \
5  "If you would like to fine-tune a model on a SQuAD task, you may " + \
6  "leverage the examples/question-answering/run_squad.py script."
```

4. 測試 2 筆資料。

```
1  # 測試 2 筆
2  result = nlp(question="What is extractive question answering?", context=context)
3  print(f"Answer: '{result['answer']}', score: {round(result['score'], 4)}",
4        f", start: {result['start']}, end: {result['end']}")
5
6  print()
7
8  result = nlp(question="What is a good example of a question answering dataset?",
9             context=context)
10 print(f"Answer: '{result['answer']}', score: {round(result['score'], 4)}",
11       f", start: {result['start']}, end: {result['end']}")
```

執行結果：非常準確，通常是從訓練資料中節錄一段文字當作回答。

```
Answer: 'the task of extracting an answer from a text given a question', score: 0.6226 , start: 33, end: 94
Answer: 'SQuAD dataset', score: 0.5053 , start: 146, end: 159
```

5. 結合分詞(Tokenizer)：可自訂分詞器，斷句會比較準確，參閱「Using tokenizers from Tokenizers」(https://huggingface.co/transformers/fast_tokenizers.html)，下面使用預設的分詞器。

6. 載入分詞器(Tokenizer)。

```
1  from transformers import AutoTokenizer, TFAutoModelForQuestionAnswering
2  import tensorflow as tf
3
4  # 結合分詞器(Tokenizer)
5  tokenizer = AutoTokenizer.from_pretrained("bert-large-uncased-whole-word-masking-finetuned-squad")
6  model = TFAutoModelForQuestionAnswering.from_pretrained("bert-large-uncased-whole-word-masking-finetuned-squad")
```

7. 載入訓練資料。

```
1  # 訓練資料
2  text = r"""
3  🤗 Transformers (formerly known as pytorch-transformers and pytorch-pretrained-bert) provides general-purpose
4  architectures (BERT, GPT-2, RoBERTa, XLM, DistilBert, XLNet…) for Natural Language Understanding (NLU) and Natural
5  Language Generation (NLG) with over 32+ pretrained models in 100+ languages and deep interoperability between
6  TensorFlow 2.0 and PyTorch.
7  """
```

8. 設定問題。

```
1  # 問題
2  questions = [
3      "How many pretrained models are available in 😊 Transformers?",
4      "What does 😊 Transformers provide?",
5      "😊 Transformers provides interoperability between which frameworks?",
6  ]
```

9. 推測答案。

```
1  # 推測答案
2  for question in questions:
3      inputs = tokenizer(question, text, add_special_tokens=True, return_tensors="tf")
4      input_ids = inputs["input_ids"].numpy()[0]
5      outputs = model(inputs)
6
7      answer_start_scores = outputs.start_logits
8      answer_end_scores = outputs.end_logits
9      answer_start = tf.argmax(answer_start_scores, axis=1).numpy()[0]
10     answer_end = (tf.argmax(answer_end_scores, axis=1) + 1).numpy()[0]
11     answer = tokenizer.convert_tokens_to_string(tokenizer.convert_ids_to_tokens
12                                 (input_ids[answer_start:answer_end]))
13
14     print(f"Question: {question}")
15     print(f"Answer: {answer}\n")
```

執行結果:非常準確。

```
Question: How many pretrained models are available in 😊 Transformers?
Answer: over 32 +

Question: What does 😊 Transformers provide?
Answer: general - purpose architectures

Question: 😊 Transformers provides interoperability between which frameworks?
Answer: tensorflow 2. 0 and pytorch
```

【範例 3】克漏字填空(Masked Language Modeling)。

程式:**12_10_BERT_填漏字.ipynb**,修改自 Transformers 官網「Summary of the tasks」的 Masked Language Modeling [23]。

1. 載入相關套件。

```
1  # 載入相關套件
2  from transformers import pipeline
```

2. 載入模型：參數須設為 fill-mask。

```
1  # 載入模型
2  nlp = pipeline("fill-mask")
```

3. 測試。

```
1  # 測試
2  from pprint import pprint
3  pprint(nlp(f"HuggingFace is creating a {nlp.tokenizer.mask_token} " + \
4             "that the community uses to solve NLP tasks."))
```

執行結果：列出前 5 名與其分數，框起來的即是填上的字。

```
[{'score': 0.17927466332912445,
  'sequence': 'HuggingFace is creating a tool that the community uses to solve '
              'NLP tasks.',
  'token': 3944,
  'token_str': ' tool'},
 {'score': 0.11349395662546158,
  'sequence': 'HuggingFace is creating a framework that the community uses to '
              'solve NLP tasks.',
  'token': 7208,
  'token_str': ' framework'},
 {'score': 0.05243542045354843,
  'sequence': 'HuggingFace is creating a library that the community uses to '
              'solve NLP tasks.',
  'token': 5560,
  'token_str': ' library'},
 {'score': 0.03493538498878479,
  'sequence': 'HuggingFace is creating a database that the community uses to '
              'solve NLP tasks.',
  'token': 8503,
  'token_str': ' database'},
 {'score': 0.028602542355656624,
  'sequence': 'HuggingFace is creating a prototype that the community uses to '
              'solve NLP tasks.',
  'token': 17715,
  'token_str': ' prototype'}]
```

4. 結合分詞(Tokenizer)。

```
1  # 載入相關套件
2  from transformers import TFAutoModelWithLMHead, AutoTokenizer
3  import tensorflow as tf
4
5  # 結合分詞器(Tokenizer)
6  tokenizer = AutoTokenizer.from_pretrained("distilbert-base-cased")
7  model = TFAutoModelWithLMHead.from_pretrained("distilbert-base-cased")
```

5. 推測答案。

```
1  # 推測答案
2  sequence = f"Distilled models are smaller than the models they mimic. " + \
3      f"Using them instead of the large versions would help {tokenizer.mask_token} " + \
4      "our carbon footprint."
5  input = tokenizer.encode(sequence, return_tensors="tf")
6  mask_token_index = tf.where(input == tokenizer.mask_token_id)[0, 1]
7  token_logits = model(input)[0]
8  mask_token_logits = token_logits[0, mask_token_index, :]
9  top_5_tokens = tf.math.top_k(mask_token_logits, 5).indices.numpy()
10 for token in top_5_tokens:
11     print(sequence.replace(tokenizer.mask_token, tokenizer.decode([token])))
```

執行結果：列出前 5 名與其分數，框起來的即是填上的字。

```
Distilled models are smaller than the models they mimic. Using them instead of the large versions would help reduce our carbon
footprint.
Distilled models are smaller than the models they mimic. Using them instead of the large versions would help increase our carbo
n footprint.
Distilled models are smaller than the models they mimic. Using them instead of the large versions would help decrease our carbo
n footprint.
Distilled models are smaller than the models they mimic. Using them instead of the large versions would help offset our carbon
footprint.
Distilled models are smaller than the models they mimic. Using them instead of the large versions would help improve our carbon
footprint.
```

【範例 4】文字生成(Text Generation)：這裡使用 GPT-2 演算法，並非 BERT，同屬於 Transformer 演算法的變形，目前已發展到 GPT-3。

程式：**12_11_ GPT2_文字生成.ipynb**，修改自 Transformers 官網 「Summary of the tasks」的 Text Generation [24]。

1. 載入相關套件。

```
1  # 載入相關套件
2  from transformers import pipeline
```

2. 載入模型：參數須設為 text-generation。

```
1  # 載入模型
2  text_generator = pipeline("text-generation")
```

3. 測試：max_length=50 表示最大生成字數，do_sample=False 表示不隨機產生，反之為 True 時，則每次生成的內容都會不同，像是聊天機器人，使用者會期望機器人表達能夠有變化，不要每次都回答一樣的答案，例如，問「How are you?」，機器人有時候回答「I am fine」，有時候回答「great」、「not bad」。

```
1  # 測試
2  print(text_generator("As far as I am concerned, I will",
3                        max_length=50, do_sample=False))
```

執行結果：每次生成的內容均相同。

[{'generated_text': 'As far as I am concerned, I will be the first to admit that I am not a fan of the idea of a "free market."
I think that the idea of a free market is a bit of a stretch. I think that the idea'}]

4. 測試：do_sample=True 表示隨機產生，每次生成內容均不同。

```
1  # 測試
2  print(text_generator("As far as I am concerned, I will",
3                        max_length=50, do_sample=True))
```

執行結果：每次生成的內容均不相同。

[{'generated_text': 'As far as I am concerned, I will not be using the name \'Archer\', even though it\'d make all of me cry!\n
\n"I\'ll wait until they leave me, you know, on this little ship, of course,'}]

5. 結合分詞(Tokenizer)：這裡使用 XLNet 演算法，而非 BERT，也屬於 Transformer 演算法的變形。

```
1  # 載入相關套件
2  from transformers import TFAutoModelWithLMHead, AutoTokenizer
3
4  # 結合分詞器(Tokenizer)
5  model = TFAutoModelWithLMHead.from_pretrained("xlnet-base-cased")
6  tokenizer = AutoTokenizer.from_pretrained("xlnet-base-cased")
```

6. 提示：針對短提示，XLNet 通常要補充說明(Padding)，因為它是針對開放式(open-ended)問題而設計的，但 GPT-2 則不用。

```
1   # 針對短提示，XLNet 通常要補充說明(Padding)
2   PADDING_TEXT = """In 1991, the remains of Russian Tsar Nicholas II and his family
3   (except for Alexei and Maria) are discovered.
4   The voice of Nicholas's young son, Tsarevich Alexei Nikolaevich, narrates the
5   remainder of the story. 1883 Western Siberia,
6   a young Grigori Rasputin is asked by his father and a group of men to perform magic.
7   Rasputin has a vision and denounces one of the men as a horse thief. Although his
8   father initially slaps him for making such an accusation, Rasputin watches as the
9   man is chased outside and beaten. Twenty years later, Rasputin sees a vision of
10  the Virgin Mary, prompting him to become a priest. Rasputin quickly becomes famous,
11  with people, even a bishop, begging for his blessing. <eod> </s> <eos>"""
12
13  # 提示
14  prompt = "Today the weather is really nice and I am planning on "
```

7. 推測答案。

```
1  # 推測答案
2  inputs = tokenizer.encode(PADDING_TEXT + prompt, add_special_tokens=False,
3                            return_tensors="tf")
4  prompt_length = len(tokenizer.decode(inputs[0], skip_special_tokens=True,
5                                       clean_up_tokenization_spaces=True))
6  outputs = model.generate(inputs, max_length=250, do_sample=True, top_p=0.95,
7                           top_k=60)
8  generated = prompt + tokenizer.decode(outputs[0])[prompt_length:]
9
10 print(generated)
```

執行結果：

```
Today the weather is really nice and I am planning on anning on getting some good photos. I need to take some long-running pict
ures of the past few weeks and "in the moment."<eop> We are on a beach, right on the coast of Alaska. It is beautiful. It is pe
aceful. It is very quiet. It is peaceful. I am trying not to be too self-centered. But if the sun doesn
```

【範例 5】命名實體識別(Named Entity Recognition, NER)：預設使用 CoNLL-2003 NER 資料集。

程式：**12_12_BERT_NER.ipynb**，修改自 Transformers 官網「Summary of the tasks」的 Named Entity Recognition [25]。

1. 載入相關套件。

```
1  # 載入相關套件
2  from transformers import pipeline
```

2. 載入模型：參數須設為 ner。

```
1  # 載入模型
2  nlp = pipeline("ner")
```

3. 測試。

```
1  # 測試資料
2  sequence = "Hugging Face Inc. is a company based in New York City. " \
3             "Its headquarters are in DUMBO, therefore very" \
4             "close to the Manhattan Bridge."
5
6  # 推測答案
7  import pandas as pd
8  df = pd.DataFrame(nlp(sequence))
9  df
```

執行結果：顯示所有實體(Entity)，word 欄位中以##開頭的，表示與其前一個詞彙結合也是一個實體，例如##gging，前一個詞彙為 Hu，即表示Hu、Hugging 均為實體。

entity 欄位有下列實體類別：

- O：非實體。
- B-MISC：雜項實體的開頭，接在另一個雜項實體的後面。
- I-MISC：雜項實體。
- B-PER：人名的開頭，接在另一個人名的後面。
- I-PER：人名。
- B-ORG：組織的開頭，接在另一個組織的後面。
- I-ORG：組織。
- B-LOC：地名的開頭，接在另一個地名的後面。
- I- LOC：地名。

	word	score	entity	index	start	end
0	Hu	0.999511	I-ORG	1	0	2
1	##gging	0.989597	I-ORG	2	2	7
2	Face	0.997970	I-ORG	3	8	12
3	Inc	0.999376	I-ORG	4	13	16
4	New	0.999341	I-LOC	11	40	43
5	York	0.999193	I-LOC	12	44	48
6	City	0.999341	I-LOC	13	49	53
7	D	0.986336	I-LOC	19	79	80
8	##UM	0.939624	I-LOC	20	80	82
9	##BO	0.912139	I-LOC	21	82	84
10	Manhattan	0.983919	I-LOC	29	113	122
11	Bridge	0.992424	I-LOC	30	123	129

4. 結合分詞(Tokenizer)。

```
1  # 載入相關套件
2  from transformers import TFAutoModelForTokenClassification, AutoTokenizer
3  import tensorflow as tf
4
5  # 結合分詞器(Tokenizer)
6  model_name = "dbmdz/bert-large-cased-finetuned-conll03-english"
7  model = TFAutoModelForTokenClassification.from_pretrained(model_name)
8  tokenizer = AutoTokenizer.from_pretrained("bert-base-cased")
```

5. 測試。

```
1   # NER 類別
2   label_list = [
3       "O",        # 非實體
4       "B-MISC",   # 雜項實體的開頭，接在另一雜項實體的後面
5       "I-MISC",   # 雜項實體
6       "B-PER",    # 人名的開頭，接在另一人名的後面
7       "I-PER",    # 人名
8       "B-ORG",    # 組織的開頭，接在另一組織的後面
9       "I-ORG",    # 組織
10      "B-LOC",    # 地名的開頭，接在另一地名的後面
11      "I-LOC"     # 地名
12  ]
13
14  # 測試資料
15  sequence = "Hugging Face Inc. is a company based in New York City. " \
16             "Its headquarters are in DUMBO, therefore very" \
17             "close to the Manhattan Bridge."
18
19  # 推測答案
20  tokens = tokenizer.tokenize(tokenizer.decode(tokenizer.encode(sequence)))
21  inputs = tokenizer.encode(sequence, return_tensors="tf")
22  outputs = model(inputs)[0]
23  predictions = tf.argmax(outputs, axis=2)
24  print([(token, label_list[prediction]) for token, prediction in
25         zip(tokens, predictions[0].numpy())])
```

執行結果：代碼説明可參照程式碼的第 3~11 行。

```
[('[CLS]', 'O'), ('Hu', 'I-ORG'), ('##gging', 'I-ORG'), ('Face', 'I-ORG'), ('Inc', 'I-ORG'), ('.', 'O'), ('is', 'O'), ('a',
'O'), ('company', 'O'), ('based', 'O'), ('in', 'O'), ('New', 'I-LOC'), ('York', 'I-LOC'), ('City', 'I-LOC'), ('.', 'O'), ('It
s', 'O'), ('headquarters', 'O'), ('are', 'O'), ('in', 'O'), ('D', 'I-LOC'), ('##UM', 'I-LOC'), ('##BO', 'I-LOC'), (',', 'O'),
('therefore', 'O'), ('very', 'O'), ('##c', 'O'), ('##lose', 'O'), ('to', 'O'), ('the', 'O'), ('Manhattan', 'I-LOC'), ('Bridge',
'I-LOC'), ('.', 'O'), ('[SEP]', 'O')]
```

【範例 6】文字摘要(Text Summarization)：從篇幅較長的文章中整理出摘要，測試的資料集是 CNN 和 Daily Mail 媒體刊登的文章。

程式：**12_13_文字摘要.ipynb**，修改自 Transformers 官網「Summary of the tasks」的 Summarization [26]。

1. 載入相關套件。

```
1   # 載入相關套件
2   from transformers import pipeline
```

2. 載入模型：參數須設為 summarization。

```
1   # 載入模型
2   summarizer = pipeline("summarization")
```

3. 測試。

```
1   # 測試資料
2   ARTICLE = """ New York (CNN) When Liana Barrientos was 23 years old, she got married in Westchester County, New York.
3   A year later, she got married again in Westchester County, but to a different man and without divorcing her first husband.
4   Only 18 days after that marriage, she got hitched yet again. Then, Barrientos declared "I do" five more times, sometimes onl
5   In 2010, she married once more, this time in the Bronx. In an application for a marriage license, she stated it was her "fir
6   Barrientos, now 39, is facing two criminal counts of "offering a false instrument for filing in the first degree," referring
7   2010 marriage license application, according to court documents.
8   Prosecutors said the marriages were part of an immigration scam.
9   On Friday, she pleaded not guilty at State Supreme Court in the Bronx, according to her attorney, Christopher Wright, who de
10  After leaving court, Barrientos was arrested and charged with theft of service and criminal trespass for allegedly sneaking
11  Annette Markowski, a police spokeswoman. In total, Barrientos has been married 10 times, with nine of her marriages occurrin
12  All occurred either in Westchester County, Long Island, New Jersey or the Bronx. She is believed to still be married to four
13  Prosecutors said the immigration scam involved some of her husbands, who filed for permanent residence status shortly after
14  Any divorces happened only after such filings were approved. It was unclear whether any of the men will be prosecuted.
15  The case was referred to the Bronx District Attorney\'s Office by Immigration and Customs Enforcement and the Department of
16  Investigation Division. Seven of the men are from so-called "red-flagged" countries, including Egypt, Turkey, Georgia, Pakis
17  Her eighth husband, Rashid Rajput, was deported in 2006 to his native Pakistan after an investigation by the Joint Terrorism
18  If convicted, Barrientos faces up to four years in prison.  Her next court appearance is scheduled for May 18.
19  """
20
21  # 推測答案
22  print(summarizer(ARTICLE, max_length=130, min_length=30, do_sample=False))
```

執行結果：摘要內容還算看得懂。

```
[{'summary_text': ' Liana Barrientos, 39, is charged with two counts of "offering a false instrument for filing in the first de
gree" In total, she has been married 10 times, with nine of her marriages occurring between 1999 and 2002 . At one time, she wa
s married to eight men at once, prosecutors say .'}]
```

4. 結合 Tokenizer：T5 是 Google Text-To-Text Transfer Transformer 的模
   型，它提供一個框架可以使用多種的模型、損失函數、超參數，來進
   行不同的任務(Tasks)，例如翻譯、語意接受度檢查、相似度比較、文
   字摘要等，詳細說明可參閱「Exploring Transfer Learning with T5: the
   Text-To-Text Transfer Transformer」[27]。

```
1   # 載入相關套件
2   from transformers import TFAutoModelWithLMHead, AutoTokenizer
3   model = TFAutoModelWithLMHead.from_pretrained("t5-base")
4
5   # 結合分詞器(Tokenizer)
6   tokenizer = AutoTokenizer.from_pretrained("t5-base")
7   # T5 uses a max_length of 512 so we cut the article to 512 tokens.
8   inputs = tokenizer.encode("summarize: " + ARTICLE, return_tensors="tf",
9                             max_length=512)
10  outputs = model.generate(inputs, max_length=150, min_length=40,
11                           length_penalty=2.0, num_beams=4, early_stopping=True)
12  [tokenizer.decode(i) for i in outputs]
```

執行結果：T5 最多只可輸入 512 個詞彙，故將多餘的文字截斷，產生的摘
要也還可以看得懂。

```
['<pad> prosecutors say the marriages were part of an immigration scam. if convicted, barrientos faces two criminal counts of
"offering a false instrument for filing in the first degree" she has been married 10 times, nine of them between 1999 and 200
2.']
```

【**範例 7**】翻譯(Translation)功能：使用 WMT English to German dataset(英翻德資料集)。

程式：**12_14_T5_翻譯.ipynb**，修改自 Transformers 官網「Summary of the tasks」的 Translation [28]。

1. 載入相關套件。

```
1  # 載入相關套件
2  from transformers import pipeline
```

2. 載入模型：參數設為 translation_en_to_de 表示英翻德。

```
1  # 載入模型
2  translator = pipeline("translation_en_to_de")
```

3. 測試。

```
1  # 測試資料
2  text = "Hugging Face is a technology company based in New York and Paris"
3  print(translator(text, max_length=40))
```

執行結果：

```
[{'translation_text': 'Hugging Face ist ein Technologieunternehmen mit Sitz in New York und Paris.'}]
```

4. 結合 Tokenizer。

```
1   # 載入相關套件
2   from transformers import TFAutoModelWithLMHead, AutoTokenizer
3   model = TFAutoModelWithLMHead.from_pretrained("t5-base")
4
5   # 結合分詞器(Tokenizer)
6   tokenizer = AutoTokenizer.from_pretrained("t5-base")
7   text = "translate English to German: Hugging Face is a " + \
8          "technology company based in New York and Paris"
9   inputs = tokenizer.encode(text, return_tensors="tf")
10  outputs = model.generate(inputs, max_length=40, num_beams=4, early_stopping=True)
11  [tokenizer.decode(i) for i in outputs]
```

執行結果：

```
['<pad> Hugging Face ist ein Technologieunternehmen mit Sitz in New York und Paris.']
```

12-9-2 Transformers 套件效能微調

Transformers 也有提供效能微調的功能，可參閱 Transformers 官網「Training and fine-tuning」[29]的網頁説明，我們現在就來練習整個程序。Transformers 效能微調可使用下列三種方式：

1. TensorFlow v2。
2. PyTorch。
3. Transformers 的 Trainer。

【範例 7】直接以 Transformers 的 Trainer 進行效能微調。

程式：12_15_BERT_Train.ipynb，修改自[30]。

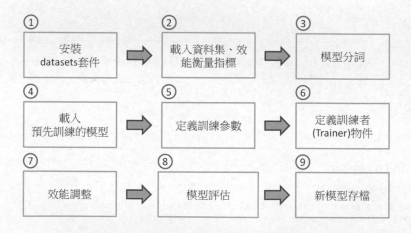

1. 安裝 datasets 套件：GLUE Benchmark(https://gluebenchmark.com/tasks) 包含許多任務(Task)與測試資料集。

   pip install datasets

2. 定義 GLUE 所有任務(Task)。

```
1  # 任務(Task)
2  GLUE_TASKS = ["cola", "mnli", "mnli-mm", "mrpc", "qnli", "qqp", "rte", "sst2", "stsb", "wnli"]
```

3. 指定任務為 cola。

```
1  # 指定任務為 cola
2  task = "cola"
3  # 預先訓練模型
4  model_checkpoint = "distilbert-base-uncased"
5  # 批量
6  batch_size = 16
```

4. 載入資料集、效能衡量指標：每個資料集有不同的效能衡量指標。

```
1  import datasets
2
3  actual_task = "mnli" if task == "mnli-mm" else task
4  # 載入資料集
5  dataset = datasets.load_dataset("glue", actual_task)
6  # 載入效能衡量指標
7  metric = datasets.load_metric('glue', actual_task)
```

5. 顯示 dataset 資料內容：dataset 資料型態為 DatasetDict，可參考 Transformers 官網的「DatasetDict 說明文件」(https://huggingface.co/docs/datasets/package_reference/main_classes.html#datasetdict)。

```
1  # 顯示 dataset 資料內容
2  dataset
```

執行結果：訓練資料 8551 筆，驗證資料 1043 筆，測試資料 1063 筆。

```
DatasetDict({
    train: Dataset({
        features: ['sentence', 'label', 'idx'],
        num_rows: 8551
    })
    validation: Dataset({
        features: ['sentence', 'label', 'idx'],
        num_rows: 1043
    })
    test: Dataset({
        features: ['sentence', 'label', 'idx'],
        num_rows: 1063
    })
})
```

6. 顯示第一筆內容。

```
1  # 顯示第一筆內容
2  dataset["train"][0]
```

執行結果：為正面/負面的情緒分析資料。

```
{'idx': 0,
 'label': 1,
 'sentence': "Our friends won't buy this analysis, let alone the next one we propose."}
```

7. 定義隨機抽取資料函數。

```
1  import random
2  import pandas as pd
3  from IPython.display import display, HTML
4
5  # 隨機抽取資料函數
6  def show_random_elements(dataset, num_examples=10):
7      picks = []
8      for _ in range(num_examples):
9          pick = random.randint(0, len(dataset)-1)
10         while pick in picks:
11             pick = random.randint(0, len(dataset)-1)
12         picks.append(pick)
13
14     df = pd.DataFrame(dataset[picks])
15     for column, typ in dataset.features.items():
16         if isinstance(typ, datasets.ClassLabel):
17             df[column] = df[column].transform(lambda i: typ.names[i])
18     display(HTML(df.to_html()))
```

8. 隨機抽取 10 筆資料查看。

```
1  # 隨機抽取10筆資料查看
2  show_random_elements(dataset["train"])
```

執行結果：

	idx	label	sentence
0	2722	unacceptable	A bicycle lent to me.
1	6537	unacceptable	Who did you arrange for to come?
2	1451	unacceptable	The cages which we donated wire for the convicts to build with are strong.
3	3119	acceptable	Cynthia munched on peaches.
4	3399	acceptable	Jackie chased the thief.
5	4705	acceptable	Nina got Bill elected to the committee.
6	1942	unacceptable	Every student who ever goes to Europe ever has enough money.
7	5889	acceptable	Bob gave Steve the syntax assignment.
8	4162	acceptable	This Government have been more transparent in the way they have dealt with public finances than any previous government.
9	1261	acceptable	I know two men behind me.

9. 顯示效能衡量指標。

```
1  # 顯示效能衡量指標
2  metric
```

執行結果： 包含準確率(Accuracy)、F1、Pearson 關聯度(Correlation)、
Spearman 關聯度(Correlation)、Matthew 關聯度(Correlation)。

```
Metric(name: "glue", features: {'predictions': Value(dtype='int64', id=None),
ge: """
Compute GLUE evaluation metric associated to each GLUE dataset.
Args:
    predictions: list of predictions to score.
        Each translation should be tokenized into a list of tokens.
    references: list of lists of references for each translation.
        Each reference should be tokenized into a list of tokens.
Returns: depending on the GLUE subset, one or several of:
    "accuracy": Accuracy
    "f1": F1 score
    "pearson": Pearson Correlation
    "spearmanr": Spearman Correlation
    "matthews_correlation": Matthew Correlation
```

10. 產生兩筆隨機亂數，測試效能衡量指標。

```
1  # 產生兩筆隨機亂數，測試效能衡量指標
2  import numpy as np
3
4  fake_preds = np.random.randint(0, 2, size=(64,))
5  fake_labels = np.random.randint(0, 2, size=(64,))
6  metric.compute(predictions=fake_preds, references=fake_labels)
```

11. 模型分詞：前置處理以利測試，可取得生字表(Vocabulary)，設定
    use_fast=True 就能夠快速處理。

```
1  from transformers import AutoTokenizer
2
3  # 分詞
4  tokenizer = AutoTokenizer.from_pretrained(model_checkpoint, use_fast=True)
```

12. 測試兩筆資料，進行分詞。

```
1  # 測試兩筆資料，進行分詞
2  tokenizer("Hello, this one sentence!", "And this sentence goes with it.")
```

13. 定義任務的資料集欄位。

```
1  # 任務的資料集欄位
2  task_to_keys = {
3      "cola": ("sentence", None),
4      "mnli": ("premise", "hypothesis"),
5      "mnli-mm": ("premise", "hypothesis"),
6      "mrpc": ("sentence1", "sentence2"),
7      "qnli": ("question", "sentence"),
8      "qqp": ("question1", "question2"),
9      "rte": ("sentence1", "sentence2"),
10     "sst2": ("sentence", None),
11     "stsb": ("sentence1", "sentence2"),
12     "wnli": ("sentence1", "sentence2"),
13 }
```

14. 測試第一筆資料。

```
1  # 測試第一筆資料
2  sentence1_key, sentence2_key = task_to_keys[task]
3  if sentence2_key is None:
4      print(f"Sentence: {dataset['train'][0][sentence1_key]}")
5  else:
6      print(f"Sentence 1: {dataset['train'][0][sentence1_key]}")
7      print(f"Sentence 2: {dataset['train'][0][sentence2_key]}")
```

執行結果：Our friends won't buy this analysis, let alone the next one we propose. 。

15. 測試 5 筆資料分詞。

```
1  # 測試 5 筆資料分詞
2  def preprocess_function(examples):
3      if sentence2_key is None:
4          return tokenizer(examples[sentence1_key], truncation=True)
5      return tokenizer(examples[sentence1_key], examples[sentence2_key], truncation=True)
6
7  preprocess_function(dataset['train'][:5])
```

執行結果：

{'input_ids': [[101, 2256, 2814, 2180, 1005, 1056, 4965, 2023, 4106, 1010, 2292, 2894, 1996, 2279, 2028, 2057, 16599, 1012, 102], [101, 2028, 2062, 18404, 2236, 3989, 1998, 1045, 1005, 1049, 3228, 2039, 1012, 102], [101, 2028, 2062, 18404, 2236, 3989, 2030, 1045, 1005, 1049, 3228, 2039, 1012, 102], [101, 1996, 2062, 2057, 2817, 16025, 1010, 1996, 13675, 16103, 2121, 2027, 2131, 1012, 102], [101, 2154, 2011, 2154, 1996, 8866, 2024, 2893, 14163, 8024, 3771, 1012, 102]], 'attention_mask': [[1, 1, 1, 1, 1, 1, 1, 1, 1, 1, 1, 1, 1, 1, 1, 1, 1, 1, 1], [1, 1, 1, 1, 1, 1, 1, 1, 1, 1, 1, 1, 1, 1], [1, 1, 1, 1, 1, 1, 1, 1, 1, 1, 1, 1, 1, 1], [1, 1, 1, 1, 1, 1, 1, 1, 1, 1, 1, 1, 1, 1, 1], [1, 1, 1, 1, 1, 1, 1, 1, 1, 1, 1, 1, 1]]}

16. 載入預先訓練的模型。

```
1  from transformers import AutoModelForSequenceClassification, TrainingArguments, Trainer
2
3  # 載入預先訓練的模型
4  num_labels = 3 if task.startswith("mnli") else 1 if task=="stsb" else 2
5  model = AutoModelForSequenceClassification.from_pretrained(model_checkpoint, num_labels=num_labels)
```

17. 定義訓練參數：可參閱 Transformers 官網的「TrainingArguments 説明文件」https://huggingface.co/transformers/main_classes/trainer.html#transformers. TrainingArguments。

```
1  # 定義訓練參數
2  metric_name = "pearson" if task == "stsb" else "matthews_correlation" \
3                      if task == "cola" else "accuracy"
4
5  args = TrainingArguments(
6      "test-glue",
7      evaluation_strategy = "epoch",
8      learning_rate=2e-5,
9      per_device_train_batch_size=batch_size,
10     per_device_eval_batch_size=batch_size,
11     num_train_epochs=5,
12     weight_decay=0.01,
13     load_best_model_at_end=True,
14     metric_for_best_model=metric_name,
15 )
```

18. 定義效能衡量指標計算的函數。

```
1  # 定義效能衡量指標計算的函數
2  def compute_metrics(eval_pred):
3      predictions, labels = eval_pred
4      if task != "stsb":
5          predictions = np.argmax(predictions, axis=1)
6      else:
7          predictions = predictions[:, 0]
8      return metric.compute(predictions=predictions, references=labels)
```

19. 定義訓練者(Trainer)物件：參數包含額外增加的訓練資料。

```
1  # 定義訓練者(Trainer) 物件
2  validation_key = "validation_mismatched" if task == "mnli-mm" else \
3                  "validation_matched" if task == "mnli" else "validation"
4
5  trainer = Trainer(
6      model,
7      args,
8      train_dataset=encoded_dataset["train"],
9      eval_dataset=encoded_dataset[validation_key],
10     tokenizer=tokenizer,
11     compute_metrics=compute_metrics
12 )
```

20. 在預先訓練好的模型基礎上繼續訓練，即是效能調整，在筆者的 PC 上至少訓練了 20 小時。

```
1  trainer.train()
```

執行結果：

Epoch	Training Loss	Validation Loss	Matthews Correlation	Runtime	Samples Per Second
1	0.519900	0.484644	0.437994	301.078100	3.464000
2	0.352600	0.519489	0.505773	299.051900	3.488000
3	0.231000	0.538032	0.556475	1863.316700	0.560000
4	0.180900	0.733648	0.515271	241.590500	4.317000
5	0.130700	0.787703	0.538738	242.532000	4.300000

訓練時間統計：

TrainOutput(global_step=2675, training_loss=0.27276652897629783, metrics={'train_runtime': 57010.5155, 'train_samples_per_secon
d': 0.047, 'total_flos': 356073036950940.0, 'epoch': 5.0, 'init_mem_cpu_alloc_delta': 757764096, 'init_mem_gpu_alloc_delta': 26
8953088, 'init_mem_cpu_peaked_delta': 273670144, 'init_mem_gpu_peaked_delta': 0, 'train_mem_cpu_alloc_delta': -935870464, 'trai
n_mem_gpu_alloc_delta': 1077715968, 'train_mem_cpu_peaked_delta': 1757851648, 'train_mem_gpu_peaked_delta': 21298176})

21. 模型評估。

```
1  # 模型評估
2  trainer.evaluate()
```

執行結果：

```
{'eval_loss': 0.5380318760871887,
 'eval_matthews_correlation': 0.5564748164739529,
 'eval_runtime': 229.9053,
 'eval_samples_per_second': 4.537,
 'epoch': 5.0,
 'eval_mem_cpu_alloc_delta': 507904,
 'eval_mem_gpu_alloc_delta': 0,
 'eval_mem_cpu_peaked_delta': 0,
 'eval_mem_gpu_peaked_delta': 20080128}
```

22. 新模型存檔：未來就能透過 from_pretrained()載入此效能調整後的模型進行預測。

```
1  # 模型存檔
2  trainer.save_model('./cola')
```

23. 新資料預測。

```
1   # 預測
2   class SimpleDataset:
3       def __init__(self, tokenized_texts):
4           self.tokenized_texts = tokenized_texts
5
6       def __len__(self):
7           return len(self.tokenized_texts["input_ids"])
8
9       def __getitem__(self, idx):
10          return {k: v[idx] for k, v in self.tokenized_texts.items()}
11
12  texts = ["Hello, this one sentence!", "And this sentence goes with it."]
13  tokenized_texts = tokenizer(texts, padding=True, truncation=True)
14  new_dataset = SimpleDataset(tokenized_texts)
15  trainer.predict(new_dataset)
```

執行結果：每筆以最大值作為預測結果。

```
PredictionOutput(predictions=array([[-0.55236566,  0.32417056],
    [-1.5994813 ,  1.4773667 ]], dtype=float32), label_ids=None, metrics={'test_runtime': 4.0388, 'test_samples_per_second':
0.495, 'test_mem_cpu_alloc_delta': 20480, 'test_mem_gpu_alloc_delta': 0, 'test_mem_cpu_peaked_delta': 0, 'test_mem_gpu_peaked_d
elta': 609280})
```

24. 之後可進行參數調校，這邊筆者就不繼續往下做了，要不然的話，機器應該會燒壞。

12-9-3　後續努力

以上只就官方的文件與範例介紹，Transformers 套件的功能非常多，要熟悉完整功能，尚待讀者後續努力地實驗，BERT 的變形不少，這些變形統稱為「BERTology」，預先訓練的模型可參閱「官網 Pretrained models」(https://huggingface.co/transformers/pretrained_models.html)，有的提供輕量型模型，如 ALBERT、TinyBERT，也有的提供更完整的模型，像是 GPT-3，號稱有 1,750 億個參數，更多內容可參閱「AI 趨勢周報第 142 期報導」[31]。

Transformer 架構的出現已經完全顛覆了 NLP 的發展，過往的 RNN/LSTM 模型雖然仍然可以拿來應用，但是，遵循 Transformer 架構的模型，它們在準確率上確實有比較明顯的優勢，因此，推測後續的研究方向應該會逐漸轉移到 Transformer 架構上，而且它不只可應用於 NLP，也開始將觸角伸向影像辨識領域，由此可見 Transformers 套件日益重要，詳情可參閱「AI 趨勢周報第 167 期報導」[32]。

12-10 總結

這一章我們介紹了處理自然語言的相關模型與其演進,包括 RNN、LSTM、GRU、注意力機制、Transformer、BERT 等,同時也實作許多範例,像是情緒分析(Sentiment Analysis)、神經機器翻譯(NMT)、字句相似度的比對、問答系統、文字摘要、命名實體識別(NER)、時間序列(Time Series)預測等。相信各位對於 NLP 應用應該有基本的認識,若要能靈活應用,還是需要找些專案或題目實作,畢竟魔鬼藏在細節內。提醒一下,由於目前 BERT 系列的模型在準確度方面已經超越 RNN/LSTM/GRU,所以如果是專案應用,建議應優先採用 BERT 模型。

參考資料 (References)

[1] Christopher Olah,《Understanding LSTM Networks》, 2015
     (https://colah.github.io/posts/2015-08-Understanding-LSTMs/)
[2] keras 官網 LSTM 的説明
     (https://keras.io/zh/layers/recurrent/)
[3] Jason Brownlee,《Time Series Prediction with LSTM Recurrent Neural
     Networks in Python with Keras》, 2016
     (https://machinelearningmastery.com/time-series-prediction-lstm-
     recurrent-neural-networks-python-keras)
[4] Junyoung Chung、 Caglar Gulcehre、KyungHyun Cho、Yoshua
     Bengio,《Empirical Evaluation of Gated Recurrent Neural Networks on
     Sequence Modeling》, 2014
     (https://arxiv.org/abs/1412.3555)
[5] Michael Phi,《Illustrated Guide to LSTM's and GRU's: A step by step
     explanation》, 2018

(https://towardsdatascience.com/illustrated-guide-to-lstms-and-gru-s-a-step-by-step-explanation-44e9eb85bf21)

[6] Alexandre Xavier,《Predicting stock prices with LSTM》, 2019
(https://medium.com/neuronio/predicting-stock-prices-with-lstm-349f5a0974d4)

[7] 陳昭明,《演算法交易(Algorithmic Trading) 實作》, 2021
(https://ithelp.ithome.com.tw/articles/10255111)

[8] 張俊林博客,《深度學習中的注意力機制(2017 版)》, 2017
(https://blog.csdn.net/malefactor/article/details/78767781)

[9] Meng Lee,《淺談神經機器翻譯 & 用 Transformer 與 TensorFlow 2 英翻中》, 2019
(https://leemeng.tw/neural-machine-translation-with-transformer-and-tensorflow2.html)

[10] TensorFlow 官網所提供的範例「Neural machine translation with attention」
(https://www.tensorflow.org/tutorials/text/nmt_with_attention)

[11] Andrej Karpathy,《The Unreasonable Effectiveness of Recurrent Neural Networks》, 2015
(http://karpathy.github.io/2015/05/21/rnn-effectiveness/)

[12] Ashish Vaswani、Noam Shazeer、Niki Parmar,《Attention Is All You Need》, 2017
(https://arxiv.org/pdf/1706.03762.pdf)

[13] Raimi Karim,《Illustrated: Self-Attention》, 2019
(https://towardsdatascience.com/illustrated-self-attention-2d627e33b20a)

[14] Jay Alammar,《The Illustrated Transformer》, 2018
(http://jalammar.github.io/illustrated-transformer/)

[15] GeeksforGeeks,《Self-attention in NLP》, 2020
(https://www.geeksforgeeks.org/self-attention-in-nlp/)

[16] Jason Brownlee,《A Gentle Introduction to Calculating the BLEU Score for Text in Python》, 2019

(https://machinelearningmastery.com/calculate-bleu-score-for-text-python/)

[17] Jacob Devlin、Ming-Wei Chang、Kenton Lee、Kristina Toutanova,《BERT: Pre-training of Deep Bidirectional Transformers for Language Understanding》, 2018

(https://arxiv.org/abs/1810.04805)

[18] Rani Horev,《BERT Explained: State of the art language model for NLP》, 2018

(https://towardsdatascience.com/bert-explained-state-of-the-art-language-model-for-nlp-f8b21a9b6270)

[19] BERT GitHub

(https://github.com/google-research/bert)

[20] Transformers GitHub

(https://github.com/huggingface/transformers)

[21] Transformers「Quick tour」

(https://huggingface.co/transformers/quicktour.html)

[22] Transformers 官網「Summary of the tasks」的 Extractive Question Answering

(https://huggingface.co/transformers/task_summary.html#extractive-question-answering)

[23] Transformers 官網「Summary of the tasks」的 Masked Language Modeling

(https://huggingface.co/transformers/task_summary.html#masked-language-modeling)

[24] Transformers 官網「Summary of the tasks」的 Text Generation

(https://huggingface.co/transformers/task_summary.html#text-generation)

[25] Transformers 官網「Summary of the tasks」的 Named Entity Recognition
(https://huggingface.co/transformers/task_summary.html#named-entity-recognition)

[26] Transformers 官網「Summary of the tasks」的 Summarization
(https://huggingface.co/transformers/task_summary.html#summarization)

[27] Adam Roberts、Staff Software Engineer、Colin Raffel 等人，《Exploring Transfer Learning with T5: the Text-To-Text Transfer Transformer》, 2020
(https://ai.googleblog.com/2020/02/exploring-transfer-learning-with-t5.html)

[28] Transformers 官網「Summary of the tasks」的 Translation
(https://huggingface.co/transformers/task_summary.html#translation)

[29] Transformers 官網「Training and fine-tuning」
(https://huggingface.co/transformers/training.html#tensorflow)

[30] (https://colab.research.google.com/github/huggingface/notebooks/blob/master/examples/text_classification.ipynb)

[31] 王若樸，《AI 趨勢周報第 142 期：推理能力新突破！OpenAI 新作 GPT-f 能自動證明數學定理》, 2020
(https://www.ithome.com.tw/news/140030)

[32] 王若樸，《AI 趨勢周報第 167 期：臉書新模型融合自監督和 Transformer，不需標註資料還能揪出複製圖》, 2021
(https://www.ithome.com.tw/news/144208)

聊天機器人(ChatBot)

這幾年 NLP 的應用範圍相當廣泛，好比說聊天機器人(ChatBot)，幾乎每一家企業都有這方面的需求，從售前支援(Pre-sale)、銷售(Sales)到售後服務(Post Services)等方面，用途十分多元，而支援系統功能的技術則涵蓋了 NLP、NLU、NLG，既要能解析對話(NLP)、理解問題 (NLU)，又要能回答得體、幽默、周全(NLG)，技術範圍幾乎整合了上一章所有的範例。另外，如果能結合語音辨識，用說話代替打字，這樣不論身處何時何地，人們都能夠更方便地用手機與機器溝通，或是結合其他的軟/硬體，例如社群軟體、智慧音箱等，使得電腦可以更貼近使用者的需求，提供人性化的服務，以往只能在電影裡看見的各種科技場景正逐漸在我們的日常生活中成真。

圖 13.1 ChatBot 商業應用

話說回來，要開發一個功能完善的 ChatBot，除了技術之外，更要有良好的規劃與設計作為基礎，當中有那些重要的「眉角」呢？現在就跟大家來一窺究竟。

13-1 ChatBot 類別

廣義來說，ChatBot 不一定要具備 AI 的功能，只要能自動回應訊息，基本上就稱為 ChatBot。通常一說到聊天機器人，大家直覺都會想到蘋果公司的 Siri，它可以跟使用者天南地北的聊天，不管是天氣、金融、音樂、生活資訊都難不倒它，但是，對於一般中小企業而言，這樣的功能並不能帶來商機，他們需要更直接的支援功能，因此我們把 ChatBot 分為以下類別：

1. 不限話題的機器人：可以與人自由自在的閒聊，包括公開資訊的查詢與應答，比如溫度、股市、播放音樂等，也包含日常寒暄，不需要精準的答案，只要有趣味性、即時回覆。

2. 任務型機器人：例如專家系統，具備特定領域的專業知識，服務範圍像是醫療、駕駛、航行、加密文件的解密等，著重在複雜的演算法或規則式(Rule based)的推理，需給予精準的答案，但不求即時的回覆。

3. 常見問答集(Frequently Asked Questions, FAQ)：客服中心將長年累積的客戶疑問集結成知識庫，當客戶詢問時，可快速搜尋，找出相似的問題，並將對應的處理方式回覆給客戶，答覆除了要求正確性與話術之外，也講究內容是否淺顯易懂和詳實周延，避免重複而空泛的回答，引發客戶不耐與不滿。

4. 資訊檢索：利用全文檢索的功能，搜尋關鍵字的相關資訊，比如 Google 搜尋，不需要完全精準的資訊，也不要單一的答案，而是提供所有可能的答案，由使用者自行作進一步判斷。

5. 資料庫應用：藉由 SQL 指令來查詢、篩選或統計資料，例如，旅館訂房、餐廳訂位、航班查詢/訂位、報價等，這是最傳統的需求，但如果能結合 NLP，讓輸出入介面更友善，例如語音輸入/輸出，可以引爆新一波的商機。

以上這五種類別的 ChatBot 各有不同的訴求，功能設計方向也因而有所差異，所以，在開工之前，務必要先搞清楚老闆要的機器人是哪一種，免得到時候開發出來，老闆才跟你說「這不是我要的」，那就欲哭無淚了。

13-2 ChatBot 設計

上一節談到的 ChatBot 種類非常多元，如果就每一種應用都詳細介紹的話 (雖然筆者很想)，應該可以再寫一本書了，所以本節僅針對共同的關鍵功能進行説明。

ChatBot 的規劃要點如下：

1. 訂定目標：根據規劃的目標，選擇適合的 ChatBot 類別，可以是多種類別的混合體。

2. 收集應用案例(Use Case)：收集應用的各種狀況和場景，整理成案例，以航空機票的銷售來舉例，就包括了每日空位查詢、旅程推薦、訂票、付款、退換票等，分析每個案例的現況與導入 ChatBot 後的場景與優點。

3. 提供的內容：現在行銷是內容為王(Content is king)的時代，有內容的資訊才能吸引人潮並帶來錢潮，這就是大家常聽到的內容行銷(Content Marketing)。因此，要評估哪些資訊是有效的、又該如何生產、並以何種方式呈現(Video、Podcast、部落文等)。

4. 挑選開發平台，有下列四種方式供選擇：

   - 套裝軟體：現在已有許多廠商提供某些行業別的解決方案，像是金融、保險等各行各業，技術也從傳統的 IVR 順勢轉為 ChatBot，提供更便利的使用介面。

   - ChatBot 平台：許多大型系統廠商都有提供 ChatBot 平台，他們利用獨有的 NLP 技術以及大量的 NER 資訊，整合各種社群軟體，使用者只要直接設定，就可以在雲端使用 ChatBot 並享有相關的服務。廠商包括 Google DialogFlow、微軟的 QnA Maker。

   - 開發工具：許多廠商提供開發工具，方便工程師快速完成一個 ChatBot，例如 Microsoft Bot Framework、Wit.ai 等，可參閱「10 Best Chatbot Development Frameworks to Build Powerful Bots」[1]，另外 Google、Amazon 智慧音箱也都有提供 SDK。

- 自行構建：可以利用套件加速開發，像是 TextBlob、Gemsim、SpaCy、Transforms 等 NLP 函數庫，或是 Rasa、ChatterBot 等 ChatBot Open Source。

5. 佈署平台：可選擇雲端或本地端，雲端可享有全球服務、或以微服務的方式運作，以使用次數計費，可節省初期的高資本支出，因此，若 ChatBot 不是資料庫交易類別的話，有越來越多企業採用雲端方案。

6. 用戶偏好(Preference)與面貌(Profile)：考量要儲存哪些與業務相關的用戶資訊。

ChatBot 的術語定義如下：

1. 技能(Skill)：例如銀行的技能包含存提款、定存、換匯、基金購買、房貸等，每一個應用都稱為一種技能。

2. 意圖(Intent)：技能中每一種對談的用意，例如，技能是旅館訂房，意圖則是有查詢某日是否有空的雙人房、訂房、換日期、退房、付款等。

3. 實體(Entity)：關鍵的人事時地物，利用前面所提的「命名實體識別」(NER)找出實體，每一個意圖可指定必要的實體，例如，旅館訂房必須指定日期、房型、住房天數、身分證字號等。

4. 例句(Utterance)：因為不同的人表達同一意圖會有各種不同表達方式，所以需要收集大量的例句，訓練 ChatBot，例如「我要訂 3/21 雙人房」、「明天雙人房一間」等。

5. 行動(Action)：所需資訊均已收集完整後，即可作出回應(Response)與相關的動作，例如訂房，若已確定日期、房型、住房天數、身分證字號後，即可採取行動，為客人保留房間，並且回應客戶「訂房成功」。

6. 開場白(Opening Message)：例如歡迎詞(Welcome)、問候語(Greeting)等，通常要有一些例句供隨機使用，避免一成不變，流於枯燥。

對話設計有些注意事項如下：

1. 對話管理：有兩種處理方式，有限狀態機(Finite-State Machine, FSM)和槽位填充(Slot Filling)。

   - 有限狀態機(FSM)：傳統的自動語音應答系統(IVR)大多採取這種方式，事先設計問題順序，確認每一個問題都得到適當的回答，才會進到下一狀態，如果中途出錯，就退回到前一狀態重來，銀行ATM 操作、電腦報修專線等也都是這種設計方式。

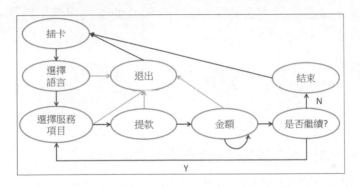

圖 13.2　ATM 提款的有限狀態機(FSM)

   - 槽位填充(Slot Filling)：有限狀態機的缺點是必須按順序回答問題，並且每次只能回答一個資訊，而且要等到系統唸完問題才能回答，對嫻熟的使用者來說會很不耐，若能引進 NLP 技術，就可以讓使用者用自然對談的方式提供資訊，例如「我要訂 3/21 雙人房」，客戶說一句話，系統就能夠直接處理，若發現資訊有欠缺，系統再詢問欠缺的資訊即可，與真人客服對談一樣，不必像往常一樣，「國語請按 1，台語請按 2，客語請按 3，英語請按 4」，只是訂個房還要過五關斬六將。

2. 整合社群媒體，譬如 Line、Facebook Messenger、Twitter 等，使用者不需額外安裝軟體，且不用教學，直接在對話群組加入官方帳號，即可開始與 ChatBot 對話。

3. 人機整合：ChatBot 設計千萬不能原地打轉，重複問相同的問題，必須設定跳脫條件，一旦察覺對話不合理，就應停止或轉由客服人員處理，避免引起使用者不快，造成反效果，使用有限狀態機設計方式，常會發生這種錯誤，若狀態已重複兩次以上，就可能是 bug。幾年前，微軟聊天機器人 Tay，推出後不到 24 小時，就因為學會罵人、講髒話，導致微軟緊急將她下架，就是一個血淋淋的案例。

除了技術層面之外，ChatBot 也稱為「Conversational AI」，因此對話的過程，需注意使用者的個資保護，包含像是對話檔案的存取權、對話中敏感資訊的保全，並且讓使用者清楚知道 ChatBot 的能力與應用範圍。

13-3 ChatBot 實作

這一節先以自行建置 ChatBot 的出發點，看看幾個範例。

【範例】NLP 加上相似度比較，製作簡單 ChatBot。
程式：13_01_simple_chatbot.ipynb。

1. 載入相關套件。

```
1  # 載入相關套件
2  import spacy
3  import json
4  import random
5  import pandas as pd
```

2. 載入訓練資料：資料來自「Learn to build your first chatbot using NLTK
   & Keras」(https://data-flair.training/blogs/python-chatbot-project)。

```
1   # 訓練資料
2   data_file = open('./chatbot_data/intents.json').read()
3   intents = json.loads(data_file)
4
5   intent_list = []
6   documents = []
7   responses = []
8
9   # 讀取所有意圖、例句、回應
10  for i, intent in enumerate(intents['intents']):
11      # 例句
12      for pattern in intent['patterns']:
13          # adding documents
14          documents.append((pattern, intent['tag'], i))
```

```
15
16          # adding classes to our class list
17          if intent['tag'] not in intent_list:
18              intent_list.append(intent['tag'])
19
20      # 回應(responses)
21      for response in intent['responses']:
22          responses.append((i, response))
23
24  responses_df = pd.DataFrame(responses, columns=['no', 'response'])
25
26  print(f'例句個數:{len(documents)}, intent個數:{len(intent_list)}')
27  responses_df
```

執行結果：例句個數有 47 個，意圖(intent)個數有 9 個。

3. 載入詞向量。

```
1  # 載入詞向量
2  nlp = spacy.load("en_core_web_md")
```

4. 定義前置處理函數：去除停用詞、詞形還原。

```
1   from spacy.lang.en.stop_words import STOP_WORDS
2
3   # 去除停用詞函數
4   def remove_stopwords(text1):
5       filtered_sentence =[]
6       doc = nlp(text1)
7       for word in doc:
8           if word.is_stop == False: # 停用詞檢查
9               filtered_sentence.append(word.lemma_) # lemma_ : 詞形還原
10      return nlp(' '.join(filtered_sentence))
11
12  # 結束用語
13  def say_goodbye():
14      tag = 1 # goodbye 項次
15      response_filter = responses_df[responses_df['no'] == tag][['response']]
16      selected_response = response_filter.sample().iloc[0, 0]
17      return selected_response
18
19  # 結束用語
20  def say_not_understand():
21      tag = 3 # 不理解的項次
22      response_filter = responses_df[responses_df['no'] == tag][['response']]
23      selected_response = response_filter.sample().iloc[0, 0]
24      return selected_response
```

5. 測試：相似度比較，為防止選出的問題相似度過低，可訂定相似度下限，低於下限即呼叫 say_not_understand()，回覆「我不懂你的意思, 請再輸入一次」，高於下限，才回答問題。

```
1  # 測試
2  prob_thread =0.6 # 相似度下限
3  while True:
4      max_score = 0
5      intent_no = -1
6      similar_question = ''
7
8      question = input('請輸入:\n')
9      if question == '':
10         break
11
12     doc1 = remove_stopwords(question)
13
14     # 比對：相似度比較
15     for utterance in documents:
16         # 兩語句的相似度比較
17         doc2 = remove_stopwords(utterance[0])
18         if len(doc1) > 0 and len(doc2) > 0:
19             score = doc1.similarity(doc2)
20             # print(utterance[0], score)
21         # else:
22             # print('\n', utterance[0],'\n')
23
24         if score > max_score:
25             max_score = score
26             intent_no = utterance[2]
27             similar_question = utterance[1] +', '+utterance[0]
28
```

```
29     # 若找到相似問題，且高於相似度下限，才回答問題
30     if intent_no == -1 or max_score < prob_thread:
31         print(say_not_understand())
32     else:
33         print(f'你問的是：{similar_question}')
34         response_filter = responses_df[responses_df['no'] == intent_no][['response']]
35         # print(response_filter)
36         selected_response = response_filter.sample().iloc[0, 0]
37         # print(type(selected_response))
38         print(f'回答：{selected_response}')
39
40 # say goodbye!
41 print(f'回答：{say_goodbye()}')
```

將回答轉成 Pandas DataFrame，便於篩選與抽樣(sample)，針對相同問題，可作不同的回覆。

執行結果：經過程式調校後，回應的結果還蠻令人滿意的。

```
請輸入:
hello
你問的是 : greeting, Hello
回答 : Hello, thanks for asking
請輸入:
How you could help me
你問的是 : options, What help you provide?
回答 : I can guide you through Adverse drug reaction list, Blood pressure tracking, Hospitals and Pharmacies
請輸入:
Adverse drug reaction
你問的是 : adverse_drug, How to check Adverse drug reaction?
回答 : Navigating to Adverse drug reaction module
請輸入:
blood pressure result
你問的是 : blood_pressure_search, Show blood pressure results for patient
回答 : Patient ID?
請輸入:
123
我不懂你的意思, 請再輸入一次.
請輸入:
pharmacy
你問的是 : pharmacy_search, Find me a pharmacy
回答 : Please provide pharmacy name
請輸入:
hospital
你問的是 : hospital_search, Hospital lookup for patient
回答 : Please provide hospital name or location
請輸入:

回答 : Bye! Come back again soon.
```

利用前一章所學的知識，只要短短數十行的程式碼，就可以完成一個具體而微的 ChatBot，當然，它還可以再加強的地方還很多，例如：

1. 中文語料庫測試。
2. 視覺化介面：可以利用 Streamlit、Flask、Django 等套件製作網頁，提供使用者測試。
3. 整合社群軟體：例如 LINE，直接在手機上測試。
4. 使用更完整的語料庫，測試 ChatBot 效能：目前使用 SpaCy 的分詞速度有點慢，應該是詞向量的轉換和前置處理花了一些時間，可以改用 NLTK 試試看。
5. 整合資料庫：例如查詢資料庫，檢查旅館是否有空房、保留訂房等。
6. 利用 NER 萃取實體：有了人事時地物的資訊，可進一步整合資料庫。

13-4 ChatBot 工具套件

網路上有許多的 ChatBot 工具套件,技術架構也很多樣化,筆者測試了一些套件如下:

1. ChatterBot(https://github.com/gunthercox/ChatterBot/tree/3eccceddd2a14eccaaeff12df7fa68513a464a00):採配接器模式(Adapter Pattern),是一個可擴充式的架構,支援多語系。

2. ChatBotAI(https://github.com/ahmadfaizalbh/Chatbot):以樣板(Template)語法訂定各式的樣板,接著再訂定變數嵌入樣板中,除了原本內建的樣板外,也可以自訂樣板和變數,來擴充 ChatBot 的功能。

3. Rasa(https://rasa.com/):以 Markdown 格式訂定意圖(Intent)、故事(Story)、回應(Response)、實體(Entity)與對話管理等功能,使用者可以編輯各個組態檔(*.yml),重新訓練後,就可以提供 ChatBot 使用。

13-4-1 ChatterBot 實作

ChatterBot 採配接器模式(Adapter Pattern),內建多種配接器(Adapters),主要分為兩類:Logic adapters、Storage adapters,也能自製配接器,是一個擴充式的架構,也支援多語系。它本身並沒有 NLP 的功能,只是單純的文字比對功能。

【範例 1】ChatterBot 測試。

程式:13_02_ChatterBot_test.ipynb。

1. 載入相關套件。

```
1  # 載入相關套件
2  from chatterbot import ChatBot
3  from chatterbot.trainers import ListTrainer
```

2. 訓練:將後一句作為前一句的回答,例如,使用者輸入 Hello 後,ChatBot 則回答「Hi there!」,它會使用到 NLTK 的語料庫。

```
1   # 訓練資料
2   chatbot = ChatBot("QA")
3
4   # 將後一句作為前一句的回答
5   conversation = [
6       "Hello",
7       "Hi there!",
8       "How are you doing?",
9       "I'm doing great.",
10      "That is good to hear",
11      "Thank you.",
12      "You're welcome."
13  ]
14
15  trainer = ListTrainer(chatbot)
16
17  trainer.train(conversation)
```

3. 簡單測試。

```
1   # 測試
2   response = chatbot.get_response("Good morning!")
3   print(f'回答：{response}')
```

執行結果：由於「Good morning!」不在訓練資料中，所以 ChatBot 就從過往的對話中隨機抽一筆資料出來回答。

4. 測試另一句在訓練資料中的句子：ChatBot 通常會回答後一句，偶而會回答過往的對話。

```
1   # 測試
2   response = chatbot.get_response("Hi there")
3   print(f'回答：{response}')
```

5. 加入內建的配接器(Adapters)。

- MathematicalEvaluation：數學式運算，檢視原始碼後發現它是使用 mathparse 函數庫(https://github.com/gunthercox/mathparse)。
- TimeLogicAdapter：有關時間的函數。
- BestMatch：從設定的句子中找出最相似的句子。

```
 1  bot = ChatBot(
 2      'Built-in adapters',
 3      storage_adapter='chatterbot.storage.SQLStorageAdapter',
 4      logic_adapters=[
 5          'chatterbot.logic.MathematicalEvaluation',
 6          'chatterbot.logic.TimeLogicAdapter',
 7          'chatterbot.logic.BestMatch'
 8      ],
 9      database_uri='sqlite:///database.sqlite3'
10  )
```

storage_adapter 參數指定對話記錄儲存在 SQL 資料庫或 MongoDB。

6. 測試時間的問題：問現在的時間。

```
 1  # 時間測試
 2  response = bot.get_response("What time is it?")
 3  print(f'回答：{response}')
```

問法可檢視原始碼 time_adapter.py。

```
self.positive = kwargs.get('positive', [
    'what time is it',
    'hey what time is it',
    'do you have the time',
    'do you know the time',
    'do you know what time it is',
    'what is the time'
])

self.negative = kwargs.get('negative', [
    'it is time to go to sleep',
    'what is your favorite color',
    'i had a great time',
    'thyme is my favorite herb',
    'do you have time to look at my essay',
    'how do you have the time to do all this'
    'what is it'
])
```

執行結果：回答「The current time is 04:37 PM」。

7. 數學式測試。

```
 1  # 算術式測試
 2  # 7 + 7
 3  response = bot.get_response("What is 7 plus 7?")
 4  print(f'回答：{response}')
 5
 6  # 8 - 7
```

```
 7  response = bot.get_response("What is 8 minus 7?")
 8  print(f'回答：{response}')
 9
10  # 50 * 100
11  response = bot.get_response("What is 50 * 100?")
12  print(f'回答：{response}')
13
14  # 50 * (85 / 100)
15  response = bot.get_response("What is 50 * (85 / 100)?")
16  print(f'回答：{response}')
```

執行結果：

```
回答：7 plus 7 = 14
回答：8 minus 7 = 1
回答：50 * 100 = 5000
回答：50 * ( 85 / 100 ) = 42.50
```

8. 加入自訂的配接器(Adapters)：自訂配接器為 my_adapter.py，類別名稱為 MyLogicAdapter。

```
 1  bot = ChatBot(
 2      'custom_adapter',
 3      storage_adapter='chatterbot.storage.SQLStorageAdapter',
 4      logic_adapters=[
 5          'my_adapter.MyLogicAdapter',
 6          'chatterbot.logic.MathematicalEvaluation',
 7          'chatterbot.logic.BestMatch',
 8      ],
 9      database_uri='sqlite:///database.sqlite3'
10  )
```

9. 測試自訂配接器。

```
 1  # 測試自訂配接器
 2  response = bot.get_response("我要訂位")
 3  print(f'回答：{response}')
```

執行結果：會回答「訂位日期、時間及人數？」或「哪一天？幾點？人數呢？」，這是程式中隨機指定的。

自訂配接器必須實現三個函數：

- __init__：初始化物件。
- can_process：訂定何種問題由此配接器處理，筆者設定的條件為 statement.text.find('訂位') >= 0。

- process：處理回答的函數。

以上範例是一個很簡單的架構，雖然沒有太多 NLP 的功能，提問的句子還必須與訓練資料完全相同，但是，它提供一個可擴充式的架構，讓讀者可以利用各自的配接器開發技能，再結合相似度辨識，找出意圖最相似的問題，最後作出對應的回答，程式就更堪用了。

13-4-2 chatbotAI 實作

chatbotAI 同時提供 AI 的功能與對話管理的架構，可以透過 REST API 與社群軟體進行整合，比方 Facebook Messenger，並且以樣板語言為主，提供學習(Learn)、 記憶(Memory)設定、條件判斷(Conditional Switch)、主題式對話(Topic-based Conversation Handling)等功能。

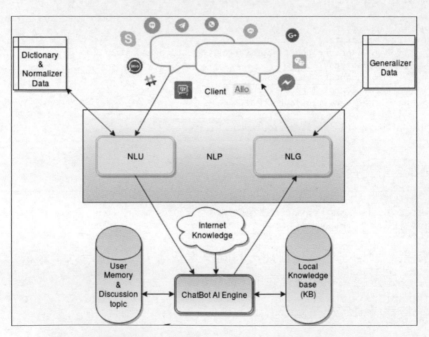

圖 13.3 chatbotAI 的架構圖，圖片來源：chatbotAI 官網[2]

【範例 2】chatbotAI 測試。

程式：13_03_ chatbotAI_test.ipynb。

1. 載入相關套件。

```
1  # 載入相關套件
2  from chatbot import demo
```

2. 功能展示。

```
1  # 功能展示
2  demo()
```

執行結果：>後面為使用者輸入的問題，一般問題的回答還算得體，要結束的話需輸入 quit。

```
Hi, how are you?
> fine
  Nice to know that you are fine. What else?
> what is your name?
My name is ChatBot
> are you male?
I may be male? -- what do you think?
> how old are you?
Perhaps you can answer your own question.
> are you a kid?
Perhaps you believe I am a kid.
> should I get out tomorrow?
Why do you ask that?
> weather
Let's change focus a bit... Tell me about your family.
> I have one brother
Now that you have one brother, what will you do next?
> quit
Thank you for talking with me.
```

3. 使用維基百科作為語料庫，需安裝 wikipedia 套件。

   pip install wikipedia

   這個套件可搜尋維基百科的資料，輸入關鍵字後，就可進行多功能的查尋，相關使用説明可參考 wikipedia 套件的官網(https://github.com/goldsmith/Wikipedia)。

4. 註冊可接收的關鍵字及負責回應的模組為「維基百科」。

```
1  # 載入相關套件
2  from chatbot import Chat, register_call
3  import wikipedia
4
5  # 註冊可接收的關鍵字及負責回應的模組
```

```
 6  @register_call("whoIs")
 7  def who_is(session, query):
 8      try:
 9          # 回應
10          return wikipedia.summary(query)
11      # 例外處理
12      except Exception:
13          for new_query in wikipedia.search(query):
14              try:
15                  return wikipedia.summary(new_query)
16              except Exception:
17                  pass
18      return "I don't know about "+query
```

5. 指定樣板，開始對話。樣本檔案內容如下：

```
{% block %}
    {% client %}(Do you know about|what is|who is|tell me about) (?P<query>.*){% endclient
    {% response %}{% call whoIs: %query %}{% endresponse %}
{% endblock %}
```

- client：使用者。
- response：ChatBot 的回應。
- (Do you know about|what is|who is|tell me about)：可接收的問句開頭。
- call whoIs: %query：指定註冊的 whoIs 模組回應。

```
1  # 第一個問題
2  first_question="Hi, how are you?"
3
4  # 使用的樣板
5  Chat("chatbot_data/Example.template").converse(first_question)
```

執行結果：詢問一些比較專業的問題，都可以應答無礙。

```
Hi, how are you?
> fine
  Nice to know that you are fine. What else?
> what is tensor
In mathematics, a tensor is an algebraic object that describes a (multilinear) relationship between sets of algebraic objects r
elated to a vector space. Objects that tensors may map between include vectors and scalars, and even other tensors. Tensors can
take several different forms – for example: scalars and vectors (which are the simplest tensors), dual vectors, multilinear map
s between vector spaces, and even some operations such as the dot product. Tensors are defined independent of any basis, althou
gh they are often referred to by their components in a basis related to a particular coordinate system.
Tensors are important in physics because they provide a concise mathematical framework for formulating and solving physics prob
lems in areas such as mechanics (stress, elasticity, fluid mechanics, moment of inertia, ...), electrodynamics (electromagnetic
tensor, Maxwell tensor, permittivity, magnetic susceptibility, ...), or general relativity (stress-energy tensor, curvature ten
sor, ... ) and others. In applications, it is common to study situations in which a different tensor can occur at each point of
an object; for example the stress within an object may vary from one location to another. This leads to the concept of a tensor
field. In some areas, tensor fields are so ubiquitous that they are often simply called "tensors".
Tensors were conceived in 1900 by Tullio Levi-Civita and Gregorio Ricci-Curbastro, who continued the earlier work of Bernhard R
iemann and Elwin Bruno Christoffel and others, as part of the absolute differential calculus. The concept enabled an alternativ
e formulation of the intrinsic differential geometry of a manifold in the form of the Riemann curvature tensor.
> tell me about chatbot
Kuki, formerly known as Mitsuku, is a chatbot created from Pandorabots AIML technology by Steve Worswick. It is a five-time win
ner of a Turing Test competition called the Loebner Prize (in 2013, 2016, 2017, 2018, and 2019), for which it holds a world rec
ord. Kuki is available to chat via an online portal, and on Facebook Messenger, Twitch group chat, Telegram and Kik Messenger,
and was available on Skype, but was removed by its developer.
```

6. 使用中文關鍵字發問。

```
1  first_question="你好嗎?"
2  Chat("chatbot_data/Example.template").converse(first_question)
```

執行結果：中文關鍵字(who is 蔡英文)能夠正確回答。

```
你好嗎?
> 好
I see.
> who is 蔡英文
Tsai Ing-wen (born 31 August 1956) is a Taiwanese politician and academic serving as the seventh president of Taiwan, since 201
6. A member of the Democratic Progressive Party, Tsai is the first female president of Taiwan. She has served as Chair of the D
emocratic Progressive Party (DPP) since 2020, and previously from 2008 to 2012 and 2014 to 2018.
Tsai studied law and international trade, and later became a law professor at Soochow University School of Law and National Che
ngchi University after earning an LLB from National Taiwan University and an LLM from Cornell Law School. She later studied law
at the London School of Economics and Political Science, with her thesis titled "Unfair trade practices and safeguard actions",
and was awarded a Ph.D. in law from the University of London. In 1993, as an independent (without party affiliation), she was a
ppointed to a series of governmental positions, including trade negotiator for WTO affairs, by the then-ruling Kuomintang (KMT)
and was one of the chief drafters of the special state-to-state relations doctrine of President Lee Teng-hui.
After DPP President Chen Shui-bian took office in 2000, Tsai served as Minister of the Mainland Affairs Council throughout Che
n's first term as a non-partisan. She joined the DPP in 2004 and served briefly as a DPP-nominated at-large member of the Legis
lative Yuan. From there, she was appointed Vice Premier under Premier Su Tseng-chang until the cabinet's mass resignation in 20
07. She was elected and assumed DPP leadership in 2008, following her party's defeat in the 2008 presidential election. She res
igned as chair after losing the 2012 presidential election.
```

7. 記憶(memory)模組定義：以下使用變數來記憶一個字串或累計值，例如訪客人數，並且使用 key/value 進行儲存。

```
1   # 記憶(memory)模組定義
2   @register_call("increment_count")
3   def memory_get_set_example(session, query):
4       # 一律轉成小寫
5       name=query.strip().lower()
6       # 取得記憶的次數
7       old_count = session.memory.get(name, '0')
8       new_count = int(old_count) + 1
9       # 設定記憶次數
10      session.memory[name]=str(new_count)
11      return f"count  {new_count}"
```

8. 記憶(memory)設定測試。

```
1   # 記憶(memory)測試
2   chat = Chat("chatbot_data/get_set_memory_example.template")
3   chat.converse("""
4   Memory get and set example
5
6   Usage:
7     increment <name>
8     show <name>
9     remember <name> is <value>
10    tell me about <name>
11
12  example:
13    increment mango
14    show mango
15    remember sun is red hot star in our solar system
16    tell me about sun
17  """)
```

chat.converse()：內含用法説明。

- increment <name>：變數值加 1。
- show <name>：顯示變數值。
- remember <name> is <value>：記憶變數與對應值。
- tell me about <name>：顯示變數對應值。

執行結果：

```
> increment mango
count  1
> increment mango
count  2
> show mango
2
> remember sun is red hot star in our solar system
I will remember sun is red hot star in our solar system
> tell me about sun
sun is red hot star in our solar system
> remember PLG 5/2 比賽結果 is 夢想家勝
I will remember plg 5/2 比賽結果 is 夢想家勝
> tell me about plg 5/2
I don't know about plg 5/2
> tell me about plg 5/2比賽結果
I don't know about plg 5/2比賽結果
> tell me about plg 5/2 比賽結果
plg 5/2 比賽結果 is 夢想家勝
```

chatbotAI 也提供一個擴充性的架構，可透過註冊的模組和樣板，以外掛的方式銜接各種技能。

13-4-3 Rasa 實作

Rasa 是一個 Open Source 的工具軟體，也有付費版本，相當多的文章有提到它。它以 Markdown 格式訂定意圖(Intent)、故事(Story)、回應(Response)、實體(Entity)以及對話管理等功能，使用者可以編輯各個組態檔(*.yml)，重新訓練過後，就可以提供給 ChatBot 來使用。

安裝過程有點挫折，依照 Rasa 官網指示操作的話，會出現錯誤，正確的安裝指令如下：

1. Windows 作業系統

```
pip install rasa --ignore-installed ruamel.yaml --user
```

** --user 參數：會讓 Rasa 被安裝在使用者目錄下。若不加此選項，則會出現權限不足的錯誤訊息，表示 Python site-packages 目錄不容安裝。

2. Linux/Mac 作業系統

```
pip install rasa --ignore-installed ruamel.yaml
```

在 Windows 作業系統下，Rasa 安裝成功後，程式會放在 c:\users\<user_name>\appdata\roaming\python\python38\scripts\。接著，測試步驟如下：

1. 新增一個專案：

```
c:\users\<user_name>\appdata\roaming\python\python38\scripts\rasa.exe init --no-prompt
```

產生一個範例專案，子目錄和檔案列表如下：

會依據以上專案檔案，同時進行訓練，完成後建立模型檔，儲存在 models 子目錄內。

data 子目錄內有幾個重要的檔案：

- nlu.yml：NLU 訓練資料，包含各類的意圖(Intent)和例句(Utterance)。
- rules.yml：包含各項規則的意圖和行動。
- stories.yml：包含各項故事情節，描述多個意圖和行動的順序。

根目錄的檔案：

- domain.yml：包含 Bot 各項的回應(Response)。
- config.yml：NLU 訓練的管線(Pipeline)與策略(Policy)。

2. 測試：

```
c:\users\<user_name>\appdata\roaming\python\python38\scripts\rasa.e
xe shell
```

對話過程如下：並沒有太大的彈性，必須完全照著 nlu.yml 問問題。

```
Bot loaded. Type a message and press enter (use '/stop' to exit):
Your input ->  hello
Hey! How are you?
Your input ->  I am fine
Great, carry on!
Your input ->  what is you name
I am a bot, powered by Rasa.
Your input ->  I am disappointed
Here is something to cheer you up:
Image: https://i.imgur.com/nGF1K8f.jpg
Did that help you?
Your input ->  yes
Great, carry on!
Your input ->  great
Great, carry on!
Your input ->  bye
Bye
Your input ->  /stop
2021-05-04 22:15:07 INFO     root  - Killing Sanic server now.
```

3. 故事情節視覺化：

```
c:\users\<user_name>\appdata\roaming\python\python38\scripts\rasa.e
xe visualize
```

執行結果：對應 stories.yml 的內容。

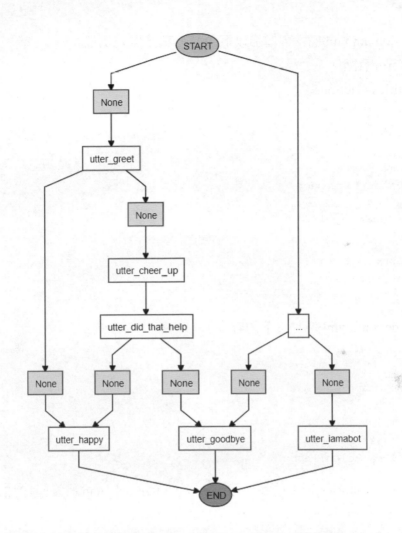

4. 訓練模型：可以修改上述的.yml 檔案後，重新訓練模型。

```
c:\users\<user_name>\appdata\roaming\python\python38\scripts\rasa.e
xe train
```

【**範例 3**】建立自訂的行動(Custom action)。以下新增一個行動，詢問姓名，並單純回答「Hello <name>」。

1. 安裝 Rasa SDK

```
pip install rasa_core_sdk
```

2. 在 domain.yml 檔案內增加以下內容，請參閱範例檔：

- inform 意圖。
- entities、actions。

```
entities:
  - name

actions:
  - action_save_name
```

- responses 增加以下內容：

```
utter_welcome:
- text: "Welcome!"
utter_ask_name:
- text: "What's your name?"
```

3. 在 data\nlu.yml 增加以下內容：

```
- intent: inform
  examples: |
    - my name is [Michael](name)
    - [Philip](name)
    - [Michelle](name)
    - [Mike](name)
    - I'm [Helen](name)
```

4. 在 data\ stories.yml 的每一段 action: utter_ask_name 後面增加以下內容：

```
- intent: inform
  entities:
  - name: "name"
- action: action_save_name
```

5. 在 endpoints.yml 解除註解：

```
action_endpoint:
 url: http://localhost:5055/webhook
```

6. 增加一段 action 處理程式。

```python
from typing import Any, Text, Dict, List
from rasa_sdk import Action, Tracker
from rasa_sdk.executor import CollectingDispatcher

class ActionSaveName(Action):
    def name(self):
        return "action_save_name"
    def run(self, dispatcher: CollectingDispatcher,
        tracker: Tracker,
        domain: Dict[Text, Any]) -> List[Dict[Text, Any]]:

        name = \
        next(tracker.get_latest_entity_values("name"))
        dispatcher.utter_message(text=f"Hello, {name}!")
        return []
```

7. 啟動 action 程式：

```
c:\users\<user_name>\appdata\roaming\python\python38\scripts\rasa.e
xe run actions
```

8. 重新訓練模型。

```
c:\users\<user_name>\appdata\roaming\python\python38\scripts\rasa.e
xe train
```

9. 測試：

```
c:\users\<user_name>\appdata\roaming\python\python38\scripts\rasa.e
xe shell
```

對話過程如下，確實有回應「Hello <name>」：

Rasa 比較像傳統的 AIML ChatBot，是以問答例句當作訓練資料，算是相對僵硬的方式，且需要大量的人力維護，但好處是可以精準控制回答的內容。

13-5 Dialogflow 實作

現在已經有許多廠商都推出成熟的 ChatBot 產品，只要經過適當的設定，就可以上線了，例如 Google Dialogflow、Microsoft QnA Maker、Azure Bot Service、IBM Watson Assistant 等。以下我們以 DialogFlow 為例介紹整個流程。

依據 Dialogflow 的官網說明[3]，它是一個 NLU 平台，可將對話功能整合至網頁、手機、語音回應介面，而輸入/輸出介面可以是文字或語音。就筆者實驗結果，它主要是以槽位填充(Slot Filling)為出發點，並搭配完整的 NER 功能，例如時間，可輸入 today、tomorrow、right now，系統會自動轉換為日期，另外全世界的城市也能辨識，算是一個可輕易上手的產品。

它有兩個版本：Dialogflow CX、Dialogflow ES，前者為進階版本，後者為標準版，可免費試用，我們用免費版本測試。兩者功能的比較表可參閱：https://cloud.google.com/dialogflow/docs/editions。

Dialogflow 的術語定義如下：

1. Agent：即 ChatBot，每個公司可建立多個 ChatBot，各司其職。

2. 意圖(Intent)：與之前定義相同，但更細緻，包含：

- 訓練的片語(Training phrases)：定義使用者表達意圖的片語，不必列舉所有可能的片語，Dialogflow 有內建的機器學習智能，會自動加入類似的片語。

- 行動(Action)：ChatBot 接收到意圖後採取的行動。

- 參數(Parameter)：定義槽位填充所需的資訊，包括必填或選填的參數，Dialogflow 可以從使用者的表達中找出對應的實體(Entity)。

- 回應(Response)：行動完畢後，回應使用者的文字或語音。

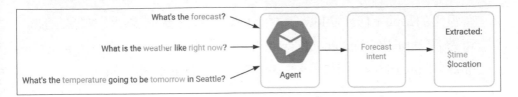

圖 13.4 Dialogflow 可從意圖中找出時間和地點，圖片來源：Dialogflow 的官網說明[3]

3. 實體(Entity)：Dialogflow 已內建許多系統實體(System Entity)類別，包括日期、時間、顏色、email 等，還包括多國語系的實體，詳情可參閱 https://cloud.google.com/dialogflow/es/docs/reference/system-entities。

4. 上下文(Context)：如下圖，Dialogflow 從第一句話中察覺意圖是「查詢帳戶資訊」(CheckingInfo)，接著會問「何種資訊」，使用者回答「帳戶餘額」後，Dialogflow 即將餘額告訴使用者。Dialogflow 會先辨識意圖，再根據缺乏的資訊進一步詢問，直到所有資訊都滿足為止，才會將答案回覆給使用者，這就是槽位填充(Slot Filling)的機制。

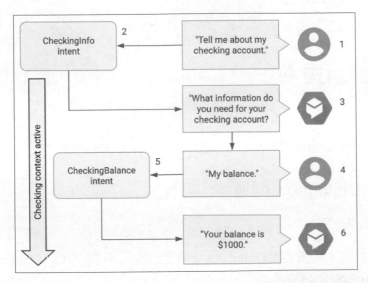

圖 13.5 Dialogflow 上下文對話的流程，圖片來源：Dialogflow 的官網說明[3]

5. 追問意圖(Follow-up intent)：可依據使用者的回答定義不同的回答方式，以追問意圖，透過此功能可以建立有限狀態機，對於複雜的流程有很大的幫助，Dialogflow 一樣有內建追問意圖的識別，如 Yes/No，可參閱 https://cloud.google.com/dialogflow/es/docs/reference/follow-up-intent-expressions，例如，yes，回答 sure、exactly 也可以。

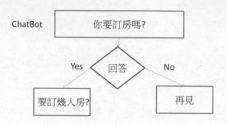

6. 履行(Fulfillment)：ChatBot 除了回應文字之外，也能夠與資料庫或社群軟體整合，開發者可以撰寫一個服務，整合各種軟硬體。

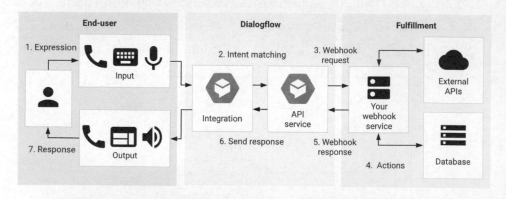

圖 13.6 履行(Fulfillment)，圖片來源：Dialogflow 的官網說明[3]

13-5-1 Dialogflow 開通

Dialogflow 不須安裝，只需開通，程序如下。

1. 開通：首先要申請 Gmail 帳號，接著再申請 GCP (Google Cloud Platform) 雲端服務。

2. 新增 GCP 的專案：參閱「Dialogflow Quickstart: Setup」(https://cloud.google.com/dialogflow/es/docs/quick/setup)， 點擊「Go to project selector」按鈕。

3. 授權專案使用 Dialogflow API：點擊「Enable the API」按鈕。

4. 建立服務帳戶：點擊「Create a service account」按鈕。

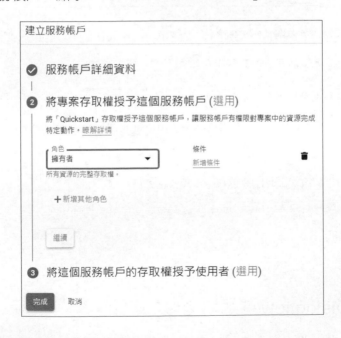

5. 新增金鑰(key)：點擊「Email account」按鈕。

「Quickstart」專案的服務帳戶

服務帳戶代表的是 Google Cloud 服務身分，例如在 Compute Engine VM 上所執行的程式

可使用機構政策來保護服務帳戶，並封鎖載有風險的服務帳戶功能，例如自動授予身分與

⇒ 篩選　輸入屬性名稱或值

☐　電子郵件

☐　📧 michael@quickstart-1590025178539.iam.gserviceaccount.com

詳細資料　　權限　　金鑰　　指標　　記錄

金鑰

Add a new key pair or upload a public key certificate from an existing key pair. Please note that public certificates need to be in RSA_X509_PEM format. Learn more about upload key formats

Block service account key creation using organization policies.
Learn more about setting organization policies for service accounts

新增金鑰 ▾

6. 設定環境變數，指向金鑰路徑。在 Powershell 內輸入如下：

```
$env:GOOGLE_APPLICATION_CREDENTIALS="<金鑰路徑>"
```

　　例如：`"F:\0_AI\Books\以 100 張圖理解深度學習\quickstart.json"`

7. 安裝 Cloud SDK：依照 https://cloud.google.com/sdk/docs/install 的指示。

8. 測試 SDK：會列印出金鑰(是一堆亂碼)。

```
gcloud auth application-default print-access-token
```

9. 安裝 Dialogflow client library：

```
pip install google-cloud-dialogflow
```

13-5-2 Dialogflow 基本功能

至此開通完成，即可進入 Dialogflow 設定相關畫面。

1. 建立 Agent：瀏覽 https://dialogflow.cloud.google.com/#/newAgent，輸入相關資訊，支援多語系，包括繁體中文，按「Create」即可。

2. 內建意圖：Dialogflow 會預先建立 2 個意圖。
   - Default Fallback Intent：不理解使用者的意圖時，會歸屬於此，通常會要求使用者再輸入其他用語。
   - Default Welcome Intent：Agent 的歡迎詞。

3. 建立意圖：點擊「Create Intent」按鈕，輸入意圖名稱，並點擊「Add Training Phrases」超連結，就可輸入多組問句與回應。

- 回應：點擊「Add Response」超連結

- 存檔：點擊「Save」按鈕。

4. 測試：存檔，並確定訓練完成的訊息出現之後，即可在畫面右側測試，亦可直接用語音輸入。

- 輸入「What is your name?」。

- 輸入「Hello」。

5. 參數：輸入的例句如果包含內建的實體(Entity)，則會被解析出來，當作參數，可進一步設定參數屬性。點選畫面左側 Intent 旁的「+」。

例如輸入「I know English」，Dialogflow 偵測到「English」是內建的語言 Entity(@sys.language)，系統自動新增一個參數(parameter)。

可針對參數設定屬性：

- Required：是否必要輸入。
- Parameter Name：參數名稱。
- Entity：選擇 Entity 類別，可修改為其他類別。
- Value：參數的名稱，回應(Response)可以此名稱取得參數值。
- Is List：參數值是否為 List，即一參數含多個值。
- Prompts：若輸入的問句或回答欠缺此參數，Agent 會顯示此提示，詢問使用者。

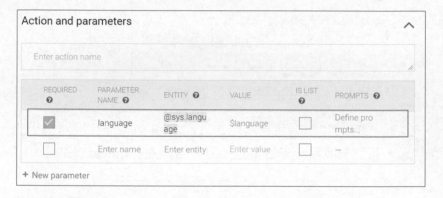

- 輸入回應(Response)：「Wow! I didn't know you knew $language.」，其中$language 會自使用者的問句取得變數值。
- 存檔。

6. 測試：輸入「I speak english」，回應的$language = english。

7. 可建立自訂的實體(Entity)：點選畫面左側 Entities 旁的「+」。

■ 輸入 Entity 和同義字(Synonym)。

■ 存檔。

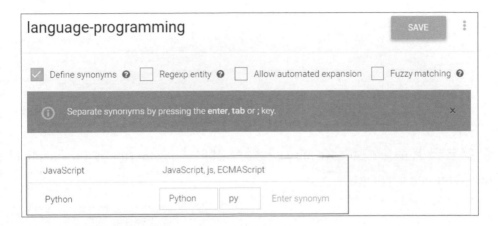

8. 使用 english 的實體(Entity)：在原來的「set-language」Entity，輸入例句(Training phrase)。

■ I know javascript.

■ I write the logic in python.

■ double click「javascript」，選取「language-programming」。

■ 輸 入 回 應 (Response)：「 $language-programming is an excellent programming language.」。

■ 存檔。

9. 測試：輸入「you know js?」。

參數 language 若勾選 Required，會出現「what is the language?」。

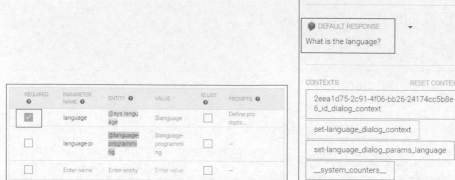

全部參數若都不勾選 Required，則會出現「JavaScript is an excellent programming language.」。

10. 追問意圖(Follow-up intent)：若需考慮上文回答，可追問詳細意圖。

- 測試：修改回應為「Wow! I didn't know you knew $language. How long have you known $language?」。
- 加追問意圖：點選畫面左側「intents」，將滑鼠移至「set-language」，會出現「Add follow-up intent」，點擊即可，會增加「set-language - custom」追問意圖。

 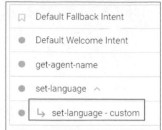

點擊「set-language - custom」，輸入例句(Training phrase)：

- 3 years
- about 4 days
- for 5 years

壽命(Lifespan)：一般意圖的預設壽命為 5 個對話，追問意圖的預設壽命則為 2 個對話，超過 20 分鐘，所有意圖均不保留，即相關的對話狀態會被重置。

測試：加入回應「I can't believe you've known #set-language-followup. language for $duration!」。

- 「#」：意圖。
- 「.」：參數值。

11. 測試追問意圖：

輸入「I know French」。

輸入「for 5 years」。

13-5-3 履行(Fulfillment)

履行(Fulfillment)是在蒐集完整資訊後採取的行動，撰寫程式完成商業邏輯和交易，可與社群軟體、硬體整合。Dialog 提供兩種履行類型：

1. Webhook：撰寫一個網頁服務(Web service)，Dialog 透過 POST 請求送給 Webhook，並接收回應。

2. Inline Editor：透過 GCP 建立 Cloud Functions，使用 Node.js 執行環境，這是比較簡單的方式，不過如果是正式的專案開發，還是要選擇 Webhook。

【範例】Inline Editor 須建立 GCP 付費帳號，才能使用，以下針對 Webhook 實作。

1. 建立一個新的意圖：輸入兩個例句「order a room in Tainan at 2021/02/05」、「I want a double room in Taipei at 2021-01-01」。

2. 參數「geo-city」、「date-time」均設為必要欄位。

3. 履行(Fulfillment)：啟用「Enable webhook call for this intent」。
4. 撰寫 Webhook 程式：可使用多種語言撰寫，這裡我們使用 Python 加上 Flask 套件，撰寫 Web 程式，完整程式請參考 dialogflow\webhook\ app.py，程式碼後續說明。
5. 程式必須佈署到網際網路上，Dialogflow 才能存取到 app.py。可以使用 ngrok.exe 將內部網址對應到外部網址，這樣就可以先在本機測試，等 到測試成功後，再將程式佈署到 Heroku 或其他網站測試，Heroku 是免 費的網站佈署平台，試用後可升級為付費帳戶。
6. 啟動 app.py：預設網址為 http://127.0.0.1:5000/。

python app.py

7. 執行下列指令，取得對應的 https 網址。

ngrok http 5000

8. 接著設定 Fulfillment：點選畫面左側的 Fulfillment 旁的「+」，啟用 Webhook，並設定上一步驟所取得的 https 網址，後面須加上 /webhook，點擊下方「Save」按鈕即可。

9. 測試：在畫面右側輸入「order a room in Tainan at 2021/02/05」。

10. 再查看 dialogflow\webhook\test.db SQLite 資料庫，就可以看到每重複執行一次，tainan/2021-02-05 的訂房數(room_count)就會加 1。而輸入不同的城市或日期則會新增一筆記錄。SQLite 資料庫可使用 SQLitespy.exe 或其他工具軟體開啟。

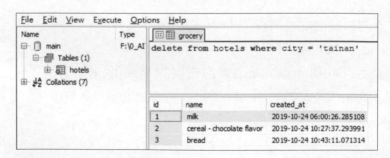

dialogflow\webhook\app.py，程式碼説明如下。

1. 安裝套件。

pip install flask

pip install sqlalchemy

2. 載入相關套件。

```
from flask import Flask, request, jsonify, make_response
from sqlalchemy import create_engine
```

3. 宣告 Flask 物件。

```
app = Flask(__name__)
```

4. 定義函數：必須為 @app.route('/webhook', methods=['POST'])。可取得
   請求、意圖、entity 等。

```
@app.route('/webhook', methods=['POST'])
def hotel_booking():
    # 取得請求
    req = request.get_json(force=True)
    # 取得意圖 set-language
    intent = req.get('queryResult').get('intent')['displayName']
    # 取得 entity
    entityCity = req.get('queryResult').get('parameters')["geo-city"].lower()
    entityDate = req.get('queryResult').get('parameters')["date-time"].lower()
    entityDate = entityDate[:10].replace('/', '-')
```

5. 開啟資料庫連線。

```
# 開啟資料庫連線
engine = create_engine('sqlite:///test.db', convert_unicode=True)
con = engine.connect()
```

6. 更新資料庫：先根據城市、日期查詢資料，若記錄存在，則訂房數加
   1，反之，則新增一筆新的記錄，最後傳回 OK 訊息。

```
if intent == 'booking':
    # 根據城市、日期查詢
    sql_cmd = f"select room_count from  hotels "
    sql_cmd += f"where city = '{entityCity}' and order_date = '{entityDate}'"
    result = con.execute(sql_cmd)
    list1 = result.fetchall()

    # 增修記錄
    if len(list1) > 0: # 訂房數加 1
        sql_cmd = f"update hotels set 'room_count' = {list1[-1][-1]+1} "
        sql_cmd += f"where city = '{entityCity}' and order_date = '{entityDate}'"
        result = con.execute(sql_cmd)
    else: # 新增一筆記錄
        sql_cmd = "insert into hotels('city', 'order_date', 'room_count')"
        sql_cmd += f" values ('{entityCity}', '{entityDate}', 1)"
        result = con.execute(sql_cmd)

    # 回應
    response = f'{entityCity}, {entityDate} OK.'
    return make_response(jsonify({ 'fulfillmentText': response }))
```

7. 以上只是示範程式，在實際情況中，我們必須做例外處理，包括程式碼錯誤、意圖、Entity 檢查等。

Dialogflow 還可以整合語音交換機、社群媒體、Spark 等，詳情可參閱「Dialogflow Integrations 説明」(https://dialogflow.cloud.google.com/#/agent/get-agent-name-ojw9/integrations)。另外，Dialogflow 也內建許多應用程式，可參閱「Dialogflow Prebuilt Agents 説明」(https://dialogflow.cloud.google.com/#/agent/get-agent-name-ojw9/prebuiltAgents/)。

以上 Dialogflow 介紹的概念，也幾乎算是業界的標準了。

13-6 結語

對於大部分的企業來説，聊天機器人的實用度相當高，可以提供售前支援、銷售甚至是售後服務等多方面的功能，但如何整合既有的流程及系統，使得 ChatBot 能無縫接軌，員工也能快速上手，是建置系統的一大課題，最後，設置時也不要忘記系統自我學習的使命，讓系統隨著服務的經驗，越來越聰明。

參考資料 (References)

[1] Adnan Rehan,《10 Best Chatbot Development Frameworks to Build Powerful Bots》, 2020

(https://geekflare.com/chatbot-development-frameworks/)

[2] ChatBotAI GitHub

(https://github.com/ahmadfaizalbh/Chatbot)

[3] Dialogflow 的官網説明

(https://cloud.google.com/dialogflow/docs)

近幾年在影像、語音等的「自然使用者介面」(Natural User Interface, NUI)有了突破性的發展，譬如 Apple Facc ID 以臉部辨識登入，手機、智慧音箱可以使用語音輸入，這類操作方式大幅降低輸入的難度，尤其是中老年人。根據統計，人們講話的速度約每分鐘 150~200 字，而打字輸入大概只有每 60 字/分，如果能提高語音辨識的能力，語音輸入就會逐漸取代鍵盤打字了，此外，鍵盤在攜帶方便性與親和力來說也遠不及語音。由此可見，要消弭人類與機器之間的隔閡，語音辨識是相當重要的關鍵技術，接下來我們就來好好認識它的發展。

回歸現實，語音辨識並不簡單，必須要克服下列挑戰：

1. 說話者的個別差異：包括口音、音調的高低起伏，像是男性和女性的音頻差異就很大。
2. 環境噪音：各種環境會有不同的背景音源，因此辨識前必須先去除噪音。
3. 語調的差異：人在不同的情緒下，講話的語調會有所不同，比如悲傷時講話速度可能較慢，聲音較小而低沉，反之，興奮時，講話速度快，聲音較大。

光是「No」一個簡單的詞，不同的人說就有各式各樣的聲波，如下圖：

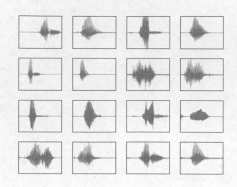

圖 14.1 一千種 No 的聲波之部分擷取，圖片來源：哥倫比亞大學語音辨識課程講義[1]

因此，要能辨識不同人的聲音，電腦必須先對收到的訊號做前置處理，之後，才能再運用各種演算法和資料庫進行辨識，而這過程中所需的基礎知識包括：

- 訊號處理(Signal processing)。
- 機率與統計(Probability and statistics)。
- 語音和語意學(Phonetics; linguistics)。
- 自然語言處理(Natural language processing, NLP)。
- 機器學習與深度學習。

看到這裡大家應該有點頭痛了，所以筆者試著以簡馭繁，將焦點放在實作上。

14-1 語音基本認識

以說話為例，人類以胸腔壓縮和變換嘴唇、舌頭形狀的方式，對空氣產生壓縮與伸張的效果，形成聲波，然後以每秒大約 340 公尺的速度在空氣中傳播，當此聲波傳遞到另一個人的耳朵時，耳膜就會感受到一伸一壓的壓力訊號，接著內耳神經再將此訊號傳遞到大腦，並由大腦解析與判讀，分辨此訊號的意義，詳細說明可參閱「Audio Signal Processing and Recognition」[2]。

聲音的訊號通常如下圖一般是不規則的,所以必須先經過數位化,才能交由電腦處理,作法是每隔一段時間衡量振幅,得到一個數字,這個過程稱為「取樣」(Sampling),如圖 14.3,之後,再把所有數字記錄下來變成數位音訊,而這個過程就是所謂的將類比(Analog)訊號轉為數位(Digital)訊號。

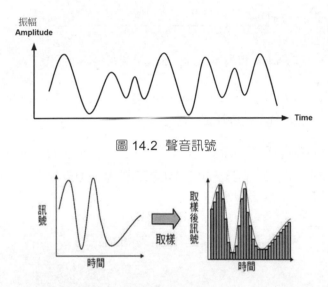

圖 14.2 聲音訊號

圖 14.3 聲音取樣(Sampling),圖片來源:國立臺灣大學普通物理實驗室[3]

訊號可由波形的振幅(Amplitude)、頻率(Frequency)及相位(Phase)來表示:

1. 振幅:指波的高度,可以形容聲音的大小。
2. 頻率:為一秒波動的週期數,可以形容聲音的高低。
3. 相位(Phase):描述訊號波形變化的度量,通常以度(角度)作為單位,也稱為相角或相。當訊號波形以週期的方式變化,波形循環一周即為 360 度。

https://gfycat.com/ickyfilthybobolink 有一個小動畫說明振幅、頻率及相位所代表的意義,簡單易懂,值得一看。

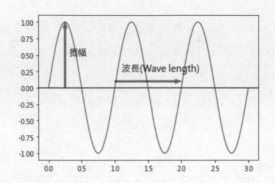

圖 14.4 振幅(Amplitude)與波長(Wave length)

可參閱程式 **14_01_ 振幅 (Amplitude)** 、 **頻率 (Frequency)** 及相位 **(Phase).ipynb**，作一簡單的測試。

頻率是以赫茲(Hz)為單位，赫茲(Hz)為訊號每秒振動的週期數，通常人耳可以聽到的頻率約在 20 Hz to 20 kHz 的範圍內，但隨著年齡的增長，人們會對高頻信號越來越不敏感。

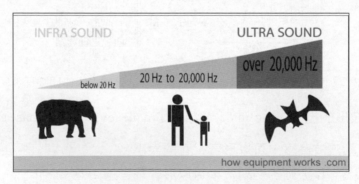

圖 14.5 動物可聽見的音頻範圍，圖片來源：「Audio Signal Processing」[4]

根據奈奎斯特(Nyquist)定理，重建訊號是取樣頻率(Sample Rate)的一半，以傳統電話為例，通常接收的音頻約為 4K，因此取樣頻率通常是 8K，其他常見裝置的取樣頻率如下：

- 網路電話：16K。
- CD：單聲道為 22.05K，立體(雙)聲道為 44.1K。

- DVD：單聲道為 48K，藍光 DVD 為 96K。
- 其他裝置的取樣頻率可參閱「Sampling (signal processing)維基百科」(https://en.wikipedia.org/wiki/Sampling_(signal_processing)#Sampling_rate)。

將訊號轉為數字時，若數字的精度不足會造成更多的損失，因此，也可分為 8、16、32 位元不同的整數精度，這個過程稱為「量化」(Quantization)，傳統電話採 8 位元，網路電話採 16 位元。

訊號經過數位化後，通常會把它存檔或透過網路傳輸給另一端，接著再把數位訊號轉回類比訊號，即可原音重現，如下圖所示，這時就牽涉到訊號壓縮的問題，如何以最小的資料量儲存或傳輸，就是所謂的編碼(Encoding)機制。

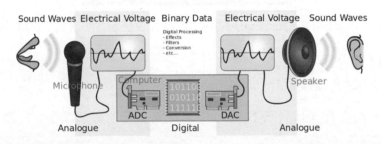

圖 14.6 訊號的數位化與重現，圖片來源：「File:CPT-Sound-ADC-DAC.svg」[5]

常見的編碼方式有：

1. 脈衝編碼調變(Pulse-code modulation,PCM)：直接將每一個取樣的振幅存檔或傳輸至對方，這種編碼方式效率不高。
2. 非線性 PCM(Non-linear PCM)：因人類對高頻信號較不敏感，故可以把高頻信號以較低精度編碼，反之，低頻信號採較高精度，可降低編碼量。
3. 可調變 PCM(Adaptive PCM)：由於訊號片段高低不一，因此不必統一編碼，可以將訊號切成很多段，並把每一段都個別編碼，進行正規化(Regularization)後，再作 PCM 編碼。

最常見的語音檔應該是 wav 檔，它支援各式的精度與編碼，最常見的是 16 位元精度與 PCM 編碼。

以上的過程可由「示波器」(Oscilloscope)觀察，如下圖，也可以直接以程式實作。

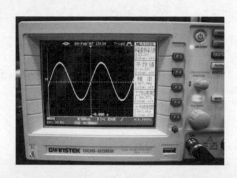

圖 14.7 示波器(Oscilloscope)，圖片來源：國立臺灣大學普通物理實驗室[3]

【範例 1】音檔解析。

程式：14_02_音檔解析.ipynb。

1. 載入相關套件：Jupyter Notebook 本身就有支援影像顯示、語音播放。

```
1  # 載入相關套件
2  import IPython
```

2. 播放音檔(wav)：檔案來源為 https://github.com/maxifjaved/sample-files，autoplay 設定為 True 時，執行即會自動播放，不須另外按 PLAY 鍵。

```
1  # 檔案來源:https://github.com/maxifjaved/sample-files
2  wav_file = './audio/WAV_1MG.wav'
3
4  # autoplay=True:自動播放，不須按 PLAY 鍵
5  IPython.display.Audio(wav_file, autoplay=True)
```

執行結果：可中止，顯示檔案長度有 33 秒。

▶ 0:02 / 0:33 ━━━━━━━━━ ◀ ⋮

3. 取得音檔的屬性：可使用 Python 內建的模組 wave，取得音檔的屬性，相關說明可參閱 wave 說明文件(https://docs.python.org/3/library/wave.html)。

```
1  # 取得音檔的屬性
2  import wave
3
4  f=wave.open(wav_file)
5  print(f'取樣頻率={f.getframerate()}, 幀數={f.getnframes()}, ' +
6      f'聲道={f.getnchannels()}, 精度={f.getsampwidth()}, ' +
7      f'檔案秒數={f.getnframes() / f.getframerate():.2f}')
8  f.close()
```

執行結果：

取樣頻率=8000, 幀數=268237, 聲道=2, 精度=2, 檔案秒數=33.53。

- getframerate()：取樣頻率。
- getnframes()：音檔總幀數。
- getnchannels()：聲道。
- getsampwidth()：量化精度。
- 檔案秒數 = 音檔總幀數 / 取樣頻率。

4. 使用 PyAudio 函數庫串流播放：每讀一個區塊，就立即播放。PyAudio 在 Windows 作業系統下不能使用 pip install PyAudio 順利安裝，請直接至「Unofficial Windows Binaries for Python Extension Packages」(https://www.lfd.uci.edu/~gohlke/pythonlibs/#pyaudio) 下載 PyAudio‑0.2.11‑cp38‑cp3**[X]**‑win_amd64.whl，再執行 pip install PyAudio‑0.2.11‑cp38‑cp3**[X]**‑win_amd64.whl。

```
1   # 使用 PyAudio 串流播放
2   import pyaudio
3
4   def PlayAudio(filename, seconds=-1):
5       # 定義串流區塊大小(stream chunk)
6       chunk = 1024
7
8       # 開啟音檔
9       f = wave.open(filename,"rb")
10
11      # 初始化 PyAudio
12      p = pyaudio.PyAudio()
13
```

```
14      # 開啟串流
15      stream = p.open(format = p.get_format_from_width(f.getsampwidth()),
16                      channels = f.getnchannels(), rate = f.getframerate(), output = True)
17
18      # 計算每秒區塊數
19      sample_count_per_second = f.getframerate() / chunk
20
21      # 計算總區塊數
22      if seconds > 0 :
23          total_chunk = seconds * sample_count_per_second
24      else:
25          total_chunk = (f.getnframes() / (f.getframerate() * f.getnchannels())) \
26                          * sample_count_per_second
27
28      print(f'每秒區塊數={sample count per second}, 總區塊數={total chunk}')
```

```
29
30      # 每次讀一區塊
31      data = f.readframes(chunk)
32      no=0
33      while data:
34          # 播放區塊
35          stream.write(data)
36          data = f.readframes(chunk)
37          no+=1
38          if seconds > 0 and no > total_chunk :
39              break
40
41      # 關閉串流
42      stream.stop_stream()
43      stream.close()
44
45      # 關閉 PyAudio
46      p.terminate()
```

5. 呼叫函數播放。

```
1  # 播放音檔
2  PlayAudio(wav_file, -1)
```

執行結果：每秒區塊數=7.8125, 總區塊數=130.97509765625。

6. 繪製波形：由於多聲道 wav 檔案格式是交錯儲存的，故先說明比較單純的單聲道 wav 檔案讀取。

```
1  # 繪製波形
2  import numpy as np
3  import wave
4  import sys
5  import matplotlib.pyplot as plt
6
7  # 單聲道繪製波形
8  def DrawWavFile_mono(filename):
```

```
9    # 開啟音檔
10   f = wave.open(filename, "r")
11
12   # 字串轉換整數
13   signal = f.readframes(-1)
14   signal = np.frombuffer(signal, np.int16)
15   fs = f.getframerate()
16
17   # 非單聲道無法解析
18   if f.getnchannels() == 1:
19       Time = np.linspace(0, len(signal) / fs, num=len(signal))
20
21       # 繪圖
22       plt.figure(figsize=(12,6))
23       plt.title("Signal Wave...")
24       plt.plot(Time, signal)
25       plt.show()
26   else:
27       print('非單聲道無法解析')
```

7. 測試。

```
1  wav_file = './audio/down.wav'
2  DrawWavFile_mono(wav_file)
```

執行結果：

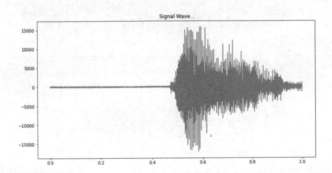

8. 多聲道繪製波形函數。

```
1   # 多聲道繪製波形
2   def DrawWavFile_stereo(filename):
3       # 開啟音檔
4       with wave.open(filename,'r') as wav_file:
5           # 字串轉換整數
6           signal = wav_file.readframes(-1)
7           signal = np.frombuffer(signal, np.int16)
8
9           # 為每一聲道準備一個 list
10          channels = [[] for channel in range(wav_file.getnchannels())]
11
```

```
12          # 將資料放入每個 list
13          for index, datum in enumerate(signal):
14              channels[index % len(channels)].append(datum)
15
16          # 計算時間
17          fs = wav_file.getframerate()
18          Time=np.linspace(0, len(signal)/len(channels)/fs,
19                           num=int(len(signal)/len(channels)))
20
21          f, ax = plt.subplots(nrows=len(channels), ncols=1,figsize=(10,6))
22          for i, channel in enumerate(channels):
23              if len(channels)==1:
24                  ax.plot(Time,channel)
25              else:
26                  ax[i].plot(Time,channel)
```

9. 測試。

```
1  wav_file = './audio/WAV_1MG.wav'
2  DrawWavFile_stereo(wav_file)
```

執行結果： 雙聲道分別如下圖。

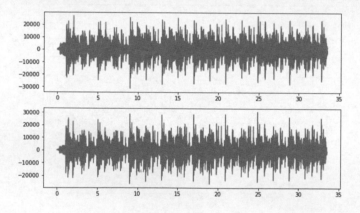

10. 將前面的單、多聲道函數整合在一起。

```
1  # 多聲道繪製波形
2  def DrawWavFile(wav_file):
3      f=wave.open(wav_file)
4      channels = f.getnchannels() # 聲道
5      f.close()
6
7      if channels == 1:
8          DrawWavFile_mono(wav_file)
9      else:
10         DrawWavFile_stereo(wav_file)
```

11. 測試。

```
1   wav_file = './audio/down.wav'
2   DrawWavFile(wav_file)
3   wav_file = './audio/WAV_1MG.wav'
4   DrawWavFile(wav_file)
```

執行結果：

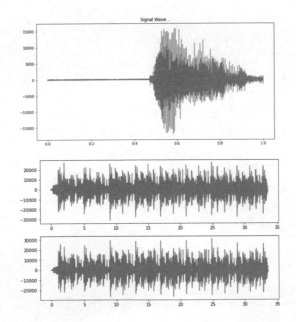

12. 產生音檔：以隨機亂數產生音檔，亂數介於(-32767, 32767)之間。

```
1   # 產生音檔
2   import wave, struct, random
3
4   sampleRate = 44100.0 # 取樣頻率
5   duration = 1.0 # 秒數
6
7   wav_file = './audio/random.wav'
8   obj = wave.open(wav_file,'w')
9   obj.setnchannels(1) # 單聲道
10  obj.setsampwidth(2)
11  obj.setframerate(sampleRate)
12  for i in range(99999):
13      value = random.randint(-32767, 32767)
14      data = struct.pack('<h', value) # <h : short, big-endian
15      obj.writeframesraw(data)
16  obj.close()
17
18  IPython.display.Audio(wav_file, autoplay=True)
```

執行結果：產生音檔 random.wav，並播放。

13. 取得音檔的屬性。

```
1  # 取得音檔的屬性
2  import wave
3
4  f=wave.open(wav_file)
5  print(f'取樣頻率={f.getframerate()}, 幀數={f.getnframes()}, ' +
6        f'聲道={f.getnchannels()}, 精度={f.getsampwidth()}, ' +
7        f'檔案秒數={f.getnframes() / f.getframerate():.2f}')
8  f.close()
```

執行結果：取樣頻率=44100, 幀數=99999, 聲道=1, 精度=2, 檔案秒數
=2.27，與設定一致。

14. 雙聲道音檔轉換為單聲道。

```
1  # 雙聲道音檔轉換為單聲道
2  import numpy as np
3
4  wav_file = './audio/WAV_1MG.wav'
5  # 開啟音檔
6  with wave.open(wav_file,'r') as f:
7      # 字串轉換整數
8      signal = f.readframes(-1)
9      signal = np.frombuffer(signal, np.int16)
10
11     # 為每一聲道準備一個 list
12     channels = [[] for channel in range(f.getnchannels())]
13
14     # 將資料放入每個 list
15     for index, datum in enumerate(signal):
16         channels[index % len(channels)].append(datum)
17
18     sampleRate = f.getframerate() # 取樣頻率
19     sampwidth = f.getsampwidth()
20
21 wav_file_out = './audio/WAV_1MG_mono.wav'
22 obj = wave.open(wav_file_out,'w')
23 obj.setnchannels(1) # 單聲道
24 obj.setsampwidth(sampwidth)
25 obj.setframerate(sampleRate)
26 for data in channels[0]:
27     obj.writeframesraw(data)
28 obj.close()
```

15. 測試。

```
1  # 取得音檔的屬性
2  import wave
3
4  f=wave.open(wav_file)
5  print(f'取樣頻率={f.getframerate()}, 幀數={f.getnframes()}, ' +
6        f'聲道={f.getnchannels()}, 精度={f.getsampwidth()}, ' +
7        f'檔案秒數={f.getnframes() / f.getframerate():.2f}')
8  f.close()
```

執行結果：取樣頻率=8000, 幀數=268237, 聲道=2, 精度=2, 檔案秒數 =33.53。

除了讀取檔案之外，要如何才能直接從麥克風接收音訊或是錄音存檔呢？我們馬上就來看看該怎麼做。

【範例 2】麥克風接收音訊與錄音存檔。

程式：14_03_錄音.ipynb。

1. SpeechRecognition 套件有提供麥克風收音的功能，並支援語音辨識，首先要進行安裝：

   pip install SpeechRecognition

2. 另外，文字轉語音(Text To Speech, TTS)的技術也已非常成熟，故一併安裝 pyttsx3 套件，下面程式碼會使用到：

```
pip install pyttsx3
```

3. 麥克風收音，並進行語音辨識。

```
1  # 載入相關套件
2  import speech_recognition as sr
3  import pyttsx3
```

4. 列出電腦中的說話者(Speaker)。

```
1   # 列出電腦中的說話者(Speaker)
2   speak = pyttsx3.init()
3   voices = speak.getProperty('voices')
4   for voice in voices:
5       print("Voice:")
6       print(" - ID: %s" % voice.id)
7       print(" - Name: %s" % voice.name)
8       print(" - Languages: %s" % voice.languages)
9       print(" - Gender: %s" % voice.gender)
10      print(" - Age: %s" % voice.age)
```

執行結果：注意有些説話者擅長説英文或中文，不過筆者實際測試後發現，其實他們兩種語言都可以講。

```
Voice:
  - ID: HKEY_LOCAL_MACHINE\SOFTWARE\Microsoft\Speech\Voices\Tokens\TTS_MS_ZH-TW_HANHAN_11.0
  - Name: Microsoft Hanhan Desktop - Chinese (Taiwan)
  - Languages: []
  - Gender: None
  - Age: None
Voice:
  - ID: HKEY_LOCAL_MACHINE\SOFTWARE\Microsoft\Speech\Voices\Tokens\TTS_MS_EN-US_ZIRA_11.0
  - Name: Microsoft Zira Desktop - English (United States)
  - Languages: []
  - Gender: None
  - Age: None
```

5. 指定説話者：每台電腦安裝的説話者均不同，請以 ID 從中指定一位。

```
1  # 指定説話者
2  speak.setProperty('voice', voices[0].id)
```

6. 麥克風收音：含文字轉語音(Text To Speech, TTS)，程式會等到持續靜默一段時間(預設是 0.8 秒)後才結束。詳細可參閱 SpeechRecognition 官方説明(https://pypi.org/project/SpeechRecognition/2.1.2/)。

```
1   # 麥克風收音
2   r = sr.Recognizer()
3   with sr.Microphone() as source:
4       # 文字轉語音
5       speak.say('請説話...')
6       # 等待説完
7       speak.runAndWait()
8
9       #降噪
10      r.adjust_for_ambient_noise(source)
11      # 麥克風收音
12      audio = r.listen(source)
```

7. 錄音存檔。

```
1  # 錄音存檔
2  wav_file = "./audio/record.wav"
3  with open(wav_file, "wb") as f:
4      f.write(audio.get_wav_data(convert_rate=16000))
```

8. 語音辨識：需以參數 language 指定要辨識的語系。

```
1  # 語音辨識
2  try:
3      text=r.recognize_google(audio, language='zh-tw')
4      print(text)
5  except e:
6      pass
```

筆者唸了一段新聞，內容如下：

「受鋒面影響，北台灣今天下午大雨特報，有些道路甚至發生積淹，曾文水庫上游也傳來好消息。」

執行結果如下：

「受封面影響北台灣今天下午大雨特報有些道路甚至發曾記股曾文水庫上游也傳來好消息。」

結果大部份是對的，錯誤的文字均為同音異字。

9. 檢查輸出檔：播放錄音。

```
1  import IPython
2
3  # autoplay=True：自動播放，不須按 PLAY 鍵
4  IPython.display.Audio(wav_file, autoplay=True)
```

10. 取得音檔的屬性。

```
1  # 取得音檔的屬性
2  # https://docs.python.org/3/library/wave.html
3  import wave
4
5  f=wave.open(wav_file)
6  print(f'取樣頻率={f.getframerate()}, 幀數={f.getnframes()}, ' +
7        f'聲道={f.getnchannels()}, 精度={f.getsampwidth()}, ' +
8        f'檔案秒數={f.getnframes() / (f.getframerate() * f.getnchannels()):.2f}')
9  f.close()
```

執行結果：取樣頻率=16000, 幀數=173128, 聲道=1, 精度=2, 檔案秒數=10.82，與設定一致。

11. 讀取音檔，轉為 SpeechRecognition 音訊格式，再進行語音辨識。

```
1  import speech_recognition as sr
2
3  # 讀取音檔，轉為音訊
4  r = sr.Recognizer()
5  with sr.WavFile(wav_file) as source:
6      audio = r.record(source)
7
8  # 語音辨識
9  try:
10      text=r.recognize_google(audio, language='zh-tw')
11      print(text)
12  except e:
13      pass
```

執行結果如下，與麥克風來源一致。

「受封面影響北台灣今天下午大雨特報有些道路甚至發曾記殷曾文水庫上游也傳來好消息。」

12. 顯示所有可能的辨識結果及信賴度。

```
1  # 顯示所有可能的辨識結果及信賴度
2  dict1=r.recognize_google(audio, show_all=True, language='zh-tw')
3  for i, item in enumerate(dict1['alternative']):
4      if i == 0:
5          print(f"信賴度={item['confidence']}, {item['transcript']}")
6      else:
7          print(f"{item['transcript']}")
```

執行結果：

信賴度=0.89820588,

所有可能的辨識結果：

受封面影響飛台灣今天下午大雨特報有些道路甚至發曾記殷曾文水庫上游野傳來好消息
受封面影響飛台灣今天下午大雨特報有些道路甚至發生技菸曾文水庫上游野傳來好消息
受封面影響飛台灣今天下午大雨特報有些道路甚至發生記菸曾文水庫上游野傳來好消息
受封面影響飛台灣今天下午大雨特報有些道路甚至發生氣菸曾文水庫上游野傳來好消息
受封面影響飛台灣今天下午大雨特報有些道路甚至發生記燕曾文水庫上游野傳來好消息

▌14-2 語音前置處理

另外，有一個非常棒的語音處理套件不得不提，那就是 Librosa，它可以將音訊做進一步的解析和轉換，我們會在後面實作相關功能，更多內容請參閱 Librosa 說明文件(https://librosa.org/doc/latest/tutorial.html)。

在開始測試之前，還有一些關於音訊的概念需要我們先了解。由於音訊通常是一段不規則的波形，很難分析，因此，學者提出「傅立葉轉換」(Fourier transform)，可以把不規則的波形變成多個規律的正弦波形(Sinusoidal)相加，如下圖所示。

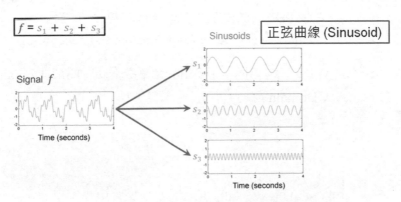

圖 14.8　傅立葉轉換(Fourier transform)

圖片來源：「Introduction Basic Audio Feature Extraction」[6]

每個正弦波形可以被表示為：

$$s_{(A,\,\omega,\,\varphi)}(t) = A \cdot \sin(2\pi(\omega t - \varphi))$$

其中：

- A：振幅
- ω：頻率
- φ：相位

例如下圖，可以觀察到振幅、頻率、相位是如何影響正弦波形的。

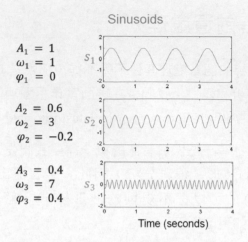

圖 14.9　正弦波形的振幅、頻率與相位

圖片來源：「Introduction Basic Audio Feature Extraction」[6]

轉換後的波形振幅和頻率均相同，原來的 X 軸為時域(Time Domain)就轉為頻域（Frequency Domain）。

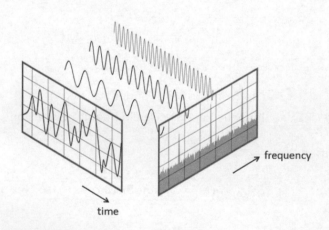

圖 14.10 傅立葉轉換後，時域(Time Domain)轉為頻域（Frequency Domain）

圖片來源：「Audio Data Analysis Using Deep Learning with Python (Part 1)」[7]

不同的頻率混合在一起稱之為頻譜(Spectrum)，而繪製的圖表就稱為頻譜圖(Spectrogram)，通常 X 軸為時間，Y 軸為頻率，可以從圖表中觀察到各種頻率的能量。

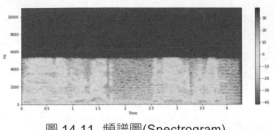

圖 14.11 頻譜圖(Spectrogram)

為了方便做語音辨識，與處理影像一樣，我們會對音訊進行特徵萃取 (Feature Extraction)，目前有 FBank(Filter Banks)、MFCC（Mel-frequency Cepstral Coefficients）兩種，特徵抽取前須先對聲音做前置處理：

1. 分幀：通常每幀是 25ms，幀與幀之間重疊 10ms，避免邊界信號的遺漏。
2. 信號加強：針對高頻信號做加強，使信號更清楚。
3. 加窗(Window)：目的是消除各個幀的兩端信號可能不連續的現象，常用的窗函數有方窗、漢明窗(Hamming window)等。有時候為了考慮上下文，會將相鄰的幀合併成一個幀，這種處理方式稱為「幀疊加」(Frame Stacking)。

去除雜訊(denoising or noise reduction)。

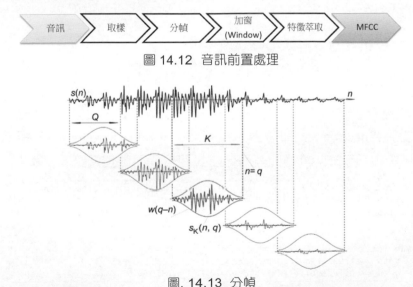

圖 14.12 音訊前置處理

圖. 14.13 分幀

在計算頻譜時，會將以上的前置處理，包含分幀、加窗、離散傳立葉轉換（Discrete Fourier Transform, DFT）合併為一個步驟，稱為短時傳立葉轉換(Short-Time Fourier Transform, STFT)，SciPy 有支援此一功能，函數名稱為 stft。

【範例 3】頻譜圖(Spectrogram)即時顯示。

程式：14_05_spectrogram.py，由於是以動畫呈現，無法在 Jupyter Notebook 上展示，故以 Python 檔案執行。另外，14_04_waves.py 可顯示即時的波形。這兩支程式均源自「Python audio spectrum analyzer」[8]。

1. 載入相關套件。

```
2   import pyaudio
3   import struct
4   import matplotlib.pyplot as plt
5   import numpy as np
6   from scipy import signal
```

2. 開啟麥克風，設定收音相關參數。

```
8    # 宣告麥克風變數
9    mic = pyaudio.PyAudio()
10
11   # 參數設定
12   FORMAT = pyaudio.paInt16 # 精度
13   CHANNELS = 1 # 單聲道
14   RATE = 48000 # 取樣頻率
15   INTERVAL = 0.32 # 緩衝區大小
16   CHUNK = int(RATE * INTERVAL) # 接收區塊大小
17
18   # 開啟麥克風
19   stream = mic.open(format=FORMAT, channels=CHANNELS, rate=RATE,
20              input=True, output=True, frames_per_buffer=CHUNK)
```

3. 頻譜圖(Spectrogram)即時顯示：呼叫 signal.spectrogram()，顯示頻譜圖，設定顯示滿 100 個圖表即停止，可依需要彈性調整。

```
22   # 設定X/Y軸標籤
23   plt.ylabel('Frequency [Hz]')
24   plt.xlabel('Time [sec]')
25
26   i=0
27   while i < 100: # 顯示100次即停止
28       data = stream.read(CHUNK, exception_on_overflow=False)
29       data = np.frombuffer(data, dtype='b')
```

```
30
31      # 繪製頻譜圖
32      f, t, Sxx = signal.spectrogram(data, fs=CHUNK)
33      dBS = 10 * np.log10(Sxx)
34      plt.clf()
35      plt.pcolormesh(t, f, dBS)
36      plt.pause(0.001)
37      i+=1
```

執行結果：

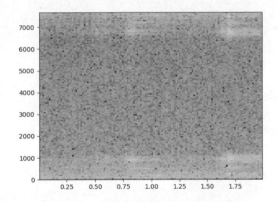

4. 關閉所有裝置。

```
39   # 關閉所有裝置
40   stream.stop_stream()
41   stream.close()
42   mic.terminate()
```

【**範例 3**】音訊前置處理：利用 Librosa 函數庫了解音訊的前置處理程序。
程式：14_06_音訊前置處理.ipynb。

1. 載入相關套件。

```
1   # 載入相關套件
2   import IPython
3   import pyaudio
4   import struct
5   import matplotlib.pyplot as plt
6   import numpy as np
7   from scipy import signal
8   import librosa
9   import librosa.display # 一定要加
10  from IPython.display import Audio
```

2. 載入檔案：呼叫 librosa.load()，傳回資料與取樣頻率。

可設定參數如下：

- hq=True，表示載入時採高品質模式(high-quality mode) 。
- sr=44100，指定取樣頻率。
- res_type='kaiser_fast'，表示快速載入檔案。

```
1  # 檔案來源: https://github.com/maxifjaved/sample-files
2  wav_file = './audio/WAV_1MG.wav'
3
4  # 載入檔案
5  data, sr = librosa.load(wav_file)
6  print(f'取樣頻率={sr}, 總樣本數={data.shape}')
```

執行結果：取樣頻率=22050, 總樣本數=(739329,)。

3. 繪製波形。

```
1  # 繪製波形
2  librosa.display.waveplot(data, sr)
```

執行結果：

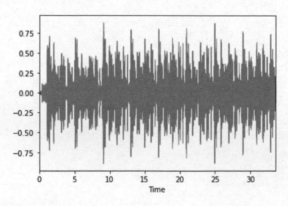

4. 顯示頻譜圖：先呼叫 melspectrogram()取得梅爾係數(Mel)，再呼叫 power_to_db()轉為分貝(db)，最後呼叫 specshow()顯示頻譜圖。

```
1  # 載入頻譜圖
2  spec = librosa.feature.melspectrogram(y=data, sr=sr)
3
4  # 顯示頻譜圖
5  db_spec = librosa.power_to_db(spec, ref=np.max,)
6  librosa.display.specshow(db_spec,y_axis='mel', x_axis='s', sr=sr)
7  plt.colorbar()
```

執行結果：

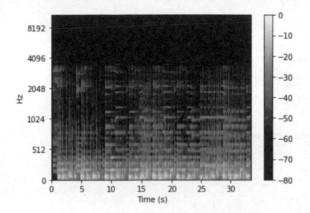

5. 存檔：v0.8 版本後已不支援 librosa.output.write_wav 函數了，須改用 soundfile 套件。

```
1  # 存檔
2  sr = 22050 # sample rate
3  T = 5.0    # seconds
4  t = np.linspace(0, T, int(T*sr), endpoint=False) # time variable
5  x = 0.5*np.sin(2*np.pi*220*t)# pure sine wave at 220 Hz
6
7  #playing generated audio
8  Audio(x, rate=sr) # load a NumPy array
9
10 # v0.8後已不支援
11 # librosa.output.write_wav('generated.wav', x, sr) # writing wave file in .wav format
12 import soundfile as sf
13 sf.write('./audio/generated.wav', x, sr, 'PCM_24')
```

6. 接著進行特徵萃取的實作，可作為深度學習模型的輸入。

7. 短時傅立葉變換(Short-time Fourier transform)：包括分幀、加窗、離散傅立葉轉換，合併為一個步驟。

```
1  # Short-time Fourier transform
2  # return complex matrix D[f, t], which f is frequency, t is time (frame).
3  D = librosa.stft(data)
4  print(D.shape, D.dtype)
```

- 傳回一個矩陣 D，其中包含頻率、時間。
- 執行結果：(1025, 1445) complex64。

8. MFCC：參數 n_mfcc 可指定每秒要傳回幾個 MFCC frame，通常是
   13、40 個。

```
1  # mfcc
2  mfcc = librosa.feature.mfcc(y=data, sr=sr, n_mfcc=40)
3  mfcc.shape
```

執行結果：(40, 1445)。

9. Log-Mel Spectrogram。

```
1  # Log-Mel Spectrogram
2  melspec = librosa.feature.melspectrogram(data, sr, n_fft=1024,
3                                           hop_length=512, n_mels=128)
4  logmelspec = librosa.power_to_db(melspec)
5  logmelspec.shape
```

執行結果：(128, 1445)。

10. 接著說明 Librosa 內建音訊載入的方法。

```
1  # librosa 內建音訊列表
2  librosa.util.list_examples()
```

執行結果：

```
AVAILABLE EXAMPLES
--------------------------------------------------------------------
brahms          Brahms - Hungarian Dance #5
choice          Admiral Bob - Choice (drum+bass)
fishin          Karissa Hobbs - Let's Go Fishin'
nutcracker      Tchaikovsky - Dance of the Sugar Plum Fairy
trumpet         Mihai Sorohan - Trumpet loop
vibeace         Kevin MacLeod - Vibe Ace
```

11. 載入 Librosa 預設的內建音訊檔。

```
1  # 載入 librosa 內建音訊
2  y, sr = librosa.load(librosa.util.example_audio_file())
3  print(f'取樣頻率={sr}, 總樣本數={y.shape}')
```

執行結果：取樣頻率=22050, 總樣本數=(1355168,)。

12. 播放：利用 IPython 模組播放音訊。

```
1  # 播放
2  Audio(y, rate=sr, autoplay=True)
```

13. 指定 ID 載入內建音訊檔。

```
1  # hq=True：high-quality mode
2  # 布拉姆斯 匈牙利舞曲
3  y, sr = librosa.load(librosa.example('brahms', hq=True))
4  print(f'取樣頻率={sr}, 總樣本數={y.shape}')
5  Audio(y, rate=sr, autoplay=True)
```

14. 音訊處理與轉換：Librosa 支援多種音訊處理與轉換功能，我們逐一來實驗。

15. 重取樣(Resampling)：從既有的音訊重取樣，通常是從高品質的取樣頻率，透過重取樣，轉換為較低取樣頻率的資料。

```
1  # 重取樣
2  sr_new = 11000
3  y = librosa.resample(y, sr, sr_new)
4
5  print(len(y), sr_new)
6
7  Audio(y, rate=sr_new, autoplay=True)
```

16. 將和音與敲擊音分離(Harmonic/Percussive Separation)：呼叫 librosa. effects.hpss()可將和音與敲擊音分離，從敲擊音可以找到音樂的節奏(Tempo)。

```
1   # 和音與敲擊音分離(Harmonic/Percussive Separation)
2   y_h, y_p = librosa.effects.hpss(y)
3   spec_h = librosa.feature.melspectrogram(y_h, sr=sr)
4   spec_p = librosa.feature.melspectrogram(y_p, sr=sr)
5   db_spec_h = librosa.power_to_db(spec_h,ref=np.max)
6   db_spec_p = librosa.power_to_db(spec_p,ref=np.max)
7
8   plt.subplot(2,1,1)
9   librosa.display.specshow(db_spec_h, y_axis='mel', x_axis='s', sr=sr)
10  plt.colorbar()
11
12  plt.subplot(2,1,2)
13  librosa.display.specshow(db_spec_p, y_axis='mel', x_axis='s', sr=sr)
14  plt.colorbar()
15
16  plt.tight_layout()
```

執行結果：

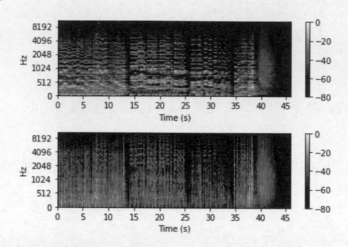

17. 取得敲擊音每分鐘出現的樣本數。

```
1  # 敲擊音每分鐘出現的樣本數
2  print(librosa.beat.tempo(y, sr=sr))
```

執行結果：143.5546875。

18. 可分別播放和音與敲擊音。

```
1  # 播放和音(harmonic component)
2  IPython.display.Audio(data=y_h, rate=sr)
```

19. 繪製色度圖(Chromagram)：chroma 為半音(semitones)，可萃取音準 (pitch)資訊。

```
1   # 萃取音準(pitch)資訊
2   chroma = librosa.feature.chroma_cqt(y=y_h, sr=sr)
3   plt.figure(figsize=(18,5))
4   librosa.display.specshow(chroma, sr=sr, x_axis='time', y_axis='chroma', vmin=0, vmax=1)
5   plt.title('Chromagram')
6   plt.colorbar()
7
8   plt.figure(figsize=(18,5))
9   plt.title('Spectrogram')
10  librosa.display.specshow(chroma, sr=sr, x_axis='s', y_axis='chroma', )
```

執行結果：y 軸顯示 12 個半音，pitch 是有週期的循環。

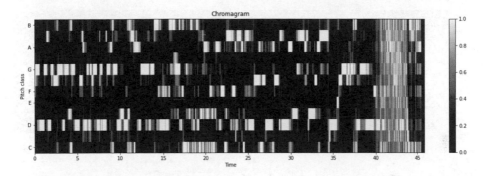

20. 可任意分離頻譜，例如將頻譜分為 8 個成份(Component)，以非負矩陣
    分解法(NMF)分離頻譜，NMF 類似主成分分析(PCA)。

```
1  # 將頻譜分為 8 個成份(Component)
2  # Short-time Fourier transform
3  D = librosa.stft(y)
4
5  # Separate the magnitude and phase
6  S, phase = librosa.magphase(D)
7
8  # Decompose by nmf
9  components, activations = librosa.decompose.decompose(S, n_components=8, sort=True)
```

21. 顯示成份(Component)與 Activations。

```
1  # 顯示成份(Component)與 Activations
2  plt.figure(figsize=(12,4))
3
4  plt.subplot(1,2,1)
5  librosa.display.specshow(librosa.amplitude_to_db(np.abs(components)
6                          , ref=np.max), y_axis='log')
7  plt.xlabel('Component')
8  plt.ylabel('Frequency')
9  plt.title('Components')
10
11 plt.subplot(1,2,2)
12 librosa.display.specshow(activations, x_axis='time')
13 plt.xlabel('Time')
14 plt.ylabel('Component')
15 plt.title('Activations')
16
17 plt.tight_layout()
```

執行結果：X 軸為 8 個成份。

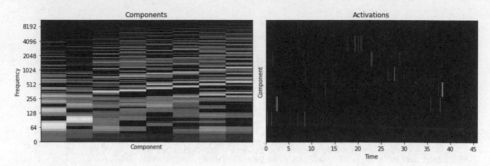

22. 再以分離的 Components 與 Activations 重建音訊。

```
1  # 以 Components 與 Activations 重建音訊
2  D_k = components.dot(activations)
3
4  # invert the stft after putting the phase back in
5  y_k = librosa.istft(D_k * phase)
6
7  # And playback
8  print('Full reconstruction')
9
10 IPython.display.Audio(data=y_k, rate=sr)
```

執行結果：播放與原曲一致，這部分的功能可用於音樂合成或修改。

23. 只以第一 Component 與 Activation 重建音訊：播放與原曲大相逕庭。

```
1  # 只以第一 Component 與 Activation 重建音訊
2  k = 0
3  D_k = np.multiply.outer(components[:, k], activations[k])
4
5  # invert the stft after putting the phase back in
6  y_k = librosa.istft(D_k * phase)
7
8  # And playback
9  print('Component #{}'.format(k))
10
11 IPython.display.Audio(data=y_k, rate=sr)
```

24. Pre-emphasis：用途為高頻加強，前面說過，人類對高頻信號較不敏感，所以能利用此技巧，補強音訊裡高頻的部分。

```
1  # Pre-emphasis：強調高頻的部分
2  import matplotlib.pyplot as plt
3
4  y, sr = librosa.load(wav_file, offset=30, duration=10)
```

```
5
6  y_filt = librosa.effects.preemphasis(y)
7
8  # 比較原音與修正的音訊
9  S_orig = librosa.amplitude_to_db(np.abs(librosa.stft(y)), ref=np.max)
10 S_preemph = librosa.amplitude_to_db(np.abs(librosa.stft(y_filt)), ref=np.max)
11
12 # 繪圖
13 plt.subplot(2,1,1)
14 librosa.display.specshow(S_orig, y_axis='log', x_axis='time')
15 plt.title('Original signal')
16
17 plt.subplot(2,1,2)
18 librosa.display.specshow(S_preemph, y_axis='log', x_axis='time')
19 fig=plt.title('Pre-emphasized signal')
20
21 plt.tight_layout()
```

執行結果：可以很明顯看到高頻已被補強。

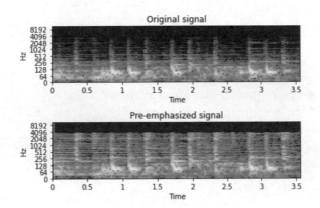

25. 常態化(Normalization)：在餵入機器學習之前，我們通常會先進行特徵
    縮放，除了能提高準確率外，也能加快優化求解的收斂速度，具體方
    式就是直接使用 SciKit-Learn 的 minmax_scale 函數即可。

```
1  # 常態化
2  from sklearn.preprocessing import minmax_scale
3
4  wav_file = './audio/WAV_1MG.wav'
5  data, sr = librosa.load(wav_file, offset=30, duration=10)
6
7  plt.subplot(2,1,1)
8  librosa.display.waveplot(data, sr=sr, alpha=0.4)
9
10 plt.subplot(2,1,2)
11 fig = plt.plot(minmax_scale(data), color='r')
```

執行結果：

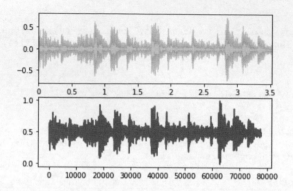

除了 Librosa 套件之外，也有 python_speech_features 套件，提供讀取音檔特徵的功能，包括 MFCC/FBank 等，安裝指令如下：

```
pip install python_speech_features
```

【範例 4】特徵萃取 MFCC、Filter bank 向量。

程式：14_07_python_speech_features.ipynb。

1. 載入相關套件。

```
1  # 載入相關套件
2  import matplotlib.pyplot as plt
3  from scipy.io import wavfile
4  from python_speech_features import mfcc, logfbank
```

2. 載入音樂檔案。

```
1  # 載入音樂檔案
2  sr, data = wavfile.read("./audio/WAV_1MG.wav")
```

3. 讀取 MFCC、Filter bank 特徵。

```
1  # 讀取 MFCC、Filter bank 特徵
2  mfcc_features = mfcc(data, sr)
3  filterbank_features = logfbank(data, sr)
4
5  # Print parameters
6  print('MFCC 維度:', mfcc_features.shape)
7  print('Filter bank 維度:', filterbank_features.shape)
```

執行結果：

- MFCC 維度: (6705, 13)。
- Filter bank 維度: (6705, 26)。

4. MFCC、Filter bank 繪圖。

```
1  # 繪圖
2  plt.subplot(2,1,1)
3  mfcc_features = mfcc_features.T
4  plt.imshow(mfcc_features, cmap=plt.cm.jet,
5      extent=[0, mfcc_features.shape[1], 0, mfcc_features.shape[0]], aspect='auto')
6  plt.title('MFCC')
7
8  plt.subplot(2,1,2)
9  filterbank_features = filterbank_features.T
10 plt.imshow(filterbank_features, cmap=plt.cm.jet,
11     extent=[0, filterbank_features.shape[1], 0, filterbank_features.shape[0]], aspect='auto')
12 plt.title('Filter bank')
13 plt.tight_layout()
14 plt.show()
```

執行結果：

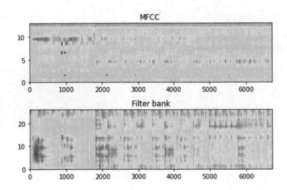

有關音訊的轉換還有另一個選擇，可以自 http://ffmpeg.org/download.html 下載 ffmpeg 工具程式，它支援的功能非常多，包括裁剪、取樣頻率、編碼等，詳細說明可參閱 ffmpeg 官網(http://ffmpeg.org/documentation.html)。以下是將 input.wav 轉為 output.wav，並改變取樣頻率、聲道、編碼：

```
ffmpeg.exe -i output.wav -ar 44100 -ac 1 -acodec pcm_s16le
output.wav
```

14-3 語音相關的深度學習應用

了解前面音訊處理與轉換的內容後，我們算是做好熱身了，接下來，就要正式實作幾個深度學習相關的應用。

【範例 1】音樂曲風的分類。我們利用上一節特徵萃取的 MFCC 向量，餵入到 CNN 模型，就可以分類曲風了。

程式：14_08_音樂曲風分類.ipynb。

資料集：MARSYAS GTZAN Genre Collection

(http://marsyas.info/downloads/datasets.html)，提供各種不同大小的資料集，本範例採用最大的資料集(http://opihi.cs.uvic.ca/sound/genres.tar.gz)，共有 10 個類別，每個類別各有 100 首歌，每首歌的長度均為 30 秒。

10 個類別分別為：

- Blues
- Classical
- Country
- Disco
- Hiphop
- Jazz
- Metal
- Pop
- Reggae
- Rock

程式修改自「Audio Data Analysis Using Deep Learning with Python (Part 2)」[9]。

1. 載入相關套件。

```
1  # 載入相關套件
2  import pandas as pd
3  import numpy as np
4  import matplotlib.pyplot as plt
5  import librosa
6  import librosa.display
7  import os
8  import pathlib
9  import csv
10 import tensorflow as tf
11 from tensorflow.keras import layers
```

2. 載入音樂檔案：檔案目錄如下。

呼叫 librosa.load()載入音樂檔案。

呼叫 librosa.feature.mfcc()將資料轉為 MFCC 向量。

CNN 輸入需要 4 維資料(筆數、寬度、高度、顏色)，因此，將訓練資料轉為 4 維。

```python
1  # 載入音樂檔案
2  genres = 'blues classical country disco hiphop jazz metal pop reggae rock'.split()
3
4  X = None
5  y = []
6  for i, g in enumerate(genres):
7      pathlib.Path(f'./GTZAN/genres//{g}').mkdir(parents=True, exist_ok=True)
8      for filename in os.listdir(f'./GTZAN/genres/{g}'):
9          songname = f'./GTZAN/genres/{g}/{filename}'
10         data, sr = librosa.load(songname, mono=True, duration=25)
11         mfcc = librosa.feature.mfcc(y=data, sr=sr, n_mfcc=40)
12         # print(data.shape, mfcc.shape)
13         if X is None:
14             X = mfcc.reshape(1, 40, -1, 1)
15         else:
16             X = np.concatenate((X, mfcc.reshape(1, 40, -1, 1)), axis=0)
17         y.append(i)
18
19 print(X.shape, len(y))
```

執行結果：$(1000, 40, 1077, 1)$。

3. 特徵縮放：資料常態化。

```python
1  # 常態化
2  X_norm = (X - X.min(axis=0)) / (X.max(axis=0) - X.min(axis=0))
```

4. 資料切割為訓練資料和測試資料。

```
1  # 資料切割
2  from sklearn.model_selection import train_test_split
3  y = np.array(y)
4  X_train, X_test, y_train, y_test = train_test_split(X_norm, y, test_size=.2)
5  X_train.shape, X_test.shape
```

執行結果：((800, 40, 1077, 1), (200, 40, 1077, 1))，切割為訓練資料 800 筆、測試資料 200 筆。

5. CNN 模型：大部份的文章使用 AveragePooling2D 取代 MaxPooling2D，不過筆者實測後發現，兩者效果並沒有太大差異，所以這裡兩種模型都列出。

```
1  # CNN 模型
2  input_shape = X_train.shape[1:]
3  model = tf.keras.Sequential(
4      [
5          tf.keras.Input(shape=input_shape),
6          layers.Conv2D(32, kernel_size=(3, 3), activation="relu"),
7          layers.MaxPooling2D(pool_size=(2, 2)),
8          layers.Conv2D(64, kernel_size=(3, 3), activation="relu"),
9          layers.MaxPooling2D(pool_size=(2, 2)),
10         layers.Flatten(),
11         layers.Dropout(0.5),
12         layers.Dense(10, activation="softmax"),
13     ]
14 )
```

```
1  # CNN 模型
2  input_shape = X_train.shape[1:]
3  model = tf.keras.Sequential(
4      [
5          tf.keras.Input(shape=input_shape),
6          layers.Conv2D(32, kernel_size=(3, 3), strides=(2, 2), activation="relu"),
7          layers.AveragePooling2D(pool_size=(2, 2)),
8          layers.Conv2D(64, kernel_size=(3, 3), strides=(2, 2), activation="relu"),
9          layers.AveragePooling2D(pool_size=(2, 2)),
10         layers.Conv2D(64, kernel_size=(3, 3), padding='same'),
11         layers.AveragePooling2D(pool_size=(2, 2)),
12         layers.Flatten(),
13         layers.Dropout(0.5),
14         layers.Dense(64, activation="relu"),
15         layers.Dropout(0.5),
16         layers.Dense(10, activation="softmax"),
17     ]
18 )
```

6. 模型訓練、評分。

```
1  # 設定優化器(optimizer)、損失函數(loss)、效能衡量指標(metrics)的類別
2  model.compile(optimizer='adam',
3                 loss='sparse_categorical_crossentropy',
4                 metrics=['accuracy'])
5
6  # 模型訓練
7  history = model.fit(X_train, y_train, epochs=20, validation_split=0.2)
8
9  # 評分(Score Model)
10 score=model.evaluate(X_test, y_test, verbose=0)
11
12 for i, x in enumerate(score):
13     print(f'{model.metrics_names[i]}: {score[i]:.4f}')
```

執行結果：訓練 20 週期，準確率為 46%，並不高。

7. 對訓練過程的準確率繪圖。

```
1  # 對訓練過程的準確率繪圖
2  plt.rcParams['font.sans-serif'] = ['Microsoft JhengHei']
3  plt.rcParams['axes.unicode_minus'] = False
4
5  plt.figure(figsize=(8, 6))
6  plt.plot(history.history['accuracy'], 'r', label='訓練準確率')
7  plt.plot(history.history['val_accuracy'], 'g', label='驗證準確率')
8  plt.legend()
```

執行結果：訓練準確率高達 99.89%，而驗證準確率只有 52.50%，表示模型有過度擬合(Overfitting)的現象。測試準確率只有 46.00%，也偏低。

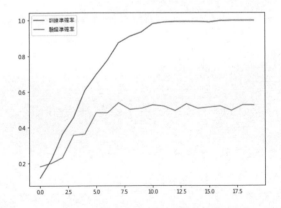

「Music Genre Recognition using Convolutional Neural Networks (CNN) — Part 1」[10]一文提到兩個改善的方向：

1. 將音樂資料分段(Segmentation)：每一段視為一筆資料，使用 pydub 套件呼叫 AudioSegment.from_wav()載入檔案，即可切割。

2. 資料增補 (Data Augmentation)：將音樂儲存為 png 檔，再利用 ImageDataGenerator 類別進行資料增補，產生更多的資料。

綜合多篇文章的實驗結果，資料增補並無太大助益，但資料分段的方法卻能大幅提高準確率，訓練 70 週期後，訓練準確率高達 99.57%，而驗證準確率也達到了 89.03%。筆者針對資料分段進行的實驗如下：

1. 資料分段：將每個檔案分為 10 段，訓練資料變成有 10000 筆。

```
1   # 載入音樂檔案
2   X = None
3   y = []
4   for i, g in enumerate(genres):
5       pathlib.Path(f'./GTZAN/genres//{g}').mkdir(parents=True, exist_ok=True)
6       for filename in os.listdir(f'./GTZAN/genres/{g}'):
7           songname = f'./GTZAN/genres/{g}/{filename}'
8           data, sr = librosa.load(songname, mono=True, duration=25)
9           try:
10              if i == 0:
11                  segment_length = int(data.shape[0] / 10)
12              for j in range(10):
13                  segment = data[j * segment_length: (j+1) * segment_length]
14                  # print(segment.shape)
15                  mfcc = librosa.feature.mfcc(y=segment, sr=sr, n_mfcc=40)
16                  # print(data.shape, mfcc.shape)
17                  if X is None:
18                      X = mfcc.reshape(1, 40, -1, 1)
19                  else:
20                      X = np.concatenate((X, mfcc.reshape(1, 40, -1, 1)), axis=0)
21                  y.append(i)
22          except:
23              print(i)
24              raise Exception('')
25  print(X.shape, len(y))
```

2. 之後的處理程序都一樣，不再贅述，實驗結果如下：

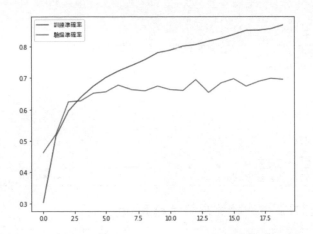

執行結果：訓練準確率高達 87.12%，而驗證準確率為 69.44%，測試準確率也有 67.80%，模型有了明顯改善。如果要再提高，可參閱上文，進行資料增補。

接著再來看另一個範例，Google 有收集一個短指令的資料集，包括常用的詞彙，譬如 stop、play、up、down、right、left 等，共有 30 個類別，如果辨識率很高的話，我們就能應用到各種場域，像是玩遊戲、控制簡報等。

【範例 2】短指令辨識。一樣使用 MFCC 向量，餵入 CNN 模型，即可進行分類。同時筆者也會結合錄音實測，示範如何控制錄音與訓練樣本一致 (Alignment)，其中有些技巧將在後面説明。

程式：14_09_短指令辨識.ipynb。

資料集：Google's Speech Commands Dataset (http://download.tensorflow.org/data/speech_commands_test_set_v0.02.tar.gz)。

每個檔案長度約為 1 秒，無雜音。它也附一個有雜音的目錄，與無雜音的檔案混合在一起，增加辨識的困難度。

1. 載入相關套件。

```
1   # 載入相關套件
2   import pandas as pd
3   import numpy as np
4   import matplotlib.pyplot as plt
5   import librosa
6   import librosa.display
7   import os
8   import pathlib
9   import csv
10  import tensorflow as tf
11  from tensorflow.keras import layers
```

2. 任選一個檔案測試，該檔案發音為 happy。

```
1   # 任選一檔案測試，發音為 happy
2   train_audio_path = './GoogleSpeechCommandsDataset/data/'
3   data, sr = librosa.load(train_audio_path+'happy/0ab3b47d_nohash_0.wav')
4
5   # 繪製波形
6   librosa.display.waveplot(data, sr)
7   print(data.shape)
```

執行結果：資料長度為 19757。

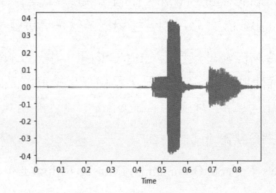

3. 再選一個檔案測試，該檔案發音也為 happy。

```
1   # 任選一檔案測試，發音為 happy
2   train_audio_path = './GoogleSpeechCommandsDataset/data/'
3   data, sr = librosa.load(train_audio_path+'happy/0b09edd3_nohash_0.wav')
4
5   # 繪製波形
6   librosa.display.waveplot(data, sr)
7   print(data.shape)
```

執行結果：與上圖相比較，同樣有兩段振幅較大的聲波，此代表 happy 的兩個音節，但是，因為錄音時每個人的起始發音點不同，因此，波形有很大差異，必須收集夠多的訓練資料才能找出共同的特徵。

資料長度為 22050，與上一筆音檔的資料長度亦不相同。

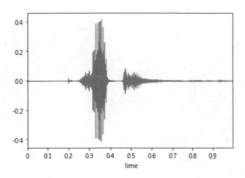

4. 可以播放看看，觀察其中的差異。

```
1  # 播放
2  from IPython.display import Audio
3
4  Audio(data, rate=sr)
```

5. 取得音檔的屬性：每個檔案的長度不等，但都接近 1 秒。

```
1   # 取得音檔的屬性
2   import wave
3
4   wav_file = train_audio_path+'happy/0ab3b47d_nohash_0.wav'
5   f=wave.open(wav_file)
6   print(f'取樣頻率={f.getframerate()}, 幀數={f.getnframes()}, ' +
7        f'聲道={f.getnchannels()}, 精度={f.getsampwidth()}, ' +
8        f'檔案秒數={f.getnframes() / f.getframerate():.2f}')
9   f.close()
10
11  nchannels2 = f.getnchannels()
12  sample_rate2 = f.getframerate()
13  sample_width2 = f.getsampwidth()
```

執行結果：取樣頻率=16000, 幀數=14336, 聲道=1, 精度=2, 檔案秒數=0.90。

6. 只取三個短指令測試：只放入三個子目錄的檔案於測試目錄內。

```
1  # 取得子目錄名稱
2  labels=os.listdir(train_audio_path)
3  labels
```

執行結果：['bed', 'cat', 'happy']。

7. 統計子目錄的檔案數。

```
1   # 子目錄的檔案數
2   no_of_recordings=[]
3   for label in labels:
4       waves = [f for f in os.listdir(train_audio_path + '/'+ label) if f.endswith('.wav')]
5       no_of_recordings.append(len(waves))
6
7   # 繪圖
8   plt.rcParams['font.sans-serif'] = ['Microsoft JhengHei']
9   plt.rcParams['axes.unicode_minus'] = False
10
11  plt.figure(figsize=(10,6))
12  index = np.arange(len(labels))
13  plt.bar(index, no_of_recordings)
14  plt.xlabel('指令', fontsize=12)
15  plt.ylabel('檔案數', fontsize=12)
16  plt.xticks(index, labels, fontsize=15, rotation=60)
17  plt.title('子目錄的檔案數')
18  print(f'檔案數={no_of_recordings}')
19  plt.show()
```

執行結果：檔案數目略有不同。

檔案數=[1713, 1733, 1742]。

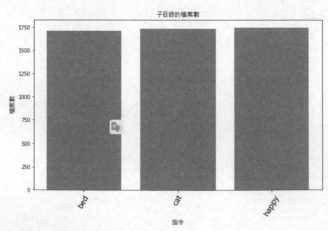

8. 載入音樂檔案：

■ 掃描每個目錄的檔案。

■ 以 librosa 載入音檔，由於每個音檔的長度不一，因此，使用 np.pad 與 np.resize 來統一長度為 1 秒。

■ 最後將每個類別的檔案合併儲存為 npy 檔案格式，之後如果要重新測試，就可直接載入，省去一一讀取和解析檔案的時間。

```
1  # 載入音樂檔案
2  TOTAL_FRAME_COUNT = 16000 # 每個檔案統一的幀數
3  duration_of_recordings=[]
4  all_wave = []
5  y = []
6  for i, label in enumerate(labels):
7      waves = [f for f in os.listdir(train_audio_path + '/'+ label) if f.endswith('.wav')]
8      class_wave=None
9      for wav in waves:
10         # 載入音樂檔案
11         samples, sample_rate = librosa.load(train_audio_path + '/' + label + '/' + wav
12                                 ,sr=None , res_type='kaiser_fast')
13         duration_of_recordings.append(float(len(samples)/sample_rate))
14         # 長度不足，右邊補 0
15         if len(samples) < TOTAL_FRAME_COUNT :
16             samples = np.pad(samples,(0, TOTAL_FRAME_COUNT-len(samples)),'constant')
17         elif len(samples) > TOTAL_FRAME_COUNT :
18             samples = np.resize(samples, TOTAL_FRAME_COUNT)
19
20         if class_wave is None:
21             class_wave = samples.reshape(1, -1)
22         else:
23             class_wave = np.concatenate([class_wave, samples.reshape(1, -1)], axis=0)
24         y.append(i)
25
26     all_wave.append(class_wave)
27     print(class_wave.shape)
28     np.save('./GoogleSpeechCommandsDataset/' + label + '.npy', class_wave) # 存成 npy
29 fig = plt.hist(np.array(duration_of_recordings))
```

執行結果：以直方圖的方式統計檔案的長度，發現到大部份的音檔是接近 1 秒鐘，但有少數檔案的時間長度非常短，事實上這會影響訓練的準確度，故統一長度有其必要性。

X 軸為音檔長度，Y 軸為筆數。

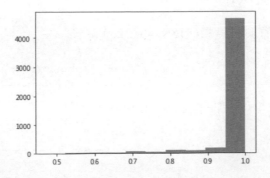

9. 載入 npy 檔案的測試。

```
1  # 載入npy檔案
2  train_audio_path2 = './GoogleSpeechCommandsDataset/'
3  npy_files = [f for f in os.listdir(train_audio_path2) if f.endswith('.npy')]
4  # npy_files
5  all_wave = []
6  y = []
7  no=0
8  for i, label in enumerate(npy_files):
9      class_wave = np.load(train_audio_path2+label)
10     all_wave.append(class_wave)
11     print(class_wave.shape)
12     no+=class_wave.shape[0]
13     y.extend(np.full(class_wave.shape[0], i))
```

執行結果：載入 npy 檔案的速度相當快，一瞬間就將所有音檔全部載入了。

10. 特徵縮放：MFCC 會有標準化的效果，後面會有測試，因此這裡不需進行特徵縮放。

11. 計算每個音檔的 MFCC，同時轉換成 CNN 模型的輸入格式。

```
1  # 計算 MFCC
2  MFCC_COUNT = 40
3  X = None
4  for class_wave in all_wave:
5      for data in class_wave:
6          mfcc = librosa.feature.mfcc(y=data, sr=len(data), n_mfcc=MFCC_COUNT)
7          # print(data.shape, mfcc.shape)
8          if X is None:
9              X = mfcc.reshape(1, MFCC_COUNT, -1, 1)
10         else:
11             X = np.concatenate((X, mfcc.reshape(1, MFCC_COUNT, -1, 1)), axis=0)
12     print(X.shape)
13 print(X.shape, len(y))
```

執行結果：輸入資料維度為(5188, 40, 32, 1)。

12. 資料切割。

```
1  # 資料切割
2  from sklearn.model_selection import train_test_split
3  y = np.array(y)
4  X_train, X_test, y_train, y_test = train_test_split(X, y, test_size=.2)
5  X_train.shape, X_test.shape
```

13. 建立 CNN 模型，與前一個範例相同，除了最後一層，類別數量改為 3。

```python
1   # CNN 模型
2   input_shape = X_train.shape[1:]
3   model = tf.keras.Sequential(
4       [
5           tf.keras.Input(shape=input_shape),
6           layers.Conv2D(32, kernel_size=(3, 3), activation="relu"),
7           layers.MaxPooling2D(pool_size=(2, 2)),
8           layers.Conv2D(64, kernel_size=(3, 3), activation="relu"),
9           layers.MaxPooling2D(pool_size=(2, 2)),
10          layers.Flatten(),
11          layers.Dropout(0.5),
12          layers.Dense(len(labels), activation="softmax"),
13      ]
14  )
```

14. 模型訓練與評分。

```python
1   # 設定優化器(optimizer)、損失函數(loss)、效能衡量指標(metrics)的類別
2   model.compile(optimizer='adam',
3                 loss='sparse_categorical_crossentropy',
4                 metrics=['accuracy'])
5
6   # 模型訓練
7   history = model.fit(X_train, y_train, epochs=20, validation_split=0.2)
8
9   # 評分(Score Model)
10  score=model.evaluate(X_test, y_test, verbose=0)
11
12  for i, x in enumerate(score):
13      print(f'{model.metrics_names[i]}: {score[i]:.4f}')
```

執行結果：準確度高達 97.88%。

15. 對訓練過程的準確率繪圖。

```python
1   # 對訓練過程的準確率繪圖
2   plt.rcParams['font.sans-serif'] = ['Microsoft JhengHei']
3   plt.rcParams['axes.unicode_minus'] = False
4
5   plt.figure(figsize=(8, 6))
6   plt.plot(history.history['accuracy'], 'r', label='訓練準確率')
7   plt.plot(history.history['val_accuracy'], 'g', label='驗證準確率')
8   fig = plt.legend(prop={"size":22})
```

執行結果：

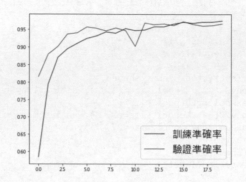

16. 定義一個預測函數，功能包括：
- 統一檔案長度，右邊補 0，過長則截掉。
- 轉為 MFCC。
- 以 MFCC 預測詞彙。
- 過程中顯示原始波形和 MFCC 散佈圖(Scatter Chart)。

```
1   # 預測函數
2   def predict(file_path):
3       samples, sr = librosa.load(file_path, sr=None, res_type='kaiser_fast')
4       # 繪製波形
5       librosa.display.waveplot(samples, sr)
6       plt.show()
7
8       # 右邊補 0
9       if len(samples) < TOTAL_FRAME_COUNT :
10          samples = np.pad(samples,(0, TOTAL_FRAME_COUNT-len(samples)),'constant')
11      elif len(samples) > TOTAL_FRAME_COUNT :
12          # 取中間一段
13          oversize = len(samples) - TOTAL_FRAME_COUNT
14          samples = samples[int(oversize/2):int(oversize/2)+TOTAL_FRAME_COUNT]
15
16      # 繪製波形
17      librosa.display.waveplot(samples, sr)
18      plt.show()
19
20      # 驗證 mfcc 是否需要標準化
21      mfcc = librosa.feature.mfcc(y=samples, sr=sr, n_mfcc=MFCC_COUNT)
22      for i in range(mfcc.shape[1]):
23          plt.scatter(x=range(mfcc.shape[0]), y=mfcc[:, i].reshape(-1))
24      X_pred = mfcc.reshape(1, *mfcc.shape, 1)
25
26      print(X_pred.shape, samples.shape)
27      # 預測
28      prob = model.predict(X_pred)
29      return np.around(prob, 2), labels[np.argmax(prob)]
```

17. 任選一個檔案進行預測，該檔案發音為 bed。

```
1  # 任選一檔案測試，該檔案發音為 bed
2  train_audio_path = './GoogleSpeechCommandsDataset/data/'
3  predict(train_audio_path+'bed/0d2bcf9d_nohash_0.wav')
```

執行結果：[[0.99, 0. , 0.01]]，正確判斷為 bed，下面第二張圖，除了起始的訊號為很大的負值以外，其他都介於(-100, 100)之間。

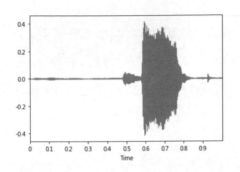 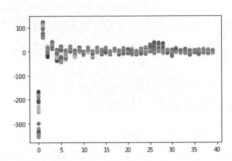

18. 再任選一個檔案測試，該檔案發音為 cat。

```
1  # 任選一檔案測試，該檔案發音為 cat
2  predict(train_audio_path+'cat/0ac15fe9_nohash_0.wav')
```

執行結果：[[0., 1., 0.]]，正確判斷為 cat。

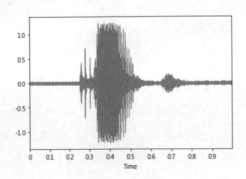

19. 接著再任選一個檔案測試，該檔案發音為 happy。

```
1  # 任選一檔案測試，該檔案發音為 happy
2  predict(train_audio_path+'happy/0ab3b47d_nohash_0.wav')
```

執行結果：[[0., 0. , 1.]]，正確判斷為 happy。

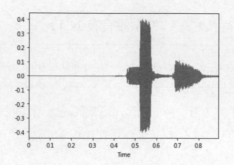

20. 自行錄音測試：筆者開發一個錄音程式 14_10_record.py，用法如下：

```
python 14_10_record.py GoogleSpeechCommandsDataset/happy.wav
```

最後的參數為存檔的檔名。程式錄音長度設為 2 秒，再利用上述 predict 函數，截取中間 1 秒鐘的音訊。

21. 測試：該檔案發音為 happy。

```
1  # 測試，該檔案發音為 happy
2  predict('./GoogleSpeechCommandsDataset/happy.wav')
```

執行結果：[[1., 0., 0.]]，正確判斷為 happy，下面第一張圖為原錄音波形，下面第二張圖為截取中段的波形。

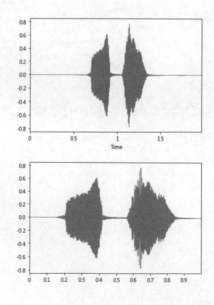

22. 再錄音測試 cat。

```
1  # 測試，該檔案發音為 cat
2  predict('./GoogleSpeechCommandsDataset/cat.wav')
```

執行結果：[[0. , 1., 0.]]，正確判斷為 cat。

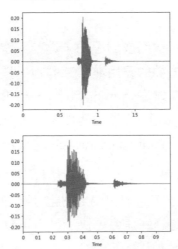

23. 再錄音測試 bed。

```
1  # 測試，該檔案發音為 bed
2  predict('./GoogleSpeechCommandsDataset/bed.wav')
```

執行結果：[[1., 0., 0.]]，正確判斷為 bed。

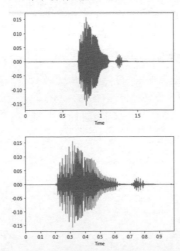

總結來說，如果用訓練資料進行測試的話，準確率都還不錯，但若是自行錄音則準確率就差強人意了，可能原因有兩點，第一是筆者發音欠佳，第二是錄音的處理方式與訓練方式不同，因此，建議還是要自己收集訓練資料為宜，也建議讀者發揮創意，多做實驗，筆者光這個範例就花了好幾天，才將結果弄得差強人意。

以上是參酌多篇文章後修改而成的程式，之前筆者有發表類似的程式在部落格上，許多網友對這方面的應用非常感興趣，提出很多的問題，讀者可參閱「Day 25：自動語音辨識(Automatic Speech Recognition) -- 觀念與實踐」[11]，此外，Kaggle 上也有一個關於這個資料集的競賽，可參閱「TensorFlow Speech Recognition Challenge」[12]。

筆者另外還做了一個實驗，以「支援向量機」(Support Vector Machine, SVM)演算法與「主成分分析」(PCA)，對上述資料訓練和預測。測試結果與 CNN 模型相同，只是訓練速度更快，讀者可參閱 14_11_SVM_短指令辨識.ipynb，處理程序與 14_09_短指令辨識.ipynb 類似，細節不再贅述。由於 SVM 輸入只接受 2 維，而 MFCC 本身即為 2 維，須 reshape 為一維，又變數過多，因此使用 PCA 降維，使其符合輸入條件。

上述實驗只能辨識短指令，假使要辨識一句話，或者更長的一段話，那就力有未逮了，因為講話的方式千變萬化，很難收集到完整的資料來訓練，所以解決的辦法則是把辨識目標再細化，以音節或音素(Phoneme)為單位，再使用語言模型，並考慮上下文才能精準預測，下一節我們就來探討相關的技術。

14-4 自動語音辨識(Automatic Speech Recognition)

自動語音辨識(Automatic Speech Recognition, ASR)的目標是將人類的語音轉換為數位信號，之後電腦就可進一步理解說話者的意圖，並做出對應的行動，譬如指令操控，應用於簡報上/下頁控制、車輛和居家裝置開關的控制、產生字幕與演講稿等。

英文的詞彙有數萬個，假如要進行分類，模型會很複雜，需要很長的訓練時間，準確率也不會太高，因為，相似音太多，因此自動語音辨識多改以「音素」（Phoneme）為預測目標，依據維基百科[13]的說明如下：

「音素」（Phoneme），又稱「音位」，是人類語言中能夠區別意義的最小聲音單位，一個詞彙由一至多個「音節」所組成，每個音節又由一至多個「音段」所組成，音素類似音段，但音素定義是要能區分語義。

舉例來說，bat 由 3 個音素/b/、/ae/、/t/所組成，連接這些音素，就是 bat 的拼音(pronunciation)，然後按照拼音就可以猜測到一個英文詞彙(Word)，當然，有可能發生同音異字的狀況，這時就必須依靠上下文做進一步的推測了。例如下圖，Human 單字被切割成多個音素 HH、Y、UW、M、AH、N。

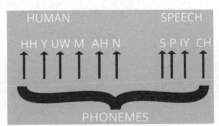

圖 14.14 音素辨識示意圖，圖片來源：Indian Accent Speech Recognition (https://anandai.medium.com/indian-accent-speech-recognition-2d433eb7edac)

各種語言的音素列表可參考「 Amazon Polly 支援語言的音素」(https://docs.aws.amazon.com/zh_tw/polly/latest/dg/ref-phoneme-tables-shell.

html)，以英文/美國(en-US)為例，音素列表主要包括母音和子音，共約 40~50 個，而中文則另外包含聲調（一聲、二聲、三聲、四聲和輕聲）。

自動語音辨識的流程可分成下列四個步驟：

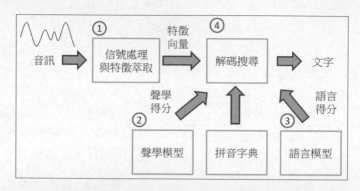

圖 14.15 自動語音辨識(Automatic Speech Recognition, ASR)架構

1. 信號處理與特徵萃取：先將音訊進行傅立葉轉換、去雜音等前置處理，接著轉為特徵向量，比如 MFCC、LPC(Linear Predictive Coding)。
2. 聲學模型(Acoustic Model)：透過特徵向量，轉換成多個音素，再將音素組合成拼音，然後至拼音字典(Pronunciation Dictionary)裡比對，找到對應的詞彙與得分。
3. 語言模型(Language Model)：依據上一個詞彙，猜測目前的詞彙，事先以 n-gram 為輸入，訓練模型，之後套用此模型，計算一個語言的得分。
4. 解碼搜尋(Decoding Search)：根據聲學得分和語言得分來比對搜尋出最有可能的詞彙。

經典的 GMM-HMM 演算法是過去數十年來語音辨識的主流，直到 2014 年 Google 學者使用雙向 LSTM，以 CTC(Connectionist Temporal Classification)為目標函數，將音訊轉成文字，深度學習演算法就此涉足這個領域，不過，目前大部份的工具箱依然以 GMM-HMM 演算法為主，因此，我們還是要先來認識 GMM-HMM 的運作原理。

自動語音辨識的流程可用貝式定理(Bayes' Theorem)來表示：

$$W = \arg\max P(W|O) \qquad (14.1)$$

$$W = \arg\max \frac{P(O|W)P(W)}{P(O)} \qquad (14.2)$$

$$W = \arg\max P(O|W)P(W) \qquad (14.3)$$

其中：

1. W 就是我們要預測的詞彙(W_1、W_2、W_3...)，O 是音訊的特徵向量。

2. P(W|O)：已知特徵向量，預測各個詞彙的機率，故以 argmax 找到獲得最大機率的 W，即為辨識的詞彙。

3. 14.2 公式的 P(O)不影響 W，可省略，故簡化為 14.3 公式。

4. P(O|W)：通常就是聲學模型，以高斯混合模型(Gaussian Mixture Model, GMM)演算法建構。

5. P(W)：語言模型，以隱藏式馬可夫模型(Hidden Markov Model, HMM)演算法建構。

高斯混合模型(Gaussian Mixture Model, 以下簡稱 GMM)是一種非監督式的演算法，假設樣本是由多個常態分配混合而成的，則演算法會利用最大概似法(MLE)推算出母體的統計量(平均數、標準差)，進而將資料分成多個集群(Clusters)。應用到聲學模型，就是以特徵向量作為輸入，算出每個詞彙的可能機率。

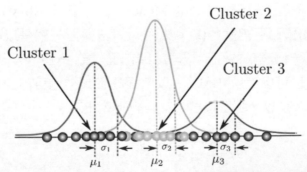

圖 14.16　一維高斯混合模型(Gaussian Mixture Model, GMM)示意圖

圖片來源：「Gaussian Mixture Models Explained」[14]

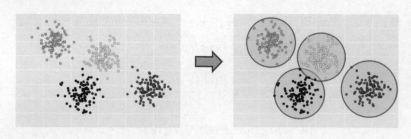

圖 14.17 二維高斯混合模型(Gaussian Mixture Model, GMM)示意圖

以聲學模型推測出多個音素後，就可以比對拼音字典(Pronunciation Dictionary)，找到相對的詞彙。

```
下雨  x ia4 ii v3
今天  j in1 t ian1
会  h ui4
北京  b ei3 j ing1
去  q v4
吗  m a1
天气  t ian1 q i4
怎么样  z en3 m o5 ii ang4
旅游  l v3 ii ou2
明天  m ing2 t ian1
的  d e5
还是  h ai2 sh i4
中  zh ong1 #1
忠  zh ong1 #2
```

圖 14.18 拼音字典(Pronunciation Dictionary)示意圖，左邊是詞彙，右邊是對照的音素，最後兩個詞彙同音，故標示#1、#2。圖片來源：「語音辨識系列 2--基於 WFST 解碼器_u012361418 的博客-程式師宅基地」[15]

隱藏式馬可夫模型(Hidden Markov Model，以下簡稱 HMM)係利用前面的狀態(k-1、k-2)預測目前狀態(k)。應用到語言模型，就是以前面的詞彙為輸入，預測下一個詞彙的可能機率，即前面提到 NLP 的 n-gram 語言模型。聲學模型也可以使用 HMM，以前面的音素推測目前的音素，例如一個字的拼音為首是ㄅ，那麼接著ㄆ就絕對不會出現。

$$P(\boldsymbol{w}) = \prod_{k=1}^{K} P(w_k|w_{k-1}, \ldots, w_1)$$

又比方，and、but、cat 三個詞彙，採用 bi-gram 模型，我們就要根據上一個詞彙預測下一個詞彙的可能機率，並取其中機率最大者。

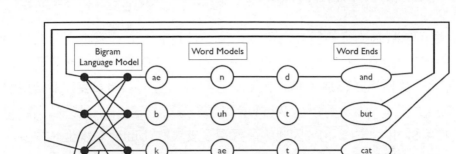

圖 14.19 bi-gram 語言模型結合 HMM 的示意圖
圖片來源：「愛丁堡大學語音辨識課程」第 11 章[16]

最後，綜合 GMM 和 HMM 模型所得到的分數，再藉由解碼的方式搜尋最有可能的詞彙。小型的詞彙集可採用「維特比解碼」(Viterbi Decoding)進行精確搜索(Exact Search)，但大詞彙連續語音辨識就會遭遇困難，所以一般會改採用「光束搜尋」(Beam Search)、「加權的有限狀態轉換機」(Weighted Finite State Transducers, WFST)或其他演算法，如要獲得較完整的概念可參閱「現階段大詞彙連續語音辨識研究之簡介」[17]一文的說明。

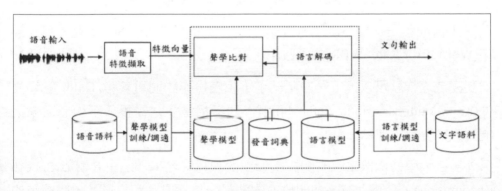

圖 14.20 大詞彙連續語音辨識的流程

▌14-5 自動語音辨識實作

Kaldi 是目前較為流行的語音辨識工具箱,它囊括了上一節所介紹的聲學模型、語言模型及解碼搜尋的相關函數庫實踐,由 Daniel Povey 等研究人員所開發,原始程式碼為 C++,其安裝程序較繁複,必須安裝許多公用程式和第三方工具,雖然可以在 Windows 作業系統上安裝,但是許多測試步驟均使用 Shell 腳本(*.sh),因此,最好還是安裝在 Linux 環境上,由於筆者手上並沒有相關設備,只好跳過這方面的實作,請讀者見諒。

以下僅列舉 Kaldi 的相關資源,供各位參考:

1. 下載 git clone https://github.com/kaldi-asr/kaldi.git kaldi --origin upstream
2. 安裝 NVidia GPU 卡驅動程式和 CUDA Toolkit。
3. 安裝 Kaldi

- cd kaldi
- 檢查相關的開發工具是否已安裝,若有缺少,必須補齊。extras/check_dependencies.sh
- 安裝第三方工具,包括 OpenFst、CUB、Sclite、Sph2pipe、IRSTLM/SRILM/Kaldi_lm 語言模型工具、OpenBLAS/MKL 矩陣運算函數庫。
- 使用 configure 設定組態,編譯 Kaldi。

其中 OpenFst 是加權有限狀態轉換器(WFST)的函數庫,Sclite 是計算錯誤率(Word Error Rate, WER)的函數庫。

安裝完成之後就可以進行一些測試了,相關操作說明可參考 Kaldi 官網文件(http://kaldi-asr.org/doc/index.html),內容相當多,需投入不少心力來研讀,大家要有心理準備。

另外,各大學的電機系都有開設整學期的課程,名稱即為語音辨識(Speech Recognition),網路上有許多開放性的教材,包括影片和投影片,有興趣的讀者可搜尋一下。

1. 台灣大學李琳山教授(http://speech.ee.ntu.edu.tw/DSP2019Spring/)。
2. 哥倫比亞大學 EECS E6870
   (https://www.ee.columbia.edu/~stanchen/fall12/e6870/outline.html)。
3. 愛丁堡大學 AUTOMATIC SPEECH RECOGNITION
   (https://www.inf.ed.ac.uk/teaching/courses/asr/lectures-2019.html)。

最後，語音辨識必須要收集大量的訓練資料，而 OpenSLR 就有提供非常多可免費下載的資料集和軟體，詳情可參考官網說明 (https://www.openslr.org/resources.php)，其中也有中文的語音資料集 CN-Celeb，它包含了 1000 位著名華人的三十萬條語音，VoxCeleb 則是知名人士的英語語音資料集，其他較知名的資料集還有：

1. TIMIT(https://catalog.ldc.upenn.edu/LDC93S1)：美式英語資料集。
2. LibriSpeech：電子書的英語朗讀資料集，可在 OpenSLR 下載。
3. 維基百科語音資料集(https://en.wikipedia.org/wiki/Wikipedia:
   WikiProject_Spoken_Wikipedia)。

▌ 14-6 結語

語音除了可以拿來語音辨識之外，還有許多其他方面的應用，譬如：

- 聲紋辨識：從講話的聲音分辨是否為特定人，屬於生物識別技術 (Biometrics)的一種，可應用在登入(Sign in)、犯罪偵測、智慧家庭等領域。
- 聲紋建模：模擬或創造特定人的聲音唱歌或講話，例如 Siri。
- 相似性比較：未來也許能夠使用語音搜尋，類似現在的文字、圖像搜尋。
- 音樂方面的應用：比方曲風辨識、模擬歌手的聲音唱歌、編曲/混音等。

各位讀者看到這裡，應該能深刻了解到，文字、影像、語言辨識是整個人工智慧應用的三大基石，不論是自駕車、機器人、ChatBot、甚至是醫療診斷通通都是建構在這些基礎技術之上。如同前文提到，現今的第三波 AI 浪潮之所以不會像前兩波一樣後繼無力、不了了之，有一部份原因是要歸功於這些基本技術的開發，這就像蓋房子的地基，技術的研發與應用實踐並進，才不會空有理論，最後成為空中樓閣。相信在未來這三大基石還會有更進一步的發展，而到時候又會有新的技能被解放，我們能透過 AI 完成的任務也更多了，換言之，我們學習的腳步永遠不會有停止的一天，筆者與大家共勉之。

參考資料 (References)

[1] Michael Picheny、Bhuvana Ramabhadran、Stanley F. Chen,《Lecture 1 Introduction/Signal Processing, Part I》, 2012
(https://www.ee.columbia.edu/~stanchen/fall12/e6870/slides/lecture1.pdf)

[2] Roger Jang (張智星),《Audio Signal Processing and Recognition (音訊處理與辨識)》, 2005
(http://mirlab.org/jang/books/audioSignalProcessing/audioIntro.asp?language=chinese)

[3] 國立臺灣大學普通物理實驗室官網關於示波器使用教學
(https://web.phys.ntu.edu.tw/gphyslab/modules/tinyd2/index8803.html?id=7)

[4] Pema Grg,《Audio Signal Processing》, 2020
(https://blog.ekbana.com/audio-signal-processing-f7e86d415489)

[5] File:CPT-Sound-ADC-DAC.svg Wikimedia Commons
(https://commons.wikimedia.org/wiki/File:CPT-Sound-ADC-DAC.svg)

[6]　Vincent Koops,《Introduction Basic Audio Feature Extraction》, 2017
　　　(http://www.cs.uu.nl/docs/vakken/msmt/lectures/SMT_B_Lecture5_DSP
　　　_2017.pdf)

[7]　Nagesh Singh Chauhan,《Audio Data Analysis Using Deep Learning
　　　with Python (Part 1)》, 2020
　　　(https://www.kdnuggets.com/2020/02/audio-data-analysis-deep-learning-
　　　python-part-1.html)

[8]　Henry Haefliger,《Python audio spectrum analyzer》, 2019
　　　(https://medium.com/quick-code/python-audio-spectrum-analyser-
　　　6a3c54ad950)

[9]　Nagesh Singh Chauhan,《Audio Data Analysis Using Deep Learning
　　　with Python (Part 2)》, 2020
　　　(https://www.kdnuggets.com/2020/02/audio-data-analysis-deep-learning-
　　　python-part-2.html)

[10] Kunal Vaidya,《Music Genre Recognition using Convolutional Neural
　　　Networks (CNN) — Part 1》, 2020
　　　(https://towardsdatascience.com/music-genre-recognition-using-
　　　convolutional-neural-networks-cnn-part-1-212c6b93da76)

[11] 陳昭明,《Day 25：自動語音辨識(Automatic Speech Recognition) -- 觀
　　　念與實踐》, 2018
　　　(https://ithelp.ithome.com.tw/articles/10195763)

[12] kaggle 官網「TensorFlow Speech Recognition Challenge」
　　　(https://www.kaggle.com/c/tensorflow-speech-recognition-challenge)

[13] 維基百科關於音素的說明
　　　(https://zh.wikipedia.org/wiki/音位)

[14] Oscar Contreras Carrasco,《Gaussian Mixture Models Explained》, 2019
　　　(https://towardsdatascience.com/gaussian-mixture-models-explained-
　　　6986aaf5a95)

[15] 心學-知行合一,《語音辨識系列 2--基於 WFST 解碼器_u012361418 的博客-程式師宅基地》, 2019
(http://www.cxyzjd.com/article/u012361418/90289912)

[16] 「愛丁堡大學語音辨識課程」第 11 章
(http://www.inf.ed.ac.uk/teaching/courses/asr/lectures-2019.html)

[17] 陳柏琳,《現階段大詞彙連續語音辨識研究之簡介》, 2005
(http://berlin.csie.ntnu.edu.tw/Berlin_Research/Manuscripts/2005_ACLC
LP-Newsletter_%E7%8F%BE%E9%9A%8E%E6%AE%B5%E5%A4%
A7%E8%A9%9E%E5%BD%99%E9%80%A3%E7%BA%8C%E8%AA
%9E%E9%9F%B3%E8%BE%A8%E8%AD%98%E7%A0%94%E7%A9
%B6%E4%B9%8B%E7%B0%A1%E4%BB%8B_Final.pdf)

| 第五篇 |

強化學習(Reinforcement Learning)

強化學習(Reinforcement Learning, RL)相關的研究少說也有數十年的歷史了，但與另外兩類機器學習相比，較不受矚目，直到 2016 年以強化學習為理論基礎的 AlphaGo，先後擊敗世界圍棋冠軍李世乭、柯潔等人，備受世界矚目，強化學習才因此一炮而紅，學者專家紛紛投入研發，接下來我們就來探究其原理與應用。

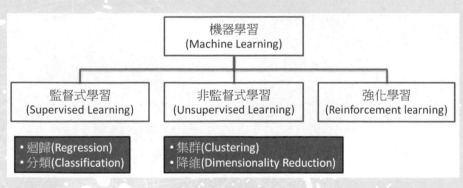

圖 15.1 機器學習分類

強化學習是指機器與環境的互動過程中，人類不必直接提供解決方案，而是透過電腦不斷的嘗試與錯誤中學習，稱為試誤法(Trial and Error)，自我學習一段時間後，電腦就可以找到最佳的行動策略。打個比方，就像是訓練狗兒接飛盤一樣，人們不會教狗如何接飛盤，而是由飼主不斷地拋出飛盤讓狗練習，如果牠成功接到飛盤，就給予食物獎勵，反之就不給獎勵，經過反覆練習後，毛小孩就能練就一身好功夫了。因此，強化學習並不是單一階段(One Step)的演算法，而是多階段、反覆求解，類似梯度下降法的求解過程。

圖 15.2　訓練狗兒接飛盤，如果成功接住飛盤，就給予食物當獎勵

根據維基百科[1]的概述，強化學習涉及到的學術領域相當多，包括博弈論(Game Theory)、自動控制、作業研究、資訊理論、模擬優化、群體智慧

(Swarm Intelligence)、統計學以及遺傳演算法等,同時它的應用領域也是包山包海,例如:

- 下棋、電玩遊戲策略(game playing)。
- 製造/醫療/服務機器人的控制策略(Robotic motor control)。
- 廣告投放策略 (Ad-placement optimization)。
- 金融投資交易策略 (Stock market trading strategies)
- 運輸路線的規劃 (Transportation Routing)。
- 庫存管理策略 (Inventory Management)、生產排程 (Production scheduling)。

甚至於殘酷的戰爭,只要應用場域是能在「模擬」環境下 Trial and Error,並且需要人工智慧提供行動的決策輔助,都是強化學習可以發揮的領域。

15-1 強化學習的基礎

強化學習的理論基礎為「馬可夫決策過程」 (Markov Decision Processes,MDP),主要是指所有的行動決策都會基於當時所處的「狀態」及行動後會帶來的「獎勵」所影響,而狀態與獎勵是由「環境」所決定的,因此就形成以下的示意圖。

- 代理人行動後,環境會依據行動更新狀態,並給予獎勵。
- 代理人觀察所處的狀態及之前的行動,決定下次的行動。

圖 15.3 馬可夫決策過程的示意圖

各個專有名詞的定義如下：

- 代理人或稱智能體(Agent)：也就是實際行動的主人翁，比如遊戲中的玩家(Player)、下棋者、機器人、金融投資者、接飛盤的狗等，他主要的任務是與環境互動，並根據當時的狀態(State)與之前得到的獎勵或懲罰，來決定下一步的行動。代理人可能不只一個，如果有多個，狀況就會複雜許多，這稱為「多代理人」(Multi-Agent)。
- 環境(Environment)：根據代理人的行動(Action)，給予立即的獎勵或懲罰，一律稱為獎勵(Reward)，也會決定代理人所處的狀態。
- 狀態(State)：指代理人所處的狀態，譬如圍棋的棋局、遊戲中玩家/敵人/寶物的位置、能力和金額。有時候代理人只能觀察到局部的狀態，例如，撲克牌遊戲 21 點(Black Jack)，莊家有一張牌蓋牌，玩家是看不到的，所以，狀態也被稱為觀察(Observation)。
- 行動(Action)：代理人依據環境所提示的狀態與獎勵而做出的決策。

整個過程就是代理人與環境互動的過程，可以以行動軌跡(trajectory)來表示：

$$\{S_0, A_0, R_1, S_1, A_1, R_2, S_2, ..., S_t, A_t, R_{t+1}, S_{t+1}, A_{t+1}, R_{t+2}, S_{t+2}\}$$

其中

- S：狀態(State)。
- A：行動(Action)。
- R：獎勵(Reward)。
- S_t：t 是時間點。
- 行動軌跡：行動(A_0)後，代理人會得到獎勵(R_1)、狀態(S_1)，之後再採取下一步的行動(A_1)，不停循環(A_t, R_{t+1}, S_{t+1}……)直至終點或中途比賽失敗/勝利為止。

馬可夫決策過程(以下簡稱 MDP)的假設是代理人的行動決策是「累計獲得的獎勵最大化」，也稱為「報酬」(Return)。注意，並不是指每一種狀態下的最大獎勵，比方說，下棋時我們會為了誘敵進入陷阱，而故意犧牲某些棋子,以求得最後的勝利,同樣道理,MDP 是追求長期的最大報酬,而非每一步驟的最大獎勵(短期利益)。強化學習類似優化求解,目標函數是報酬,希望找到報酬最大化時應採取的行動策略(Policy),對比於神經網路求解,神經網路目標是最小化損失函數,對各神經元的權重求解。

MDP 是由「馬可夫獎勵過程」(Markov Reward Process, MRP)加上「行動轉移矩陣」(Action Transition matrix)所組成,接著我們依序講解這兩個概念。

在說明 MRP 之前,先從 MRP 的基礎開始講起,馬可夫過程(Markov Process, MP),也稱為「馬可夫鏈」(Markov Chain),主要內容為描述狀態之間的轉換,例如,假設天氣的變化狀態有兩種,晴天和雨天,下圖就是一個典型的馬可夫鏈的狀態轉換圖:

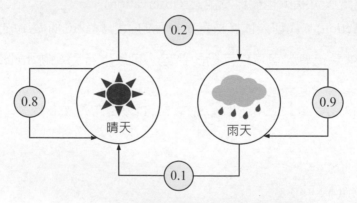

圖 15.4 馬可夫鏈(Markov Chain)的示意圖

也可以用表格來表示,稱為狀態轉換矩陣(State Transition matrix):

	晴天	雨天
晴天	0.8	0.2
雨天	0.1	0.9

圖 15.5 狀態轉換矩陣(State Transition matrix)

上面圖表要表達的資訊如下：

- 今天是晴天，明天也是晴天的機率：0.8。
- 今天是晴天，明天是雨天的機率：0.2。
- 今天是雨天，明天是晴天的機率：0.1。
- 今天是雨天，明天也是雨天的機率：0.9。

也就是説，明日天氣會受今日天氣的影響，換言之，下一個狀態出現的機率會受到目前狀態的影響，符合這種特性的模型就稱為「馬可夫性質」(Markov Property)，即目前狀態(S_t)只受前 1 個狀態(S_{t-1})影響，與之前的狀態(S_{t-2}，S_{t-3}，...)無關，也可以擴展受 n 個狀態(S_{t-1}, S_{t-2}，S_{t-3}，...，S_{t-n})影響，類似時間序列。

因此，我們可從上面圖表推測出 n 天後出現晴天或雨天的機率，例如：

- 今天是晴天，「後天」是晴天的機率 = 0.8 x 0.8 + 0.2 x 0.1 = 0.66。
- 今天是雨天，「後天」是晴天的機率 = 0.1 x 0.8 + 0.9 x 0.1 = 0.17。

再擴充一下，馬可夫鏈加上獎勵(Reward)，就稱為「馬可夫獎勵過程」(Markov Reward Process, MRP)，即：

Markov Process + Reward = Markov Reward Process

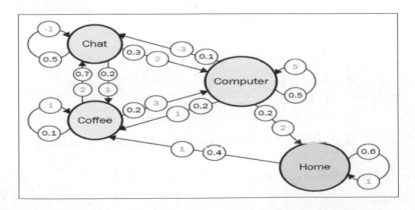

圖 15.6 馬可夫獎勵過程(Markov Reward Process, MRP)，每個轉換除了機率外，還有獎勵資訊

上圖是學生作息的狀態轉換，狀態包括聊天(Chat)、喝咖啡(Coffee)、玩電腦(Computer)和在家(Home)，我們依據轉換機率和獎勵，可算出每個狀態的期望值，來表達每個狀態的價值：

$$V(chat) = -1 * 0.5 + 2 * 0.3 + 1 * 0.2 = 0.3$$

$$V(coffee) = 2 * 0.7 + 1 * 0.1 + 3 * 0.2 = 2.1$$

$$V(home) = 1 * 0.6 + 1 * 0.4 = 1.0$$

$$V(computer) = 5 * 0.5 + (-3) * 0.1 + 1 * 0.2 + 2 * 0.2 = 2.8$$

理解 MRP 後，那「行動轉移矩陣」(Action Transition matrix)又是什麼呢？舉例來說，走迷宮時，玩家往上/下/左/右走的機率可能不相等，又如玩剪刀石頭布時，每個人的猜拳偏好都不盡相同，假使第一次雙方平手，第二次出手可能就會參考第一次的結果來出拳，因此出剪刀/石頭/布的機率又會有所改變，這就是「行動轉移矩陣」，它在每個狀態的轉移矩陣值可能都不一樣，如下圖。

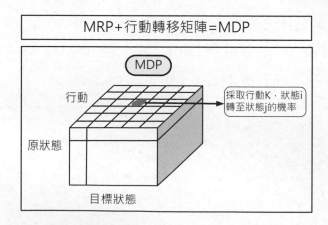

圖 15.6 馬可夫決策過程(Markov Decision Process, MDP)

我們可將行動轉移矩陣理解為策略(Policy)，若策略是固定的常數，MDP就等於 MRP，但是，通常我們面臨的環境是多變的，策略不會一成不變，因此總而言之，**強化學習的目標就是在 MDP 的機制下，要找出最佳的行動策略，而目的是希望獲得最大的報酬。**

15-2 強化學習模型

接下來用數學式建立強化學習模型，將馬可夫決策過程的示意圖轉為數學符號(第二張圖)。

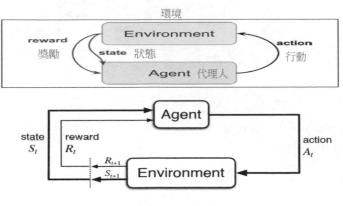

圖 15.7　強化學習模型

行動軌跡(trajectory)如下：

$$\{S_0, A_0, R_1, S_1, A_1, R_2, S_2, ..., S_t, A_t, R_{t+1}, S_{t+1}, A_{t+1}, R_{t+2}, S_{t+2}...\}$$

1. 狀態轉移機率：達到狀態 S_{t+1} 的機率如下。

$$p(S_{t+1}| S_t, A_t, S_{t-1}, A_{t-1}, S_{t-2}, A_{t-2}, S_{t-3}, A_{t-3}...)$$

依據「馬可夫性質」的假設，S_{t+1} 只與前一個狀態(S_t)有關，上式簡化為：

$$p(S_{t+1}| S_t, A_t)$$

2. 報酬(Return)：就以走迷宮當例子，到達終點時，所累積的獎勵總和稱為報酬，下式為從 t 時間點走到終點(T)的累積獎勵。

$$G_t = R_{t+1} + R_{t+2} + R_{t+3} + ... + R_T = \sum_{k=1}^{T} R_{t+k}$$

例 1. 以下圖走迷宮為例，目標是以最短路徑到達終點，故設定每走一步獎勵為-1，即可算出每一個位置的報酬，計算方法是由終點倒推回起點，結果如下圖中的數字所示。

圖 15.8 計算迷宮每一個位置的報酬

3. 折扣報酬(Discount Return)：模型目標是追求報酬最大化，若迷宮很大的話要考慮的獎勵(R_i)個數也會很多，所以為了簡化模型，將每個時間的獎勵乘以一個小於 1 的折扣因子(γ)，讓越久遠的獎勵越不重要，避免要考慮太多的狀態，類似複利的概念，報酬公式修正為：

$$G_t = R_{t+1} + \gamma R_{t+2} + \gamma^2 R_{t+3} + ... + \gamma^{T-1} R_T = \sum_{k=1}^{T} \gamma^{k-1} R_{t+k}$$

$$又\ G_{t+1} = R_{t+2} + \gamma R_{t+3} + \gamma^2 R_{t+4} + ... + \gamma^{T-2} R_T$$

$$故\ G_t = R_{t+1} + \gamma G_{t+1}$$

4. 狀態值函數(State Value Function)：以圖 15.8 迷宮為例，玩家所在的位置就是狀態，所以，圖中的數字(報酬)即是狀態值，代表每個狀態的價值。又加上要考量從起點走到終點的路徑可能不只一種，故狀態值函數是每條路徑報酬的期望值(平均數)。

例 2. 舉另一個迷宮遊戲為例，規則如下，起點為(1, 1)，終點為(4, 3)或(4, 2)，走到(4, 3)獎勵為 1，走到(4, 2)獎勵為 -1，每走一步獎勵均為 -0.04。

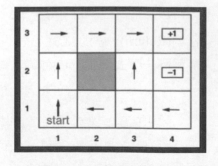

圖 15.9 另一個迷宮遊戲

假設有三種走法：

①
$(1,1) \rightarrow (1,2) \rightarrow (1,3) \rightarrow (1,2) \rightarrow (1,3) \rightarrow (2,3) \rightarrow (3,3) \rightarrow (4,3)$
-0.04　-0.04　　-0.04　　-0.04　　-0.04　　-0.04　　-0.04　　+1

②
$(1,1) \rightarrow (1,2) \rightarrow (1,3) \rightarrow (2,3) \rightarrow (3,3) \rightarrow (3,2) \rightarrow (3,3) \rightarrow (4,3)$
-0.04　-0.04　　-0.04　　-0.04　　-0.04　　-0.04　　-0.04　　+1

③
$(1,1) \rightarrow (2,1) \rightarrow (3,1) \rightarrow (3,2) \rightarrow (4,2)$
-0.04　-0.04　　-0.04　　-0.04　　-1

三條走法的報酬計算如下：

(1) 1 - 0.04 x 7 = 0.72

(2) 1 - 0.04 x 7 = 0.72

(3) -1 - 0.04 x 4 = -1.16

(1, 1)狀態期望值 = (0.72 + 0.72 + (-1.16)) / 3= 0.28 / 3 ≒ 0.09

狀態值函數以 $V_\pi(s)$為表示，其中 π 為特定策略。

$$V_\pi(s) = E(G|S{=}s)$$

5. Bellman 方程式：特定策略下採取各種行動的機率，狀態值函數的公式如下。

$$v_\pi(s) = \sum_a \pi(a|s) \sum_{s'} \mathcal{P}_{ss'}^a \left[\mathcal{R}_{ss'}^a + \gamma v_\pi(s') \right]$$

其中：

- s'：下一狀態。
- $\pi(a|s)$：採取特定策略時，在狀態 s 採取行動 a 的機率。
- $P_{ss'}^a$：為狀態轉移機率，即在狀態 s 採取行動 a，會達到狀態 s'的機率。
- 後面[]的公式係依據 $G_t = R_{t+1} + \gamma G_{t+1}$ 轉換而來。

Bellman 方程式讓我們可以從下一狀態的獎勵/狀態值函數推算出目前狀態的值函數。這裡讀者可能會愣一下,在目前狀態下怎麼會知道下一個狀態的值函數?不要忘了,強化學習是以嘗試錯誤(Trial and Error)的方式進行訓練,我們可以從之前的訓練結果推算目前回合的值函數,譬如,要算第 50 回合的值函數,可以從 1~49 回合推算每一狀態的期望值。

6. 行動值函數(Action Value Function):採取某一行動的值函數,類似狀態值函數,公式如下。

$$q_\pi(s, a) = \sum_{s'} \mathcal{P}_{ss'}^a \left[\mathcal{R}_{ss'}^a + \gamma \sum_{a'} \pi(a'|s') \, q_\pi(s', a') \right]$$

同樣可以從下一狀態的獎勵/行動值函數推算目前行動值函數。

7. 狀態值函數與行動值函數的關係可以由「倒推圖」(Backup Diagram)來表示。

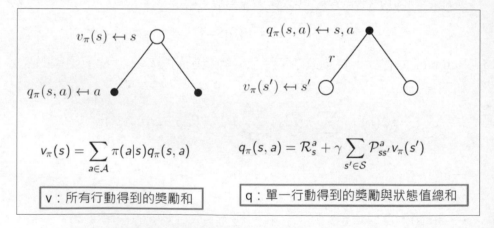

圖 15.10 狀態值函數與行動值函數的關係(1)

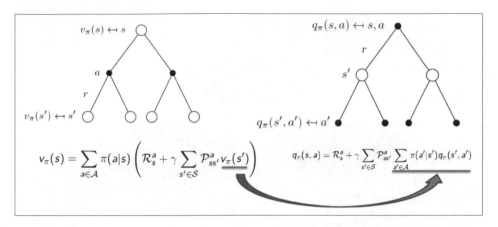

圖 15.11 狀態值函數與行動值函數的關係(2)

從倒推圖可以看出行動值函數的公式來源。

依照上述公式行動,不斷更新所有狀態值函數,之後在每一個狀態下,即以最大化狀態或行動值函數為準則,採取行動,以獲取最大報酬。

15-3 簡單的強化學習架構

這一節我們把前面剛學到熱騰騰的理論統整一下,就能完成一個初階的程式。回顧前面談到的強化學習機制如下:

圖 15.12 強化學習機制

藉由強化學習機制，我們可以制定程式架構如下：

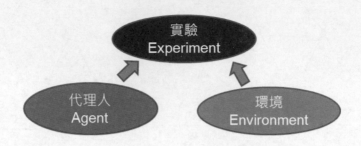

圖 15.13 強化學習程式架構

採取物件導向設計(OOP)，程式架構大致分為三個步驟，共有兩個類別，如下所示：

1.　環境(Environment)：比如迷宮、遊戲或圍棋，它會給予玩家獎勵並負責狀態轉換，若是單人遊戲，環境還要擔任玩家的對手，像是電腦圍棋。職責(方法)列舉如下。

- 初始化(Init)：需定義狀態空間(State Space)、獎勵(Reward)辦法、行動空間(Action Space)、狀態轉換(State Transition definition)。

- 重置(Reset)：每一回合(Episode)結束時，須重新開始，重置所有變數。

- 步驟(Step)：代理人行動後，驅動行動軌跡的下一步，而環境就會隨之更新狀態、給予獎勵，並判斷回合是否結束。

- Render (渲染)：更新顯示的畫面。

2.　代理人(Agent)：即玩家，職責(方法)列舉如下。

- 行動(Act)：代理人依據既定的策略與目前所處的狀態，採取行動，例如上、下、左、右。

- 通常如果要訂定特殊的策略，會使用繼承代理人類別(Agent class)，在衍生的類別中，覆寫行動函數(Act)，並撰寫策略邏輯。

- 進行實驗時，先建立兩個類別物件，觸發環境各項方法，等待玩家行動。

接下來，我們就依照上述架構來開發程式。

【範例 1】建立簡單的迷宮遊戲：共有 5 個位置，玩家一開始站中間位置，每走一步扣分 0.2，走到左端點得-1 分，走到右端點得 1 分，走到左右端點該回合即結束。

起點

程式：RL_15_01_simple_game.py。

1. 載入相關套件。

```
1  # 載入相關套件
2  import random
```

2. 建立環境類別。

- __init__：初始化，每回合結束後比賽重置(Reset)。
- get_observation：傳回狀態空間(State Space)，本遊戲假設有五種:1、2、3、4、5。
- get_actions：傳回行動空間(Action Space)，本遊戲假設有兩種:-1、1，只能往左或往右。
- is_done：判斷比賽回合是否結束。
- step：觸發下一步，根據傳入的行動，更新狀態，並計算獎勵。

```
4   # 環境類別
5   class Environment:
6       def __init__(self): # 初始化
7           self.poistion = 3 # 玩家一開始站中間位置
8
9       def get_observation(self):
10          # 狀態空間(State Space)，共有5個位置
11          return [i for i in range(1, 6)]
12
13      def get_actions(self):
14          return [-1, 1] # 行動空間(Action Space)
15
16      def is_done(self): # 判斷比賽回合是否結束
17          # 是否走到左右端點
18          return self.poistion == 1 or self.poistion == 5
```

15-13

```
20        # 步驟
21        def step(self, action):
22            # 是否回合已結束
23            if self.is_done():
24                raise Exception("Game is over")
25
26            # 隨機策略，任意行動，並給予獎勵(亂數值)
27            action = random.choice(self.get_observation())
28            self.position += action
29            if self.poistion == 1:
30                reward = -1
31            elif self.position == 5:
32                reward = 1
33            else:
34                reward = -0.2
35
36            return action, reward
```

3. 建立代理人類別：主要是訂定行動策略，本範例採取隨機策略。

```
37   # 代理人類別
38   class Agent:
39       # 初始化
40       def __init__(self):
41           pass
42
43       def action(self, env):
44           # 取得狀態
45           current_obs = env.get_observation()
46           # 隨機行動
47           return random.choice(env.get_actions())
```

4. 定義好環境與代理人功能後，就可以進行實驗了。

```
50   if __name__ == "__main__":
51       # 建立實驗，含環境、代理人物件
52       env = Environment()
53       agent = Agent()
54
55       # 進行實驗
56       total_reward=0  # 累計報酬
57       while not env.is_done():
58           # 採取行動
59           action = agent.action(env)
60
61           # 更新下一步
62           state, reward = env.step(action)
63
64           # 計算累計報酬
65           total_reward += reward
66
67       # 顯示累計報酬
68       print(f"累計報酬: {total_reward:.4f}")
```

執行：python RL_15_01_simple_game.py。

執行結果：累計報酬: -1.2，由於採隨機策略，因此，每次結果均不相同。

【範例 2】呼叫 RL_15_01_simple_game.py，執行 10 回合。

程式：RL_15_02_simple_game_test.py。

1. 載入相關套件。

```
1  # 載入相關套件
2  from RL_15_01_simple_game import Environment, Agent
```

2. 建立實驗，包含環境、代理人物件。

```
4  # 建立實驗，含環境、代理人物件
5  env = Environment()
6  agent = Agent()
```

3. 進行實驗。

```
8   # 進行實驗
9   for _ in range(10):
10      env.__init__()  # 重置
11      total_reward=0  # 累計報酬
12      while not env.is_done():
13          # 採取行動
14          action = agent.action(env)
15          
16          # 更新下一步
17          state, reward = env.step(action)
18
19          # 計算累計報酬
20          total_reward += reward
21
22      # 顯示累計報酬
23      print(f"累計報酬: {total_reward:.4f}")
```

執行結果： 可以看到每次的結果均不相同，這就是程式在嘗試錯誤的證明。

```
累計報酬: -1.2000
累計報酬: -1.6000
累計報酬: 0.8000
累計報酬: -1.2000
累計報酬: 0.4000
累計報酬: -1.2000
累計報酬: -1.2000
累計報酬: -1.2000
累計報酬: -1.6000
累計報酬: 0.4000
```

【範例 3】以狀態值函數最大者為行動依據，執行 10 回合，並將程式改成較有彈性，允許更多的節點。

程式：RL_15_03_simple_game_with_state_value.ipynb。

1. 載入相關套件。

```
1  # 載入相關套件
2  import numpy as np
3  import random
```

2. 參數設定。

```
1  # 參數設定
2  NODE_COUNT = 5          # 節點數
3  NORMAL_REWARD = -0.2 # 每走一步扣分 0.2
```

3. 建立環境類別：增加以下函數。

- update_state_value：由終點倒推，更新狀態值函數。
- get_observation：取得狀態值函數的期望值，以狀態值函數最大者為行動依據。

```
1  # 環境類別
2  class Environment():
3      # 初始化
4      def __init__(self):
5          # 儲存狀態值函數,[0]:不用
6          self.state_value = np.full((NODE_COUNT+1), 0.0)
7          self.state_value[1]=-1
8          self.state_value[NODE_COUNT]=1
9
10         # 更新次數,[0]:不用
11         self.state_value_count = np.full((NODE_COUNT+1), 0)
12         self.state_value_count[1]=1
13         self.state_value_count[NODE_COUNT]=1
14
15     # 初始化
16     def reset(self):
17         self.position = int((1+NODE_COUNT) / 2)  # 玩家一開始站中間位置
18         self.trajectory=[] # 行動軌跡
19
20     def get_states(self):
21         # 狀態空間(State Space),共有5個位置
22         return [i for i in range(1, 6)]
23
24     def get_actions(self):
25         return [-1, 1] # 行動空間(Action Space)
```

```
26
27      def is_done(self): # 判斷比賽回合是否結束
28          # 是否走到左右端點
29          return self.poistion == 1 or self.poistion == NODE_COUNT
```

```
31      # 步驟
32      def step(self, action):
33          # 是否回合已結束
34          if self.is_done():
35              raise Exception("Game is over")
36
37          self.trajectory.append(self.poistion)
38          self.poistion += action
39          if self.poistion == 1:
40              reward = -1
41          elif self.poistion == NODE_COUNT:
42              reward = 1
43          else:
44              reward = NORMAL_REWARD
45
46          return action, reward
47
48      def update_state_value(self, final_value):
49          # 倒推，更新狀態值函數
50          for i in range(len(self.trajectory)-1, -1, -1):
51              final_value += NORMAL_REWARD
52              self.state_value[self.trajectory[i]] += final_value
53              self.state_value_count[self.trajectory[i]] += 1
54
55      # 取得狀態值函數期望值
56      def get_observation(self):
57          np.seterr(divide='ignore',invalid='ignore')
58          mean1 = self.state_value / self.state_value_count
59          return mean1
```

4. 代理人類別：比較左右相鄰的節點，以狀態值函數最大者為行動方
   向，如果兩個的狀態值一樣大，就隨機選擇一個。

```
1   # 代理人類別
2   class Agent():
3       # 初始化
4       def __init__(self):
5           pass
6
7       def action(self, env):
8           # 取得狀態值函數期望值
9           state_value = env.get_observation()
10
11          # 以左/右節點狀態值函數大者為行動依據，如果兩個狀態值一樣大，隨機選擇一個
12          if state_value[env.poistion-1] > state_value[env.poistion+1]:
13              return -1
14          if state_value[env.poistion-1] < state_value[env.poistion+1]:
15              return 1
16          else:
17              return random.choice([-1, 1])
```

5. 建立實驗，包含環境、代理人物件。

```
1  # 建立實驗，含環境、代理人物件
2  env = Environment()
3  agent = Agent()
4
5  # 進行實驗
6  total_reward_list = []
7  for _ in range(10):
8      env.reset()  # 重置
9      total_reward=0  # 累計報酬
10     while not env.is_done():
11         # 採取行動
12         action = agent.action(env)
13
14         # 更新下一步
15         state, reward = env.step(action)
16         #print(state, reward)
17         # 計算累計報酬
18         total_reward += reward
19
20     print('trajectory', env.trajectory)
21     env.update_state_value(total_reward)
22     total_reward_list.append(round(total_reward, 2))
23
24  # 顯示累計報酬
25  print(f"累計報酬: {total_reward_list}")
```

執行：python RL_15_02_simple_game_test.py。

執行結果： 可以看出每次的結果均不相同，基本上訓練了兩次之後，接下來的每一次都會往右走，這表示已經找到最佳策略，就是一直往右走。

```
trajectory [3, 2, 3, 4, 3, 2]
trajectory [3, 4]
trajectory [3, 4]
trajectory [3, 4]
trajectory [3, 4]
trajectory [3, 4]
trajectory [3, 4]
trajectory [3, 4]
trajectory [3, 4]
trajectory [3, 4]
累計報酬: [-2.0, 0.8, 0.8, 0.8, 0.8, 0.8, 0.8, 0.8, 0.8, 0.8]
```

6. 繪圖。

```
1  # 繪圖
2  import matplotlib.pyplot as plt
3
4  plt.figure(figsize=(10,6))
5  plt.plot(total_reward_list)
```

執行結果： 可以看出訓練兩次過後，即可找到最佳策略，就是一直往右走。

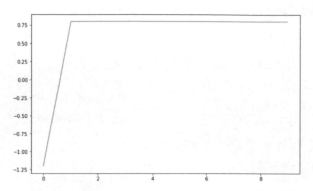

把節點數(NODE_COUNT)參數放大為 11，執行結果：可以看出訓練三次後，就找到最佳策略，也是一直往右走。

```
trajectory [6, 5, 4, 5, 4, 3, 2, 3, 4, 3, 2, 3, 4, 3, 4, 5, 4, 3, 4, 3, 2, 3, 4, 5, 6, 5, 6, 5, 4, 3, 2]
trajectory [6, 5, 6, 5, 6, 7, 8, 9, 10, 9, 10]
trajectory [6, 7, 8, 9, 10]
trajectory [6, 7, 8, 9, 10]
trajectory [6, 7, 8, 9, 10]
trajectory [6, 7, 8, 9, 10]
trajectory [6, 7, 8, 9, 10]
trajectory [6, 7, 8, 9, 10]
trajectory [6, 7, 8, 9, 10]
累計報酬: [-7.0, -1.0, 0.2, 0.2, 0.2, 0.2, 0.2, 0.2, 0.2, 0.2]
```

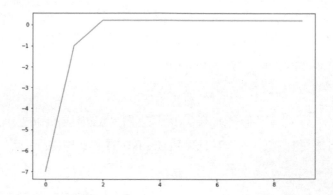

結論：以狀態值函數最大者為行動依據，訓練模型，果然可以找到最佳解，模型就是每個狀態的值函數，之後實際上線時即可載入模型執行。

15-4 Gym 套件

根據前面的練習，可以瞭解到強化學習是在各種環境中尋找最佳策略，因此，為了節省開發者的時間，網路上有許多套件設計了各式各樣的環境，供大家實驗，也能藉由動畫來展示訓練過程，例如 Gym、Amazon SageMaker 等套件。

以下就來介紹 Gym 套件的用法，它是 OpenAI 開發的學習套件，提供數十種不同的遊戲，Gym 官網(https://gym.openai.com/)首頁上有展示一些遊戲畫面。不過請讀者留意，有些遊戲在 Windows 作業環境並不能順利安裝，因為 Gym 是以 gcc 撰寫的。

安裝指令如下：

```
pip install gym
```

假使有要修改 Gym 的程式碼，可直接從 GitHub 下載原始程式碼後，再自行安裝：

```
git clone https://github.com/openai/gym
cd gym
pip install -e .
```

如需安裝全部遊戲，可執行以下指令，注意，這只能在 Linux 環境下執行，而且必須先安裝相關軟體工具，請參考 Gym GitHub 說明(https://github.com/openai/gym#installing-everything)。

```
pip install -e '.[all]'
```

在 Windows 作業環境下僅能加裝 Atari 遊戲，Atari 為 1967 年開發的遊戲機，擁有幾十種的遊戲，例如打磚塊(Breakout)、桌球(Pong)等：

```
pip install -e '.[atari]'
```

圖 15.14 Atari 遊戲機

Gym 提供的環境分為四類：

1. 經典遊戲(Classic control)和文字遊戲(Toy text)：屬於小型的環境，適合初學者開發測試。

2. 演算法類(Algorithmic)：像是多位數的加法、反轉順序等，這對電腦來說非常簡單，但使用強化學習方式求解的話，是一大挑戰。

3. Atari：Atari 遊戲機內的一些遊戲。

4. 2D and 3D 機器人(Robot)：機器人模擬環境，有些是要付費的，可免費試用 30 天。可惜在 Windows 作業環境安裝會有問題，還好網路上有些文章有提到解決方案，需要的讀者可以 google 一下或參考「Install OpenAI Gym with Box2D and Mujoco in Windows 10」[2]。

接下來先認識一下 Gym 的架構，各位可能會覺得有點眼熟，這是由於前面的範例刻意模仿 Gym 的設計架構，因此兩者的環境類別有些類似。

Gym 的環境類別包括以下方法：

1. reset()：比賽一開始或回合結束時，呼叫此方法重置環境。

2. step(action)：傳入行動，觸動下一步，回傳下列資訊。
   - observation：環境更新後的狀態。
   - reward：行動後得到的獎勵。
   - done：布林值，True 表示比賽回合結束，False 表示比賽回合進行中。
   - info：為字典(dict)的資料型態，通常是除錯訊息。

3. render()：渲染，即顯示更新後的畫面。

4. close()：關閉環境。

【範例 4】Gym 入門。

程式：RL_15_04_Gym.ipynb。

1. 載入相關套件。

```
1  # 載入相關套件
2  import gym
3  from gym import envs
```

2. 顯示已註冊的遊戲環境。

```
1  # 已註冊的遊戲
2  all_envs = envs.registry.all()
3  env_ids = [env_spec.id for env_spec in all_envs]
4  print(env_ids)
```

執行結果：有數百種的遊戲，種類依每台機器安裝的情形而有所不同。

```
['Copy-v0', 'RepeatCopy-v0', 'ReversedAddition-v0', 'ReversedAddition3-v0', 'DuplicatedInput-v0', 'Reverse-v0', 'CartPole-v
0', 'CartPole-v1', 'MountainCar-v0', 'MountainCarContinuous-v0', 'Pendulum-v0', 'Acrobot-v1', 'LunarLander-v2', 'LunarLanderC
ontinuous-v2', 'BipedalWalker-v3', 'BipedalWalkerHardcore-v3', 'CarRacing-v0', 'Blackjack-v0', 'KellyCoinflip-v0', 'KellyCoin
flipGeneralized-v0', 'FrozenLake-v0', 'FrozenLake8x8-v0', 'CliffWalking-v0', 'NChain-v0', 'Roulette-v0', 'Taxi-v3', 'Guessing
Game-v0', 'HotterColder-v0', 'Reacher-v2', 'Pusher-v2', 'Thrower-v2', 'Striker-v2', 'InvertedPendulum-v2', 'InvertedDoublePen
dulum-v2', 'HalfCheetah-v2', 'HalfCheetah-v3', 'Hopper-v2', 'Hopper-v3', 'Swimmer-v2', 'Swimmer-v3', 'Walker2d-v2', 'Walker2d
-v3', 'Ant-v2', 'Ant-v3', 'Humanoid-v2', 'Humanoid-v3', 'HumanoidStandup-v2', 'FetchSlide-v1', 'FetchPickAndPlace-v1', 'Fetch
Reach-v1', 'FetchPush-v1', 'HandReach-v0', 'HandManipulateBlockRotateZ-v0', 'HandManipulateBlockRotateZTouchSensors-v0', 'Han
dManipulateBlockRotateZTouchSensors-v1', 'HandManipulateBlockRotateParallel-v0', 'HandManipulateBlockRotateParallelTouchSenso
rs-v0', 'HandManipulateBlockRotateParallelTouchSensors-v1', 'HandManipulateBlockRotateXYZ-v0', 'HandManipulateBlockRotateXYZT
ouchSensors-v0', 'HandManipulateBlockRotateXYZTouchSensors-v1', 'HandManipulateBlockFull-v0', 'HandManipulateBlock-v0', 'Hand
ManipulateBlockTouchSensors-v0', 'HandManipulateBlockTouchSensors-v1', 'HandManipulateEggRotate-v0', 'HandManipulateEggRotate
TouchSensors-v0', 'HandManipulateEggRotateTouchSensors-v1', 'HandManipulateEggFull-v0', 'HandManipulateEgg-v0', 'HandManipula
```

3. 任意載入一個環境，例如木棒台車(CartPole)遊戲，並顯示行動空間、
   狀態空間/最大值/最小值。

```
1  # 載入 木棒台車(CartPole) 遊戲
2  env = gym.make("CartPole-v1")
3
4  # 環境的資訊
5  print(env.action_space)
6  print(env.observation_space)
7  print('observation_space 範圍：')
8  print(env.observation_space.high)
9  print(env.observation_space.low)
```

執行結果： 行動空間有兩個離散值(0:往左，1:往右)、狀態空間為 Box 資
料型態，表示連續型變數，維度大小為 4，分別代表台車位置(Cart
Position)、台車速度(Cart Velocity)、木棒角度(Pole Angle)及木棒速度(Pole
Velocity At Tip)，另外顯示 4 項資訊的最大值和最小值。

```
Discrete(2)
Box(4,)
observation_space 範圍:
[4.8000002e+00 3.4028235e+38 4.1887903e-01 3.4028235e+38]
[-4.8000002e+00 -3.4028235e+38 -4.1887903e-01 -3.4028235e+38]
```

4. 載入「打磚塊」(Breakout)遊戲，顯示環境的資訊。

```
1  # 載入 打磚塊(Breakout) 遊戲
2  env = gym.make("Breakout-v0")
3
4  # 環境的資訊
5  print(env.action_space)
6  print(env.observation_space)
7  print('observation_space 範圍 : ')
8  print(env.observation_space.high)
9  print(env.observation_space.low)
```

執行結果：相較於木棒台車，打磚塊就複雜許多，官網並未提供相關資訊，必須從 GitHub 下載原始程式碼，再觀看程式說明。或是查看安裝目錄，Atari 程式安裝在 anaconda3\Lib\site-packages\atari_py 目錄下，Breakout 原始程式為 ale_interface\src\games\supported\Breakout.cpp。

```
Discrete(4)
Box(210, 160, 3)
observation_space 範圍:
[[[255 255 255]
  [255 255 255]
  [255 255 255]
  ...
  [255 255 255]
  [255 255 255]
  [255 255 255]]

 [[255 255 255]
  [255 255 255]
  [255 255 255]
  ...
  [255 255 255]
  [255 255 255]
  [255 255 255]]
```

5. 實驗木棒台車(CartPole)遊戲。

```
1  # 載入 木棒台車(CartPole) 遊戲
2  env = gym.make("CartPole-v1")
3
4  # 重置
5  observation = env.reset()
6  # 將環境資訊寫入日誌檔
7  with open("CartPole_random.log", "w", encoding='utf8') as f:
8      # 執行 1000 次行動
9      for _ in range(1000):
10         # 更新畫面
11         env.render()
12         # 隨機行動
13         action = env.action_space.sample()
14         # 觸動下一步
15         observation, reward, done, info = env.step(action)
16         # 寫入資訊
17         f.write(f"action={action}, observation={observation}," +
18                 f"reward={reward}, done={done}, info={info}\n")
19         # 比賽回合結束，重置
20         if done:
21             observation = env.reset()
22  env.close()
```

執行結果：以隨機的方式行動，可以看到台車時而前進，時而後退。

CartPole_random.log 日誌檔的部份內容如下：

```
action=0, observation=[ 0.02797028 -0.203182    0.00185102  0.28630734],reward=1.0, done=F
action=1, observation=[ 0.02390664 -0.00808649  0.00757717 -0.00579121],reward=1.0, done=F
action=1, observation=[ 0.02374491  0.18692597  0.00746134 -0.29607385],reward=1.0, done=F
action=0, observation=[ 0.02748343 -0.00830155  0.00153987 -0.00104711],reward=1.0, done=F
action=0, observation=[ 0.0273174  -0.20344555  0.00151892  0.29212127],reward=1.0, done=F
action=1, observation=[ 2.32484899e-02 -8.34528672e-03  7.36135014e-03 -8.22245954e-05],re
action=1, observation=[ 0.02308158  0.18667032  0.00735971 -0.2904335 ],reward=1.0, done=F
action=0, observation=[ 0.02681499 -0.00855579  0.00155104  0.00456148],reward=1.0, done=F
action=0, observation=[ 0.02664387 -0.20369996  0.00164227  0.29773338],reward=1.0, done=F
action=1, observation=[ 0.02256988 -0.00860145  0.00759693  0.00556884],reward=1.0, done=F
action=0, observation=[ 0.02239785 -0.20383153  0.00770831  0.30063898],reward=1.0, done=F
```

木棒台車(CartPole)的遊戲規則說明如下：

1. 可控制台車往左(0)或往右(1)。
2. 每走一步得一分。

3. 台車一開始定位在中心點,平衡桿是直立(upright)的,在行駛中要保持平衡。

4. 遊戲結束的要件,三擇一:
   - 平衡桿偏差超過 12℃ ➔ 敗。
   - 離中心點 2.4 單位 ➔ 勝。
   - 兩個版本,v0:行動超過 200 步,v1:行動超過 500 步 ➔ 勝。

5. 如果連續 100 回合的平均報酬超過 195 步,即視為解題成功。

Step 函數傳回的內容:

1. observation:環境更新後狀態。
   - 台車位置(Cart Position)。
   - 台車速度(Cart Velocity)。
   - 木棒角度(Pole Angle)。
   - 木棒速度(Pole Velocity At Tip)。

2. reward:行動後得到的獎勵。

3. done:布林值,True 表示比賽回合結束,False 表示比賽回合進行中。

4. info:為字典(dict)的資料型態,通常是除錯訊息,本遊戲均不傳回資訊。

【範例 5】木棒台車(CartPole)實驗。

程式:RL_15_05_CartPole.ipynb。

1. 載入相關套件。

```
1  # 載入相關套件
2  import gym
3  from gym import envs
```

2. 設定比賽回合數。

```
1  # 參數設定
2  no = 50          # 比賽回合數
```

3. 實驗：採隨機行動。

```
 1  # 載入 木棒台車(CartPole) 遊戲
 2  env = gym.make("CartPole-v0")
 3
 4  # 重置
 5  observation = env.reset()
 6  all_rewards=[] # 每回合總報酬
 7  total_rewards = 0
 8
 9  while no > 0:    # 執行 50 比賽回合數
10      # 隨機行動
11      action = env.action_space.sample()
12      # 觸動下一步
13      observation, reward, done, info = env.step(action)
14      # 累計報酬
15      total_rewards += reward
16
17      # 比賽回合結束，重置
18      if done:
19          observation = env.reset()
20          all_rewards.append(total_rewards)
21          total_rewards = 0
22          no-=1
23
24  env.close()
```

執行結果：如右表所示，結果相當慘烈，沒有一回合走超過 200 步，全軍覆滅，顯示對於強化學習而言這個遊戲很有挑戰性。

4. 基於上述實驗，傳統的作法會針對問題提出對策，例如：

- 台車距離中心點大於 2.4 單位就算輸了，所以設定每次行動採一左一右的方式，盡量不偏離中心點。

- 由於台車的平衡桿角度偏差 12 度以上也算輸，故而設定平衡桿角度若偏右 8 度以上，就往右前進，直到角度偏右小於 8 度為止。

- 反之，偏左也是類似處理。

回合	報酬	結果
0	56.0	Loss
1	12.0	Loss
2	37.0	Loss
3	23.0	Loss
4	18.0	Loss
5	23.0	Loss
6	16.0	Loss
7	12.0	Loss
8	16.0	Loss
9	24.0	Loss
10	11.0	Loss
11	58.0	Loss
12	25.0	Loss
13	15.0	Loss
14	12.0	Loss
15	25.0	Loss
16	19.0	Loss
17	36.0	Loss
18	12.0	Loss
19	10.0	Loss
20	13.0	Loss
21	19.0	Loss
22	31.0	Loss
23	49.0	Loss
24	17.0	Loss
25	13.0	Loss
26	15.0	Loss
27	23.0	Loss
28	24.0	Loss

5. 首先建立 Agent 類別，撰寫 act 函數實現以上邏輯。

```
1  import math
2
3  # 參數設定
4  left, right = 0, 1  # 台車行進方向
5  max_angle = 8        # 偏右8度以上，就往右前進，偏左也是同樣處理
```

```
1  class Agent:
2      # 初始化
3      def __init__(self):
4          self.direction = left
5          self.last_direction=right
6
7      # 自訂策略
8      def act(self, observation):
9          # 台車位置、台車速度、平衡桿角度、平衡桿速度
10         cart_position, cart_velocity, pole_angle, pole_velocity = observation
11
12         ...
13         行動策略：
14         1. 設定每次行動採一左一右，盡量不離中心點。
15         2. 平衡桿角度偏右8度以上，就往右前進，直到角度偏右小於8度。
16         3. 反之，偏左也是同樣處理。
17         ...
18         if pole_angle < math.radians(max_angle) and \
19             pole_angle > math.radians(-max_angle):
20             self.direction = (self.last_direction + 1) % 2
21         elif pole_angle >= math.radians(max_angle):
22             self.direction = right
23         else:
24             self.direction = left
25
26         self.last_direction = self.direction
27
28         return self.direction
```

6. 以 agent.act(observation)取代 env.action_space.sample()。

```
1  no = 50           # 比賽回合數
2
3  # 載入 木棒台車(CartPole) 遊戲
4  env = gym.make("CartPole-v0")
5
6  # 重置
7  observation = env.reset()
8  all_rewards=[] # 每回合總報酬
9  total_rewards = 0
10 agent = Agent()
11 while no > 0:     # 執行 50 比賽回合數
12     # 行動
13     action = agent.act(observation) #env.action_space.sample()
14     # 觸動下一步
15     observation, reward, done, info = env.step(action)
```

```
16      # 累計報酬
17      total_rewards += reward
18
19      # 比賽回合結束，重置
20      if done:
21          observation = env.reset()
22          all_rewards.append(total_rewards)
23          total_rewards = 0
24          no-=1
25
26  env.close()
```

7. 顯示執行結果：雖然比起隨機行動的方式，改良後的報酬增加很多，
   但結果還是都失敗了。

回合	報酬	結果
0	97.0	Loss
1	71.0	Loss
2	112.0	Loss
3	129.0	Loss
4	96.0	Loss
5	78.0	Loss
6	116.0	Loss
7	82.0	Loss
8	84.0	Loss
9	105.0	Loss
10	80.0	Loss
11	62.0	Loss
12	71.0	Loss
13	145.0	Loss
14	72.0	Loss
15	135.0	Loss
16	127.0	Loss
17	78.0	Loss
18	100.0	Loss
19	87.0	Loss
20	93.0	Loss
21	55.0	Loss
22	95.0	Loss
23	66.0	Loss
24	156.0	Loss
25	79.0	Loss
26	48.0	Loss
27	101.0	Loss
28	77.0	Loss

上述的解法還有一個缺點，不具通用性，這個策略就算在木棒台車有效，
也不能套用到其他遊戲上，也沒有自我學習的能力，無法隨著訓練次數的
增加，使模型更加聰明與準確。

木棒台車是一款很簡單的遊戲，不過，要能成功解題並不如想像中容易，
筆者應該以前面講到的「狀態值函數最大化」為策略來進行實驗，但很不

幸的,如此又會碰到另一些難題:

1. 狀態非單一變數,而是四個變數,包括台車的位置和速度、木棒的角度和速度。
2. 這四個變數都是連續性變數,然而計算狀態值函數是針對每一狀態,因此狀態空間必須是離散的,僅有限的個數才能倒推計算出狀態值函數。

以上兩個問題可以使用下列技巧處理:

1. 四個變數混合列舉出所有組合。
2. 將連續性變數分組,變成有限組別。

這裡我們先不急著進行實驗,因為後續可以搭配更好的演算法來解決上述問題,所以這題先擱在一邊,待會再回頭搞定它。

▌ 15-5 Gym 擴充功能

雖然 Gym 有很多環境供開發者挑選,但萬一還是沒找到完全符合需求的環境該怎麼辦?這時可以利用擴充功能 Wrapper,客製化預設的環境,包括:

1. 修改 step()回傳的狀態:預設只會回傳最新的狀態,我們可以利用 ObservationWrapper 達成此一功能。
2. 修改獎勵值:可以利用 RewardWrapper 修改預設的獎勵值。
3. 修改行動值:可以利用 ActionWrapper 餵入行動值。

【範例 6】ActionWrapper 示範:執行木棒台車遊戲,原先台車固定往左走,但加上 ActionWrapper 後,會有 1/10 的機率採隨機行動。

程式:RL_15_06_Action_Wrapper_test.py。

1. 載入相關套件。

```
1  import gym
2  import random
```

2. 建立一個 RandomActionWrapper 類別，繼承 gym.ActionWrapper 基礎
   類別，覆寫 action()方法。epsilon 變數為隨機行動的機率，每次行動時
   (step)都會呼叫 RandomActionWrapper 的 action()。

```
4  # 繼承 gym.ActionWrapper 基礎類別
5  class RandomActionWrapper(gym.ActionWrapper):
6      def __init__(self, env, epsilon=0.1):
7          super(RandomActionWrapper, self).__init__(env)
8          self.epsilon = epsilon # 隨機行動的機率
9
10     def action(self, action):
11         # 隨機亂數小於 epsilon，採取隨機行動
12         if random.random() < self.epsilon:
13             print("Random!")
14             return self.env.action_space.sample()
15         return action
```

3. 實驗：step(0)表示固定往左走，但卻會被 RandomActionWrapper 的
   action()所攔截，結果如下，偶而會出現隨機行動。

```
18 if __name__ == "__main__":
19     env = RandomActionWrapper(gym.make("CartPole-v0"))
20
21     for _ in range(50):
22         env.reset()
23         total_reward = 0.0
24         while True:
25             env.render()
26             # 固定往左走
27             print("往左走!")
28             obs, reward, done, _ = env.step(0)
29             total_reward += reward
30             if done:
31                 break
32
33         print(f"報酬: {total_reward:.2f}")
34     env.close()
```

執行結果：

```
往左走!
往左走!
Random!
往左走!
報酬: 9.00
往左走!
往左走!
往左走!
往左走!
往左走!
往左走!
往左走!
往左走!
往左走!
往左走!
往左走!
報酬: 10.00
```

【範例 7】使用 wrappers.Monitor 錄影：可將訓練過程存成多段影片檔 (mp4)。

程式：RL_15_07_Record_test.py。

1. 載入相關套件。

```
1  # 載入相關套件
2  import gym
```

2. 錄影：呼叫 Monitor()。

```
4  # 載入環境
5  env = gym.make("CartPole-v0")
6
7  # 錄影
8  env = gym.wrappers.Monitor(env, "recording", force=True)
```

3. 實驗。

```
10  # 實驗
11  for _ in range(50):
12      total_reward = 0.0
13      obs = env.reset()
14
15      while True:
16          env.render()
17          action = env.action_space.sample()
18          obs, reward, done, _ = env.step(action)
19          total_reward += reward
```

```
20        if done:
21            break
22
23    print(f"報酬: {total_reward:.2f}")
24
25 env.close()
26
```

執行結果：會將錄影結果存在 recording 資料夾。

以上只是列舉兩個簡單的範例，讀者有興趣可參閱官網說明。

15-6 動態規劃(Dynamic Programming)

依據前面談到的 Bellman 方程式：

$$v_\pi(s) = \sum_a \pi(a|s) \sum_{s'} \mathcal{P}^a_{ss'} \left[\mathcal{R}^a_{ss'} + \gamma v_\pi(s') \right]$$

如果上述的行動轉移機率(π)、狀態轉移機率(P)均為已知，亦即環境是明確的(deterministic)，我們就可以利用 Bellman 方程式計算出狀態值函數、行動值函數，以反覆的方式求解，這種解法稱為「動態規劃」 (Dynamic Programming, DP)，類似程式 RL_15_03_simple_game_with_state_value. ipynb，不過，更為細緻一點。

動態規劃的概念，是將大問題切分成小問題，然後逐步解決每個小問題，由於每個小問題都很類似，因此整合起來就能解決大問題。譬如「費波那契數列」（Fibonacci），其中 $F_n=F_{n-1}+F_{n-2}$，要計算整個數列的話，可以設計一個 $F_n=F_{n-1}+F_{n-2}$ 函數(小問題)，以遞迴的方式完成整個數列的計算(大問題)。

【範例 8】費波那契數列計算。
程式：RL_15_08_Fibonacci_Calculation.ipynb。

```
1   def fibonacci(n):
2       if n == 0 or n ==1:
3           return n
4       else:
5           return fibonacci(n-1)+fibonacci(n-2)
6
7   list1=[]
8   for i in range(2, 20):
9       list1.append(fibonacci(i))
10  print(list1)
```

執行結果：

[1, 2, 3, 5, 8, 13, 21, 34, 55, 89, 144, 233, 377, 610, 987, 1597, 2584, 4181]

每個小問題的計算結果都會被儲存下來，作為下個小問題的計算基礎。

強化學習將問題切成兩個步驟：

1. 策略評估(Policy Evaluation)：當玩家走完一回合後，就可以更新所有狀態的值函數，這就稱為策略評估，即將所有狀態重新評估一次，也稱為「預測」(prediction)。

2. 策略改善(Policy Improvement)：依照策略評估的最新狀態，採取最佳策略，以改善模型，也稱為「控制」(control)，通常都會依據最大的狀態值函數行動，我們稱之為「貪婪」(Greedy)策略，但其實它是有缺陷的，後面會再詳加討論。

最後，將兩個步驟合併，循環使用，即是所謂的策略循環或策略疊代(Policy Iteration)，如下圖所示。

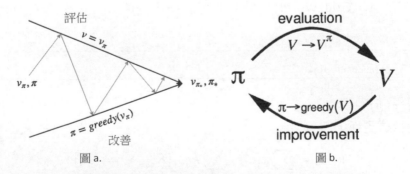

圖 15.15　策略循環或策略疊代(Policy Iteration)

圖 a.表示先走一回合後，進行評估，接著依評估結果採貪婪策略行動，再評估，一直循環下去，直到收斂為止。

圖 b.強調經過行動策略 π 後，狀態值函數會由原來的 V 變成 V_π，之後再以 V_π 作為行動的依據。

【**範例 9**】以下圖的迷宮為例，執行策略評估(Policy Evaluation)。

程式：RL_15_09_Policy_Evaluation.ipynb，修改自 Denny Britz 網站[3]，他 是 解 答 「 Reinforcement Learning ： An Introduction 」 一 書 (http://incompleteideas.net/book/the-book-2nd.html)的習題，該書可說是強化學習的聖經。

圖 15.16 Grid World 迷宮，左上角為起點，右下角為終點

1. 載入相關套件。

```
1  # 載入相關套件
2  import numpy as np
3  from lib.envs.gridworld import GridworldEnv
```

2. 建立 Grid World 迷宮環境：程式為 lib\envs\gridworld.py，定義遊戲規則，注意，Grid World 在網路上有許多不同的版本，遊戲規則略有差異。

```
1  # 環境
2  env = GridworldEnv()
```

3. 策略評估函數：
依下列公式更新。

$$v_\pi(s) = \sum_a \pi(a|s) \sum_{s'} \mathcal{P}_{ss'}^a \left[\mathcal{R}_{ss'}^a + \gamma v_\pi(s') \right]$$

參數說明如下：

- discount_factor：折扣因子，預設為 1，即無折扣。
- theta：差異容忍值，前後兩次狀態值函數的最大差異小於此數值，即停止更新。

P 均為 1，表示狀態轉移機率均等，定義在 gridworld.py 中。

```
1   # 策略評估函數
2   def policy_eval(policy, env, discount_factor=1.0, theta=0.00001):
3       # 狀態值函數初始化
4       V = np.zeros(env.nS)
5       while True:
6           delta = 0
7           # 更新每個狀態值的函數
8           for s in range(env.nS):
9               v = 0
10              # 計算每個行動後的狀態值函數
11              for a, action_prob in enumerate(policy[s]):
12                  # 取得所有可能的下一狀態值
13                  for  prob, next_state, reward, done in env.P[s][a]:
14                      # 狀態值函數公式，依照所有可能的下一狀態值函數加總
15                      v += action_prob * prob * (reward + discount_factor * V[next_state])
16              # 比較更新前後的差值，取最大值
17              delta = max(delta, np.abs(v - V[s]))
18              V[s] = v
19          # 若最大差值 < 門檻值，則停止評估
20          if delta < theta:
21              break
22      return np.array(V)
```

4. 呼叫策略評估函數：採隨機策略，往上/下/左/右走的機率(π)均等。

```
1   # 隨機策略，機率均等
2   random_policy = np.ones([env.nS, env.nA]) / env.nA
3   # 評估
4   v = policy_eval(random_policy, env)
```

5. 顯示狀態值函數。

```
1   print("狀態值函數:")
2   print(v)
3   print("")
4
5   print("4x4 狀態值函數:")
6   print(v.reshape(env.shape))
7   print("")
```

6. 驗證答案是否正確：與書中的答案相對照。

```
1  # 驗證答案是否正確
2  expected_v = np.array([0, -14, -20, -22, -14, -18, -20, -20, -20, -20, -18, -14, -22, -20, -14, 0])
3  np.testing.assert_array_almost_equal(v, expected_v, decimal=2)
```

完成策略循環中的評估後，接下來再結合策略改善。

【範例 10】以上述的迷宮為例，執行策略循環(Policy Iteration)。

程式：RL_15_10_Policy_Iteration.ipynb，修改自 Denny Britz 網站。

1. 載入相關套件，前 3 步驟與上例相同。

```
1  # 載入相關套件
2  import numpy as np
3  from lib.envs.gridworld import GridworldEnv
```

2. 建立 Grid World 迷宮環境：程式在 lib\envs\gridworld.py，定義遊戲規
   則，注意，Grid World 在網路上有許多不同的版本，遊戲規則略有差
   異。

```
1  # 環境
2  env = GridworldEnv()
```

3. 策略評估函數：

```
1  # 策略評估函數
2  def policy_eval(policy, env, discount_factor=1.0, theta=0.00001):
3      # 狀態值函數初始化
4      V = np.zeros(env.nS)
5      while True:
6          delta = 0
7          # 更新每個狀態值的函數
8          for s in range(env.nS):
9              v = 0
10             # 計算每個行動後的狀態值函數
11             for a, action_prob in enumerate(policy[s]):
12                 # 取得所有可能的下一狀態值
13                 for  prob, next_state, reward, done in env.P[s][a]:
14                     # 狀態值函數公式，依照所有可能的下一狀態值函數加總
15                     v += action_prob * prob * (reward + discount_factor * V[next_state])
16             # 比較更新前後的差值，取最大值
17             delta = max(delta, np.abs(v - V[s]))
18             V[s] = v
19         # 若最大差值 < 門檻值，則停止評估
20         if delta < theta:
21             break
22     return np.array(V)
```

4. 定義策略改善函數：

■ one_step_lookahead：依下列公式計算每一種行動的值函數。

$$q_\pi(s, a) = \sum_{s'} \mathcal{P}^a_{ss'} \left[\mathcal{R}^a_{ss'} + \gamma \sum_{a'} \pi(a'|s') \, q_\pi(s', a') \right]$$

■ 一開始採隨機策略，進行策略評估，計算狀態值函數。

■ 呼叫 one_step_lookahead，計算下一步的行動值函數，找出最佳行動，並更新策略轉移機率(π)。

■ 直到已無較佳的行動策略，則回傳策略與狀態值函數。

```
1   # 策略改善函數
2   def policy_improvement(env, policy_eval_fn=policy_eval, discount_factor=1.0):
3       # 計算行動值函數
4       def one_step_lookahead(state, V):
5           A = np.zeros(env.nA)
6           for a in range(env.nA):
7               for prob, next_state, reward, done in env.P[state][a]:
8                   A[a] += prob * (reward + discount_factor * V[next_state])
9           return A
10
11      # 剛開始採隨機策略，往上/下/左/右走的機率(π)均等
12      policy = np.ones([env.nS, env.nA]) / env.nA
13
14      while True:
15          # 策略評估
16          V = policy_eval_fn(policy, env, discount_factor)
17
18          # 若要改變策略，會設定 policy_stable = False
19          policy_stable = True
20
21          for s in range(env.nS):
22              # 依 P 選擇最佳行動
23              chosen_a = np.argmax(policy[s])
24
25              # 計算下一步的行動值函數
26              action_values = one_step_lookahead(s, V)
27              # 選擇最佳行動
28              best_a = np.argmax(action_values)
```

```
30              # 貪婪策略：若有新的最佳行動，修改行動策略
31              if chosen_a != best_a:
32                  policy_stable = False
33              policy[s] = np.eye(env.nA)[best_a]
34
35          # 如果已無較佳行動策略，則回傳策略及狀態值函數
36          if policy_stable:
37              return policy, V
```

5. 執行策略循環。

```
1  # 執行策略循環
2  policy, v = policy_improvement(env)
```

6. 顯示結果。

```
1  # 顯示結果
2  print("策略機率分配:")
3  print(policy)
4  print("")
5
6  print("4x4 策略機率分配 (0=up, 1=right, 2=down, 3=left):")
7  print(np.reshape(np.argmax(policy, axis=1), env.shape))
8  print("")
9
10 print("狀態值函數:")
11 print(v)
12 print("")
13
14 print("4x4 狀態值函數:")
15 print(v.reshape(env.shape))
16 print("")
```

執行結果：分別得到 π、P，也就是模型的參數。

```
策略機率分配:
[[1. 0. 0. 0.]
 [0. 0. 0. 1.]
 [0. 0. 0. 1.]
 [0. 0. 1. 0.]
 [1. 0. 0. 0.]
 [1. 0. 0. 0.]
 [1. 0. 0. 0.]
 [0. 0. 1. 0.]
 [1. 0. 0. 0.]
 [1. 0. 0. 0.]
 [0. 1. 0. 0.]
 [0. 0. 1. 0.]
 [1. 0. 0. 0.]
 [0. 1. 0. 0.]
 [0. 1. 0. 0.]
 [1. 0. 0. 0.]]

4x4 策略機率分配 (0=up, 1=right, 2=down, 3=left):
[[0 3 3 2]
 [0 0 0 2]
 [0 0 1 2]
 [0 1 1 0]]
```

狀態值函數:
[0. -1. -2. -3. -1. -2. -3. -2. -2. -3. -2. -1. -3. -2. -1. 0.]

4x4 狀態值函數:
[[0. -1. -2. -3.]
 [-1. -2. -3. -2.]
 [-2. -3. -2. -1.]
 [-3. -2. -1. 0.]]

7. 驗證答案是否正確：與書中的答案相對照。

```
1  # 驗證答案是否正確
2  expected_v = np.array([ 0, -1, -2, -3, -1, -2, -3, -2, -2, -3, -2, -1, -3, -2, -1,  0])
3  np.testing.assert_array_almost_equal(v, expected_v, decimal=2)
```

8. 因此，參考策略機率分配，從左上角走到右下角的最佳路徑如下：

$$
\begin{bmatrix} 0 & 3 & 3 & 2 \\ 0 & 0 & 0 & 2 \\ 0 & 0 & 1 & 2 \\ 0 & 1 & 1 & 0 \end{bmatrix}
$$

▌ 15-7 值循環(Value Iteration)

採用策略循環時，在每次策略改善前，必須先作一次策略評估，執行迴圈，更新所有狀態值函數，直至收斂，非常耗費時間，所以，考量到狀態值函數與行動值函數的更新十分類似，乾脆將其二者合併，以改善策略循環的缺點，稱之為值循環(Value Iteration)。

【範例 11】以上述的迷宮為例，使用值循環(Value Iteration)。
程式：RL_15_11_Value_Iteration.ipynb，修改自 Denny Britz 網站。

1. 載入相關套件，前 2 步驟與上例相同。

```
1  # 載入相關套件
2  import numpy as np
3  from lib.envs.gridworld import GridworldEnv
```

2. 建立 Grid World 迷宮環境。

```
1  # 環境
2  env = GridworldEnv()
```

3. 定義值循環函數：直接以行動值函數取代狀態值函數，將策略評估函數與策略改善函數合而為一。

```python
 1  # 值循環函數
 2  def value_iteration(env, theta=0.0001, discount_factor=1.0):
 3      # 計算行動值函數
 4      def one_step_lookahead(state, V):
 5          A = np.zeros(env.nA)
 6          for a in range(env.nA):
 7              for prob, next_state, reward, done in env.P[state][a]:
 8                  A[a] += prob * (reward + discount_factor * V[next_state])
 9          return A
10
11      # 狀態值函數初始化
12      V = np.zeros(env.nS)
13      while True:
14          delta = 0
15          # 更新每個狀態值的函數
16          for s in range(env.nS):
17              # 計算下一步的行動值函數
18              A = one_step_lookahead(s, V)
19              best_action_value = np.max(A)
20              # 比較更新前後的差值，取最大值
21              delta = max(delta, np.abs(best_action_value - V[s]))
22              # 更新狀態值函數
23              V[s] = best_action_value
24          # 若最大差值 < 門檻值，則停止評估
25          if delta < theta:
26              break
```

```python
28      # 一開始採隨機策略，往上/下/左/右走的機率(π)均等
29      policy = np.zeros([env.nS, env.nA])
30      for s in range(env.nS):
31          # 計算下一步的行動值函數
32          A = one_step_lookahead(s, V)
33          # 選擇最佳行動
34          best_action = np.argmax(A)
35          # 永遠採取最佳行動
36          policy[s, best_action] = 1.0
37
38      return policy, V
```

4. 執行值循環。

```python
 1  # 執行值循環
 2  policy, v = value_iteration(env)
```

5. 顯示結果：與策略循環結果相同。

```
1   # 顯示結果
2   print("策略機率分配:")
3   print(policy)
4   print("")
5
6   print("4x4 策略機率分配 (0=up, 1=right, 2=down, 3=left):")
7   print(np.reshape(np.argmax(policy, axis=1), env.shape))
8   print("")
9
10  print("狀態值函數:")
11  print(v)
12  print("")
13
14  print("4x4 狀態值函數:")
15  print(v.reshape(env.shape))
16  print("")
```

執行結果：分別得到 π、P，也就是模型的參數。

```
策略機率分配:
[[1. 0. 0. 0.]
 [0. 0. 0. 1.]
 [0. 0. 0. 1.]
 [0. 0. 1. 0.]
 [1. 0. 0. 0.]
 [1. 0. 0. 0.]
 [1. 0. 0. 0.]
 [0. 0. 1. 0.]
 [1. 0. 0. 0.]
 [1. 0. 0. 0.]
 [0. 1. 0. 0.]
 [0. 0. 1. 0.]
 [1. 0. 0. 0.]
 [0. 1. 0. 0.]
 [0. 1. 0. 0.]
 [1. 0. 0. 0.]]

4x4 策略機率分配 (0=up, 1=right, 2=down, 3=left):
[[0 3 3 2]
 [0 0 0 2]
 [0 0 1 2]
 [0 1 1 0]]

狀態值函數:
[ 0. -1. -2. -3. -1. -2. -3. -2. -2. -3. -2. -1. -3. -2. -1.  0.]

4x4 狀態值函數:
[[ 0. -1. -2. -3.]
 [-1. -2. -3. -2.]
 [-2. -3. -2. -1.]
 [-3. -2. -1.  0.]]
```

6. 驗證答案是否正確：與書中的答案相對照。

```
1   # 驗證答案是否正確
2   expected_v = np.array([ 0, -1, -2, -3, -1, -2, -3, -2, -2, -3, -2, -1, -3, -2, -1,  0])
3   np.testing.assert_array_almost_equal(v, expected_v, decimal=2)
```

7. 因此，參考策略機率分配，從左上角走到右下角的最佳路徑如下：

$$\begin{bmatrix} 0 & 3 & 3 & 2 \\ 0 & 0 & 0 & 2 \\ 0 & 0 & 1 & 2 \\ 0 & 1 & 1 & 0 \end{bmatrix}$$

動態規劃的優缺點如下：

1. 適合定義明確的問題，即策略轉移機率、狀態轉移機率均為已知的狀況。

2. 適合中小型的模型，狀態空間不超過百萬個，比方說圍棋，狀態空間 $=3^{19 \times 19}=1.74 \times 10^{172}$，狀態值函數更新就會執行太久。另外，可能會有大部份的路徑從未走過，導致樣本代表性不足，進而造成維數災難(Curse of Dimensionality)。

15-8 蒙地卡羅(Monte Carlo)

我們在玩遊戲時，通常不會知道策略轉移機率、狀態轉移機率，其他應用領域也是如此，這稱為無模型(Model Free)學習，在這樣的情況下，動態規劃就無法派上用場。因此，有學者就應用蒙地卡羅(Monte Carlo, MC)演算法，透過模擬的方式估計出轉移機率。

根據維基百科[4]描述，它命名的由來相當有趣，二戰時期，美國研發核武器的團隊發明了此計算方式，而發明人之一的「斯塔尼斯拉夫·烏拉姆」，因為他的叔叔經常在摩納哥的蒙地卡羅賭場輸錢，故而將其方法取名為蒙地卡羅。蒙地卡羅演算法主要就是使用亂數生成器產生數據，進而估計實際問題的答案。

【範例 12】以蒙地卡羅演算法求圓周率(π)。

程式：RL_15_11_Value_Iteration.ipynb。

1. 載入相關套件。

```
1  # 載入相關套件
2  import random
```

2. 計算圓周率(π)：

如下圖，假設有一個正方形與圓形，圓形半徑為 r，則圓形面積為 πr^2，正方形面積為$(2r)^2=4r^2$。

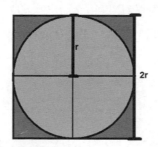

在正方形的範圍內隨機產生亂數一千萬個點，計算落在圓形內的點數。

落在圓形內的點數 / 一千萬 ≒ $\pi r^2 / 4r^2 = \pi/4$

化簡後 π= 4 x (落在圓形內的點數) / 一千萬。

```
1   # 模擬一千萬次
2   run_count=10000000
3   list1=[]
4
5   # 在 X:(-0.5, -0.5)，Y:(0.5, 0.5) 範圍內產生一千萬個點
6   for _ in range(run_count):
7       list1.append([random.random()-0.5, random.random()-0.5])
8
9   in_circle_count=0
10  for i in range(run_count):
11      # 計算在圓內的點，即 (X^2 + Y^2 <= 0.5 ^ 2)，其中 半徑=0.5
12      if list1[i][0]**2 + list1[i][1]**2 <= 0.5 ** 2:
13          in_circle_count+=1
14
15  # 正方形面積：寬高各為2r，故面積=4*(r**2)
16  # 圓形面積：pi * (r ** 2)
17  # pi = 圓形點數 / 正方形點數
18  pi=(in_circle_count/run_count) * 4
19  pi
```

執行結果：得到 π= 3.1418028，與正確答案 3.14159 相去不遠，如果模擬更多的點，就會更相近。

從以上的例子延伸，假如轉移機率未知，我們也可以利用蒙地卡羅演算法估計轉移機率，利用隨機策略去走迷宮，根據結果計算轉移機率。

為了避免無聊，我們換另一款遊戲「21 點撲克牌」(Blackjack)實驗，讀者如不熟悉遊戲規則，可參照維基百科的規則說明 https://zh.wikipedia.org/wiki/二十一點。

【範例 13】實驗 21 點撲克牌(Blackjack)之策略評估。
程式：RL_15_13_Blackjack_Policy_Evaluation.ipynb，修改自 Denny Britz 網站。

1. 載入相關套件。

```
1  # 載入相關套件
2  import numpy as np
3  from lib.envs.blackjack import BlackjackEnv
4  from lib import plotting
5  import sys
6  from collections import defaultdict
7  import matplotlib
8
9  matplotlib.style.use('ggplot') # 設定繪圖的風格
```

2. 建立環境：程式為 lib\envs\blackjack.py。

```
1  # 環境
2  env = BlackjackEnv()
```

3. 試玩：採用的策略為如果玩家手上點數超過(>=)20 點，才不補牌(stick)，反之都跟莊家要一張牌(hit)，策略並不合理，通常超過 16 點就不補牌了，若考慮更周延的話，會再視莊家的點數，才決定是否補牌，讀者可自行更改策略，觀察實驗結果的變化。採用這個不合理的策略，是要測試各種演算法是否能有效提升勝率。

```
1   # 試玩
2   def print_observation(observation):
3       score, dealer_score, usable_ace = observation
4       print(f"玩家分數: {score} (是否持有A: {usable_ace}), 莊家分數: {dealer_score}")
5
6   def strategy(observation):
7       score, dealer_score, usable_ace = observation
8       # 超過20點，不補牌(stick)，否則都跟莊家要一張牌(hit)
9       return 0 if score >= 20 else 1
10
11  # 試玩 20 次
12  for i_episode in range(20):
13      observation = env.reset()
14      # 開始依策略玩牌，最多 100 步驟，中途分出勝負即結束
15      for t in range(100):
16          print_observation(observation)
17          action = strategy(observation)
18          print(f'行動: {["不補牌", "補牌"][action]}')
19          observation, reward, done, _ = env.step(action)
20          if done:
21              print_observation(observation)
22              print(f"輸贏分數: {reward}\n")
23              break
```

執行結果：讀者若不熟悉玩法，可以觀察下列過程。

```
玩家分數: 21 (是否持有A: True), 莊家分數: 9
行動: 不補牌
玩家分數: 21 (是否持有A: True), 莊家分數: 9
輸贏分數: 0

玩家分數: 15 (是否持有A: False), 莊家分數: 3
行動: 補牌
玩家分數: 17 (是否持有A: False), 莊家分數: 3
行動: 補牌
玩家分數: 22 (是否持有A: False), 莊家分數: 3
輸贏分數: -1

玩家分數: 20 (是否持有A: True), 莊家分數: 6
行動: 不補牌
玩家分數: 20 (是否持有A: True), 莊家分數: 6
輸贏分數: 1

玩家分數: 20 (是否持有A: False), 莊家分數: 1
行動: 不補牌
玩家分數: 20 (是否持有A: False), 莊家分數: 1
輸贏分數: 0
```

4. 定義策略評估函數：主要是透過既定策略計算狀態值函數。

■ 實驗 1000 回合，記錄玩牌的過程，然後計算每個狀態的平均值函數。

■ 狀態為 Tuple 資料型態，內含:玩家的總點數 0~31、莊家亮牌的點數 1~11(A)、玩家是否拿 A，維度大小為(32, 11, 2)，其中 A 可為 1 點或 11 點，由持有者自行決定。

15-45

- 行動只有兩種：0為不補牌，1為補牌。
- 注意，在一回合中每個狀態有可能被走過兩次以上，例如，一開始持有 A、5，玩家視 A 為 11 點，加總為 16 點，後來補牌後抽到 10 點，改視 A 為 1 點，加總也是 16 點，故 16 點這個狀態被經歷兩次。所以，在計算狀態值函數時，有兩種方式，「首次訪問」 (First Visit)及「每次訪問」(Every Visit)。首次訪問只計算第一次訪問時的報酬，而每次訪問則計算所有訪問的平均報酬，本程式採用首次訪問。

```
1   # 策略評估函數
2   def policy_eval(policy, env, num_episodes, discount_factor=1.0):
3       returns_sum = defaultdict(float)    # 記錄每一個狀態的報酬
4       returns_count = defaultdict(float)  # 記錄每一個狀態的訪問個數
5       V = defaultdict(float) # 狀態值函數
6
7       # 實驗 N 回合
8       for i_episode in range(1, num_episodes + 1):
9           # 每 1000 回合顯示除錯訊息
10          if i_episode % 1000 == 0:
11              print(f"\r {i_episode}/{num_episodes}回合.", end="")
12              sys.stdout.flush() # 清除畫面
13
14          # 回合(episode)資料結構為陣列，每一項目含 state, action, reward
15          episode = []
16          state = env.reset()
17          # 開始依策略玩牌，最多 100 步驟，中途分出勝負即結束
18          for t in range(100):
19              action = policy(state)
20              next_state, reward, done, _ = env.step(action)
21              episode.append((state, action, reward))
22              if done:
23                  break
24              state = next_state
```

```
26          # 找出走過的所有狀態
27          states_in_episode = set([tuple(x[0]) for x in episode])
28          # 計算每一狀態的值函數
29          for state in states_in_episode:
30              # 找出每一步驟內的首次訪問(First Visit)
31              first_occurence_idx = next(i for i,x in enumerate(episode)
32                                         if x[0] == state)
33              # 算累計報酬(G)
34              G = sum([x[2]*(discount_factor**i) for i,x in
35                      enumerate(episode[first_occurence_idx:])])
36              # 計算狀態值函數
37              returns_sum[state] += G
38              returns_count[state] += 1.0
39              V[state] = returns_sum[state] / returns_count[state]
40
41      return V
```

5. 訂定策略：與前面策略相同。

```
1   # 採相同策略
2   def sample_policy(observation):
3       score, dealer_score, usable_ace = observation
4       # 超過20點，不補牌(stick)，否則都跟莊家要一張牌(hit)
5       return 0 if score >= 20 else 1
```

6. 分別實驗 10,000 與 500,000 回合。

```
1   # 實驗 10000 回合
2   V_10k = policy_eval(sample_policy, env, num_episodes=10000)
3   plotting.plot_value_function(V_10k, title="10,000 Steps")
4
5   # 實驗 500,000 回合
6   V_500k = policy_eval(sample_policy, env, num_episodes=500000)
7   plotting.plot_value_function(V_500k, title="500,000 Steps")
```

執行結果：顯示各狀態的值函數，分成持有 A(比較容易獲勝)及未持有 A。可以看到當玩家持有的分數很高時，勝率會明顯提升。下列彩色圖表可參考程式執行結果。

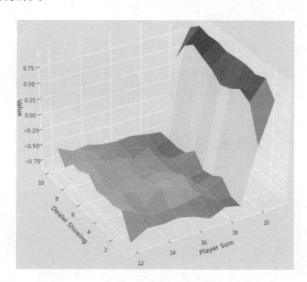

持有 A，但分數不高時，勝率也有明顯提升。

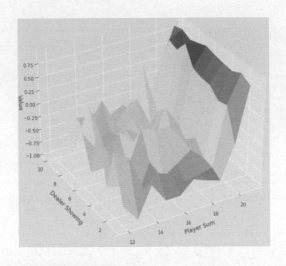

實驗 500,000 回合後的表現更明顯。

接著進行值循環，結合策略評估與策略改善。另外，還要介紹一個新的策略 ε-greedy，之前策略改善時都是採用貪婪(greedy)策略，它有一個弱點，就是一旦發現最大值函數的路徑後，貪婪策略會一直走相同的路徑，這樣便失去了找到更好路徑的潛在機會，舉例來說，家庭聚餐時，都會選擇最好的美食餐廳，若為了不踩雷，每次都去之前最好吃的餐廳用餐，那新開的餐廳就永遠沒機會被發現了，這就是所謂的「探索與利用」(Exploration and Exploitation)，而 ε-greedy 所採取的方式就是除了採最佳路徑之外，還保留一個比例去探索，不走既有的老路。以下我們就嘗試這種新策略與蒙地卡羅演算法結合。

圖 15.17 探索與利用(Exploration and Exploitation)

【範例 14】實驗 21 點撲克牌(Blackjack)之值循環。

程式：RL_15_14_Blackjack_Value_Iteration.ipynb，修改自 Denny Britz 網站。

1. 載入相關套件。

```
1  # 載入相關套件
2  import numpy as np
3  from lib.envs.blackjack import BlackjackEnv
4  from lib import plotting
5  import sys
6  from collections import defaultdict
7  import matplotlib
8
9  matplotlib.style.use('ggplot') # 設定繪圖的風格
```

2. 建立環境：程式為 lib\envs\blackjack.py。

```
1  # 環境
2  env = BlackjackEnv()
```

3. 定義 ε-greedy 策略：若 ε=0.1，則 10 次行動有 1 次採隨機行動。

```
1   # ε-greedy策略
2   def make_epsilon_greedy_policy(Q, epsilon, nA):
3       def policy_fn(observation):
4           # 每個行動的機率初始化，均為 ε / n
5           A = np.ones(nA, dtype=float) * epsilon / nA
6           best_action = np.argmax(Q[observation])
7           # 最佳行動的機率再加 1 - ε
8           A[best_action] += (1.0 - epsilon)
9           return A
10      return policy_fn
```

4. 定義值循環函數：與上例的程式邏輯幾乎相同，主要差異是將狀態值函數改為行動值函數。

```
1   # 值循環函數
2   def value_iteration(env, num_episodes, discount_factor=1.0, epsilon=0.1):
3       returns_sum = defaultdict(float)     # 記錄每一個狀態的報酬
4       returns_count = defaultdict(float)   # 記錄每一個狀態的訪問個數
5       Q = defaultdict(lambda: np.zeros(env.action_space.n)) # 行動值函數
6
7       # 採用 ε-greedy策略
8       policy = make_epsilon_greedy_policy(Q, epsilon, env.action_space.n)
9
10      # 實驗 N 回合
11      for i_episode in range(1, num_episodes + 1):
```

```
12          # 每 1000 回合顯示除錯訊息
13          if i_episode % 1000 == 0:
14              print(f"\r {i_episode}/{num_episodes}回合.", end="")
15              sys.stdout.flush() # 清除畫面
16
17          # 回合(episode)資料結構為陣列，每一項目含 state, action, reward
18          episode = []
19          state = env.reset()
20          # 開始依策略玩牌，最多 100 步驟，中途分出勝負即結束
21          for t in range(100):
22              probs = policy(state)
23              action = np.random.choice(np.arange(len(probs)), p=probs)
24              next_state, reward, done, _ = env.step(action)
25              episode.append((state, action, reward))
26              if done:
27                  break
28              state = next_state
```

```
30          # 找出走過的所有狀態
31          sa_in_episode = set([(tuple(x[0]), x[1]) for x in episode])
32          for state, action in sa_in_episode:
33              # (狀態, 行動)組合初始化
34              sa_pair = (state, action)
35              # 找出每一步驟內的首次訪問(First Visit)
36              first_occurence_idx = next(i for i,x in enumerate(episode)
37                                         if x[0] == state and x[1] == action)
38              # 算累計報酬(G)
39              G = sum([x[2]*(discount_factor**i) for i,x in
40                       enumerate(episode[first_occurence_idx:])])
41              # 計算行動值函數
42              returns_sum[sa_pair] += G
43              returns_count[sa_pair] += 1.0
44              Q[state][action] = returns_sum[sa_pair] / returns_count[sa_pair]
45
46      return Q, policy
```

5. 執行值循環。

```
1   # 執行值循環
2   Q, policy = value_iteration(env, num_episodes=500000, epsilon=0.1)
```

6. 顯示執行結果。

```
1   # 顯示結果
2   V = defaultdict(float)
3   for state, actions in Q.items():
4       action_value = np.max(actions)
5       V[state] = action_value
6   plotting.plot_value_function(V, title="Optimal Value Function")
```

執行結果：上個範例只有當玩家的分數接近 20 分的時候，值函數特別高，然而，這個策略即使在低分時也有不差的表現，勝率比起上例明顯提

升。下列彩色圖表可參考程式執行結果。

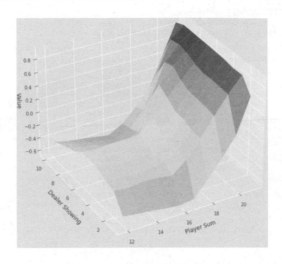

再來看另一種想法，目前為止的值循環在策略評估與策略改良上，均採用同一策略，即 ε-greedy，而這次兩者則各自採用不同策略，即策略評估時採用隨機策略，盡可能走過所有路徑，在策略改良時，改採貪婪策略，盡量求勝。所以，採用同一策略，我們稱為 On-policy，採用不同策略則稱為 Off-policy。

【**範例 15**】實驗 21 點撲克牌(Blackjack)之 Off-policy 值循環。
程式：RL_15_15_Blackjack_Off_Policy.ipynb，修改自 Denny Britz 網站。

1. 載入相關套件。

```
1  # 載入相關套件
2  import numpy as np
3  from lib.envs.blackjack import BlackjackEnv
4  from lib import plotting
5  import sys
6  from collections import defaultdict
7  import matplotlib
8
9  matplotlib.style.use('ggplot') # 設定繪圖的風格
```

2. 建立環境：程式為 lib\envs\blackjack.py。

```
1  # 環境
2  env = BlackjackEnv()
```

3. 定義隨機策略在評估時使用。

```
1  # 隨機策略
2  def create_random_policy(nA):
3      A = np.ones(nA, dtype=float) / nA
4      def policy_fn(observation):
5          return A
6      return policy_fn
```

4. 定義貪婪(greedy)策略在改良時使用。

```
1   # 貪婪(greedy)策略
2   def create_greedy_policy(Q):
3       def policy_fn(state):
4           # 每個行動的機率初始化，均為 0
5           A = np.zeros_like(Q[state], dtype=float)
6           best_action = np.argmax(Q[state])
7           # 最佳行動的機率 = 1
8           A[best_action] = 1.0
9           return A
10      return policy_fn
```

5. 定義值循環策略，使用重要性加權抽樣。

- 重要性加權抽樣(Weighted Importance Sampling)：以值函數大小作為隨機抽樣比例的分母。

- 依重要性加權抽樣，值函數公式如下：

$$Q(S_t, A_t) \leftarrow Q(S_t, A_t) + \frac{W}{C(S_t,A_t)} \left[G - Q(S_t, A_t) \right]$$

```
1   # 定義值循環策略，使用重要性加權抽樣
2   def mc_control_importance_sampling(env, num_episodes, behavior_policy, discount_f
3       Q = defaultdict(lambda: np.zeros(env.action_space.n)) # 行動值函數
4       # 重要性加權抽樣(weighted importance sampling)的累計分母
5       C = defaultdict(lambda: np.zeros(env.action_space.n))
6
7       # 在策略改良時，採貪婪策略
8       target_policy = create_greedy_policy(Q)
9
10      # 實驗 N 回合
11      for i_episode in range(1, num_episodes + 1):
12          # 每 1000 回合顯示除錯訊息
13          if i_episode % 1000 == 0:
14              print(f"\r {i_episode}/{num_episodes}回合.", end="")
15              sys.stdout.flush() # 清除畫面
16
17          # 回合(episode)資料結構為陣列，每一項目含 state, action, reward
18          episode = []
19          state = env.reset()
```

```
20          # 開始依策略玩牌，最多 100 步驟，中途分出勝負即結束
21          for t in range(100):
22              # 評估時採用隨機策略
23              probs = behavior_policy(state)
24              # 以值函數大小作為隨機抽樣比例的分母
25              action = np.random.choice(np.arange(len(probs)), p=probs)
26              next_state, reward, done, _ = env.step(action)
27              episode.append((state, action, reward))
28              if done:
29                  break
30              state = next_state
```

```
32          G = 0.0 # 報酬初始化
33          W = 1.0 # 權重初始化
34          # 找出走過的所有狀態
35          for t in range(len(episode))[::-1]:
36              state, action, reward = episode[t]
37              # 累計報酬
38              G = discount_factor * G + reward
39              # 累計權重
40              C[state][action] += W
41              # 更新值函數，公式參見書籍
42              Q[state][action] += (W / C[state][action]) * (G - Q[state][action])
43              # 已更新完畢，即跳出迴圈
44              if action != np.argmax(target_policy(state)):
45                  break
46              # 更新權重
47              W = W * 1./behavior_policy(state)[action]
48
49      return Q, target_policy
```

6. 執行值循環：評估時採用隨機策略。

```
1  # 執行值循環
2  random_policy = create_random_policy(env.action_space.n)
3  Q, policy = mc_control_importance_sampling(env, num_episodes=500000,
4                                      behavior_policy=random_policy)
```

7. 顯示執行結果。

```
1  # 顯示結果
2  V = defaultdict(float)
3  for state, actions in Q.items():
4      action_value = np.max(actions)
5      V[state] = action_value
6  plotting.plot_value_function(V, title="Optimal Value Function")
```

執行結果：玩家分數在低分時勝率也明顯提升。下列彩色圖表可參考程式執行結果。

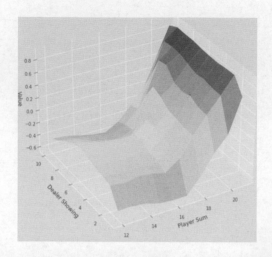

這一節我們學會了運用蒙地卡羅演算法，還有探索與利用、On/Off Policy，使模型勝率提高了不少，這些概念不只可以應用在蒙地卡羅演算法上，也能套用到後續其他的演算法，讀者可視專案不同的需求來選擇。

15-9 時序差分(Temporal Difference)

蒙地卡羅演算法必須先完成一些評估後，才能計算值函數，接著依據值函數計算出狀態轉移機率，有以下缺點：

1. 每個回合必須走到終點，才能夠倒推每個狀態的值函數。
2. 假使狀態空間很大的話，還是一樣要走到終點，才能開始下行動決策，速度實在太慢，例如圍棋，根據統計，每下一盤棋平均約需 150 手，而且圍棋共有 $3^{19 \times 19} \fallingdotseq 1.74 \times 10^{172}$ 個狀態，就算使用探索也很難測試到每個狀態，計算值函數。

於是學者提出時序差分(Temporal Difference, TD)演算法，透過邊走邊計算值函數的方式，解決上述問題。

值函數更新公式如下：

$$v(s_t) = v(s_t) + \alpha[r_{t+1} + \gamma v(s_{t+1}) - v(s_t)]$$

值函數每次加上下一狀態值函數與目前狀態值函數的差額，以目前的行動產生的結果代替 Bellman 公式的期望值，另外，再乘以學習率(Learning Rate)α。這種走一步更新一次的作法稱為 TD(0)，如果是走 n 步更新一次的作法稱則為 TD(λ)。

以倒推圖比較動態規劃、蒙地卡羅及時序差分演算法的作法。

- 動態規劃：逐步搜尋所有的下一個可能狀態，計算值函數期望值。
- 蒙地卡羅：試走多個回合，再以回推的方式計算值函數期望值。
- 時序差分：每走一步更新一次值函數。

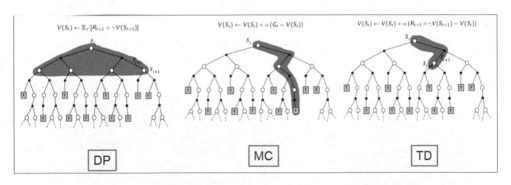

圖 15.18 動態規劃、蒙地卡羅及時序差分演算法的倒推圖

時序差分有兩個種類的演算法：

- SARSA 演算法：On Policy 的時序差分。
- Q-learning 演算法：Off Policy 的時序差分。

先介紹 SARSA 演算法，它的名字是行動軌跡中 5 個元素的縮寫 s_t, a_t, r_{t+1}, s_{t+1}, a_{t+1}，如下圖，意謂著每走一步更新一次：

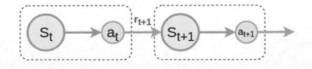

圖 15.19 SARSA：s_t, a_t, r_{t+1}, s_{t+1}, a_{t+1}

接著我們就來進行演算法的實作，再介紹一款新遊戲 Windy Gridworld，與原來的 Grid World 有些差異，變成 10x7 個格子，第 4,5,6,9 行的風力 1 級，第 7,8 行的風力 2 級，會把玩家往上吹 1 格和 2 格，而起點(x)與終點(T)的位置如下：

```
o o o o o o o o o o
o o o o o o o o o o
o o o o o o o o o o
x o o o o o o T o o
o o o o o o o o o o
o o o o o o o o o o
o o o o o o o o o o
```

【範例 16】實驗 Windy Gridworld 之 SARSA 策略。

程式：RL_15_16_SARSA.ipynb，修改自 Denny Britz 網站。

1. 載入相關套件。

```
1   # 載入相關套件
2   import gym
3   import itertools
4   import matplotlib
5   import numpy as np
6   import pandas as pd
7   import sys
8   from collections import defaultdict
9   from lib.envs.windy_gridworld import WindyGridworldEnv
10  from lib import plotting
11
12  matplotlib.style.use('ggplot') # 設定繪圖的風格
```

2. 建立 Windy Gridworld 環境。

```
1   # 建立環境
2   env = WindyGridworldEnv()
```

3. 試玩，一律往右走。

```
1   # 試玩
2   print(env.reset()) # 重置
3   env.render()        # 更新畫面
4
5   print(env.step(1)) # 走下一步
6   env.render()        # 更新畫面
7
8   print(env.step(1)) # 走下一步
9   env.render()        # 更新畫面
```

```
10
11  print(env.step(1))  # 走下一步
12  env.render()         # 更新畫面
13
14  print(env.step(1))  # 走下一步
15  env.render()         # 更新畫面
16
17  print(env.step(1))  # 走下一步
18  env.render()         # 更新畫面
19
20  print(env.step(1))  # 走下一步
21  env.render()         # 更新畫面
```

執行結果：

30：第 30 個點，表示起始點為第 3 列第 0 行，索引值從 0 開始算。

```
o   o   o   o   o   o   o   o   o   o

o   o   o   o   o   o   o   o   o   o

o   o   o   o   o   o   o   o   o   o

x   o   o   o   o   o   o   T   o   o

o   o   o   o   o   o   o   o   o   o

o   o   o   o   o   o   o   o   o   o

o   o   o   o   o   o   o   o   o   o
```

移至第 3 列第 1 行，獎勵為-1。
(31, -1.0, False, {'prob': 1.0})

```
o   o   o   o   o   o   o   o   o   o

o   o   o   o   o   o   o   o   o   o

o   o   o   o   o   o   o   o   o   o

o   x   o   o   o   o   o   T   o   o

o   o   o   o   o   o   o   o   o   o

o   o   o   o   o   o   o   o   o   o

o   o   o   o   o   o   o   o   o   o
```

...

(33, -1.0, False, {'prob': 1.0})
(24, -1.0, False, {'prob': 1.0}) ➔ 往上吹一格
(15, -1.0, False, {'prob': 1.0}) ➔ 往上吹一格

4. 定義 ε-greedy 策略：與前面相同。

```
1   # 定義 ε-greedy策略
2   def make_epsilon_greedy_policy(Q, epsilon, nA):
3       def policy_fn(observation):
4           # 每個行動的機率初始化，均為 ε / n
5           A = np.ones(nA, dtype=float) * epsilon / nA
6           best_action = np.argmax(Q[observation])
7           # 最佳行動的機率再加 1 - ε
8           A[best_action] += (1.0 - epsilon)
9           return A
10      return policy_fn
```

5. 定義 SARSA 策略：走一步算一步，然後採 ε-greedy 策略，決定行動。

```
1   # 定義 SARSA 策略
2   def sarsa(env, num_episodes, discount_factor=1.0, alpha=0.5, epsilon=0.1):
3       # 行動值函數初始化
4       Q = defaultdict(lambda: np.zeros(env.action_space.n))
5       # 記錄 所有回合的長度及獎勵
6       stats = plotting.EpisodeStats(
7           episode_lengths=np.zeros(num_episodes),
8           episode_rewards=np.zeros(num_episodes))
9
10      # 使用 ε-greedy策略
11      policy = make_epsilon_greedy_policy(Q, epsilon, env.action_space.n)
12
13      # 實驗 N 回合
14      for i_episode in range(num_episodes):
15          # 每 100 回合顯示除錯訊息
16          if (i_episode + 1) % 100 == 0:
17              print(f"\r {(i_episode + 1)}/{num_episodes}回合.", end="")
18              sys.stdout.flush() # 清除畫面
19
20          # 開始依策略實驗
21          state = env.reset()
22          action_probs = policy(state)
23          action = np.random.choice(np.arange(len(action_probs)), p=action_probs)
```

```
25          # 每次走一步就更新狀態值
26          for t in itertools.count():
27              # 走一步
28              next_state, reward, done, _ = env.step(action)
29
30              # 選擇下一步行動
31              next_action_probs = policy(next_state)
32              next_action = np.random.choice(np.arange(len(next_action_probs))
33                                          , p=next_action_probs)
34
35              # 更新長度及獎勵
36              stats.episode_rewards[i_episode] += reward
37              stats.episode_lengths[i_episode] = t
38
```

```
39              # 更新狀態值
40              td_target = reward + discount_factor * Q[next_state][next_action]
41              td_delta = td_target - Q[state][action]
42              Q[state][action] += alpha * td_delta
43
44              if done:
45                  break
46
47              action = next_action
48              state = next_state
49
50      return Q, stats
```

6. 執行 SARSA 策略 200 回合。

```
1  # 執行 SARSA 策略
2  Q, stats = sarsa(env, 200)
```

7. 顯示執行結果。

```
1  # 顯示結果
2  fig = plotting.plot_episode_stats(stats)
```

執行結果：共有三張圖表。

每一回合走到終點的距離：剛開始的時候要走很多步才會到終點，不過執行到大約第 50 回合後就逐漸收斂了，每回合幾乎都相同步數。

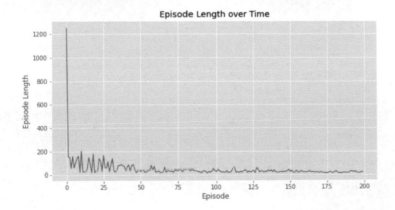

每一回合的報酬：每回合獲得的報酬越來越高。

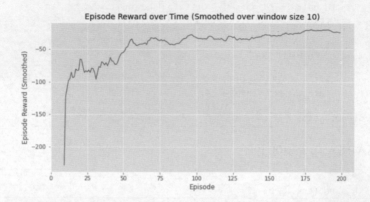

累計的步數與回合對比：呈現曲線上揚的趨勢，即每回合到達終點的步數越來越少。

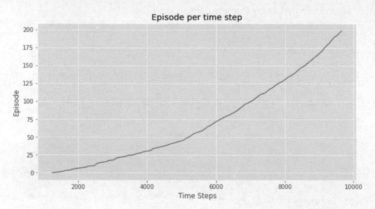

接著說明時序差分的第二種演算法 Q-learning，它與 SARSA 的差別是採取 Off Policy，評估時使用 ε-greedy 策略，改良時選擇 greedy 策略。再介紹另一款遊戲 Cliff Walking，同樣也是迷宮，最下面一排除了起點(x)與終點(T)之外，其他都是陷阱(C)，踩到陷阱即 Game Over。

```
o o o o o o o o o o o o
o o o o o o o o o o o o
o o o o o o o o o o o o
x C C C C C C C C C C T
```

【**範例 17**】實驗 Cliff Walking 之 Q-learning 策略。

程式：RL_15_17_Q_learning.ipynb，修改自 Denny Britz 網站。

1. 載入相關套件。

```
1  # 載入相關套件
2  import gym
3  import itertools
4  import matplotlib
5  import numpy as np
6  import pandas as pd
7  import sys
8  from collections import defaultdict
9  from lib.envs.cliff_walking import CliffWalkingEnv
10 from lib import plotting
11
12 matplotlib.style.use('ggplot') # 設定繪圖的風格
```

2. 建立 Cliff Walking 環境。

```
1  # 建立環境
2  env = CliffWalkingEnv()
```

3. 試玩：隨便走。

```
1  # 試玩
2  print(env.reset()) # 重置
3  env.render()        # 更新畫面
4
5  print(env.step(0)) # 往上走
6  env.render()        # 更新畫面
7
8  print(env.step(1)) # 往右走
9  env.render()
10
11 print(env.step(1)) # 往右走
12 env.render()
13
14 print(env.step(2)) # 往下走
15 env.render()
```

執行結果：

36：第 36 個點，表示起始點為第 3 列第 0 行，但程式卻是在第 3 列第 6 行，應該是程式邏輯有問題，但不影響測試，就不除錯了。

```
36
O  O  O  O  O  O  O  O  O  O  O  O
O  O  O  O  O  O  O  O  O  O  O  O
O  O  O  O  O  O  O  O  O  O  O  O
x  C  C  C  C  C  C  C  C  C  C  T
```

走到最後一步，移至第 3 列第 2 行，走到陷阱，獎勵-100。

```
(38, -100.0, True, {'prob': 1.0})
O  O  O  O  O  O  O  O  O  O  O  O
O  O  O  O  O  O  O  O  O  O  O  O
O  O  O  O  O  O  O  O  O  O  O  O
O  C  x  C  C  C  C  C  C  C  C  T
```

4. 定義 ε-greedy 策略：與前面相同。

```
1  # 定義 ε-greedy策略
2  def make_epsilon_greedy_policy(Q, epsilon, nA):
3      def policy_fn(observation):
4          # 每個行動的機率初始化，均為 ε / n
5          A = np.ones(nA, dtype=float) * epsilon / nA
6          best_action = np.argmax(Q[observation])
7          # 最佳行動的機率再加 1 - ε
8          A[best_action] += (1.0 - epsilon)
9          return A
10     return policy_fn
```

5. 定義 Q-learning 策略：評估時採 ε-greedy 策略，改良時選擇 greedy 策略。

```
1   # 定義 Q_learning 策略
2   def q_learning(env, num_episodes, discount_factor=1.0, alpha=0.5, epsilon=0.1):
3       # 行動值函數初始化
4       Q = defaultdict(lambda: np.zeros(env.action_space.n))
5       # 記錄 所有回合的長度及獎勵
6       stats = plotting.EpisodeStats(
7           episode_lengths=np.zeros(num_episodes),
8           episode_rewards=np.zeros(num_episodes))
9
10      # 使用 ε-greedy策略
11      policy = make_epsilon_greedy_policy(Q, epsilon, env.action_space.n)
12
13      # 實驗 N 回合
14      for i_episode in range(num_episodes):
15          # 每 100 回合顯示除錯訊息
16          if (i_episode + 1) % 100 == 0:
17              print(f"\r {(i_episode + 1)}/{num_episodes}回合.", end="")
18              sys.stdout.flush() # 清除畫面
19
```

```
20          # 開始依策略實驗
21          state = env.reset()
22          # 每次走一步就更新狀態值
23          for t in itertools.count():
24              # 使用 ε-greedy策略
25              action_probs = policy(state)
26              # 選擇下一步行動
27              action = np.random.choice(np.arange(len(action_probs)), p=action_probs)
28              next_state, reward, done, _ = env.step(action)
```

```
30              # 更新長度及獎勵
31              stats.episode_rewards[i_episode] += reward
32              stats.episode_lengths[i_episode] = t
33
34              # 選擇最佳行動
35              best_next_action = np.argmax(Q[next_state])
36              # 更新狀態值
37              td_target = reward + discount_factor * Q[next_state][best_next_action]
38              td_delta = td_target - Q[state][action]
39              Q[state][action] += alpha * td_delta
40
41              if done:
42                  break
43
44              state = next_state
45
46      return Q, stats
```

6. 執行 Q-learning 策略 500 回合。

```
1  # 執行 Q-Learning 策略
2  Q, stats = q_learning(env, 500)
```

7. 顯示執行結果。

```
1  # 顯示結果
2  fig = plotting.plot_episode_stats(stats)
```

執行結果：共有三張圖表。

每一回合走到終點的距離：與 SARSA 相同，剛開始要走很多步後才會到終點，但執行到大概第 50 回合就逐漸收斂，之後的每個回合幾乎都相同步數。由於，這兩個範例的遊戲不同，不能比較 SARSA 與 Q-learning 的效能，若要比較效能，可改用同一款遊戲進行比較。

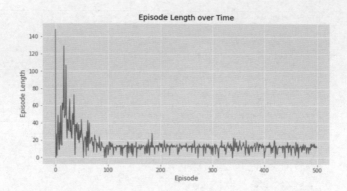

每一回合的報酬：每回合獲得的報酬越來越高，不過尚未收斂。

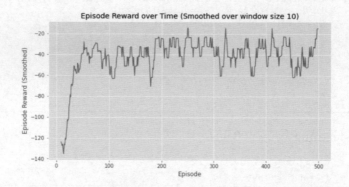

累計的步數與回合對比：呈現直線上揚的趨勢，即每回合到達終點的步數差不多，這可能是因為有陷阱的關係，應該分成勝敗兩類比較，會更清楚。

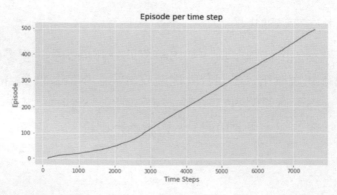

總體而言，SARSA 與 Q-learning 的比較如下：

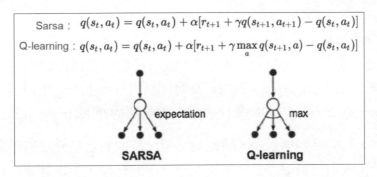

圖 15.20　SARSA 與 Q-learning 策略的比較

看了這麼多的演算法，不管是策略循環(Policy Iteration)或值循環(Value Iteration)，總體而言，它們的邏輯與神經網路優化求解，其實有那麼一點相似。

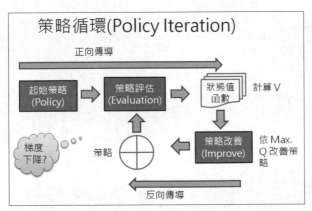

圖 15.21　策略循環與梯度下降法的比較

15-10 其他演算法

不管是使用狀態值函數或是行動值函數，前述的演算法都是建立一個陣列(List)，來記錄所有的對應值，之後就從陣列中選擇最佳的行動，所以這類演算法被稱為「表格型」(Tabular)強化學習，作法簡單直接，但只適合離散型的狀態，像木棒台車這種的連續型變數，就不適用，變通的作法有兩種：

1. 將連續型變數進行分組，轉換成離散型變數。
2. 使用機率分配或神經網路模型取代表格，以策略評估的訓練資料，來估計模型的參數(權重)，選擇行動時，就依據模型推斷出最佳預測值，而 Deep Q-learning(DQN)即是利用神經網路的 Q-learning 演算法。

另外一個研究的課題則是多人遊戲的情境，不同於之前介紹的遊戲，玩家都只有一位，現代的遊戲設計重視人際之間的交流，可以有多位玩家同時參與一款遊戲，這時就會產生協同合作或互相對抗的情境，玩家除了考慮獎勵與狀態外，也需觀察其他玩家的狀態，這種演算法稱為「多玩家強化學習」(Multi-agent Reinforcement Learning)，比如許多撲克牌遊戲都屬於多玩家遊戲。

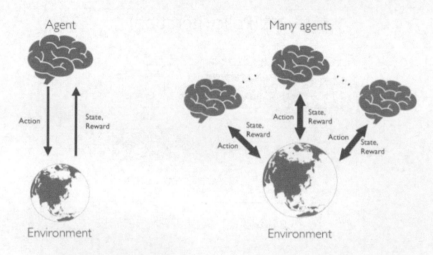

圖 15.22 單玩家與多玩家強化學習的比較，圖片來源：「An Overview of Multi-Agent Reinforcement Learning from Game Theoretical Perspective」[5]

近年來，強化學習研究的環境越來越複雜，各種演算法相繼推陳出新，可參閱維基百科強化學習的介紹(https://en.wikipedia.org/wiki/Reinforcement_learning)，如下表：

Algorithm ⬦	Description ⬦	Policy ⬦	Action Space ⬦	State Space ⬦	Operator ⬦
Monte Carlo	Every visit to Monte Carlo	Either	Discrete	Discrete	Sample-means
Q-learning	State–action–reward–state	Off-policy	Discrete	Discrete	Q-value
SARSA	State–action–reward–state–action	On-policy	Discrete	Discrete	Q-value
Q-learning - Lambda	State–action–reward–state with eligibility traces	Off-policy	Discrete	Discrete	Q-value
SARSA - Lambda	State–action–reward–state–action with eligibility traces	On-policy	Discrete	Discrete	Q-value
DQN	Deep Q Network	Off-policy	Discrete	Continuous	Q-value
DDPG	Deep Deterministic Policy Gradient	Off-policy	Continuous	Continuous	Q-value
A3C	Asynchronous Advantage Actor-Critic Algorithm	On-policy	Continuous	Continuous	Advantage
NAF	Q-Learning with Normalized Advantage Functions	Off-policy	Continuous	Continuous	Advantage
TRPO	Trust Region Policy Optimization	On-policy	Continuous	Continuous	Advantage
PPO	Proximal Policy Optimization	On-policy	Continuous	Continuous	Advantage
TD3	Twin Delayed Deep Deterministic Policy Gradient	Off-policy	Continuous	Continuous	Q-value
SAC	Soft Actor-Critic	Off-policy	Continuous	Continuous	Advantage

圖 15.23 強化學習演算法的比較

本書關於演算法的介紹就到此告一段落，想了解更多內容的讀者，可詳閱強化學習的聖經「Reinforcement Learning: An Introduction」一書 (http://incompleteideas.net/book/the-book.html)。

15-11 井字遊戲

接著我們就開始實戰吧，拿最簡單的井字遊戲(Tic-Tac-Toe)來練習，包括如何把井字遊戲轉換為環境、如何定義狀態及立即獎勵、模型存檔等。

圖 15.24 井字遊戲

【範例 18】實驗井字遊戲之 Q-learning 策略。

程式：TicTacToe_1\ticTacToe.py，程式修改自「Reinforcement Learning — Implement TicTacToe」[6]，程式撰寫成類似 Gym 的架構。

1. 載入相關套件。

```
1  # 載入相關套件
2  import numpy as np
3  import pickle
4  import os
5
6  # 參數設定
7  BOARD_ROWS = 3 # 列數
8  BOARD_COLS = 3 # 行數
```

2. 定義環境類別：與前面的範例類似，主要就是棋盤重置(reset)、更新狀態、給予獎勵及判斷輸贏，勝負未分的時候，電腦加 0.1 分，而玩家加 0.5 分，這樣設定是希望電腦能儘速贏得勝利，讀者可以試試看其他的給分方式，若勝負已定，則給 1 分。

3. 環境初始化。

```
10  # 環境類別
11  class Environment:
12      def __init__(self, p1, p2):
13          # 變數初始化
14          self.board = np.zeros((BOARD_ROWS, BOARD_COLS))
15          self.p1 = p1 # 第一個玩家
16          self.p2 = p2 # 第二個玩家
17          self.isEnd = False    # 是否結束
18          self.boardHash = None # 棋盤
19
20          self.playerSymbol = 1 # 第一個玩家使用X
21
22      # 記錄棋盤狀態
23      def getHash(self):
24          self.boardHash = str(self.board.reshape(BOARD_COLS * BOARD_ROWS))
25          return self.boardHash
```

4. 判斷輸贏：取得勝利的情況包括連成一列、一行或對角線。

```
27      # 判斷輸贏
28      def is_done(self):
29          # 連成一列
30          for i in range(BOARD_ROWS):
31              if sum(self.board[i, :]) == 3:
32                  self.isEnd = True
33                  return 1
34              if sum(self.board[i, :]) == -3:
35                  self.isEnd = True
36                  return -1
37
38          # 連成一行
39          for i in range(BOARD_COLS):
```

```
40              if sum(self.board[:, i]) == 3:
41                  self.isEnd = True
42                  return 1
43              if sum(self.board[:, i]) == -3:
44                  self.isEnd = True
45                  return -1
```

```
47          # 連成對角線
48          diag_sum1 = sum([self.board[i, i] for i in range(BOARD_COLS)])
49          diag_sum2 = sum([self.board[i, BOARD_COLS - i - 1] for i in
50                                              range(BOARD_COLS)])
51          diag_sum = max(abs(diag_sum1), abs(diag_sum2))
52          if diag_sum == 3:
53              self.isEnd = True
54              if diag_sum1 == 3 or diag_sum2 == 3:
55                  return 1
56              else:
57                  return -1
58
59          # 無空位置即算平手
60          if len(self.availablePositions()) == 0:
61              self.isEnd = True
62              return 0
63          self.isEnd = False
64          return None
```

5. 定義顯示空位置、更新棋盤、給予獎勵等函數。

```
66      # 顯示空位置
67      def availablePositions(self):
68          positions = []
69          for i in range(BOARD_ROWS):
70              for j in range(BOARD_COLS):
71                  if self.board[i, j] == 0:
72                      positions.append((i, j))
73          return positions
74
75      # 更新棋盤
76      def updateState(self, position):
77          self.board[position] = self.playerSymbol
78          # switch to another player
79          self.playerSymbol = -1 if self.playerSymbol == 1 else 1
80
81      # 給予獎勵
82      def giveReward(self):
83          result = self.is_done()
84          # backpropagate reward
85          if result == 1: # 第一玩家贏，P1加一分
86              self.p1.feedReward(1)
87              self.p2.feedReward(0)
88          elif result == -1: # 第二玩家贏，P2加一分
89              self.p1.feedReward(0)
90              self.p2.feedReward(1)
91          else: # 勝負未分，第一玩家加 0.1分，第二玩家加 0.5分
92              self.p1.feedReward(0.1)
93              self.p2.feedReward(0.5)
```

6. 棋盤重置。

```
95      # 棋盤重置
96      def reset(self):
97          self.board = np.zeros((BOARD_ROWS, BOARD_COLS))
98          self.boardHash = None
99          self.isEnd = False
100         self.playerSymbol = 1
```

7. 訓練：這是重點，本例訓練 50,000 回合後，可產生狀態值函數表，將電腦和玩家的狀態值函數表分別存檔 (policy_p1 、 policy_p2)，policy_p1 為先下子的策略模型，policy_p2 為後下子的策略模型。

```
102     # 開始訓練
103     def play(self, rounds=100):
104         for i in range(rounds):
105             if i % 1000 == 0:
106                 print("Rounds {}".format(i))
107             while not self.isEnd:
108                 # Player 1
109                 positions = self.availablePositions()
110                 p1_action = self.p1.chooseAction(positions, self.board, self.playerSymbol)
111                 # take action and upate board state
112                 self.updateState(p1_action)
113                 board_hash = self.getHash()
114                 self.p1.addState(board_hash)
115                 # check board status if it is end
116
117                 win = self.is_done()
118                 if win is not None:
119                     # self.showBoard()
120                     # ended with p1 either win or draw
121                     self.giveReward()
122                     self.p1.reset()
123                     self.p2.reset()
124                     self.reset()
125                     break
```

```
126                 else:
127                     # Player 2
128                     positions = self.availablePositions()
129                     p2_action = self.p2.chooseAction(positions, self.board, self.playerSymbol)
130                     self.updateState(p2_action)
131                     board_hash = self.getHash()
132                     self.p2.addState(board_hash)
133
134                     win = self.is_done()
135                     if win is not None:
136                         # self.showBoard()
137                         # ended with p2 either win or draw
138                         self.giveReward()
139                         self.p1.reset()
140                         self.p2.reset()
141                         self.reset()
142                         break
```

8. 比賽：與訓練邏輯類似，差別是玩家要自行輸入行動。

```
144        # 開始比賽
145        def play2(self, start_player=1):
146            is_first = True
147            while not self.isEnd:
148                # Player 1
149                positions = self.availablePositions()
150                if not (is_first and start_player==2):
151                    p1_action = self.p1.chooseAction(positions, self.board, self.playerSymbol)
152                    # take action and upate board state
153                    self.updateState(p1_action)
154                is_first = False
155                self.showBoard()
156                # check board status if it is end
157                win = self.is_done()
158                if win is not None:
159                    if win == -1 or win == 1:
160                        print(self.p1.name, " 勝!")
161                    else:
162                        print("平手!")
163                    self.reset()
164                    break
```

```
165                else:
166                    # Player 2
167                    positions = self.availablePositions()
168                    p2_action = self.p2.chooseAction(positions)
169
170                    self.updateState(p2_action)
171                    self.showBoard()
172                    win = self.is_done()
173                    if win is not None:
174                        if win == -1 or win == 1:
175                            print(self.p2.name, " 勝!")
176                        else:
177                            print("平手!")
178                        self.reset()
179                        break
```

9. 顯示棋盤目前的狀態。

```
181        # 顯示棋盤目前狀態
182        def showBoard(self):
183            # p1: x  p2: o
184            for i in range(0, BOARD_ROWS):
185                print('-------------')
186                out = '| '
187                for j in range(0, BOARD_COLS):
188                    if self.board[i, j] == 1:
189                        token = 'x'
190                    if self.board[i, j] == -1:
191                        token = 'o'
192                    if self.board[i, j] == 0:
193                        token = ' '
194                    out += token + ' | '
195                print(out)
196            print('-------------')
```

10. 電腦類別：包括電腦依最大值函數行動、比賽結束前存檔、比賽開始
前載入檔案。

11. 初始化。

```
202  # 電腦類別
203  class Player:
204      def __init__(self, name, exp_rate=0.3):
205          self.name = name
206          self.states = []   # record all positions taken
207          self.lr = 0.2
208          self.exp_rate = exp_rate
209          self.decay_gamma = 0.9
210          self.states_value = {}   # state -> value
211
212      def getHash(self, board):
213          boardHash = str(board.reshape(BOARD_COLS * BOARD_ROWS))
214          return boardHash
```

12. 電腦依最大值函數行動。

```
212      # 電腦依最大值函數行動
213      def chooseAction(self, positions, current_board, symbol):
214          if np.random.uniform(0, 1) <= self.exp_rate:
215              # take random action
216              idx = np.random.choice(len(positions))
217              action = positions[idx]
218          else:
219              value_max = -999
220              for p in positions:
221                  next_board = current_board.copy()
222                  next_board[p] = symbol
223                  next_boardHash = self.getHash(next_board)
224                  value = 0 if self.states_value.get(next_boardHash) is None \
225                             else self.states_value.get(next_boardHash)
226
227                  # 依最大值函數行動
228                  if value >= value_max:
229                      value_max = value
230                      action = p
231          # print("{} takes action {}".format(self.name, action))
232          return action
```

13. 更新狀態值函數。

```
234      # 更新狀態值函數
235      def addState(self, state):
236          self.states.append(state)
237
238      # 重置狀態值函數
239      def reset(self):
240          self.states = []
```

```
241
242        # 比賽結束，倒推狀態值函數
243        def feedReward(self, reward):
244            for st in reversed(self.states):
245                if self.states_value.get(st) is None:
246                    self.states_value[st] = 0
247                self.states_value[st] += self.lr * (self.decay_gamma * reward
248                                                    - self.states_value[st])
249                reward = self.states_value[st]
```

14. 存檔、載入檔案。

```
252        def savePolicy(self):
253            fw = open('policy_' + str(self.name), 'wb')
254            pickle.dump(self.states_value, fw)
255            fw.close()
256
257        # 載入檔案
258        def loadPolicy(self, file):
259            fr = open(file, 'rb')
260            self.states_value = pickle.load(fr)
261            fr.close()
```

15. 玩家類別：自行輸入行動。

```
267  # 玩家類別
268  class HumanPlayer:
269        def __init__(self, name):
270            self.name = name
271
272        # 行動
273        def chooseAction(self, positions):
274            while True:
275                position = int(input("輸入位置(1~9):"))
276                row = position // 3
277                col = (position % 3) - 1
278                if col < 0:
279                    row -= 1
280                    col = 2
281                # print(row, col)
282                action = (row, col)
283                if action in positions:
284                    return action
285
286        # 狀態值函數更新
287        def addState(self, state):
288            pass
289
290        # 比賽結束，倒推狀態值函數
291        def feedReward(self, reward):
292            pass
293
294        def reset(self):
295            pass
```

16. 畫圖說明輸入規則：說明如何輸入位置。

```
297    # 畫圖說明輸入規則
298    def first_draw():
299        rv = '\n'
300        no=0
301        for y in range(3):
302            for x in range(3):
303                idx = y * 3 + x
304                no+=1
305                rv += str(no)
306                if x < 2:
307                    rv += '|'
308            rv += '\n'
309            if y < 2:
310                rv += '-----\n'
311        return rv
```

執行結果：輸入位置的號碼如下。

```
1|2|3
-----
4|5|6
-----
7|8|9
```

17. 主程式：

- 若已訓練過，就不會再訓練了，要重新訓練的話可將 policy_p1、policy_p2 檔案刪除。

- 提供執行參數 2，讓玩家可先下子，指令如下，否則一律由電腦先下子。

    python ticTacToe.py 2

```
313    if __name__ == "__main__":  # 主程式
314        import sys
315        if len(sys.argv) > 1:
316            start_player = int(sys.argv[1])
317        else:
318            start_player = 1
319
320        # 產生物件
321        p1 = Player("p1")
322        p2 = Player("p2")
323        env = Environment(p1, p2)
324
325        # 訓練
326        if not os.path.exists(f'policy_p{start_player}'):
327            print("開始訓練...")
328            env.play(50000)
```

```
329            p1.savePolicy()
330            p2.savePolicy()
331
332    print(first_draw())  # 棋盤說明
333
334    # 載入訓練成果
335    p1 = Player("computer", exp_rate=0)
336    p1.loadPolicy(f'policy_p{start_player}')
337    p2 = HumanPlayer("human")
338    env = Environment(p1, p2)
339
340    # 開始比賽
341    env.play2(start_player)
```

執行結果：筆者試了幾回合，要贏電腦還蠻難的。

```
-------------
| o | o | x |
-------------
| x | x | o |
-------------
| o | x | x |
-------------
平手!
```

15-12 木棒台車(CartPole)

透過 15-4 節的實驗，我們了解到木棒台車的狀態是連續型變數，無法以陣列儲存所有狀態的值函數，而這一節我們就來看看如何使用「行動者與評論者」(Actor Critic)演算法進行木棒台車實驗。

Actor Critic 類似 GAN，主要分為兩個神經網路，Actor(行動者)在評論者 (Critic)的指導下，優化行動決策，而評論者則負責評估行動決策的好壞，並主導值函數模型的參數更新，詳細的說明可參閱「Keras 官網說明」 (https://keras.io/examples/rl/actor_critic_cartpole/)，以下的程式也來自該網頁。

程式 RL_15_18_Actor_Critic.py。
執行指令：python RL_15_18_Actor_Critic.py。
執行結果：若報酬超過 195 分，即停止，表示模型已非常成熟，部份的執行結果如下，在 763 回合成功達成目標。

```
running reward: 173.41 at episode 680
running reward: 181.18 at episode 690
running reward: 188.73 at episode 700
running reward: 186.98 at episode 710
running reward: 179.31 at episode 720
running reward: 181.53 at episode 730
running reward: 187.06 at episode 740
running reward: 190.73 at episode 750
running reward: 194.45 at episode 760
Solved at episode 763!
```

官網也提供兩段錄製的動畫,分別為訓練初期與後期的比較,可以看出後期的木棒台車行駛得相當穩定。

- 訓練初期:https://i.imgur.com/5gCs5kH.gif。
- 訓練後期:https://i.imgur.com/5ziiZUD.gif。

15-13 結論

強化學習的應用範圍越來越廣泛,像是自駕車或無人搬運車(Automated Guided Vehicle, AGV)都會在不久的將來大行其道,它們都是利用攝影機偵測前方的障礙,但更核心的部分是利用強化學習,來採取最佳行動決策。以無人搬運車為例,我們只要模仿 Grid World 遊戲的作法,將辦公室/工廠/醫院平面圖製成類似迷宮的路徑,進行模擬訓練後,將模型移植到機器人,就可以驅動機器人從 A 點送貨至 B 點。

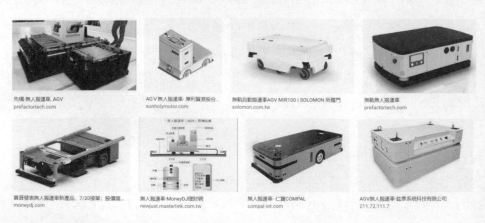

圖 15.25 各廠牌的無人搬運車(Automated Guided Vehicle, AGV)

不僅如此,在股票投資、腦部手術等其他領域,也都看得到強化學習的身影,雖然,其理論較為艱深,且需要較紮實的程式基礎,但是,它的好處是不用蒐集大量的訓練資料,也不需標記資料(Labeling)。

參考資料 (References)

[1] 維基百科關於強化學習的說明
    (https://zh.wikipedia.org/wiki/%E5%BC%BA%E5%8C%96%E5%AD%A6%E4%B9%A0)
[2] Sayan Mandal,《Install OpenAI Gym with Box2D and Mujoco in Windows 10》, 2019
    (https://medium.com/@sayanmndl21/install-openai-gym-with-box2d-and-mujoco-in-windows-10-e25ee9b5c1d5)
[3] Denny Britz Github
    (https://github.com/dennybritz/reinforcement-learning)
[4] 維基百科關於蒙地卡羅方法的說明
    (https://zh.wikipedia.org/wiki/%E8%92%99%E5%9C%B0%E5%8D%A1%E7%BE%85%E6%96%B9%E6%B3%95)
[5] Yaodong Yang、Jun Wang,《An Overview of Multi-Agent Reinforcement Learning from Game Theoretical Perspective》, 2020
    (https://arxiv.org/abs/2011.00583)
[6] Jeremy Zhang,《Reinforcement Learning — Implement TicTacToe》, 2019
    (https://towardsdatascience.com/reinforcement-learning-implement-tictactoe-189582bea542)

Note

Note